MAIN LENDING   **3 WEEK LOAN**   DCU LIBRARY

Fines are charged **PER DAY** if this item is overdue.
Check at www.dcu.ie/~library or telephone (01) 700 5183 for fine rates and
renewal regulations for this item type.
Item is subject to recall.
Remember to use the Book Bin when the library is closed.
The item is due for return on or before the latest date shown below.

Janos H. Fendler

Nanoparticles and Nanostructured Films

 **WILEY-VCH**

Further Titles of Interest

D. F. Evans, H. Wennerström
The Colloidal Domain
ISBN 1-56081-525-6

G. Schmid (Ed.)
Clusters and Colloids
From Theory to Applications
ISBN 3-527-29043-5

Advanced Materials
The leading journal in high-tech materials science
Published monthly
ISSN 0935-9648

Janos H. Fendler (Ed.)

Nanoparticles and Nanostructured Films

Preparation, Characterization and Applications

WITHDRAWN

 **WILEY-VCH**

Weinheim • New York • Chichester
Brisbane • Singapore • Toronto

Prof. Janos H. Fendler
Center for Advanced Material Processing
Clarkson University
Potsdam, NY 13699
USA

Library of Congress Card No. applied for.

A catalogue record for this book is available from the British Library.

Deutsche Bibliothek Cataloguing-in-Publication Data:

Nanoparticles and nanostructured films: preparation, characterization and applications / Janos H. Fendler (ed.). - Weinheim ; New York ; Chichester ; Brisbande ; Singapore ; Toronto : Wiley-VCH, 1998
  ISBN 3-527-29443-0

Composition: Asco Trade Typesetting Ltd., Hong Kong.
Printing: betz-druck gmbh, D-64291 Darmstadt.
Bookbinding: Wilhelm Osswald & Co, D-67433 Neustadt.
Printed in the Federal Republic of Germany.

Preface

Small is not only beautiful but also eminently useful. The virtues of working in the nanodomain are increasingly recognized by the scientific community, the technological world and even the popular press. The number of research publications in this area has been increasing exponentially. Additionally, national and international biological, physical, chemical, engineering, and materials science societies and government agencies have been organizing workshops, meetings, and symposia around some aspects of nanoparticle research with increasing frequency. This burgeoning interest is amply justified, of course, by the unique properties of nanoparticles and nanostructured materials and by the promise these systems hold as components of optical, electrical, electro-optical, magnetic, magneto-optical, and catalytic sensors and devices.

The appearance of numerous review articles and books on nanoparticle research has helped the neophyte to digest the veritable information overload. No recent overview has appeared, however, to the best of our knowledge, that focuses upon the utilization of "wet" chemical and colloid chemical methods for the preparation of nanoparticles and nanostructured films. The purpose of the present book is to fill this gap by summarizing current accomplishments in preparing and characterizing nanoparticles and nanostructured films and to point out their potential applications. Versatility, relative ease of preparation and transfer from the liquid to the solid phase, convenience of scale-up, and economy are the advantages of the chemical approach to advanced materials synthesis.

Electrochemistry has reached sufficient maturity and sophistication to be used for the layer-by-layer deposition of nanoparticles and nanoparticulate films. In Chapters 1 and 3 the state-of-the-art electrodeposition of quantum dots, superlattices, and nanocomposites is surveyed. Chemists have plenty to learn from mother nature. Much of the work on template-directed nanoparticle growth is inspired by biomineralization, the oriented growth of inorganic crystals in biomembranes. Advantage has been taken of organized surfactant assemblies that mimic the biological membranes to grow nanoparticles and nanoparticulate films. Chapters 2 and 4 highlight the growth of metallic, semiconducting, and magnetic nanoparticles under monolayers and within the confines of reverse micelles. More rigid templates have also been employed for nanoparticle preparations. This approach is illustrated for such diverse templates as opal (Chapter 13), nanoporous membranes (Chapter 10), and zeolites (Chapter 17). Chapter 7 emphasizes the use of block copolymer micelles as hosts for generating metallic nanoparticles.

The recent attention to porous silicon nanoparticles has been prompted by their demonstrated photoluminescence and electroluminescence, as well as by their promise to function as optical interconnects and chemically tunable sensors, which require passive surfaces that are stable to oxidation yet are able to conduct current efficiently. Chemical and plasma-induced silicon nanocluster formation and growth are examined in Chapters 5 and 8. The potentially important, albeit as yet unexplored fullerene nanoparticles and their two-dimensional crystal growth are surveyed in Chapter 6.

Nanoparticles themselves can be used as building blocks for two-dimensional arrays and/or three-dimensional networks. They can also be derivatized and treated as if they were simple molecules. This approach should lead to the type of hetero-supramolecular structures that are illustrated in Chapter 16. Such complex chemistries must go hand-in-hand with an improved understanding of surface and colloid chemical interactions. Some aspects of these are discussed in Chapters 11 and 12.

Exploitation of nanoparticles and nanostructured materials requires an appreciation of electron and photoelectron transfer mechanisms therein. Chapter 9 presents a well balanced view of the electron transfer processes in nanostructured semiconductor thin films while Chapter 14 discusses charge transfer at nanocrystalline metal, oxide–semiconductor interfaces and its relation to electrochromic–battery and photovoltaic–photocatalytic interfaces. Significantly, as summarized in Chapter 15, nanoparticles provide us with the possibility of monitoring, and ultimately exploiting, single electron transfer events.

An attempt has been made in the last chapter to provide the newcomer with handy "recipes" for the preparation of nanoparticles and nanostructured films as well as to summarize current accomplishments and future prospects in this intellectually fascinating and highly relevant area of research. Inevitably, current activities soon become "past achievements", and interested readers will have to acquaint themselves with the latest results as they appear in primary publications and as they are disseminated at scientific meetings. Chapter 18 also lists selected data on the properties of the most frequently used bulk semiconductors in order to permit much needed comparisons between the bulk and size-quantized materials.

I am grateful to all the contributing authors who took time from their busy schedule to write their chapters and thus to share their expertise with the scientific community. I also thank Dr. Peter Gregory and Dr. Jörn Ritterbusch, the Editors at WILEY-VCH, and their staff or initiating this project and for providing enthusiastic support throughout the various stages of publication.

October 1997   Janos H. Fendler

Contents

Contributors

B. Alperson
Department of Materials and Interfaces
The Weizmann Institute of Science
IL-76100 Rehovot
Israel

M. Antonietti
Max-Planck-Institut für Kolloid- und
    Grenzflächenforschung
Kantstraße 55
14513 Teltow-Seehof
Germany

D. Behar
Department of Materials and Interfaces
The Weizmann Institute of Science
IL-76100 Rehovot
Israel

R. A. Bley
Department of Chemistry
University of California
Davis, CA 95616
USA

L. Bronstein
The Russian Academy of Sciences
A. N. Nesmeyanov Institute of
    Organoelement Compounds
28 Vavilov St., INEOS
Moscow
Russia

S. Carrara
Institute of Biophysics
University of Genova
Via Giotto 2
16153 Genova
Italy

L. Cusack
Department of Chemistry
University College Dublin
Belfield
Dublin 4
Ireland

J. Dutta
Powder Technology Laboratory
Department of Materials Science
Swiss Federal Institute of Technology
CH-1015 Lausanne
Switzerland

J. H. Fendler
Center for Advanced Material
    Processing
Clarkson University
Potsdam, NY 13699
USA

D. Fitzmaurice
Department of Chemistry
University College Dublin
Belfield
Dublin 4
Ireland

Y. Golan
Department of Materials and Interfaces
The Weizmann Institute of Science
IL-76100 Rehovot
Israel

D. M. Guldi
Radiation Laboratory
University of Notre Dame
Notre Dame, IN 46556
USA

I. Hannus
Applied Chemistry Department
Jozsef Attila University
Rerrich Béla tér
6720 Szeged
Hungary

G. Hodes
Department of Materials and Interfaces
The Weizmann Institute of Science
IL-76100 Rehovot
Israel

H. Hofmann
Powder Technology Laboratory
Department of Materials Science
Swiss Federal Institute of Technology
CH-1015 Lausanne
Switzerland

H. Hofmeister
Powder Technology Laboratory
Department of Materials Science
Swiss Federal Institute of Technology
CH-1015 Lausanne
Switzerland

C. Hollenstein
Powder Technology Laboratory
Department of Materials Science
Swiss Federal Institute of Technology
CH-1015 Lausanne
Switzerland

J. C. Hulteen
Department of Chemistry
Colorado State University
Fort Collins, Colorado 80523
USA

J. T. Hupp
Department of Chemistry
Northwestern University
2145 Sheridan Rd.
Evanston, IL 60208-3113
USA

P. V. Kamat
Radiation Laboratory
University of Notre Dame
Notre Dame, IN 46556
USA

S. M. Kauzlarich
Department of Chemistry
University of California
Davis, CA 95616
USA

S. Kelly
Department of Physics
Brooklyn College of CUNY
Brooklyn, NY 11210
USA

I. Kiricsi
Applied Chemistry Department
Jozsef Attila University
Rerrich Béla tér
6720 Szeged
Hungary

B. I. Lemon
Department of Chemistry
Northwestern University
2145 Sheridan Rd.
Evanston, IL 60208-3113
USA

L. A. Lyon
Department of Chemistry
Northwestern University
2145 Sheridan Rd.
Evanston, IL 60208-3113
USA

X. Marguerettaz
Department of Chemistry
University College Dublin
Belfield
Dublin 4
Ireland

C. R. Martin
Department of Chemistry
Colorado State University
Fort Collins, Colorado 80523
USA

F. C. Meldrum
Department of Applied Mathematics
Research School of Physical Sciences
Australian National University
Canberra, ACT 0200
Australia

P. Mulvaney
School of Chemistry
University of Melbourne
Parkville, VIC 3052
Australia

J. B. Nagy
Laboratoire de Résonance Magnétique
   Nucléaire
Facultés Universitaires Notre-Dame
   de la Paix
61 Rue de Bruxelles
5000 Namur
Belgium

M.-P. Pileni
Laboratoire S.R.S.I.
U.R.A.C.N.R.S.

1662 Université P. et M. Curie (Paris VI)
4 Place Jussieu
F-75231 Paris Cedex 05
France

S. G. Romanov
University of Wuppertal
Dept. of Electronics
Fuhlrottstr. 10
42097 Wuppertal
Germany

I. Rubinstein
Department of Materials and Interfaces
The Weizmann Institute of Science
IL-76100 Rehovot
Israel

C. M. Sotomayor-Torres
University of Wuppertal
Dept. of Electronics
Fuhlrottstr. 10
42097 Wuppertal
Germany

J. A. Switzer
Martin E. Straumanis Hall
University of Missouri-Rolla
Rolla, MI 65401-0249
USA

M. Tomkiewicz
Department of Physics
Brooklyn College of CUNY
Brooklyn, NY 11210
USA

P. Valetsky
The Russian Academy of Sciences
A. N. Nesmeyanov Institute of
   Organoelement Compounds
28 Vavilov St., INEOS
Moscow
Russia

Y. Zhang
Department of Materials and Interfaces
The Weizmann Institute of Science
IL-76100 Rehovot
Israel

Chapter 1

Electrodeposited Quantum Dots: Size Control by Semiconductor–Substrate Lattice Mismatch

G. Hodes, Y. Golan, D. Behar, Y. Zhang, B. Alperson, and I. Rubinstein

1.1 Introduction

Semiconductor nanocrystals, and more specifically quantum dots (QDs), are the subject of a rapidly developing field. Nanocrystals can be loosely defined as crystals with dimensions up to ca. 100 nm; above this size, they are more commonly termed microcrystals. QDs are nanocrystals that display quantum size effects. While there are different quantum size effects with different size scales, the term is commonly understood to refer to nanocrystals whose dimensions are smaller than the bulk Bohr diameter of the semiconductor – typically several nm up to several tens of nm. In this size regime, as the crystal size becomes smaller the semiconductor energy levels become more separated from each other and the effective bandgap increases. This means that a material with a fixed chemical composition and crystal structure can be made to have very different optoelectronic properties solely by virtue of its physical dimensions. It is largely this property that has generated the high level of interest in the field [1–3].

Numerous methods have been applied to fabricate QDs. The commonly used gas phase techniques are molecular beam epitaxy or chemical vapor deposition of the semiconductor. The QDs are often formed by lithographically patterning quantum wells or superlattices, sometimes together with the application of a confining potential (Reference [4] contains many references to these techniques). More recently, self-assembly of QDs that form spontaneously during the deposition process has been described [5, 6]. The spontaneous formation of small islands, rather than of coherent films, is believed to be due to the mismatch strain between the substrate and epitaxially deposited semiconductor. Perfect lattice match should ideally result in a continuous film, while increasing mismatch results in smaller islands owing to the buildup of mismatch strain with increasing size, eventually causing cessation of growth when the strain energy becomes too large and nucleation of a new island becomes more favorable. As will be seen later, this principle of lattice mismatch plays a central role in the subject of this chapter.

There are also various methods based on wet chemical reactions that result in very small crystals. The most common of these is based on the reaction of two (or more) reactants in a liquid or solid phase to form a colloid of the desired semi-conductor [1]. Size control of the colloidal nanocrystals can be achieved by a number of different techniques, the most common of which are based upon temperature of formation or subsequent heat treatment and capping the nanocrystals with a strongly adsorbed layer that prevents further crystal growth [7–9].

In contrast with the gas phase techniques, which in most cases result in QDs deposited on a solid substrate, usually a different semiconductor, the wet chemical reactions form nanocrystals that most often are dispersed in a separate phase, which may be a liquid, glass, or plastic. This places major limitations on potential uses of such systems; while they may be highly suitable for optical applications, for example, they are less suitable for electronic applications, where an electronic contact is needed. In some cases, films with some electronic conductivity have been formed from dispersed semiconductor nanocrystals by subsequent processing, such as transferring Langmuir–Blodgett monolayers containing semiconductor nanocrystals onto a solid substrate [10, 11] or attaching colloids from solution to a substrate with thiol linkages [12].

Our approach to the preparation of semiconductor QDs is a chemical one that results in the direct formation of the QDs on a substrate, usually in the form of aggregated nanocrystals, but also as isolated QDs. The two techniques used by us are chemical solution deposition and electrodeposition [13]. While each method has its advantages and disadvantages, a very important advantage of the electrodeposition technique is the high degree of control over the amount of deposited material. This results from Faraday's laws – the relationship between the amount of deposition charge passed and the amount of material deposited: $(96\ 500 \times n)$ coulombs results in 1 gm-mole of deposit where n is the number of electrons passed in depositing one molecule of the deposit.

Thus, the amount of deposit formed can be measured from the deposition charge passed (the integrated current) if the number of electrons taking part in the deposition of one molecule is known. While this calculation can be modified by several factors, in particular by a less than 100% current efficiency, it allows a very high degree of control over the amount of material deposited, since very small currents and very short times (and therefore very small amounts of charge) are experimental variables that are simple to achieve. As an example of such a calculation, in the deposition of CdSe as described below (a two-electron reaction), a current density of 0.1 mA cm^{-2} for one second will theoretically result in a deposit of ca. 2×10^{-7} g, which is equivalent to a continuous layer of CdSe ca. 0.35 nm thick – about a single monomolecular layer. In practice, the deposit often does not form a continuous layer, but rather islands, and hence the usefulness of the electrodeposition technique for depositing both isolated and thick films of nanocrystals. Another possible complication, as discussed below, is that the very first step of electrodeposition on a bare substrate may be different from the deposition on a semiconductor-covered substrate.

In this chapter, we concentrate on isolated semiconductor QDs electrodeposited (ED) onto well-defined metal substrates. We show the important effect of semiconductor–substrate lattice mismatch in determining the QD size. At the same time, the epitaxial strain resulting from this mismatch is sometimes not sufficient in itself to explain the experimental results. Other factors, such as wetting of the substrate by the semiconductor and surface energies of both semiconductor and substrate, may become dominant in some cases. We have previously described in detail three systems that can be well explained by the lattice mismatch. These are CdSe/Au, Cd(Se, Te)/Au and CdSe/Pd. The important features of these systems will be described first. Next, the results of other semiconductor–substrate combinations – some of which are still preliminary – will be discussed.

While this chapter concentrates on deposition aspects, some electronic structure characterization of the QDs will be described. This is important since the interest in controlling the QD size lies in its effect on controlling the electronic energy structure, and therefore optoelectronic properties, of the deposited QDs.

1.2 The CdSe/Au System

The research described in this chapter developed from previous experiments on the electrodeposition of CdSe nanocrystals on thin-film Au substrates [14, 15]. The Au films, 35 nm thick, were evaporated onto glass or mica and then annealed in air to give a largely {111} textured Au film with individual grain sizes up to hundreds of nm. Electrodeposition of hexagonal CdSe was carried out from a hot DMSO solution of $Cd(ClO_4)_2$ (50 mM) saturated with elemental Se at elevated temperatures using a constant current, typically of 0.1 mA cm^{-2}. The amount of deposit could be readily controlled by the amount of charge passed, as described in Section 1.2; this allowed controllable and reproducible degrees of coverage of the Au substrate by the CdSe nanocrystals.

The CdSe nanocrystals were typically 4–5 nm in all three dimensions, the lateral dimensions measured by transmission electron microscopy (TEM) imaging (Figure 1.1(a)) and the vertical dimension estimated by both X-ray diffraction (XRD) and atomic force microscopy (AFM). Thus complete coverage of the substrate – a monolayer of nanocrystals – is equivalent in amount of material to a film of ca. 4–5 nm in thickness (ignoring voids between crystals, which will decrease this amount to a small degree). We will often refer to a monolayer of nanocrystals, which should not be confused with a monolayer of CdSe (ca. 0.35 nm).

Selected area electron diffraction (SAD) (Figures 1.1(b) and (c)), high-resolution transmission electron microscopy (HRTEM), and optical diffraction of HRTEM images showed that the QDs are epitaxially aligned with the Au substrate, in a $\{111\}Au\|\{00.2\}CdSe$ and $\langle100\rangle Au\|\langle11.0\rangle$ CdSe orientation relationship. We suggested that the epitaxy is due to the close lattice match between the relevant lattice spacings, $d - \{110\}_{Au}$ and $a_0(CdSe)$. The literature values of those lattice spacings are $d - \{110\}_{Au} = 0.2884$ nm and $a_0(CdSe) = 0.4299$ nm. In a ratio of 3:2, these values correspond to a mismatch of −0.6% (i.e., the CdSe lattice should be expansively strained by the Au substrate). From the Poisson relationship, this expansive strain in the plane of the substrate should result in a compressive strain in the perpendicular direction. This was indeed observed in XRD studies of these

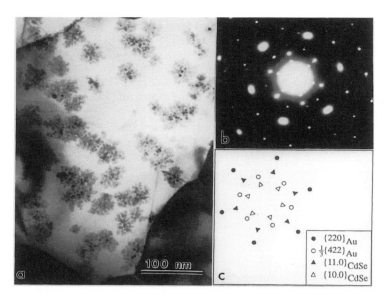

Figure 1.1. (a) TEM image of aggregates of CdSe nanocrystals ED on evaporated {111} gold. (b) SAD pattern corresponding to (a). (c) Schematic drawing of (b) showing the assignment of the relevant diffraction spots. (Reproduced by permission from Reference [15], Y. Golan et al.)

deposits, where a compression in the {00.2} spacing (the CdSe c direction) of 1.7% compared to the bulk value was observed [16].

An important question arising in any method used to prepare QDs is what controls the crystal size? Why are the CdSe QDs the size they are (4–5 nm)? In addition, the size distribution of the nanocrystals was relatively narrow [14]. This was of particular interest, both because a narrow size distribution is necessary for most potential applications and because most measured properties of the QDs are smeared out, with a resulting loss of information, by inhomogeneous size broadening. Within a wide range, the crystal size was independent of many deposition parameters that might be expected to affect size, in particular deposition temperature and current density. These parameters did strongly affect the distribution of the deposit, i.e., higher temperatures and smaller current densities/longer deposition times resulted in a greater degree of aggregation of the QDs, which could be explained by increased opportunity of the individual QDs to migrate on the Au surface at higher temperatures or over longer times (although at room temperature isolated QDs did not aggregate, even over a period of a year). However, the size of the *individual* QDs in the aggregates did not vary appreciably with these parameters.

It was suggested that the QD size and narrow size distribution was largely due to the mismatch-induced strain in the CdSe QDs [16]. As the QDs nucleate and grow, the total strain in the individual QDs grows. At some critical size, the buildup of strain will be sufficiently great that it will be energetically more favorable for renucleation to occur rather than continued growth. Such a mechanism concurs with the equivalent self-assembly of semiconductor islands deposited by vapor deposition techniques [5, 6], although both the island size and degree of mismatch in the case of the ED samples are considerably smaller than for the vapor-deposited ones.

If the above hypothesis is correct, then it should be possible to control the ED QD size by varying the semiconductor–substrate mismatch. This can be done in a number of ways, the most obvious being to vary the composition of either the semiconductor or the substrate.

1.3 Change of Semiconductor Lattice Spacing – Cd(Se, Te)/Au

CdSe and CdTe form a solid solution over the complete composition range. The predominant crystal phase changes from hexagonal (CdSe) to cubic (CdTe) approximately in the middle of the composition range, and the solid solution obeys Vegard's law, i.e., the lattice spacing varies linearly with composition from one end member to the other [17]. This means that the lattice spacing can be accurately controlled through the Se:Te ratio in the deposit. A composition of $CdSe_{0.92}Te_{0.08}$ should be very closely lattice-matched to Au (at the 2:3 ratio). Unfortunately, our deposition technique does not allow inclusion of much Te into the electrodeposit. We have been unable to dissolve Te in DMSO to any noticeable extent that allows deposition of CdTe, even at high temperatures and low current densities. However, in the presence of Se, Te does dissolve to a small extent in DMSO. By varying the amount of Se, the deposition temperature, and current density, we have been able to incorporate small amounts of Te into the CdSe – up to several percent of the Se concentration (the analyses, by X-ray photoelectron spectroscopy, were only semi-quantitative but did give an indication of the Te concentration) [18].

Figure 1.2 shows the variation of crystal size with increasing Te content, from pure CdSe (Figure 1.2(a)) up to several percent (Figure 1.2(f)). The crystal size varies from 4–5 nm (pure CdSe, Figure 1.2(a)) to ca. 20 nm (Figure 1.2(f)). As expected from our reasoning above, the crystal size increases with increasing Te concentration. The wider size distribution typically observed for the larger QDs (larger Te content) is probably due to nonhomogeneous chemical composition. The inset in Figure 1.2(c) shows a SAD pattern of a single Au grain with Cd(Se, Te) crystals verifying the epitaxy of the nanocrystals.

The height of the Cd(Se, Te) crystals, measured by XRD, also increased with increasing Te content, although only about half as much as the growth in the lateral dimensions [18]. Thus, while the pure CdSe crystals were approximately equi-dimensional, the largest crystals were more disklike (20 nm in diameter by ca. 10 nm in height). This can indicate relaxation of interfacial strain leading to crystals with a flatter shape (increased interfacial strain favors taller, thinner shapes) [19].

Another predictable consequence of the decreased strain in the Cd(Se, Te) crystals compared to pure CdSe is that the Poisson contraction in the c direction was decreased from 1.7% (CdSe) to ca. 0.6% (for several percent Te content) [18].

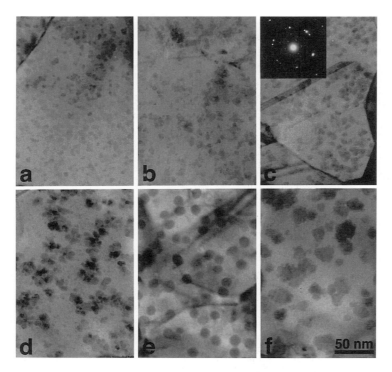

Figure 1.2. TEM images of (a) (Te free) CdSe QDs on Au; (b–f) Cd(Se, Te) QDs with increasing average size from 6 nm (b) to 18 nm (f). All the images in this figure are shown at the same magnification and were all electrodeposited using the same amount of electrical charge ($0.70\ \text{mC cm}^{-2}$). Deposition parameters are given in Reference [18]. Inset: SAD pattern obtained from (c), showing the orientation relationship with the Au substrate. The $\{hk.0\}_{Cd(Se,Te)}$ d spacings retain the hetero-epitaxial values determined by the Au substrate, for all Se/Te ratios. (Reproduced by permission from Reference [18], Y. Golan et al.)

1.4 Change of Substrate Lattice Spacing – CdSe/Pd

Pd has a similar fcc structure but a smaller lattice parameter compared with Au. Thus the corresponding lattice mismatch between CdSe and Pd is considerably larger (+4.1% for the 3:2 ratio) than for CdSe on Au with the same orientation relationship (−0.6%). Based on the mismatch strain argument, we expect very tiny nanocrystals of CdSe on this substrate.

We found that electrodeposition of CdSe on Pd, using the same experimental procedure as for deposition on Au, results in diffraction-amorphous CdSe QDs, formed on top of a thin diffraction-amorphous CdSe layer [20]. HRTEM imaging revealed, however, many regions of localized sixfold symmetry in the deposit. Digital image analysis of the HRTEM images established the presence of irregularly shaped, ordered CdSe clusters, typically 1–2.5 nm in size, surrounded by dis-

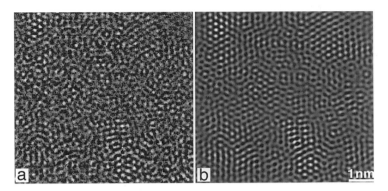

Figure 1.3. (a) Raw digitized HRTEM image of CdSe ED onto Pd. (b) Image obtained by cross correlation between the structural motif developed by the autocorrelation function of this deposit and a high-pass-filtered version of (a). (Reproduced by permission from Reference [20], Y. Golan et al.)

ordered material (Figure 1.3). These clusters were observed in the QDs as well as in the CdSe thin layer. In spite of the relatively large mismatch in this system, the ordered clusters exhibited considerable preferential orientation. Fourier analysis indicated a {111}Pd‖{00.2}CdSe, 110 Pd‖ 10.0 CdSe epitaxial relationship between the ordered CdSe clusters and the {111}Pd substrate. Thus the CdSe on Pd is rotated by 30° relative to the orientational relationship of epitaxial CdSe QDs on {111}$_{Au}$. Simulation of the CdSe lattice on top of that of Pd confirmed that this R30° orientational relationship resulted in better registry between the CdSe and Pd compared with a nonrotated relationship (and vice versa for CdSe on Au) [16].

The results of this study support the mismatch strain argument, although instead of isolated nanocrystals, the ordered clusters were surrounded by truly amorphous material. The structure of the CdSe as a disordered, amorphous phase accommodates mismatch strain (the disordered lattice can accommodate local lattice variations with the Pd), hence preventing the formation of dislocations. Thus it is understandable that a continuous layer of CdSe can form in this system, in contrast to the CdSe/Au system, where buildup of strain apparently prevents formation of such a layer.

1.5 Thicker Layers of CdSe on Au and Pd

While this chapter concentrates on isolated QDs (or at most a single layer of aggregated QDs) on metal substrates, it is important to describe the extension of such deposits to thicker films. Previous publications have described ED nanocrystalline CdS and CdSe "thick" (ca. 100 nm or more) films, using the DMSO/S(Se)/Cd^{2+} plating bath [13, 21, 22]. The crystal size was determined predominantly by the anion of the Cd salt. When CdCl$_2$ was used as the Cd source, the typical crystal size for both CdSe and CdS was ca. 5 nm, although with a fairly large size distribution, while use of Cd(ClO$_4$)$_2$ (the salt used in the above studies) resulted in considerably larger crystals – typically ca. 10 nm. It may be that this difference is caused by adsorption of Cl$^-$ on the growing crystals, resulting in termination of growth, the capping mechanism well known in colloidal growth and recently suggested by us as the cause of size control in CdTe films electrodeposited from DMSO solutions containing phosphines [23].

For CdSe deposition on Au, the epitaxy described above occurs for the first layer of CdSe crystals but not for subsequent layers. The epitaxy is normally lost with subsequent layers, as seen by a change of the single crystal spot diffraction pattern of the first layer to a polycrystalline ring diffraction pattern for thicker films [14]. Thus, in this case, CdSe deposits epitaxially onto Au but *not* onto itself. An important consequence of this loss of epitaxy is the corresponding loss in size control; the subsequent layers of CdSe are composed of larger crystals (ca. 10 nm) with a large size distribution.

Gold evaporated onto mica substrates and subjected to a long annealing treatment (250 °C for 12 hr) forms highly oriented crystals, in contrast with the textured (but azimuthally random) Au films used in all the above studies, obtained on glass or mica after a short annealing time. CdSe deposited onto such substrates retains a high (although not total, as for the first layer) degree of epitaxy, even for 15 nm-thick layers [24]. HRTEM showed the ability of a single of CdSe crystal to bridge the grain boundaries between the oriented Au crystals, in contrast to the change in orientation between two adjacent and touching CdSe crystals on adjacent Au grains that were not oriented (on an Au-on-glass substrate). In the former case, the second layer contained large (several tens of nm) CdSe crystals. This might be due to the perfect CdSe-on-CdSe epitaxy that arises from the epitaxial Au substrate. For CdSe on Pd, the situation is completely different. Since the deposit is – to a large extent – disordered, and therefore relaxed, the Pd is initially covered with a continuous film of CdSe followed by island growth as described above. The size of the islands depends on the thickness of the deposit; they grow larger in size and with a large size distribution as more CdSe is deposited [20]. Such behavior is typical for growth where no size-limiting mechanism, such as mismatch strain or capping, occurs.

1.6 Other Semiconductor–Substrate Systems

Up to now, we have discussed three semiconductor–substrate combinations that have been the subjects of intensive investigation by us, namely CdSe/Au, Cd(Se, Te)/Au and CdSe/Pd. More recently, we have been investigating other combinations; some have been studied in depth while others are at a preliminary stage. Here we will treat some aspects of these studies.

1.6.1 (Cd, Zn)Se/Au

Our investigation of this system began with a chance observation. Initial attempts to electrodeposit ZnSe, using $Zn(ClO_4)_2$ in place of $Cd(ClO_4)_2$, resulted not in ZnSe, but in a deposit that subsequent analyses showed to be Cd-rich (Cd, Zn)Se. It was later realized that the source of the Cd was the graphite anode, which had previously been used for the deposition of CdSe and clearly contained occluded Cd, in spite of being well rinsed in DMSO and water. The few control experiments we have attempted up to now, using defined mixtures of Cd and Zn perchlorates (Cd^{2+} concentrations down to 1 mM and constant Zn^{2+} concentration of 50 mM) gave only CdSe. It is clear that much smaller Cd^{2+} concentrations are required if appreciable concentrations of Zn are to be introduced into the deposit. Larger deposition current densities could also be used in principle, but this may not be practical owing to the low-saturation Se concentrations in the deposition solutions (<10 mM). Since Cd (and CdSe) deposit at considerably more positive potentials than the corresponding Zn species, codeposition of Zn with Cd requires a diffusion-limited concentration of Cd in solution. The preliminary results shown below were obtained from deposits using the Cd-contaminated $Zn(ClO_4)_2$ plating solution with a Zn^{2+} concentration of 50 mM and an estimated Cd^{2+} concentration no greater than 0.1 mM.

The plating efficiency of the (Cd, Zn)Se appears to be much lower than that of either CdSe or CDs. Figure 1.4 shows a TEM micrograph of a nominally 8 nm-thick film (plated at 1 mA cm^{-2} for 5 s at 120 °C). The coverage of the Au is still not complete, in contrast with the situation for CDs and even more for CdSe using the same deposition charge. The crystals are very tiny – ca. 2 nm. The ED pattern (inset in Figure 1.4) shows that the deposit is highly oriented.

An XRD of the above sample is shown in Figure 1.5. The height of the nanocrystals measured from the peak broadening is ca. 5 nm – more than twice the average lateral dimension. Also shown in this figure, by vertical lines, are the expected bulk peak positions of CdSe, ZnSe and $Cd_{0.78}Zn_{0.22}Se$, assuming Vegard's law is valid for this system (see the XPS results below). The peak position lies between CdSe and ZnSe and closer to the former.

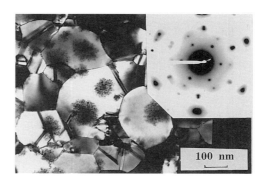

Figure 1.4. TEM image of $Cd_{0.78}Zn_{0.22}Se$ (nominal composition – see text) ED onto Au. Inset: SAD pattern of this deposit.

100 nm

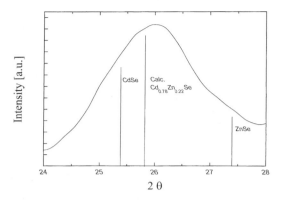

Figure 1.5. XRD spectrum of a 15 nm (nominal thickness – the actual thickness is less) $Cd_{0.78}Zn_{0.22}Se$ deposit on Au. The literature values for the positions of the CdSe and ZnSe {00.2} reflections are shown. The expected position of the corresponding $Cd_{0.78}Zn_{0.22}Se$ peak is also shown.

An XPS analysis of the deposit gave a Cd:Zn ratio of 3.5:1; from this, a composition of $Cd_{0.78}Zn_{0.22}Se$ was inferred (there was some excess Se as is normally observed for incompletely covered Au owing to chemisorption of Se on free Au). Returning to the XRD peak in Figure 1.5, we see that the peak position is shifted to a higher angle (smaller {00.2} spacing) than that expected from the above bulk position by 0.7%. This shift is less than the 1.7% contraction measured for CdSe on Au; since the nanocrystals are smaller and (almost) epitaxial, a larger contraction might be expected. However, the results are in qualitative accord with the lattice mismatch principle. There is a -2.1% mismatch between $Cd_{0.78}Zn_{0.22}Se$ and {111} Au, considerably larger than the mismatch between CdSe and Au (-0.6%). Therefore, to a first order, if we ignore effects other than lattice mismatch, a crystal size of ca. 1.5 nm could be estimated for this system. This is only slightly smaller than the ca. 2 nm measured. Both this slight discrepancy and the smaller shift of the XRD peak than anticipated suggest that the composition is somewhat more Cd rich than the value we derive from the XPS results, a distinct possibility owing to the lack of reliability of XPS elemental composition measurements in general and in particular since the technique is surface specific while the composition might not necessarily be homogeneous.

1.6.1.1 CdS/Au

CdS has a 3.9% smaller lattice than CdSe and a -4.5% mismatch with {111} Au (in the 2:3 ratio). This is not very different from the $+4.1\%$ mismatch between CdSe and Pd, although in the opposite direction (the CdS should be stretched by the gold). Based purely on a mismatch argument, therefore, we might expect the CdS/Au system to behave like the CdSe/Pd one, i.e., a largely disordered deposit, possibly with some local order. We note that consecutive electrodeposition of S and Cd on the {111} surface of Au has been shown to form an epitaxial monolayer of CdS [25] and that this epitaxy follows the same 2:3 ratio as we find for CdSe on Au [16]. Based on mismatch strain, however, we do not expect this situation to exist for much larger thicknesses.

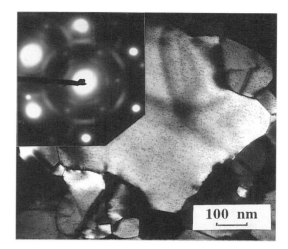

Figure 1.6. TEM image of 5 nm (nominal thickness) CdS ED onto Au. Inset: SAD pattern of the deposit.

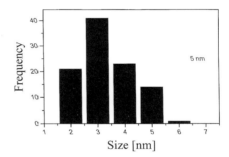

Figure 1.7. Histogram of the crystal size distribution from the CdS deposit in Figure 1.6.

In fact, the CdS/Au system shows very different behavior than CdSe/Pd and is closer to the CdSe/Au combination, in spite of the large (compared to CdSe/Au) mismatch. Figure 1.6 shows a TEM micrograph of a (nominally) 5 nm-thick CdS deposit on Au. The crystal size is typically 3–4 nm with a fairly narrow size distribution, shown in the histogram in Figure 1.7. The coverage of the substrate is not complete, suggesting a lower plating efficiency compared with CdSe, at least in these early stages of deposition. Preliminary XPS elemental analyses of the deposits suggest that little CdS is deposited in the first stages of electrodeposition, and that efficient CdS deposition occurs only at nominal thicknesses above 2 nm (compare qualitatively similar behavior for CdSe deposition discussed above). Complete coverage is obtained for a nominal thickness of ca. 8 nm. The arc-shaped SAD pattern of this CdS deposit (inset in Figure 1.6) shows that, unlike CdSe on Au, the CdS QDs are not completely epitaxial with the Au, although there is a pronounced preferred orientation. As for CdSe/Au, however, the CdS crystals show a preferred orientation with the Au lattice, unlike the 30% rotation that occurs for CdSe on Pd substrates.

The deposits exhibit increasing crystal size and gradual loss in orientation with increasing deposit thickness, as for CdSe on Au. Thus the average crystal size for a nominal 10 nm-thick film ($\leq$ two layers of CdS) is ca. 6 nm. Peak broadening of the XRD spectrum (the {00.2} planes) was used to measure the average height of the crystals, as described above for CdSe/Au. A well-defined XRD peak could only be obtained for films of nominal thickness 10 nm and greater, corresponding to slightly more than a monolayer of crystals. A coherence length (equivalent to nanocrystal height) of ca. 6 nm was obtained for the 10 nm film, in agreement with the lateral dimensions measured by TEM imaging. The position of the {00.2} XRD peak was found to be identical to that of bulk CdS, in contrast with the large contraction (-1.7%) found for the (thinner) CdSe/Au case. Since thinner (one crystal layer or less) samples could not be measured by our XRD equipment, it is not clear whether the XRD peak is dominated by the partial second layer of CdS (in this size range, larger crystal sizes often dominate the XRD pattern, even if present in appreciably smaller amounts) or, less probably, whether this really reflects a much smaller vertical strain in the CdS, compared with the CdSe nanocrystals.

The smaller crystal size of CdS/Au compared to CdSe/Au is in qualitative agreement with the lattice mismatch theory. Quantitatively, we would have expected the difference to be larger, i.e., smaller (or amorphous) CdS crystals, if we assume heteroepitaxy. There are a number of factors that could be invoked in a general manner to explain this, including the nonperfect epitaxy, different surface energy of CdS, and different chemical reactivity with the Au. Unlike the CdS/Pd system described below, however, such factors do not appear to dominate the general behavior of the system.

1.6.1.2 CdS/Pd

CdS has a $+0.17\%$ mismatch with Pd in the 2:3 ratio. Based on the mismatch principle, therefore, we expect relatively large crystals of CdS to form on Pd.

Figure 1.8 shows a Pd substrate onto which CdS (nominally 2.5 nm thick) has been deposited. Very small, slightly oblate crystals typically 2.5×2.0 nm in size and with a narrow size distribution are observed. Coverage of the Pd by the deposit is high (close observation shows the presence of similar nanocrystals, but with lower contrast, between the better-resolved dark ones): this would be expected if the height of the nanocrystals was ca. 2.5 nm. However, there are two additional factors that must be taken into account here. The first is the total absence of any diffraction pattern, either ED or XRD, for this or even thicker (up to 10 nm) layers. The second, and more important, factor is that XPS elemental analysis shows ca. 92% Pd and only 7% S and 0.6% Cd on the 2.5 nm sample (compare with corresponding values of 83, 12, and 5% for the corresponding CdS/Au deposit). This, and particularly the very low Cd concentrations, brings into question the composition of the nanocrystals seen in Figure 1.8

Regardless of the uncertainty in the composition of these nanocrystals, it is clear that the CdS/Pd system cannot be understood by the mismatch principle. The explanation for this may well lie in the XPS results. The fact that the deposition on Pd follows a very different course than either CdS on Au or CdSe on both Au and

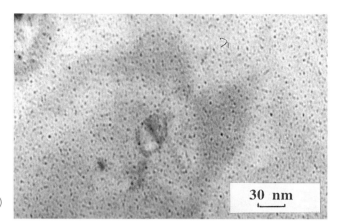

Figure 1.8. TEM image of 2.5 nm (nominal thickness) CdS ED onto Pd.

Pd, and in particular the very low Cd concentrations present even for a nominally 2.5 nm-thick film, where high coverage of the substrate by some material is apparent (Figure 1.8), suggests that this material may be sulfur in greater than monolayer coverage (although in this case, more than the measured value (by XPS) of 7% S would be expected). The relevant parameter for determining the mismatch would then be the lattice spacing of the S (disordered if the S is amorphous). Another, less likely possibility is that chemisorption of S onto Pd might distort (reconstruct) the Pd surface or cause it to become disordered.

1.6.1.3 CdSe/Au–Pd

Since the Au lattice is 0.6% larger than that of CdSe (in the 3:2 ratio) and Pd is 4.1% smaller, then an alloy of Au–Pd containing 12.8% Pd should be perfectly lattice-matched with CdSe, if we assume that the lattice spacings vary linearly with composition. A previous study of this alloy, in bulk form, showed a slight, but definite, deviation from Vegard's law in the Au-rich alloys [26]. If we take this nonlinearity into account, and assume it to hold equally for thin films (an assumption that is not always valid), the composition of perfect match should be ca. 12% Pd. We have therefore carried out a study of the effect of composition of Au–Pd alloys on the electrodeposited CdSe.

The Au–Pd alloys were vacuum-evaporated from tungsten boats using alloys that were previously made by melting Au and Pd together in a crucible by an electron beam. While it might be expected that the evaporated films would be Au rich, owing to the higher vapor pressure of Au compared with Pd, the XRD peak positions of the various alloys were close to those expected from the composition of the evaporant.

Five different alloys were studied: 2.6, 9, 13, 19, and 26% Pd in Au. According to the mismatch-strain principle, the crystal size should grow with increasing Pd concentration and then decrease again for the 19 and 26% alloys; epitaxy should be

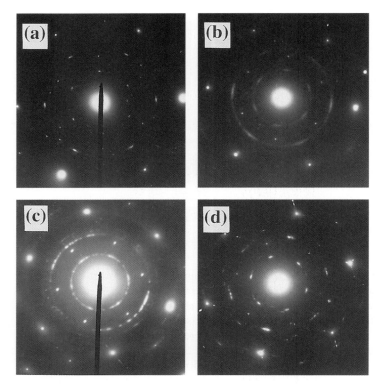

Figure 1.9. SAD patterns of CdSe (nominal thicknesses between 1.2 and 2.5 nm) ED at 0.1 mA cm^{-2} (a–c) onto Au/Pd alloys containing (a) 2.6% Pd, (b) 13% Pd, and (c) 26% Pd and at 0.01 mA cm^{-2} onto 26% Pd/Au (d).

maintained for all the alloys, although it could be partially lost for the 26% Pd substrate.

The behavior we obtained from this system was quite different from this. Under the same conditions normally used for ED CdSe on Au (in particular, at a deposition current density of 0.1 mA cm^{-2}), the degree of epitaxy was gradually reduced – although not lost entirely – with an increase in Pd concentration. Figure 1.9(a–c) shows ED patterns of three CdSe deposits, all between 1.2 and 2.5 nm nominal coverage (less than a monolayer coverage of crystals), deposited at 0.1 mA cm^{-2}. Even for the 2.6% alloy (Figure 1.9(a)), a slight deviation from the perfect epitaxy, obtained for CdSe on pure Au, is visible as arcing of the CdSe diffraction spots. For the 13% alloy (Figure 1.9(b)), where an almost perfect lattice match should occur, the loss of epitaxy is considerable. The 26% alloy sample shows what at first sight appears to be a polycrystalline pattern with some degree of orientation (Figure 1.9(c)). However, if we take into account the 30° rotation that occurs for CdSe on Pd discussed above, it is also possible that there is considerably greater orientation than at first apparent, and that some of the crystals are rotated 30° to the substrate lattice direction. Such a supposition is supported by the form of the ED pattern for the 26% Pd deposit: there is a concentration of intensity of the CdSe reflections aligned with the Au/Pd reflections, and also aligned between adjacent Au/Pd reflections, i.e., 30° rotated, with a noticeable gap between these two concentrations. This effect is only very slightly apparent with the 13% deposit and not at all with the 2.6% one. These results might indicate the presence of Au-rich and Pd-rich domains in the films.

While loss of epitaxy that increases with increasing Pd content is observed for CdSe on the Au/Pd alloys, the epitaxy is very dependent on the deposition current. Figure 1.9(d) shows a SAD of a film similar to that in Figure 1.9(c), only deposited at 0.01 mA cm^{-2} (and for ten times as long, i.e., the same number of coulombs). In this case, the degree of orientation is much better than that obtained at ten times the deposition current. This suggests that the loss of epitaxy at higher currents is due to kinetic limitations: thermodynamically, the epitaxial configuration is still preferred. However, it should be noted that the deposition efficiency is considerably reduced at these lower deposition currents. This may be due to a parasitic electrochemical reaction.

With the breakdown of the mismatch principle as the main factor in determining epitaxy for Au/Pd alloys, it is not surprising that the crystal size does not behave as expected. In fact, the variation of crystal size with substrate composition is irreproducible compared to the large reproducibility obtained with the CdSe/Au system. Typical crystal size varies from ca. 1.5–2 nm on the low side to ca. 8 nm and sometimes even larger. While there is a general trend for the Pd-rich alloys to have somewhat larger crystal sizes than the lower Pd alloys for the same deposition conditions, this is not a strong effect. The crystallite size does increase with increasing amount of deposit. This can be seen from Figures 1.10(a) and (b), which

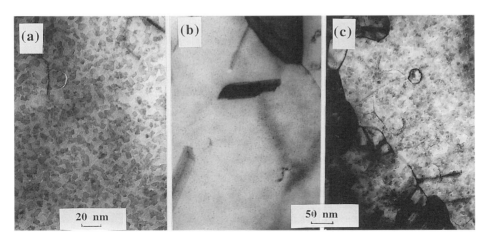

Figure 1.10. TEM images of CdSe ED onto Pd/Au. (a) 13% Pd, CdSe deposited at 0.1 mA cm^{-2} for 16 s at 90 °C; (b) as (a), but CdSe deposited for 32 s; (c) 26% Pd, CdSe deposited at 0.1 mA cm^{-2} for 16 s at 120 °C.

show the crystal size for two samples deposited under the same conditions, except that the former (typical crystal size ca. 5 nm) was deposited for half the time of the latter (ca. 2.5 nm). It is noticeable that, as for the CdSe/Pd system, the TEM images suggest complete coverage of the substrate with CdSe, particularly for the Pd-rich alloys and at high (0.1 mA cm^{-2}) deposition currents. Figure 1.10(c) shows a TEM image of a nominally 2.5 nm deposit on 26% Pd/Au. The coverage appears to be complete. It is difficult to verify this without using HRTEM, as was necessary for CdSe on Pd [20]. However, bright areas, ca. 2–4 nm in size, and usually with poor contrast, can often be seen on these samples (seen on close inspection of Figure 1.10(c)). These bright areas could be holes in the deposit. HRTEM will be carried out in the future to investigate this.

XRD has also been used in an attempt to measure the heights of the crystals (or films, as they may sometimes be). The main conclusion here is that the XRD peak becomes increasingly difficult to see with increasing Pd content. Note that for CdSe on pure Pd, no XRD peak was seen at all. However, in at least some of the alloy samples, quite large CdSe crystals were obtained, in contrast to the very tiny regions of ordered CdSe surrounded by disordered material in the CdSe on Pd. This suggests that these "large" crystals may be defected, in contrast to the defect-free CdSe on Au [15]. There is no consistent difference between the widths of the CdSe {00.2} peaks for the alloy and pure Au deposits, although the useful information that we have obtained from the alloy deposits is still too limited to be sure of this. We do see a narrowing of the peak (increase in crystal height) for lower deposition currents, particularly when very low currents $(0.01 \text{ mA cm}^{-2})$ are used. However, this behavior is also observed for CdSe on Au.

1.7 Bandgap Measurements

Our interest in the nanocrystalline semiconductors described above derived from the ability to tailor the optoelectronic properties if control of crystal size could be attained. In order to demonstrate this ability, we had first and foremost to measure the bandgaps of the semiconductors and show that this parameter is larger than for bulk material. For the ultrathin "films" on metal substrates studied by us, this could not be done easily by normal spectrophotometry, owing both to the very weak absorption of the semiconductor and to the intense absorption and reflection of the underlying metal substrate (ca. 35 nm thick). Instead we have concentrated on two techniques from which the bandgaps can be extracted.

The first is photoelectrochemical photocurrent spectroscopy, where the sample is made one of the electrodes in a suitable electrolyte. Upon illumination with super-bandgap light, a photocurrent is generated and measured as a function of illuminating wavelength. We have previously used this technique to study CdSe on Au [27]. For the first monolayer of CdSe crystals on Au, a bandgap of 1.95 eV was estimated, with the possibility of a second bandgap, due to smaller crystals, of ca. 2.2 eV. These values correspond to approximate crystal sizes of 6 and 4 nm. Since the typical crystal size in these films is 4–5 nm, and there is only a small fraction of crystals as large as 6 nm, it may be that the photocurrent onset is actually a strong sub-bandgap signal, a decided possibility in view of the very high surface area of the CdSe. Another possibility is that, if the electronic contact between CdSe and Au is good, quantization in the substrate direction is reduced or lost and the CdSe is quantized in only two directions. Increase in film thickness (equivalent to additional layers of larger crystals) results in the expected red shift in the photocurrent onset and even the clear appearance of sub-bandgap signals.

The second technique used to measure the bandgap (or rather, to estimate it, since interpretation of the results is very model specific) is conductance spectroscopy [28]. In this method, current–voltage spectroscopy is carried out using a metallized atomic force microscope (AFM) tip that ideally contacts one of the nanocrystals on the substrate. Current can only flow when resonance occurs between the Fermi level of one of the contacts and one of the semiconductor levels. A voltage region where no current flows is observed in both bias directions, and this zero current range corresponds to the bandgap. This measurement must be corrected for the relatively large charging energies for the very tiny crystals (typically 0.2 eV – this charging process can be seen as peaks in the conductance spectra (the Coulomb staircase) and measured by the spacing between consecutive peaks. As with photocurrent spectroscopy, sub-bandgap signals may occur owing to surface states, which will give an apparent bandgap value lower than the real value. Values of bandgaps ranging between 2.05 and 2.4 eV were measured (in Reference [28], the values given, between 1.9 and 2.2 eV, were in error owing to the subtraction of two, rather than only one, times the charging energy), corresponding to crystal sizes between 4.6 and 2.7 nm, respectively (using Reference [9] to correlate bandgap and size). These values agreed well with the measured size distribution of the CdSe.

Here we show the application of the same techniques to measure bandgaps for CdS nanocrystals on Au. Normalized photocurrent spectra of CdS films of various thickness are shown in Figure 1.11. The 5 and 7.5 nm (nominal thickness) samples

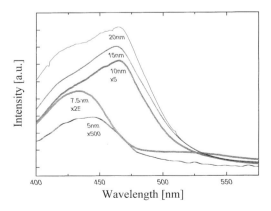

Figure 1.11. Photoelectrochemical photocurrent spectra of CdS deposits of various nominal thicknesses (shown in the figure) on Au. The electrolyte used was 0.2 M Se dissolved in 0.4 M aqueous Na_2SO_3 (sodium selenosulfate).

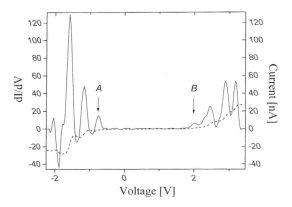

Figure 1.12. *I–V* and associated differential conductance spectra of 5 nm (nominal thickness) CdS on Au measured using a metallized AFM tip.

exhibits a clear blue shift, to ca. 2.6 eV, compared with the thicker films (10, 15, and 20 nm), where a value of 2.35–2.4 eV is estimated (the bulk bandgap of CdS is ca. 2.4 eV, the exact value varying for different literature sources). From the tight binding calculations of Lippens and Lannoo [29], an increase in bandgap of 0.2 eV corresponds to a CdS crystal size of ca. 4.5 nm; this is only slightly larger than the measured size for the 5 nm CdS sample (3–4 nm; see above), particularly since the spectrum will probably be dominated by the larger crystals in the distribution.

The second technique, conductance spectroscopy, provides considerable information on the probe–nanocrystal–substrate circuit properties, in addition to the semiconductor bandgap. Since our interest for the present purpose is to measure bandgaps, we will discuss the additional information only briefly. Figure 1.12 shows an *I–V* spectrum of 5 nm CdS on Au (less than a monolayer of crystals) and the associated conductance spectrum (d*I*/d*V–V*). As described for CdSe/Au above, the zero current region should give the bandgap plus the energy required to charge the nanocrystal, the latter measured by the peak spacing in the conductivity spectrum (assuming the peaks represent Coulomb charging). The position of the first peak for each polarity is quite often ambiguous: Since the first peaks are often small and sometimes broad, they may represent direct charge transfer into a surface state rather than into the first quantized bulk level. From the conductance spectrum in Figure 1.12, the zero current region (between the positions of peaks *A* and *B*) is 2.79 V. Subtracting from this a charging voltage of ca. 0.35 V (see below) gives a value of 2.44 eV for the bandgap. This value is considerably lower than the expected value of between 2.6 and 2.7 eV for these 3–4 nm CdS crystals. If we conveniently assume that the small peak *B* is due to transfer into surface states, the bandgap value would then be 2.71 eV, in line with the expected value. To put this measurement on a more statistical footing, a necessary step owing to the rather large variation between measurements, nine different spectra were taken over different regions of the same sample. Two extreme bandgap measurements were made: the spacing between the first peaks on either side of zero bias and the spacing between the corresponding second peaks, in both cases, less the charging energy. If we average over all the measurements, the first value should be lower than the true bandgap while the second should be higher. The average value of the former was 2.2 eV, while that of the latter was 2.8 eV, with a scatter in values over ca. 20% of the mean. Thus, while this technique is not, at present, very accurate for bandgap measurement, at least of this system, it does give values that are in the right ball-park. In particular, if we compare the values with those obtained for CdSe on Au [28] (for which more accurate bandgap values were obtained), the ability of these measurements to provide a reasonable estimation of the bandgap for very tiny amounts of the relevant semiconductor is clear.

The features, other than the zero current region, in these spectra are of great interest in themselves. The interpeak spacings are a function of nanocrystal charging and/or higher-lying levels (see Reference [4] for an in-depth treatment of these phenomena). At this point, we cannot separate these effects, although we do believe that the charging phenomenon is dominant in our measurements. On that basis, we have used the value of 0.35 V for the charging voltage, taken as the average of the negative (0.4 V) and positive (0.3 V) peak spacings. A basic difference between these measurements with CdS and our earlier ones with CdSe is the lack of symmetry of the first conductivity peaks about the zero bias for CdS, compared with the symmetrical behavior of the CdSe system. Although we do not understand this behavior, we note that the symmetry (or lack thereof) is a function of the equilibrium energy position of the semiconductor levels with respect to the metal Fermi level.

1.8 Conclusion and Speculations

The rationale for the research presented in this paper is to understand the growth processes that determine nanocrystal size and thus to control the optoelectronic properties through control of the nanocrystal size. While possible applications for electrodeposited quantum dots (or nanocrystals in general) are not a main driving force for this research at present (they may become so later), it is of interest to speculate on potential applications for these and similar materials. The first and most obvious use is a general one: the ability to tailor bandgap and energy level spacing of a semiconductor without changing its chemistry allows a wider range of materials with specific properties. This is clearly useful for devices with properties that depend on these properties. Thus light-emitting devices (lasers or electroluminescent displays) could, in principle, be tailored for the exact wavelength output desired. Particularly for lasers, a very narrow size distribution will be necessary to prevent incoherent emission of radiation. The increase in the bandgap also means a decrease in electron affinity of the quantized semiconductor. A change of electron affinity would allow, e.g., control of band offsets at semiconductor heterojunctions (this includes the junction between one semiconductor and the same semiconductor with different nanocrystal size), an important parameter in many devices such as diodes and photovoltaic cells. One of the most frequently cited potential uses of semiconductor nanocrystals is in nonlinear optics. The main reason for good nonlinear optical behavior of these materials is the concentration of the optical transitions in only a few levels. As for lasers, a narrow size distribution will be important for most possible devices, although there may be some uses where a wide size distribution is less of a problem, such as in devices where optical saturation of a subset of crystals (hole burning) is operative. The current–voltage behavior resulting from single electron charging and/or higher-level charge transport can form the basis of future electronic devices. There is still a very large gap between our experiments shown here and a useful device. Problems to be overcome include reproducibility of current–voltage characteristics and the ability to form a third contact (a gate electrode) that is capacitively coupled to the nanocrystals in order to allow fine external control of the nanocrystal energy levels with respect to the Fermi levels of the contacts. It may be possible (and certainly easier) to use an optical signal, rather than an electrical one, for this purpose. It is clear that good reproducibility and the ability to obtain a very narrow size distribution are central elements in most applications considered at present. It is hoped that our efforts to understand and control growth of semiconductor quantum dots (and nanoparticles in general) will help toward realizing these goals.

Acknowledgments

We thank Dr. Udi Meirav (Department of Physics, Weizmann Institute) for useful discussions. G. Hodes and I. Rubinstein acknowledge support of this work by the US Office of Naval Research, Grant No. N00014-93-1-1151.

References

[1] L. E. Brus, *Appl. Phys.* **1991**, *A 53*, 465–474.
[2] L. Banyai, S. W. Koch, *Semiconductor Quantum Dots*, World Scientific, Singapore, 1993.
[3] A. D. Yoffe, *Adv. Phys.*, **1993**, *42*, 173–266.
[4] U. Meirav, E. B. Foxman, *Semicond. Sci. Technol.*, **1995**, *10*, 255–284.
[5] D. J. Eaglesham, *M.* Cerullo, *Phys. Rev. Lett.*, **1990**, *64*, 1943–1946.
[6] P. M. Petroff, S. P. DenBaars, *Superlattices and Microstructures*, **1994**, *15*, 15–21.
[7] Y. Nosaka, K. Yamaguchi, H. Miyama, H. Hayashi, *Chem. Lett.*, **1988**, 605–608.
[8] M. L. Steigerwald et al., *J. Am. Chem. Soc.*, **1988**, *110*, 3046–3050.
[9] C. B. Murray, D. J. Norris, M. G. Bawendi, *J. Amer. Chem. Soc.*, **1993**, *115*, 8706–8715.
[10] X. K. Zhao, J. H. Fendler, *Chem. Mater.*, **1991**, *3*, 168–174.
[11] J. Yang, J. H. Fendler, J. T.-C., T. Laurion, *Microscopy Research and Technique*, **1994**, *27*, 403–411.
[12] V. L. Colvin, A. N. Goldstein, A. P. Alivisatos, *J. Am. Chem. Soc.*, **1992**, *114*, 5221–5230.
[13] G. Hodes, *Isr. J. Chem.*, **1993**, *33*, 95–106.
[14] Y. Golan, L. Margulis, I. Rubinstein, G. Hodes, *Langmuir*, **1992**, *8*, 749–752.
[15] Y. Golan, L. Margulis, G. Hodes, I. Rubinstein, J. L. Hutchison, *Surf. Sci.*, **1994**, *311*, L633–640.
[16] Y. Golan, G. Hodes, I. Rubinstein, *J. Phys. Chem.*, **1996**, *100*, 2220–2228.

[17] A. D. Stuckes, G. Farrell, *J. Phys. Chem. Sol.*, **1964**, *25*, 477–482.

[18] (a) Y. Golan, J. L. Hutchison, I. Rubinstein, G. Hodes, *Adv. Mater.*, **1996**, *8*, 631–633.
(b) Y. Golan, A. Hatzor, J. L. Hutchison, I Rubinstein and G. Hodes, *Isr. J. Chem.*, **1997**, *37*, 303–313.

[19] D. J. Srolovitz, M. G. Goldiner, *JOM – Journal of the Mineral Metals and Materials Society*, **1995**, *47*, 31.

[20] Y. Golan, E. Ter-Ovanesyan, Y. Manassen, L. Margulis, G. Hodes, I. Rubinstein, E. G. Bithell, J. L. Hutchison, *Surf. Sci.*, **1996**, *350*, 277–284.

[21] G. Hodes, A. Albu-Yaron, *Proc. Electrochem. Soc.*, **1988**, *88–14*, 298–303.

[22] G. Hodes, T. Engelhard, A. Albu-Yaron, A. Pettford-Long, *Mat. Res. Soc. Symp. Proc.*, **1990**, *164*, 81–86.

[23] Y. Mastai, G. Hodes, *J. Phys. Chem.*, **1997**, *101*, 2685–2690.

[24] Y. Golan, B. Alperson, J. L. Hutchison, G. Hodes, I. Rubinstein, *Adv. Mater.*, **1997**, *9*, 236–238.

[25] U. Demir, C. Shannon, *Langmuir*, **1994**, *10*, 2794–2799

[26] A. Maeland, T. B. Flanagan, *Can. J. Phys.*, **1964**, *42*, 2364–2366.

[27] B. Alperson, S. Cohen, Y. Golan, I. Rubinstein, G Hodes, *NATO ASI Series 3* (Ed.: E. Pelizzeti), Kluwer, **1995**, Vol. 12, pp. 579–590.

[28] B. Alperson, S. Cohen, I. Rubinstein, G. Hodes, *Phys. Rev. B.*, **1995**, *52*, R17017–17020.

[29] P. E. Lippens, M. Lannoo, *Phys. Rev. B*, **1989**, *39*, 10935–10942.

Chapter 2

Oriented Growth of Nanoparticles at Organized Assemblies

F. C. Meldrum

2.1 Introduction

Many techniques exist to produce oriented arrays of particles on a compatible substrate. Indeed, the epitaxial growth of one material on another, whereby sets of lattice planes in the substrate and developing phase have corresponding form, has been widely investigated to manufacture technologically important devices. The development of much novel physics, for applications in a wide range of electronic, photonic, and microwave devices has been based upon the production of superior quality thin films, whose structure can be defined at a molecular level [1–5].

Traditional methods of producing epitaxial thin films can be broadly classified as gas, liquid, and solution phase. The principal gas phase methods are molecular beam epitaxy (MBE) and metal organic vapor phase epitaxy (MOVPE), although many hybrid techniques exist. Both MBE and MOVPE provide high growth rates and can manufacture intricate structures. A further gas phase technique, atomic layer epitaxy (ALE), offers complete control at the monolayer level and uses precisely defined incremental growth at the substrate to generate a film, a monolayer thickness at a time. Liquid phase epitaxy (LPE) describes the deposition of a crystalline film onto a single crystal substrate from a supersaturated liquid metal solution and can yield superior quality epitaxial films. Epitaxial thin films can also be formed by solution phase techniques, which are experimentally straightforward and do not demand the extreme vacuum or temperature conditions of gas or liquid phase methods. Solution deposition methods encompass chemical bath, electrochemical techniques, and successive ionic layer adsorption and reaction (SILAR), which operate at ambient temperatures and pressures and use ionic solutions as precursors. Chemical solution deposition enables the preparation of films of compound semiconductors such as CdS, PbS, $PbSe$, $CdSe$, and $ZnSe$ and is based upon a controllable chemical reaction that proceeds at a rate defined by the composition of the reaction solution [6–10]. Electrochemical deposition represents one of the earliest methods for the preparation of thin films and has been applied to the epitaxial deposition of semiconductor nanoparticles on a range of substrates [11–13]. SILAR (also termed liquid phase atomic layer epitaxy (LPALE)) is the solution phase counterpart of the gas phase atomic layer epitaxy (ALE) method and employs the alternate exposure of the substrate to solutions of the reactant materials, with copious washing between the reactant solutions to remove unbound or unreacted ions [14, 15]. As a further related method, electrochemical atomic layer epitaxy (ECALE) relies upon the alternate underpotential electrochemical deposition of two elements that react to form a binary compound on the substrate [16, 17].

Epitaxial growth is necessarily determined by the structural and dimensional correspondence between the substrate and developing phase. While the majority of work has been carried out on single crystal inorganic substrates, a range of organized organic assemblies have also been investigated as surfaces to support the growth of oriented arrays of inorganic nanoparticles. Self-assembled monolayers and multilayers, Langmuir–Blodgett multilayers, and Langmuir monolayers formed at the air–water interface have all been successfully utilized. The organic matrix selects the nucleating crystal face and can further act to direct the growth of arrays of coaligned crystals. The purpose of the studies is multifold. Technologically important materials such as zeolites [18, 19] and semiconductors [20–24] can be produced of defined size, morphology, and orientation; for example, the physiochemical properties of PbS crystals have been shown to vary as a function of the particle morphology [25]. In addition, organic matrices have the potential for introducing lateral order into a crystal film. Self-assembled films can be readily patterned using standard photolithographic techniques and have been used as templates for the development of ordered crystalline arrays [26, 27].

The studies also yield information on the processes of crystal nucleation and growth. The structure of the organic substrate can be readily determined and tailored to a selected nucleating inorganic crystal face. Experimental methods such as grazing incidence angle X-ray diffraction (GID) and transmission electron microscopy (TEM) can be applied to determine the relationship between the crystal and matrix [28–31]. Particle formation in association with organic matrices can in many ways be considered to be biomimetic. In natural systems, inorganic crystal growth always takes place in association with organic membranes, to define the size, shape, and orientation of the growing crystals [32]. Although it is difficult to demonstrate whether a well-defined spatial relationship exists between the associated organic matrix and crystal phase, a number of systems have been studied in depth. In mollusc nacre, the *a* and *b* axes of the aragonite crystallographic axes are aligned with respect to the chitin fibers and *b*-sheet polypeptide chains that form the organic matrix framework, while the *c* axis lies perpendicular to the matrix plane. Such a pronounced orientational relationship strongly suggests that crystal growth is directed by epitaxial matching between the organic matrix and inorganic crystals. Thus, investigations into the influence of organic matrices on crystal growth can improve understanding of biomineralization and suggest routes to the preparation of materials with superior properties.

2.2 Oriented Crystal Growth on Self-assembled Monolayers and Multilayers

2.2.1 Growth of Zincophosphate Zeolites on Zirconium Phosphate Multilayers [18]

Ordered self-assembled monolayers, formed by chemisorption of suitable molecules onto a partner substrate, can be used as surfaces on which to deposit oriented inorganic crystal films. Organophosphonate multilayers were prepared on gold substrates by adsorption of 11-mercapto-1-undecanol, which was then phosphorylated with $POCl_3$ and 2,4,6-collidine to give the corresponding phosphate. The prepared substrate was then immersed in solutions of zirconyl chloride and 1,10-decanediylbiphosphonic acid to form a zirconium phosphate trilayer. The zincophosphate zeolites were grown on these films by placing the prepared substrate in the zeolite "reaction solution" at 7 °C for 5 hours and finally washing with water. Examination of the film after this period by scanning electron microscopy (SEM) and X-ray diffraction showed that zeolite crystals of basic octahedral morphology and triangular basal faces had grown on the film and that over 90% were nucleated on the {111} face (Figure 2.1). No alignment of the nucleation plane with respect to the substrate was observed. Control experiments in which clean gold surfaces were immersed in the zeolite reaction solution showed that the presence of the multilayer was necessary for crystal deposition. The zirconium phosphonate multilayer interacts strongly with the {111} face of small zeolite nuclei formed in the reaction solution, presumably via electrostatic and geometric matching processes. The $[-PO_3H_2]$ surfaces stabilize the zeolite {111} face, resulting in arrays of crystals deposited specifically on this face.

2.2.2 Oriented Aluminophosphate Zeolite Crystals Grown on Self-assembled Monolayers [19]

Arrays of aluminophosphate crystals were produced on zirconium phosphonate (ZrP) multilayers. The pore structure of the zeolites comprises a hexagonal array of parallel cylinders, and oriented growth of the zeolites resulted in vertical alignment of the channels with respect to the substrate. The ZrP multilayers were prepared by the methodology described above [18]. Aluminophosphate zeolite crystals were synthesized hydrothermally from AlOOH, 85% orthophosphoric acid, 1,2-diazabicyclo[2,2,2]octane, and water. The reaction temperature and fraction of water in the solution were varied and were shown to profoundly affect the orientation, morphology, and aspect ratio of the crystals. For crystal growth on the multilayers, the substrate was placed face down in the gel for 8 to 12 hours, prior to washing and drying.

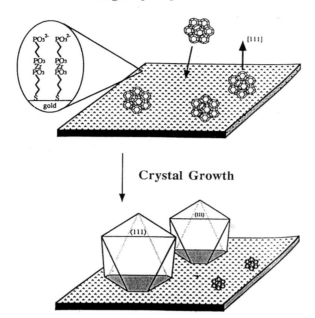

Figure 2.1. Proposed reaction scheme for the growth of zinco-phosphate molecular sieves on organophosphonate films. Reproduced with permission from Reference [18].

XRD examination of the crystals on the substrate demonstrated that the crystals were oriented with their *c* axes perpendicular to the substrate. It was considered that the {001} faces interact most strongly with the phosphonate surface, thus favoring retention of this orientation. The influence of the temperature and water content on crystal growth was investigated. Water content of $d = [H_2O]/[Al_2O_3] = 60$ and temperatures of 150 °C resulted in semispherical aggregates. Dilution produced larger hexagonal prismatic crystals with higher aspect ratios. When $d = 300$, almost all of the crystals exhibited hexagonal morphologies and {001} orientation. At further dilutions to $d = 400$ and 600, the majority of crystals were aligned with their *c* axes slightly tilted from the vertical; the misorientation probably results from the high aspect ratio of the crystals, which causes the vertical orientation to be somewhat unstable.

2.2.3 Nucleation and Growth of Oriented Ceramic Films on Self-assembled Monolayers [33]

Goethite (α-FeOOH) was deposited on self-assembled monolayers comprising sulfonate-terminated alkane silanes on silica, via the hydrolysis of iron nitrate solution at 70 °C and pH 2.1. Examination of the films as planer areas and in cross section showed that the crystals were densely packed and comparable in structure to vapor-deposited films. This suggested that film growth was by a heterogeneous mechanism rather than by adsorption of preformed colloids precipitated in the bulk solution. The films comprised individual crystallites growing perpendicular to the film surface to form columnar structures. Single crystals extended from the substrate to the top of the film. Diffraction patterns were solved as goethite, which is described by an orthorhombic unit cell where $a = 4.62$, $b = 9.95$, and $c = 3.01$. Selected area diffraction patterns primarily comprised $\{101\}$, $\{002\}$, and $\{200\}$ reflections, indicating a high degree of b axis orientation perpendicular to the substrate, and that growth occurs from "ac" planes. No specific orientation of the basal plane with respect to the self-assembled film was observed. Akageneite (β-FeOOH) films were also prepared by the same experimental procedure on substitution of $FeCl_3$ for $Fe(NO_3)_3$. The crystals were again oriented such that the [001] axis was parallel to the substrate.

The use of sulfonated polystyrene substrates in place of the self-assembled monolayers yielded the same experimental results. That the disordered polystyrene surfaces could support oriented crystal growth showed that an organized substrate was not a prerequisite for producing crystal films, aligned along a single axis. The orientational effects can not be explained in terms of epitaxial matching between the inorganic crystal and underlying substrate, and must be less specific than strict spatial relationships between surface sites and solution species. Indeed, the concentration of cations at the negatively charged sulfonate surface may play a role in the selection of the nucleation face. XPS studies demonstrated that one $FeOH^+$ species coordinates to a pair of sulfonate sites on the monolayer, which may favor nucleation of the "ac" plane of goethite, which is the crystal face bearing the highest density of Fe^{3+} cations.

2.3 Epitaxial Crystal Growth on Langmuir–Blodgett Films [34]

Multilayers of behenic acid $(CH_3(CH_2)_{20}COOH)$ were utilized as substrates on which to precipitate strontium sulfate. Standard Langmuir–Blodgett dipping techniques were used to deposit multilayers of behenic acid on coated electron microscope grids. The grids were then placed in supersaturated solutions of 1:1 $SrNO_3$:Na_2SO_4 and were incubated for varying lengths of time. The product of crystallization varied according to whether the LB film was hydrophobic or hydrophilic in character, which was controlled by the number of dipping cycles employed during the multilayer preparation. Thin, disklike crystals nucleated from a $\{010\}$ face on the hydrophobic surfaces, while open, floret-shaped crystals with needlelike outgrowths nucleated on the hydrophilic films. Diffraction data from these crystals did not match the orthorhombic $SrSO_4$ structure. In contrast, control experiments performed in the absence of an LB film produced crystals of tabular, rhombic morphologies. Interactions between the highly ordered organic surface and ions present in solution was proposed to influence crystal nucleation and growth.

2.4 Langmuir Monolayers as Templates for Epitaxial Crystal Growth

Langmuir monolayers represent the best-characterized system in which the influence of an ordered organic matrix on crystal growth has been studied. Monolayers provide a model system in which to mimic biomineralization processes [35–45], to study the fundamental processes of crystal nucleation and growth [28–31, 46–50], and to produce highly oriented arrays of technologically important materials such as semiconductors [20–25]. The system offers many attractions. Experimentally simple, the monolayer is formed at the air–water interface and crystals are precipitated from the solution; heterogeneous nucleation at the monolayer is favored over bulk precipitation. The method has been applied to a wide range of systems, such as the precipitation of NaCl, $CaCO_3$, and $BaSO_4$ from supersaturated solutions, cooling to induce ice nucleation, and diffusion of a reactant gas through the monolayer to precipitate such materials as CdS and PbS. The monolayer structure can be determined using techniques such as grazing incidence angle X-ray diffraction, cryogenic TEM, and Brewster angle microscopy and correlated with the structure of the nucleating crystals. In addition, the monolayer and associated crystals can subsequently be transferred to a solid substrate as either mono- or multilayer arrays by standard Langmuir–Blodgett dipping techniques, without disruption of the original crystal alignment.

2.4.1 Epitaxial Growth of Semiconductor Nanoparticles under Langmuir Monolayers

Langmuir monolayers have been successfully employed to direct the growth of oriented arrays of semiconductor crystals [23]. Epitaxy has been demonstrated in the systems PbS/AA [24], PbS/(AA and ODA) [22], PbSe/AA [20], and CdS/AA [21], where AA = arachidic acid and ODA = octadecylamine. A general experimental methodology was used in all cases. The selected surfactant was spread on a subphase of the metal ion solution in a Langmuir trough, and evaporation of the carrier solvent was allowed. The surfactant was then compressed to a selected surface pressure. The trough was isolated from the environment, and a known volume of reactant gas (such as H_2S or H_2Se) was injected into the container. Slow diffusion of the gas through the monolayer resulted in the precipitation of the insoluble semiconductor particles at the monolayer–solution interface. The rate of growth and size of the particles could be controlled by varying the volume of gas injected and the exposure time.

2.4.2 Formation of PbS Crystals under Arachidic Acid (AA) and Octadecylamine (ODA) Monolayers [22, 24]

Oriented films of PbS crystals were produced in association with AA, ODA, and mixed AA/ODA monolayers. The surfactant was spread on a subphase of $Pb(NO_3)_2$ and PbS precipitation induced by injection of H_2S into the environment enclosing the compressed monolayer. The crystal nucleation face and degree of selectivity was controlled by variation of the monolayer composition.

Crystallization under a 100% AA monolayer compressed to a solid state resulted in a population of crystals of equilateral triangle morphologies (Figure 2.2). The size of the crystals depended on the time of exposure to the H_2S gas. After 5 minutes the crystals exhibited sides of 29.7 nm, while 30 minutes exposure resulted in significantly larger crystals of side 60.7 nm. The orientation of the crystals was investigated by selected area diffraction and revealed the epitaxial relationship between the crystals and the monolayer. Analysis of a 2 μm area showed a single crystal pattern of sixfold symmetry, demonstrating that the PbS particles nucleated from the same crystal face and that they were oriented at angular separations of 60°. The diffraction pattern showed reflections corresponding to {220}, {422}, and {440} planes, indicative of a {111}-type nucleation face. PbS crystals were also precipitated under monolayers maintained at lower surface pressures, and epitaxial growth was observed even under gaseous state monolayers. In these cases, circular domains of aligned particles were observed. It was considered that small domains of crystallized AA provide an initial template for the nucleation of oriented PbS. The precipitation of a PbS particle then induces growth in the area of the AA domain, which in turn facilitates nucleation of further PbS particles.

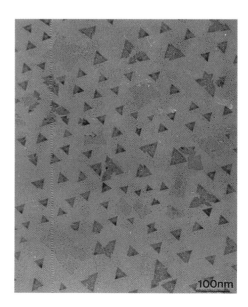

Figure 2.2. TEM image of PbS crystals produced under AA monolayers.

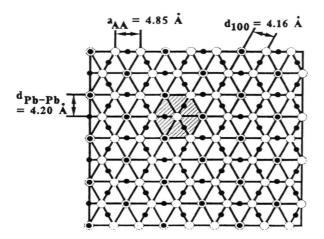

Figure 2.3. Schematic representation of the overlap between Pb^{2+} ions and AA headgroups. The full circles represent the Pb ions, the empty circles the AA headgroups. A unit cell is shaded.

The effect of the monolayer in directing crystal growth was rationalized by considering the structural match between the AA and the PbS lattices (Figure 2.3). Synchrotron X-ray studies of crystalline cadmium arachidate monolayers spread on a $CdCl_2$ subphase [47, 48] have shown that they crystallize in a hexagonal close-packed array with a lattice constant of $a = 4.85$. The surfactant molecules are in a fully extended state and possess a planer zigzag conformation. An experimental value of $a = 4.81$, as derived from surface pressure – surface area isotherms showed good agreement and was used in the analysis of the PbS system. Epitaxial growth of the PbS crystals from the monolayer resulted from the structural match between the PbS $\{111\}$ face and the monolayer. PbS is of cubic, NaCl-type structure and lattice constant $a = 5.9458$. The Pb–Pb and S–S separations of 4.20 Å in the $\{111\}$ face closely match the $d\{100\}$ spacing of 4.16 Å for AA. The spatial match is almost perfect, the mismatch being in the order of only 1%.

Doping of ODA into the basic AA monolayer to ratios of AA:ODA 5:1, 2:1, 1:1, 1:2, 1:5, and 0:1 influenced the nucleation face of the PbS crystals and the degree of specificity. The experimental results are summarized in Table 2.1. At composition AA:ODA 5:1 and $\pi = 30$ mN m^{-1}, the monolayer directed the growth of a population of equilateral triangles of side 45 nm, similar to those produced under a pure AA monolayer. Selected area diffraction showed patterns of sixfold symmetry which were solved in terms of nucleation from a $\{111\}$ plane. However, some subtle variations as compared with the pure AA monolayers were observed in that most of the triangles had indentations in the middle of their edges, and that the spots in the diffraction pattern were more elongated, indicative of less perfect alignment.

Variation of the monolayer composition to a AA:ODA ratio of 2:1 resulted in 50 nm PbS crystals with right-angle triangle morphologies for monolayers maintained at surface pressures of $\pi = 30$ mN m^{-1}. A small proportion of 100 nm irregular crystals were also observed. Selected area diffraction patterns showed twelvefold symmetry and reflections from $\{200\}$, $\{220\}$, $\{400\}$, and $\{420\}$ planes. The right-angle triangles thus nucleate from a $\{001\}$ basal plane and lie in direc-

Table 2.1. Summary of experimental observations on PbS crystallites grown under monolayers of mixed arachidic acid (AA) and octadecylamine (ODA).

AA:ODA	Electron Diffraction	Orientation	Morphology
1:0	Sixfold {220}, {420}, {420},...	Epitaxy on PbS {111}	Equilateral triangles 45 ± 9 nm
5:1	Sixfold {220}, {420}, {440},... more diffuse reflections than 1:0	Epitaxy on PbS {111}	Indented triangles 45 nm
2:1	Twelvefold {200}, {220}, {400}...	Epitaxy on PbS {001}	Irregular (> 100 nm) and some right-angle triangles (50 nm)
2:1[b]	Sixfold {111}, {200}, {222}, {311}, {400}	Epitaxy on PbS {110}	Predominately right-angle triangles
1:1	Twelvefold {200}, {222}, {311}, {400}	Epitaxy on PbS {001}	Predominately right-angle triangles
1:2	Powder-type ring {200}, {220}, {400} reflections	No Epitaxy PbS {001} texturing	Square-shaped, 80 nm, sparsely distributed
1:5	NA	NA	Crystals grown in bulk solution
0:1	NA	NA	Crystals grown in bulk solution

tions separated by 60°, so generating the apparent twelvefold symmetry. While the maintained surface pressure had little effect on the crystallization of PbS particles in the case of 100% AA monolayers, it was shown to be very important for the mixed monolayers. In the case of the 2:1 AA:ODA monolayers, crystallization under monolayers held at surface pressure $\pi = 0$ mN m^{-1} and headgroup area 23 Å^2 mol^{-1} produced mostly right-angle triangles of mean side 20 nm and angles $75 \pm 8°$. Selected area diffraction yielded patterns of sixfold symmetry and reflections corresponding to {111}, {200}, {222}, {311}, and {400} planes, consistent with a nucleation face of {110}.

At a monolayer composition of AA:ODA = 1:1 and surface pressure $\pi = 30$ mN m^{-1}, PbS crystallized in a similar manner to the AA:ODA = 2:1 situation, except that fewer irregular crystals were viewed and there was a higher percentage of right-angle triangle morphologies. Electron diffraction patterns were of sixfold symmetry, but spots were elongated, demonstrating nonideal alignment. Monolayers in which the proportion of ODA exceeded that of AA (AA:ODA = 1:2) did not support epitaxial growth of PbS crystals. At a surface pressure of $\pi = 30$ mN m^{-1}, randomly oriented crystals of rectangular shape and large size distribution were observed. Selected area diffraction patterns were of powder type in which the most intense reflections were from {200}, {220}, {400}, and {420} planes, suggesting that a {001} orientation predominated. Monolayers of higher ODA concentration to the level AA:ODA = 1:5 failed to support crystal growth. Irregular, brown particles formed in the bulk solution, and with time, precipitated out.

Depending upon the concentration ratios of the surfactants comprising the monolayer, they may be either statistically distributed or segregated into domains. The amine headgroups of ODA hydrogen bond to the carboxyl moiety in AA, so providing the structural integrity of the mixed monolayer. The surface pressure – surface area isotherms of the mixed monolayer were identical to that of pure AA for AA:ODA of 5:1 and 2:1, implying that hexagonal close packing was maintained on condition of excess AA. Experiment demonstrated that introduction of ODA into the AA matrix alters and finally destroys the lattice matching between the monolayer and PbS crystals. That epitaxial growth switched from the {111} to {001} face of PbS can be attributed to a reduced concentration of Pb^{2+} ions under a given monolayer area and an alteration in the geometric alignment of the monolayer headgroups. The possible geometric matching between PbS {001} with the monolayer is shown in Figure 2.3. A close spatial match exists between the monolayer $d\{10\} = 4.16$ and the Pb–Pb separation in the {001} plane of $d\{110\} = 4.20$.

2.4.3 Investigation of PbS Physiochemical Properties as a Function of Crystal Morphology [25]

The electrical, electrochemical, and spectroelectrical properties of the PbS nanoparticle films prepared under Langmuir monolayers were investigated as a function of the morphologies of the crystals. Three contrasting nanocrystalline films were studied. PbS-I films comprised equilateral triangle PbS precipitated under AA monolayers, PbS-II nanoparticle films were right-angle triangle PbS particles grown under mixed monolayers of AA and ODA, and PbS-III films were nonoriented and were prepared under hexadecyl phosphate monolayers. Examination of the potential-dependent absorption spectra on optically transparent conducting glass supports, and photocurrent on platinum supports, demonstrated differing behavior for PbS-I, -II, and -III.

The absorption spectra of all three films showed broad bands with maxima at 380 nm and absorption edges of about 800 nm (Figure 2.4). The structured absorp-

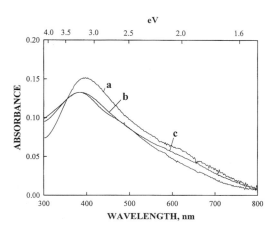

Figure 2.4. Absorption spectra of (a) PbS-I, (b) PbS-II, and (c) PbS-III.

tion spectrum and shift in the absorption edge from the bulk value of over 3000 nm demonstrates size quantization. The absorption spectra were also measured under application of potentials between +0.30 V and −0.80 V. The epitaxial (PbS-I and PbS-II) films behaved quite differently from the randomly oriented particle films (PbS-III), in that biasing to negative potentials increased the near-infrared absorption in the former case, while no change in absorption at wavelengths longer than 700 nm was observed for PbS-III. The absorption increase is likely to derive from the accumulation of trapped conduction band electrons.

The variation of the absorption at the absorption maximum with the applied potential was more complex (Figure 2.5). In the case of PbS-I (Figure 2.5(a)), A_{384} increased rapidly with a reduction in the potential from 0 V to −0.4 V, before subsequently decreasing, most rapidly in the −1.6 V to −2.0 V range. The PbS-II films also showed an increase in A_{386} on decreasing the potential from 0 V to −0.4 V (Figure 2.5(b)). The absorption then decreased in the −0.4 V to −1.2 V range, increased again from −1.3 V to −1.8 V, and finally decreased again at voltages more negative than −18 V. In contrast, the PbS-III films displayed relatively simple behavior (Figure 2.5(c)), with A_{380} decreasing linearly in the 0 V to −1.7 V range and then more rapidly below −17 V. The PbS films also showed differences in their capacitance vs. voltage (Figure 2.6) and photocurrent (Figure 2.7) behaviors. The increase in the photocurrent at negative potentials is characteristic of p-type semiconductors.

The variation of the absorbance, photocurrent, and film capacitance with applied voltage depends upon such factors as the electron concentration in the electronic bands, in the presence of charge traps and available surface states, and on the interfacial electrochemical and photochemical processes. Thus, although a complete rationalization of the spectroelectrochemical properties of the PbS nanoparticulate has not yet been realized, the results clearly demonstrate that control of crystal size, morphology, and orientation permit selection of the physical properties of the film.

2.4.4 Epitaxial Growth of Cadmium Sulfide Nanoparticles under Arachidic Acid Monolayers [21]

Arachidic acid (AA) monolayers were employed to direct the growth of CdS particles. The monolayer was spread on a $CdCl_2$ solution and precipitation induced by diffusion of H_2S through the monolayer. The effect of temperature was studied by maintaining the trough either at 20 °C or at 3–4 °C. TEM examination of the crystals produced at room temperature showed that both isolated particles and areas of densely packed crystals were present (Figure 2.8). The crystals were predominantly rod shaped, being 50–300 nm in length and 5–15 nm in width. Extensive twinning, apparent as dendritic outgrowths from the main crystal length, was present. That the crystals were highly oriented with respect to the monolayer was apparent in that they only grew in three directions at angular separations of 120°. In addition to the rods, a smaller proportion of crystals were observed with disklike morphologies. These particles were typically present as small outgrowths from the rodlike CdS. Selected area diffraction of the crystals were of single crystal

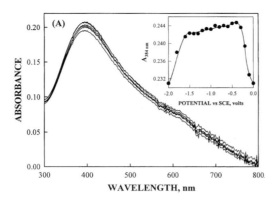

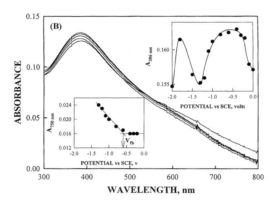

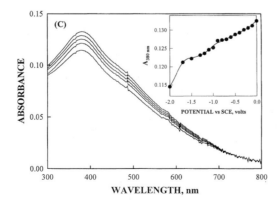

Figure 2.5. (a) Absorption spectra of PbS-I at different applied potentials. The insert shows the absorbances at 384 nm as functions of the applied potential. (b) Absorption spectra of PbS-II at different applied potentials. The inserts show the absorbances at 750 nm and at 386 nm as functions of the applied potential. (c) Absorption spectra of PbS-III at different applied potentials. The insert shows the absorbances at 380 nm as functions of the applied potential.

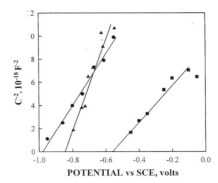

Figure 2.6. Capacitance changes of a single layer of PbS-I (●), PbS-II (▲), and PbS-III (■) on platinum electrodes in 1.0 M aqueous NaCl solution as a function of the applied potential.

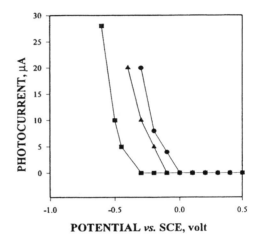

Figure 2.7. Photocurrent generated under different applied potentials from five layers of PbS-I (■), PbS-II (▲), and PbS-III (●) on platinum electrodes.

type, demonstrating oriented growth with respect to the monolayer. Analysis of the diffraction patterns showed that the CdS was of hexagonal structure (wurtzite, $a = 4.136$ and $c = 6.713$) and that reflections from the three zone axes $\langle 0001 \rangle$, $\langle 120 \rangle$, and $\langle 110 \rangle$ were present. Hence, crystals nucleated from the monolayer on $\{0001\}$, $\{01.0\}$, and $\{11.0\}$ faces. As judged from the relative intensities of the diffraction spots, the crystals displaying $\{11.0\}$ basal planes were in the minority.

Reduction in the experimental temperature to 3–4 °C modified the morphology of the CdS film. The crystals were present as a single intergrown sheet over the monolayer and appeared thinner than their room temperature counterparts. This was verified by UV–Vis spectra, which showed a shift in the absorption edge from 500 nm to 460 nm (Figure 2.9). A band gap of ≥ 520 nm is characteristic of bulk CdS, while an absorption edge of 460 nm can be attributed to particles of diameter ca. 45 Å. Selected area diffraction produced patterns identical to those obtained from the room temperature sample, demonstrating epitaxial crystal growth from the same set of crystal faces.

Comparison of the structure of the arachidic acid monolayer and the packing of

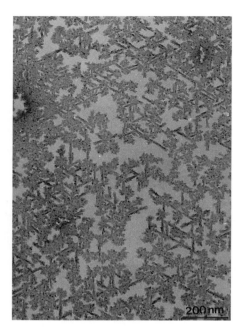

Figure 2.8. TEM image of CdS crystals produced at room temperature under AA monolayers.

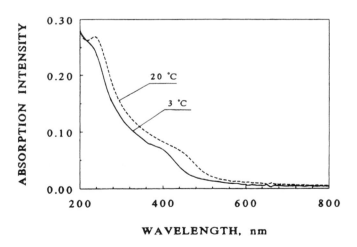

Figure 2.9. Absorption spectra of the CdS crystals produced under the AA monolayer at (a) room temperature and (b) low temperature (3–4 °C). The low-temperature spectrum has been scaled up to match the absorption at 200 nm of the room temperature sample.

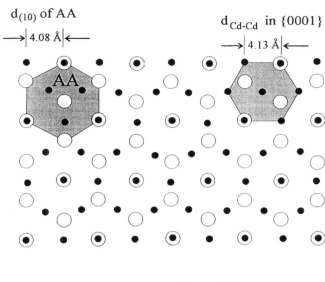

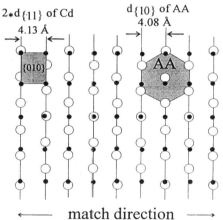

← match direction ⟶

Figure 2.10. Schematic diagrams showing: (Top) The suggested matching between the AA head-groups (empty circles) and the Cd^{2+} ions in the $\{0001\}$ plane (filled circles). The unit cells of both lattices are shaded. A good match exists between the $d\{10\}$ of the AA lattice and the nearest Cd–Cd separation, and isotropic growth is predicted. (Bottom) The possible matching in the horizontal direction to produce rod-shaped CdS.

the Cd^{2+} ions in the $\{0001\}$ and $\{01.0\}$ faces of CdS shows a good structural match (Figure 2.10). In the case of the $\{0001\}$ face, there is only a 3.6% mismatch between the AA headgroup separation, $d\{10\} = 3.98$ and the Cd ion separation $d\langle 100\rangle = 4.13$. The sixfold symmetry of the $\{0001\}$ face mimics that of the AA lattice, reasonably suggesting that isotropic crystal growth could occur to produce the disk-shaped crystals. In contrast, a lattice match between the monolayer and CdS crystals only occurs along the 100 CdS direction in the case of nucleation from the {01.0} face; the mismatch is only 3%, compared with 15% in the perpendicular direction. Thus, the rodlike morphology derives from rapid growth along the 100 direction, compared with slow growth along the perpendicular $\langle 0001 \rangle$ direction, where there is significant misfit and associated strain.

2.4.5 Epitaxial growth of PbSe Crystals under Arachidic Acid Monolayers [20]

In addition to acting as a template for the growth of PbS crystalline films, AA monolayers were shown to direct the epitaxial growth of nanosized PbSe particles. Infusion of H_2Se through an AA monolayer on a $Pb(NO_3)_2$ subphase resulted in the epitaxial growth of PbSe crystals whose size and morphologies depended on the monolayer surface pressure. At a surface pressure of $\pi = 35$ mN m^{-1} and 10 minutes exposure to the reactant gas, crystals of equilateral triangle morphologies and dimensions 50.1 nm were precipitated. AFM studies showed that the crystals had thicknesses in the order of 65 Å and that they were thicker at the corners than at the center. Selected area diffraction yielded single-crystal-type patterns with reflections corresponding to {220}, {422}, and {440} planes, which demonstrated that epitaxial crystal growth occurred from a {111} basal plane. Very slow, overnight infusion of H_2Se resulted in smaller crystals with a larger size distribution (10–20 nm).

Increase in the monolayer surface pressure to $\pi = 40$ mN m^{-1} caused a dramatic alteration in the PbSe crystal morphology. The former triangles were replaced by rods of lengths 100 nm and thicknesses 10 nm. A small number of 10 nm circular particles were also observed (Figure 2.11(a)). Selected area diffraction gave patterns of sixfold symmetry and were solved to show that the crystals nucleated from {110} planes. The rod-shaped crystals conspicuously lay in only three directions on the monolayer separated by angles of 60° (Figure 2.11(b)). Such orientation was responsible for the sixfold nature of the diffraction pattern.

The epitaxial growth of both morphological forms of PbSe can be explained in terms of geometric matching between the crystal lattice and the hcp structure of the monolayer (Figure 2.12). PbSe crystallizes with an FCC structure that has a lattice

Figure 2.11. (a) TEM micrograph of PbSe crystalline film. (b) Selected area diffraction pattern of crystals shown in (a).

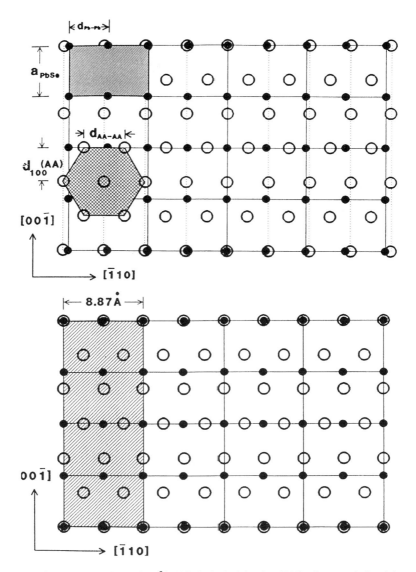

Figure 2.12. (a) Schematic representation of Pb^{2+} (filled circles) in the (110) plane and the AA lattice (open circles) The unit cells are shaded rectangular and hexagonal respectively. The mismatch between the lattices is evident. (b) Representation of the improved matching achieved by stretching the PbSe lattice along the $[(-1)10]$ direction. The strained PbSe lattice was experimentally determined as $a = 6.27$.

constant of 6.1255 Å. For the case of the triangular PbSe particles, the closest Pb–Pb separation in the {111} plane, d(Pb–Pb)⟨110⟩ = 4.33 matches the d{100} = 4.16 of AA to a mismatch of 4%. This can be compared with the corresponding 1% mismatch in the PbS experiments and may be responsible for the poorer reproducibility in the PbSe system. Growth of the rod-shaped PbSe particles can be attributed to matching of the {110} plane of PbSe to the monolayer. Mismatches of 2% and 10% occur along the [001] and [110] axes respectively. Thus, the superior overlap in the [001] direction promotes accelerated growth in this direction, compared with the [110] direction, where considerable strain occurs.

2.5 Sodium Chloride Growth under Monolayers [50]

Sodium chloride was precipitated from supersaturated solutions in the presence of positively charged (octadecylamine, ODA), negatively charged (long-chain carboxylic acids) and zwitterionic (stearoyl-R-glutamate) monolayers. Appropriate selection of the monolayer was shown to influence the crystal growth habit. NaCl precipitates from solution with a face-centered cubic structure of lattice constant $a = 5.638$ and displays {100} faces when precipitated under ambient conditions. Crystal growth in association with Langmuir monolayers selected for either {100}, {110}, or {111} faces, where the latter two are not naturally occurring growth habits.

NaCl crystals with platelike morphologies were nucleated from {100} faces under ODA monolayers. The {100} face has a limiting area of 31.8 Å^2 and exhibits an equal distribution of positive and negative ions. The monolayer bares a close spatial match, having a limiting area of 31.0 Å^2 on a subphase of 6 M NaCl. It was suggested that chloride ions penetrate between the ODA molecules, producing a similar array of charged species at the monolayer to the {100} face of the developing NaCl crystals. Nucleation from the {111} face of NaCl was preferred (70–90% of crystals) under long-chain carboxylate monolayers such as stearic acid or arachidic acid. The {111} face of NaCl expresses ions of only one type. Accumulation of Na$^+$ ions under the compressed monolayer may thus create a template for this face. A third growth habit was selected using zwitterionic monolayers of stearoyl-R-glutamate. At a subphase pH of 1–3, 70–80% of the crystals nucleated from the {110} face, while an increase in the pH to 4–6 reduced the percentage of {110} type crystals to 40–50% and resulted in the nucleation of {111} type crystals. The {110} face of NaCl exhibits alternating rows of Na$^+$ and Cl$^-$ ions. Stabilization of this face suggests that the monolayer may also have alternating rows of like charges.

The results demonstrate that direct simulation of the nucleating NaCl face is not necessary to induce oriented crystal growth. None of the monolayers studied simulated the first NaCl face, which suggests that electrostatic interactions between the monolayer and growing crystal face are also active in defining crystal growth.

2.5.1 Ice Nucleation under Aliphatic Alcohol Monolayers [28–31, 46, 49]

Compressed and uncompressed monolayers of amphiphilic alcohols were shown to be effective in promoting ice crystallization at the air–water interface. This behavior contrasts with that of water-soluble alcohols, which act as antifreeze agents. The structure of the monolayer mimics the crystal structure of ice, and oriented nucleation occurs owing to structural fitting, complementarity, or electrostatic attraction between the monolayer headgroups and the top layer of bound molecules of the developing crystals.

Monolayers of aliphatic alcohols of structure $C_nH_{2n+1}OH$ were spread on water droplets and the threshold freezing point was measured [46, 49]. Control experiments comprising pure water droplets and water droplets covered with aliphatic acid monolayers were run simultaneously to ensure that only factors involving a different structural match between the monolayer and nucleating ice crystals were considered. Alcohols with n in the range 16 to 31 were studied and the freezing point was shown to be dependent on the magnitude and parity of n. The n-odd monolayers were superior to the n-even monolayers in promoting ice nucleation. The freezing point for the n-odd series increased asymptotically from $-11\ °C$ to $-1\ °C$ on increasing the value of n from 17 to 31, while a gradual increase in the freezing point from $-14\ °C$ and leveling off at $-7.5\ °C$ occurred in the n-even series for $n = 22$ to 30. The OH headgroup in the monolayer presumably assumes a different orientation depending upon the parity of the chain, and a closer structural match between the basal plane of ice and the monolayer exists for the n-odd surfactants. This contrasts with the behavior of monolayers of the analogous aliphatic carboxylic acids $C_{n-1}H_{2n-1}COOH$, which show no correlation of the temperature at which ice was nucleated with chain length in the temperature range $-12\ °C$ to $-18\ °C$. Alcohols with fluorocarbon chains, or with a steroidal backbone, also induced ice nucleation at much lower temperatures.

The role of specific interactions between the alcohol monolayer and nucleating ice crystals was demonstrated by investigating the effect of mixed alcohol monolayers [46]. The freezing point of supercooled water was measured under monolayers of $n = 29$ and $n = 31$ molecules over a range of compositions. In all cases, freezing occurred at lower temperatures than for either of the pure alcohols. Similarly, mixtures of the $n = 28$ and $n = 30$ alcohols, which have identical threshold freezing points when in the pure state, supported significantly lower freezing points than the pure alcohols. Although the aliphatic alcohols were completely miscible, the results show that a structural perturbation occurs on introducing a second molecule into the pure monolayer and reduces the quality of the structural match between alcohol and ice crystals. This may be due to a surface roughness at the monolayer–water interface, which is caused by the difference in the hydrocarbon chain lengths.

The effect of the headgroup area on crystal nucleation was investigated by introducing ester $(CH_3(CH_2)_nCO_2(CH_2)_mOH)$ or amide $(CH_3(CH_2)_nCONH(CH_2)_mOH)$ functional groups into the hydrocarbon chain of the basic alcohol molecule [46]. For these monolayers, no dependence of the freezing point on the chain length or on the parity of the chain was observed. However, the molecules could be classified into two distinct groups whose freezing points were separated by 4 °C. These groups were distinguished by the parity of m, the number of carbon atoms connecting the functional group to the alcohol endgroup. The packing arrangement of the molecules, and thus the orientation of the headgroup, is affected by the location in the surfactant of the ester or amide groups.

The structural complementarity between the monolayers and ice crystals was determined using combined grazing incidence angle X-ray diffraction (GID) and lattice energy calculations [28, 29], and cryogenic transmission electron microscopy (TEM) [30, 31]. The orientation of ice crystals at the monolayer can be rationalized in terms of the good structural match between the monolayer and the ice lattice. The structure of the $C_{31}H_{63}OH$ monolayer was determined by GID over the temperature range +5 °C to the threshold freezing point. The monolayer comprised crystalline domains of surfactant, azimuthally randomly oriented on the water surface. The unit cell of the crystallites was rectangular, with unit cell parameters $a = 5.0$, $b = 7.5$, and $\gamma = 120°$, and an area of 18.5 Å^2. This cell shows a good match to the ab face of hexagonal ice. Under ambient conditions ice crystallizes with a hexagonal structure with lattice parameters $a = b = 4.5$, $c = 7.3$, and $\gamma = 120$. Each water molecule in the ab layer occupies 17.5 Å^2. Indeed, the highest freezing points occurred under the alcohol monolayers with molecular areas of 18.5–20.0 Å^2, which most closely approaches the unit cell area of water in ice (46). The freezing point under $C_nH_{2n+1}OH$ monolayers for n in the range $16 < n < 31$ generally increased with increasing n. This appeared to result from higher crystallinity and lateral coherence with increased length of the hydrocarbon chain. The longer-chain alcohols also exhibit a smaller molecular tilt angle and thus provide a better structural match to ice. In monolayers containing bulky chains, a lattice or structural match to the ice crystal cannot be achieved.

The GID studies yielded an explanation of the freezing point dependence on the length and parity of the hydrocarbon chains. The structure of monolayers on pure water at +5 °C of the simple alcohols $C_{31}H_{63}OH$ and $C_{30}H_{61}OH$, and of $C_{19}H_{39}CO_2(CH_2)_nOH$ for $n = 9$ and 10 were analyzed using combined GID measurements and lattice energy calculations. The hydrocarbon chains were shown to be similarly packed in both the $n = 30$ and $n = 31$ alcohol molecules. The neighboring hydrocarbon chains assumed an all-trans conformation, tilted at 9° to the b axis and were organized with a herringbone packing motif. The orientation of the CH_2OH moieties with respect to the water interface could not be determined for these molecules. The absolute orientation of the OH headgroup could be resolved for the monolayers of $C_{19}H_{39}CO_2(CH_2)_nOH$. The monolayers self-aggregated with a high degree of crystallinity, and the unit cell was similar to that of the pure alcohol, being rectangular with cell parameters $a = 5.7$ and $b = 7.5$. The OH endgroup existed in distinctly different orientations with respect to the water surface for the $n = 9$ and $n = 10$ molecules. The differing abilities of these molecules towards nucleation can be attributed to such a molecular rearrangement at the headgroup, which may affect the free energy of nucleation. A similar headgroup reorientation can also be expected to cause the differing behaviors of the $C_{31}H_{63}OH$ and $C_{30}H_{61}OH$ molecules.

Uncompressed monolayers of $C_{31}H_{63}OH$ were studied by GID in the temperature range $+6\ °C$ to freezing point. The studies showed that although some change in the molecular tilt between monolayers on water and on ice occurred, the $C_{31}H_{63}OH$ monolayer maintained its two-dimensional crystallinity upon nucleating ice from the (0001) face. Reduction in the temperature to the threshold freezing point produced a homogeneous population of ice crystallites over the monolayer surface. GID studies and X-ray powder diffractometer measurements of the monolayer conclusively demonstrated epitaxial growth of ice crystals from the monolayer. The crystals nucleated exclusively from a (0001) face [28]. A coherence length of the ice crystallites parallel to the interface of approximately 25 Å was measured, as was consistent with the extent of match between the monolayer lattice and the *ab* face of ice. A coherence length of 25 Å corresponds to a nucleus of approximately 50 water molecules and can be equated with the maximum size of the critical nucleus. The average separation of nucleation sites at the monolayer is estimated at 50–60 Å. The geometric match between the monolayer and ice lattices extends over 30–50 Å. However, as demonstrated by TEM, the ice crystals can also grow to much larger diameters [30, 31]. The final size thus appears to be defined by the growth conditions.

Cryogenic studies of the monolayer and nucleated ice crystals provided further evidence of an epitaxial match [30, 31]. Cryogenic TEM permits virtually *in situ* examination of the monolayer. The monolayer and ice crystals were transferred to electron microscope grids and were rapidly frozen by plunging the specimen into liquid ethane at its freezing point. Fast freezing ensures that the thin water layer supporting the monolayer is frozen in a vitreous state and thus that the original monolayer structure is preserved. Selected area electron diffraction techniques were applied to determine the orientational relationship between monolayers of $C_{31}H_{63}OH$ and ice crystals formed at the threshold freezing point. Patterns taken of a single ice crystal within a single domain of the monolayer showed that the ice crystals nucleated specifically from the (0001) plane

2.5.2 Kinetic Measurements of Ice Nucleation under Alcohol Monolayers [51]

The crystallization of ice under monolayers of $C_{30}H_{61}OH$ was studied in order to investigate the kinetics of nucleation. While many studies on crystallization in association with Langmuir monolayers have been performed, the majority have concerned the mechanism by which structural information is transferred from the monolayer to nascent crystal, as given by morphological and diffraction data. The induction time prior to crystallization was measured as a function of the working temperature and considered to be inversely proportional to the nucleation rate. Experiment demonstrated that the nucleation rate was increased in the presence of the monolayer and that nucleation was heterogeneous. However, the monolayer did not affect either the free energy of formation of the critical nucleus or reduce the critical nucleus size, compared with control experiments carried out in the absence of a monolayer. The results suggested that the catalytic effect of the monolayer is due to the increased number of potential nucleation sites, as opposed to the structural compatibility between the monolayer and ice crystals. Indeed, the structural match between the nuclei and monolayer was no better than with other heteronuclei present in the control situation. The effect of the monolayer dominates at higher temperatures since the heteronuclei only become active on temperature reduction. In the freezing range -3 to $-5\,°C$, the critical nucleus size was calculated to be 15–30 Å, a value consistent with the 20 determined by grazing incidence angle X-ray diffraction experiments [31] for crystallization of ice under $C_{31}H_{63}OH$ monolayers. It was thus concluded that the alcohol monolayer catalyses the nucleation of ice because the structural match between the two surfaces provides active sites for heterogeneous nucleation.

2.6 Biomineralization

Inorganic crystals of defined phase, size, morphology, and orientation are produced by biological systems. Indeed, nature has evolved to use a wide range of minerals for purposes as diverse as skeleton, magnetoreception, gravity devices, and eye lenses [32]. In all cases, control over the crystal nucleation, growth, and aggregation is regulated by organic matrices, in well-defined, spatially delineated sites [52, 53]. Precipitation can be intra- or extracellular, with greater structural control being applied in the former case. Three principal stages are considered to occur during biomineralization: supramolecular preorganization, interfacial molecular recognition, and cellular processing [54]. An organized reaction environment, such as a lipid vesicle or protein cage or an extended protein-polysaccharide network is first produced. This structure then provides a framework for mineralization, and crystal nucleation is governed by electrostatic, structural, and stereochemical complementarity. The specific orientation of crystals with respect to an underlying organic matrix has been demonstrated in a number of systems; for example, aragonite crystals in mollusc shells are aligned with respect to the β-pleated protein. The final stage of biomineralization involves a variety of constructional processes involving large-scale cellular activity. Subsequent to nucleation, particles can grow to a size limited by their reaction environment or undergo further cellular processing to give unusual sizes and shapes or be organized into elaborate assemblies.

Although the *in vivo* system is highly complex, many elements can be investigated in an artificial system. Langmuir monolayers were used as substrates on which to precipitate inorganic particles, since they constitute a simple ordered organic membrane [35]. The structure of the monolayer is readily characterized and its effect on crystals nucleated and grown at the air–water interface can be determined. The influence of Langmuir monolayers formed from a range of surfactants on the growth of calcium carbonate, barium sulfate, and calcium sulfate was investigated, and the crystal nucleation and growth processes related to the structure of the monolayer [35].

2.6.1 Growth of Calcium Carbonate under Langmuir Monolayers [37–39, 42–44]

Monolayers of either octadecylamine (ODA), stearic acid, or eicosyl sulfate were spread on supersaturated solutions of calcium bicarbonate. Calcium carbonate precipitates on passive loss of CO_2 from the solution and nucleates and grows under the monolayer as three possible polymorphs termed vaterite, calcite, and aragonite. All three polymorphs were obtained at the monolayer–solution interface under certain experimental conditions. Calcite and vaterite are described by hexagonal unit cells with lattice constants $a = 4.959$, $c = 17.002$, and $a = 7.15$, $c = 16.917$ respectively. Aragonite is orthorhombic with lattice constants $a = 4.959$, $b = 7.689$, and $c = 5.741$. Experiments were carried out under a range of solution concentrations and monolayer surface pressures. The monolayer-supported crystals were sampled over a range of reaction times and were examined by transmission electron microscopy (TEM) and scanning electron microscopy (SEM).

Crystallization under ODA monolayers from subphases of concentration $4.5\ \text{mM} < [Ca^{2+}] < 9.0\ \text{mM}$ produced oriented vaterite crystals in two contrasting forms, termed Type I and Type II respectively. At early stages of growth, the Type I crystals were hexagonal disks which nucleated from the (0001) face, with the [0001] direction lying perpendicular to the monolayer. Further growth produced a floret morphology. In contrast, Type II crystals nucleated from the (11.0) face. Early crystals were disklike, with outgrowths related by a twofold axis, while mature crystals had a complex shape and exhibited twofold symmetry.

Crystal nucleation and growth under stearic acid monolayers, unlike the ODA system, was dependent upon the subphase Ca^{2+} concentration. At $[Ca^{2+}] = 9$ mM, oriented calcite crystals were preferentially nucleated in two morphological forms. Populations of crystals were either almost exclusively Type I calcite or were a mixture of Type I and Type II. Both crystal types were platelike at early growth stages and nucleated from a $(1(-1).0)$ plane such that the $[1(-1).0]$ axis was perpendicular to the monolayer plane. Realignment during further growth resulted in the capped rhombohedral plates of the Type I crystals and a related triangular morphology in the case of the Type II crystals. When the concentration of subphase calcium was reduced to $[Ca^{2+}] = 4.5$ mM, only oriented Type I vaterite crystals, which were identical to those produced under ODA monolayers, were observed. Intermediate Ca^{2+} concentrations resulted in mixed populations of calcite and vaterite crystals.

The influence of the monolayer on calcium carbonate precipitation was considered on the basis of ion binding and on geometric and stereochemical complementarity between the monolayer and the nucleating crystal face. The results suggested that Ca^{2+} binding was necessary for the crystallization of calcite but not of vaterite. While vaterite was precipitated on both the positively charged ODA monolayers and the negatively charged stearic acid monolayers, calcite was only produced under the negatively charged surface. The polymorph selectivity under stearic acid monolayers indicated the same dependence on the Ca^{2+} concentration. At high concentrations of Ca^{2+} ions, the thermodynamically favored phase, calcite is precipitated. On reduction of the Ca^{2+} concentration, the activation energy for calcite nucleation increases until it becomes comparable with that of the metastable vaterite phase; calcite is replaced by vaterite at lower Ca^{2+} concentrations.

Stereochemical and geometric matching occurs between the stearic acid monolayer and the $(1(-1).0)$ face of calcite. The monolayer crystallizes in a pseudohexagonal lattice with interheadgroup spacings in the order of 5 Å [38], which closely approaches the carbonate–carbonate spacing of 4.69 in the $(1(-1).0)$ face of calcite. A close epitaxial match occurs in two directions. However, such a geometric complementarity cannot be the sole criterion for selection of the $(1(-1).0)$ plane since, for example, the (0001) plane offers a similar geometric match. The stereochemistry of the monolayer headgroups is also essential in directing nucleation. The orientation of the carboxylate groups in the stearic acid monolayer mimics the rows of perpendicular carbonate ions in the $(1(-1).0)$ calcite face but not in the (0001) face. In the case of vaterite nucleation, no geometric match exists and oriented crystal growth is directed by stereochemical considerations only. Both the (0001) and the (11.0) faces of vaterite contain anions that lie perpendicular to the crystal surfaces and can be mimicked by bidentate binding of the carboxylate headgroups in the stearic acid monolayers. In the case of the ODA monolayers, no Ca^{2+} binding at the monolayer takes place and only vaterite crystals were produced. The amine headgroups in the ODA monolayers also have no stereochemical equivalent in the vaterite crystal. That oriented nucleation was observed under this monolayer may be due to electrostatic binding of HCO_3^- ions orthogonal to the $-NH_3^+$ headgroups, which could provide an indirect means of stereochemical recognition.

Crystallization under monolayers of *n*-eicosyl sulfate and *n*-eicosyl phosphonate was also investigated since the monolayer headgroups exhibit a trigonal symmetry, which contrasts with that of the ODA and stearic acid monolayers [44]. The induction period prior to precipitation was reduced to under 45 minutes in the presence of the sulfate monolayer. Immature crystals were pseudohexagonal in form and developed to a trigonal pyramidal morphology. Electron diffraction demonstrated that nucleation was from a (0001) face. Monolayers of the corresponding phosphonate surfactant yielded similar results, although a reduced nucleation density was obtained. An approximate geometric matching exists between the hexagonal array of Ca^{2+} ions in the (0001) calcite face, $a = 4.96$, and the pseudohexagonal packing of the headgroups in the sulfate, $a = 5.5$, and the phosphonate, $a = 5.2$, monolayers. However, the stereochemical recognition between the monolayer and nucleating crystal is probably the overriding factor determining the orientation. The monolayer headgroups have trigonal symmetry, which mimics that of the planer carbonate ions in the (0001) face of calcite.

In the presence of Mg^{2+} ions in the subphase, aragonite crystals nucleated under monolayers of *n*-eicosanoic acid, *n*-eicosyl sulfate, and *n*-eicosyl phosphonate [44]. Crystals precipitated under the *n*-eicosanoic acid monolayers comprised an intergrown bundle of acicular crystals which were oriented with the [100] axis perpendicular to the monolayer. In the case of the sulfate and phosphonate monolayers, pseudohexagonal crystals were produced with the [001] axis perpendicular to the monolayer. The arrangement of ions in the (001) plane of aragonite is very similar to that in the (0001) calcite face. Thus, the (001) face of aragonite can be selected by the sulfate and phosphonate monolayers through the same mechanism as was the calcite (0001) face.

2.6.2 Epitaxial Growth of Barium Sulfate under Surfactant Monolayers [36, 40, 41]

Barium sulfate was precipitated from supersaturated solutions of barium sulfate, which were equimolar in barium chloride and sodium sulfate, in the presence of monolayers of *n*-eicosyl sulfate, eicosanoic acid, and $C_{20}H_{41}PO(OH)_2$ surfactants. The solution concentration was selected such that there was an induction period of over 2 hours prior to bulk precipitation. The monolayer was spread and compressed during this time. Barium sulfate is described by an orthorhombic unit cell of dimensions $a = 8.878$, $b = 5.450$, and $c = 7.152$.

In control experiments in which no monolayer was spread, rectangular tablets of $BaSO_4$ were precipitated and collected from the bottom of the trough. TEM examination showed that they were elongated along a [001] direction [36]. In the presence of sulfate monolayers (L), the induction time was reduced to less than 1 hour, and smaller crystals with larger aspect ratios and complex internal structures were precipitated. Each crystal displayed nanoscale texture comprising an interconnecting mosaic of diamond- and rectangular-shaped subunits. Electron diffraction demonstrated that the particles were single crystals, that they nucleated from the monolayer on a (100) face, and that in common with the control crystals, elongation was along the [001] direction. Crystallization under eicosanoic acid monolayers [36] occurred after a longer induction time of 1.5–2 hours and with a significantly lower nucleation density, compared with the *n*-eicosyl sulfate monolayers. The nascent crystals were elongated, asymmetric disks that nucleated on the monolayer such that elongation was along the [001] axis and the [010] direction was perpendicular to the monolayer plane. No well-defined nucleation face was observed. Growth produced mature crystals with complex, dendritic, bow-tie morphologies. The induction time under the phosphonate monolayers was reduced to less than 1 hour, and in contrast with the two previous monolayer types discussed, three different crystal forms were precipitated [40]. The majority of crystals were classified as Type I and were platelike with a central rhombic elevation. These developed from small rhombic particles, which nucleated from the monolayer on a {100} basal plane. The Type II crystals possessed a bow-tie morphology and also developed from a rhombic nucleus. Growth of two triangular plates in opposite directions from the original rhombus resulted in the final form. Diffraction from single crystals demonstrated that Type I and Type II crystals nucleated from (100) faces. The Type III crystals were polar laths approximately 300 nm long and 80–100 nm in width and were elongated along the [100] direction.

Despite the nonpolar nature of the [100] axis, one end of the crystals was rounded while the other was well defined. This suggested that nucleation may have occurred from one end while the other was directed into the solution. Although diffraction demonstrated that the laths were lying in an [010] zone, it appeared more reason-

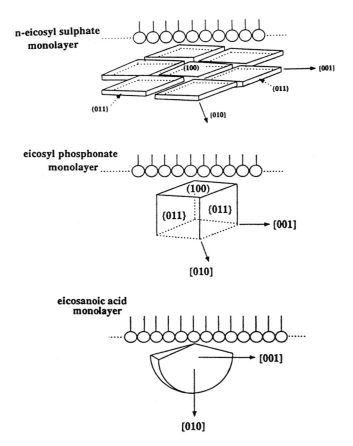

Figure 2.13. Schematic diagram illustrating the orientation and morphology of BaSO₄ crystals nucleated under compressed anionic monolayers. (Top) Eicosyl sulfate. (Middle) Eicosyl phosphonate. (Bottom) Eicosanoic acid. Reproduced with permission from Reference [41].

able that they had in fact nucleated from a (100) face, and that the [010] zone was adopted on transfer of the crystals to electron microscope grids. Thus, all three crystal types nucleated from the barite (100) face, but the Type I and Type II crystals developed from rhombic precursors, while the growth mechanism of the Type III particles was quite different (Figure 2.13).

Stereochemical complementarity between the monolayers and nucleating crystals was considered to be the principal factor determining the orientation of the barium sulfate, since no close geometric match exists. Both the sulfate and phosphonate monolayers promoted the nucleation of a (100) face despite marked differences in the dimensions of their unit cells. The tridentate symmetry of the sulfate and phosphonate headgroups at the monolayer mirrors a similar arrangement of the sulfate ions in the (100) face of barite. Binding of Ba²⁺ ions to the monolayer headgroups simulates the coordination requirements of ions in the (100) barite face, thus providing a stereochemical matching and inducing oriented nucleation. In the case of the eicosanoic acid monolayers, although the headgroup spacing is close to that of the eicosyl-phosphonate monolayers, the stereochemistry is quite different. The carboxylate headgroup has a bidendate symmetry, as opposed to the tridentate symmetry of the sulfate and phosphonate monolayers. That no geometric or stereochemical matching occurs in this system explains the experimentally shown absence of a well-defined nucleation face. The preferred [010] orientation may be due to kinetic factors as opposed to structural recognition [41].

2.6.3 Oriented Nucleation of Gypsum (CaSO$_4 \cdot$2H$_2$O) under Langmuir Monolayers [45]

The crystallization of gypsum under compressed monolayers of negatively charged (eicosanyl sulfate, eicosanyl phosphonate, and stearic acid), positively charged (octadecylamine), and polar (octadecanol) surfactants was studied and showed the importance of hydrogen bonding and ion binding on inorganic crystal nucleation. In the absence of a monolayer, control experiments yielded bunches of intergrown needles that were elongated along the c axis. Under a layer of the amorphous, partially hydroxylated polymer Formvar, the majority of crystals were deposited at the bottom of the crystallization dish. Those precipitated at the monolayer–solution interface were oriented with their {010} faces parallel to the monolayer.

In all of the systems studied, the presence of a monolayer increased the nucleation density at the air–solution interface, with the negatively charged surfactant being more effective than the polar monolayer and the positively charged octadecylamine being only slightly more effective than the control situation. The crystals were oriented specifically with their c axis either parallel or perpendicular to the surface. The relative proportions of the two crystal types depended on the nature of the monolayer headgroup. Approximately equal numbers of both orientations were obtained under charged monolayers, while the c axis was almost exclusively parallel to the surface of polar monolayers. SEM examination of large crystals showed that crystals whose c axes lay parallel to the surface nucleated off the {010} face, while those lying perpendicular were nucleated from a range of crystal planes, possibly of {001}, {103}, or {203} types.

The {010} face of gypsum is commonly exhibited in the equilibrium crystal form. However, its expression under charged and polar monolayers must also reflect influence of the monolayer since the nucleation density of this face was substantially reduced under the disordered Formvar film. The crystal structure of gypsum parallel to the {010} face comprises layers of Ca^{2+} and SO$_4{}^{2-}$ ions, interspaced with a double layer of water molecules. Thus, the {010} face can be either charged or polar in character, depending upon the section through the crystal considered. Preferential nucleation of a {010} basal plane may derive from either ions bound to the charged headgroups mimicking the crystal face or water molecules under the alcohol monolayer simulating hydrogen bonding interactions in the polar layer of the crystal. The organization of oxygen-bound calcium ions in the {010} gypsum face may also match the stereochemistry of the headgroup oxygen of the monolayer

($-SO_3^-$, $-OPO_3H^-$, $-CO_2^-$). While the former mechanism is valid for many gypsum faces, the H-bonding mechanism only holds for the {010} face. The high specificity of the polar monolayer for the {010} basal plane suggests that in the absence of strong ion-binding effects, hydrogen bonding between the hydrated crystal face and the alcohol monolayer acts to direct oriented crystal growth.

Thus, the experimental results suggest that the ordering of water molecules through directional hydrogen bonding is an important factor in the molecular recognition at the inorganic–organic interface of hydrated surfaces.

References

[1] K. L. Chopra, R. C. Kainthla, D. K. Pandya, A. P. Thakoor, in *Physics of Thin Films* (Eds: G. Hass, H. M. Francombe, J. L. Vossen), Academic Press **1982**, Vol. 12.
[2] *Atomic Layer Epitaxy* (Eds: T. Suntola, M. Simpson), Chapman and Hall, New York **1990**.
[3] *Molecular Beam Epitaxy and Heterostructures* (Eds: L. L. Chang, K. Ploog), Nato Series, Martinos Nijhoff, the Netherlands **1985**.
[4] C. J. Weisbuch, *Cryst. Growth* **1993**, *127*, 742–751.
[5] N. Inoue, *Adv. Mater.* **1993**, *5*, 192–197.
[6] N. C. Sharma, D. K. Pandya, H. K Sehgal, K. L. Chopra, *Thin Solid Films* **1979**, *59*, 157–164.
[7] I. Kaur, D. K. Pandya, K. L. Chopra, *J. Electrochem. Soc.* **1980**, *127(4)*, 943–948.
[8] S. Gorer, A. Albu-Yaron, G. Hodes, *Chem. Mater.* **1995**, *7*, 1243–1256.
[9] S. Gorer, A. Albu-Yaron, G. Hodes, *J. Phys. Chem.* **1995**, *99*, 16442–16448.
[10] D. Lincot, R. Ortega Borges, *J. Electrochem. Soc.* **1992**, *139(7)*, 1880–1889.
[11] A. S. Baranski, W. R. Fawcett, A. C. McDonald, R. M. de Nobriga, J. R MacDonald, *J. Electrochem. Soc.* **1981**, *128(5)*, 963–968.
[12] Y. Golan, G. Hodes, I. Rubinstein, *J. Phys. Chem.* **1996**, *100*, 2220–2228.
[13] Y. Golan, E. Ter-Ovanesyan, Y. Manassen, L. Margulis, G. Hodes, I. Rubinstein, E. G. Bithell, J. L. Hutchinson, *Surf. Sci.* **1996**, *350*, 277–284.
[14] Y. F. Nicolau, M. Dupuy, M. Brunel, *J. Electrochem. Soc.* **1990**, *137(9)*, 2915–2923.
[15] S. Lindroos, T. Kanniainen, M. Leskela, *Appl. Surf. Sci.* **1994**, *75*, 70–74.
[16] I. Villegas, J. L. Stickney, *J. Electrochem. Soc.* **1992**, *139(3)*, 686–694.
[17] B. W. Gregory, D. W. Suggs, J. L. Stickney, *J. Electrochem. Soc.* **1991**, *138(5)*, 1279–1284.
[18] S. Feng, T. Bein, *Nature* **1994**, *368*, 834–836.
[19] S. Feng, T. Bein, *Science* **1994**, *265*, 1839–1841.
[20] J. Yang, J. H. Fendler, T. C. Jao, T. Laurion, *Micros. Res. Tech.* **1994**, *27*, 402–411.
[21] J. Yang, F. C. Meldrum, J. H. Fendler, *J. Phys. Chem.* **1995**, *99*, 5500–5504.
[22] J. Yang, J. H. Fendler, *J. Phys. Chem.* **1995**, *99*, 5505–5511.
[23] J. H. Fendler, F. C. Meldrum, *Adv. Mater.* **1995**, *7(7)*, 607–632.
[24] X. K. Zhao, Y. Yang, L. D. McCormick, J. H. Fendler, *J. Phys. Chem.* **1992**, *96*, 9933–9939.
[25] Y. Tian, C. Wu, N. Kotov, J. H. Fendler, *Adv. Mater.* **1994**, *6(12)*, 959–962.
[26] B. C. Bunker, P. C. Rieke, B. J. Tarasevich, A. A. Campbell, G. E. Fryxell, G. L. Graff, L. Song, J. Lui, J. W. Virden, G. L. McVay, *Science* **1994**, *264*, 48–55.
[27] S. J. Potochnik, P. E. Pehrsson, D. S. Y. Hsu, J. M. Calvert, *Langmuir* **1995**, *11(6)*, 1841–1845.
[28] J. Majewski, R. Popovitz-Biro, K. Kjaer, J. Als-Nielsen, M. Lahav, L. Leiserowitz, *J. Phys. Chem.* **1994**, *98*, 4087–4093.
[29] J. Wang, F. Leveiller, D. Jacquemain, K. Kjaer, J. Als-Nielsen, M. Lahav, L. Leiserowitz, *J. Am. Chem. Soc.* **1994**, *116*, 1192–1204.

[30] J. Majewski, L. Margulis, I. Weissbuch, R. Popovitz-Biro, T. Arad, Y. Talmon, M. Lahav, L. Leiserowitz, *Adv. Mater.* **1995**, *7*, 26–35.

[31] J. Majewski, L. Margulis, L.; Jacquemain, D.; Leveiller, F.; Bohm, C.; Arad, T.; Talmon, Y.; Lahav, M.; Leiserowitz, L. *Science* **1993**, *261*, 899–902.

[32] H. A. Lowenstam, S. Weiner, in *On Biomineralization*, Oxford University Press, Oxford **1989**.

[33] B. J. Tarasevich, P. C. Rieke, J. Liu, *J. Chem. Mater.* **1996**, *8*, 292–300.

[34] N. P. Hughes, D. Heard, C. C. Perry, R. J. P. Williams, *J. Phys. D.* **1991**, *24*, 146–153.

[35] B. R. Heywood, S. Mann, *Adv. Mater.* **1994**, *6(1)*, 9–20.

[36] B. R. Heywood, S. Mann, *J. Am. Chem. Soc.* **1992**, *114*. 4681–4686.

[37] S. Mann, B. R. Heywood, S. Rajam, J. B. A. Walker, *J. Phys. D.* **1991**, *24*, 154–164.

[38] B. R. Heywood, S. Rajam, S. Mann, *J. Chem. Soc. Faraday Trans.* **1991**, *87(5)*, 735–743.

[39] S. Rajam, B. R. Heywood, J. B. A. Walker, S. Mann, *J. Chem. Soc. Faraday Trans.* **1991**, *87(5)*, 727–734.

[40] B. R. Heywood, S. Mann, *Langmuir* **1992**, *8*, 1492–1498.

[41] B. R. Heywood, S. Mann, *Adv. Mater.* **1992**, *4(4)*, 278–282.

[42] S. Mann, B. R. Heywood, S. Rajam, J. B. A. Walker, R. J. Davey, J. D. Birchall, *Adv. Mater.* **1990**, *2(5)*, 257–261.

[43] S. Mann, B. R. Heywood, S. Rajam, J. D. Birchall, *Nature* **1988**, *334*, 692–695.

[44] B. R. Heywood, S. Mann, *Chem. Mater.* **1994**, *6*, 311–318.

[45] T. Douglas, S. Mann, *Mat. Sci. Eng. C1* **1994**, 193–199.

[46] R. Popovitz-Biro, J. L. Wang, J. Majewski, E. Shavit, L. Leiserowitz, M. J. Lahav, *Am. Chem. Soc.* **1994**, *116*, 1179–1191.

[47] F. Leveiller, D. Jacquemain, M. Lahav, L. Leiserowitz, M. Deutsch, K. Kjaer, J. Als-Nielsen, *Science* **1991**, *252*, 1532–1535.

[48] F. Leveiller, C. Boehm, D. Jacquemain, H. Moehwald, L. Leiserowitz, K. Kjaer, J. Als-Nielsen, *Langmuir* **1994**, *10*, 819–829.

[49] M. Gavish, R. Popovitz-Biro, M. Lahav, L. Leiserowitz. *Science* **1990**, 250, 973–975.

[50] E. M. Landau, R. Popovitz-Biro, M. Levanon, L. Leiserowitz, M. Lahav, J. Sagiv, *J. Mol. Cryst. Liq. Cryst.* **1986**, *134*, 323–335.

[51] R. J. Davey, S. J. Maginn, R. B. Steventon, J. M. Ellery, A. V. Murrell, J. Booth, A. D. Godwin, J. E. Rout, *Langmuir* **1994**, *10*, 1673–1675.

[52] S. Mann, D. D. Archibald, J. M. Didymus, T. Douglas, B. R. Heywood, F. C. Meldrum, N. J. Reeves, *Science* **1993**, *261*, 1286–1292.

[53] S. J. Mann, *Chem. Soc. Dalton Trans.* **1993**, 1–9.

[54] S. Mann, *Nature* **1993**, *365*, 499–505.

Chapter 3

Electrodeposition of Superlattices and Nanocomposites

J. A. Switzer

3.1 Introduction

The interest in nanoscale materials stems from the fact that the properties (optical, electrical, mechanical, and chemical) are a function of the dimensions of the material. The concept here is not to see how many transistors can be squeezed onto a chip, but rather to grow materials in a nanoscale-size regime, in which some normally intrinsic property such as a semiconductor bandgap (e.g., 1.1 eV for silicon and 1.4 eV for gallium arsenide) can be tuned by simply changing the dimensions of the material. Superlattices and nanocomposites are particularly interesting subclasses of these "designer solids", since they have nanoscale confinement dimensions for electrons in the solid, yet they can by grown as large-area films or even monolithic solids. This aspect of these materials makes them easily amenable to device manufacture. In this chapter, we will briefly review the general area of nanoscale materials and follow this with a discussion of the use of electrochemistry to assemble nanoscale architectures such as superlattices and nanocomposites.

The field of nanoscale materials began about twenty years ago when Esaki and Chang reported resonant tunneling across potential barriers in nanoscale structures grown by molecular beam epitaxy [1] and Dingle reported optical verification of quantum confinement in semiconductor quantum wells [2]. A quantum well is produced when a smaller-bandgap material such as GaAs is sandwiched between a larger-bandgap material such as GaAlAs. Carriers confined in this quantum well behave like quantum mechanical particles in a box.

Since this early work, there has been explosive growth in the area of nanoscale materials. This growth has been driven both by the basic scientific interest in quantum physics and by the development of useful devices [3–5]. Examples of actual devices are the high-electron-mobility transistor, HEMT (also called the two-dimensional electron gas field effect transistor, TEGFET), and the semiconductor quantum well laser. The nanoscale dimensions produce diode lasers which operate at lower threshold currents and emit light at wavelengths that are determined by the layer thickness (bandgap engineering).

Another area of interest in the nanomaterials field is building materials that circumvent a selection rule for optical transitions in semiconductors that derives from translational symmetry. In indirect-gap semiconductors such as silicon, the valence-to-conduction-band transition is electronically forbidden, since it violates conservation of momentum. The transition does occur but with a low transition probability, since a phonon is needed to assist the transition. This the is reason that Si has not been used in optical devices (especially for light emission), whereas direct-gap materials such as GaAs perform well in LEDs and diode lasers. There is interest in using the much less expensive material silicon for this application. Zu, Lockwood, and Baribeau have observed room temperature light emission in SiO_2/Si superlattices with Si layers in the 2 nm thickness range [6].

Although the emphasis of this chapter will be on two-dimensional nanomaterials known as superlattices, there is also a strong interest in zero-dimensional materials known as quantum dots [7–10]. Research in this area has recently been reviewed by Alivisatos [10]. Most of the work in this area has been on metal chalcogenides like cadmium sulfide. This material can be produced by colloidal chemical techniques. The bandgap can be tuned from 2.5 eV for macroscopic crystals to 4.5 eV for nanocrystals. Also, the radiative lifetime for the lowest allowed optical transition can vary from tens of picoseconds to several nanoseconds. An important step in the synthesis of these materials is the passivation of the surface.

One problem with the utilization of quantum dot devices is that it is difficult to make a useful device out of the dots because of their small size. In superlattices, the dimensions of the layers are in the nanometer range, but the actual superlattice can be grown to any size. Bawendi and coworkers have grown a quantum dot system by colloidal processing that goes a long way towards solving this problem [11]. They have grown a three-dimensional semiconductor quantum dot superlattice by the self-organization of CdSe nanocrystallites. They verified that the material was a superlattice by X-ray diffraction.

A particularly beautiful example of quantum confinement is the work by Crommie, Lutz, and Eigler [12], in which the scanning tunneling microscope (STM) was used to arrange 48 iron atoms in a "quantum corral" with a radius of 7.1 nm on a single crystal copper(111) surface. STM images of this quantum corral showed standing waves from the surface electrons that were trapped in the round two-dimensional box.

3.2 Electrodeposition of Inorganic Materials

How can an electrochemist contribute to a field that is dominated by sophisticated high-vacuum deposition techniques and solid state quantum physics? The answer lies in the advantages that electrodeposition provides over vapor deposition techniques such as molecular beam epitaxy when it is applicable for a given material:

- The low processing temperatures (usually room temperature) of electrodeposition minimize interdiffusion
- One can control the film thickness by monitoring the delivered charge

- Composition and defect chemistry can be controlled
- Films can be deposited onto complex shapes
- Nonequilibrium phases can be deposited
- The driving force can be precisely controlled
- The technique is not capital intensive

There is an additional advantage of the electrochemical method that has not been explored sufficiently:

- The current-time transient following a potential step provides an *in situ* measurement of the deposition process, since the current is proportional to the deposition rate.

In this chapter we will briefly review work that has been done on the electro-deposition of nanoscale architectures, and we will cite examples that should help to crystallize some of these ideas. A common theme in all of the work that we will discuss is the use of various tricks to ensure that not only are the electrodeposited materials small but that the size distribution of the materials is very narrow. The emphasis in this chapter will be on the synthesis of nonmetallic inorganic phases in the nanometer regime. We will first outline synthesis schemes for these materials, followed by techniques for building nanometer-scale architectures.

3.2.1 Electrodeposition of Metal Chalcogenides

When you think of electrodeposition, you think of metal electroplating. There are several nonmetallic materials, however, such as conducting polymers, semiconductors, and ceramics that have been electrodeposited. We will focus our attention on compound semiconductors and ceramics. Most of the work on semiconductor deposition has been on Group II–VI compound semiconductors of metal chalcogenides, such as CdS, CdSe, CdTe, and $Hg_{1-x}Cd_xTe$ [13–16]. Both anodic and cathodic processes have been used to deposit metal chalcogenides. In the anodic process, the metal is simply electrochemically oxidized in the presence of chalcogenide ions. The technique is simple, but it does not allow the deposition of the metal chalcogenide onto substrates other than metals. The cathodic process is a true deposition process, since both components of the film are deposited from solution precursors. In this case higher-valency metal and chalcogenide ions (for example, Cd^{2+} and $HTeO_2^+$) are electrochemically reduced to the elements at the electrode surface, where they combine to form the metal chalcogenide. These reactions are summarized in Eqs. (3.1–3.5).

Cathodic

$$Cd^{2+} + 2e^- = Cd \tag{3.1}$$

$$HTeO_2^+ + 4e^- + 3H^+ = Te + 2H_2O \tag{3.2}$$

$$Cd + Te = CdTe \tag{3.3}$$

Anodic

$$Cd = Cd^{2+} + 2e^- \qquad (3.4)$$

$$Cd^{2+} + Te^{2-} = CdTe \qquad (3.5)$$

Rajeshwar has shown that the cathodic reactions shown above are a simplification of the actual deposition process [16]. The reduction of $HTeO_2^+$ according to Eq. (3.2) is thought to be the crucial step, followed by the subsequent assimilation of Cd into the Te layer in a two-electron process, as shown in Eq. (3.6) below. The Te electrodeposition is also very complex, and the six-electron reduction product, H_2Te, may also participate in the precipitation of CdTe, according to Eq. (3.7). A further complication in the deposition process is that homogeneous chemical reactions between electrogenerated species can lead to impurities in the film. The homogeneous reaction in Eq. (3.8), for instance, will lead to Te impurities.

$$Cd^{2+} + 2e^- + Te = CdTe \qquad (3.6)$$

$$Cd^{2+} + H_2Te = CdTe + 2H^+ \qquad (3.7)$$

$$2H_2Te + HTeO_2^+ = 3Te + 2H_2O + H^+ \qquad (3.8)$$

In spite of the complexity of the electrodeposition process for the metal chalcogenide semiconductors, these electrodeposited materials have been shown to compete very successfully with vapor-deposited materials for optoelectronic applications such as photovoltaic solar cells, infrared detectors, and "smart" goggles. Photovoltaic conversion efficiencies in the 8–10% range have been observed [16]. The majority of these studies have been on electrodeposited polycrystalline films. There has been recent work by Daniel Lincot et al., however, that shows that single crystal films of CdTe can be epitaxially grown onto single crystal InP substrates [17]. The deposition was carried out in an aqueous solution at pH 2 with 1 M $CdSO_4$, 5×10^{-4} M TeO_2, a deposition temperature of 85 °C, and a deposition rate of 0.7 μm hr^{-1}. Epitaxy was verified by five-circle X-ray diffraction and reflection high-energy electron diffraction (RHEED). The epitaxy was found to exhibit higher perfection when the InP was covered with a thin film (20–30 nm) of epitaxial CdS grown by chemical bath deposition.

3.2.2 Electrodeposition of Metal Oxides

Oxide ceramics can be electrodeposited using either a redox change method or the electrochemical generation of base [18, 19]. In the redox method, a metal ion or complex is placed in a solution at a pH at which the starting oxidation state is stable but the oxidized (or reduced) form of the ion undergoes hydrolysis to form the oxide. We have used the redox method to produce thallium(III) oxide [20–22], lead(IV) oxide [23], silver(II) oxide [24], copper(I) oxide [25], the oxysalt $Ag(Ag_3O_4)_2NO_3$ [23], and superlattices in the Pb–Tl–O system [24–31]. All of the redox depositions were done using anodic reactions except for the case of Cu_2O, which was deposited by electrochemically reducing a Cu(II) lactate complex in a pH range of 9–12. All of these films grew as columnar films with a very strong preferred orientation.

We have also grown oxide ceramics using the electrochemical generation of base. In this case the metal ion or complex is already in a high oxidation state, and the local pH at the electrode surface is increased by a cathodic reaction such as the reduction of water to produce hydrogen gas. We have used this technique to prepare nanoscale CeO_2 [18, 32] and ZrO_2 [33]. Mitchell and Wilcox used the base generation method to produce preshaped ceramic bodies [34], and Redepenning has used this technique to deposit brushite onto prosthetic alloys [35].

3.3 Electrodeposition of Nanophase Materials

3.3.1 Growth in Nanobeakers

We now turn our attention to the various techniques that use electrochemistry to produce materials in the nanometer regime. One approach to growing nanoscale materials is to use nanoscale beakers. Chuck Martin at Colorado State University has been exploring this area for about ten years [36]. This approach involves using the pores in nanoporous membranes as templates to prepare nanoscopic particles of the desired material. The membranes (such as anodized aluminum or track-etch polymers) have cylindrical pores of uniform diameter. When a polymer, metal, semiconductor, or carbon is synthesized electrochemically within one of these pores, a nanocylinder of the desired material is obtained. Depending on the material and the chemistry of the pore wall, this nanocylinder may be hollow (a tubule) or solid (a fibril). Martin has shown that metal nanotube membranes can also serve as ion-selective membranes [37]. The nanotube diameter can be as small as 0.8 nm, and the length of the nanotube can span the complete thickness of the membrane. These membranes show selective transport analogous to ion-exchange polymers. The ion permselectivity is thought to occur because of excess charge density that is present on the inner walls of the nanotubes. Since the sign of the excess charge can be changed potentiostatically, a metal nanotube can be either cation or anion selective, depending on the applied potential [37].

3.3.2 Scanning Probe Nanolithography

Penner and coworkers at the University of California, Irvine use a different strategy. An STM is used to modify a surface with nanometer-scale defects, so as to induce nucleation of the deposited material at these defect sites. Penner's group has produced silver pillars 10–30 nm in diameter and 4–10 nm high on an STM-modified highly ordered pyrolytic graphite surface [38]. They have also fabricated a nanometer-scale galvanic cell composed of copper and silver nanopillars [39].

3.3.3 Epitaxial Growth of Quantum Dots

Rubinstein, Hodes, and coworkers at the Weizmann Institute have produced epitaxially oriented CdSe quantum dots with diameters of about 5 nm with a controllable spatial distribution and narrow size distribution by electrodepositing the nanocrystals on evaporated gold substrates [40, 41]. An interesting feature of their work is that the size of these quantum dots is believed to be controlled by the strain that is induced by the CdSe/Au lattice mismatch.

3.3.4 Electrodeposition of Superlattices

The Switzer group entered the nanoregime in 1990, when they showed that it was possible to electrodeposit nanometer-scale ceramic superlattices based on the Pb–Tl–O system [26]. A superlattice is a crystalline multilayer structure with coherent stacking of atomic planes and periodic modulation of the structure or composition or both [28]. A schematic of a superlattice is shown at the top of Figure 3.1. A general requirement for growing a superlattice is that the layers grow epitaxially, so the lattice mismatch must be fairly small. For the Pb–Tl–O systems the mismatch is on the order of 0.3%. The layers of the superlattice do not need to be equal, as long as the structure is periodic. The bilayer thickness is called the modulation wavelength. As shown at the bottom of Figure 3.1, superlattices can be grown by pulsing either the applied current or potential in a solution containing precursors of the materials in both layers. The thickness of each layer is determined by the charge passed. The Pb–Tl–O compositional superlattices were deposited from a solution of

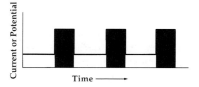

Figure 3.1. Schematic of a superlattice (top) and the current-time or potential-time waveform used to grow the superlattice (bottom). The bilayer thickness is called the modulation wavelength. Superlattices can be grown electrochemically with layers that are only a few atoms thick.

0.005 M Tl(I) and 0.1 M Pb(II) in 5 M NaOH. Thallium-rich films deposited at low overpotential, and lead-rich films deposited at higher potentials at which the deposition was mass-transport-limited in thallium.

A new type of nanoscale material called a defect chemistry superlattice was grown electrochemically at the University of Missouri-Rolla [31]. These nanometer-scale layered structures based on thallium(III) oxide were electrodeposited in a beaker at room temperature by pulsing the applied potential during deposition. The conducting metal oxide samples had layers as thin as 6.7 nm. The defect chemistry was a function of the applied overpotential: high overpotentials favored oxygen vacancies, whereas low overpotentials favored cation interstitials. The transition from one defect chemistry to another in this nonequilibrium process occurred in the same overpotential range (100–120 mV) in which the back electron transfer reaction became significant (as evidenced by deviation from linearity in a Tafel plot). The epitaxial structures have the high carrier density and low dimensionality of high-transition-temperature superconductors. They also have very interesting near-IR optical properties because of their high carrier density. The materials exhibit a plasma resonance in the near-IR at about 1500 nm, and the optical bandgap is shifted by up to 1.1 eV by the Moss–Burstein shift owing to the high free carrier density [42].

Several workers have shown that compositionally modulated metallic alloys can be electrochemically deposited if either the potential or current is cycled in a single plating solution containing salts or complexes of both metals [43–47]. The interest in these metallic multilayers stems from the enhanced mechanical properties induced by the low dimensionality and nanometer-scale dimensions [45] and recently from the discovery of the giant magnetoresistance (GMR) effect in nanoscale magnetic materials [48–50]. GMR materials have attracted attention because of their application in magnetic recording heads [51]. The general scheme for the deposition of metallic superlattices from a single plating bath is to use a solution containing a large excess of the ion with the more negative standard reduction potential. For example, in the electrodeposition of Cu/Ni superlattices, copper is present in solution at typically 1% of the total electroactive ion concentration. At low currents, pure copper is deposited. At higher currents, the copper deposition becomes mass-transport limited, and the films are essentially pure nickel. Using a different technique, Spaepen and coworkers have deposited Ni/NiP$_x$ multilayers from a dual bath, in which the electrode is moved repeatedly from one bath to another [52].

3.4 Characterization of Superlattices

3.4.1 X-ray Diffraction

How do you know if you have a superlattice? The best evidence is X-ray diffraction [26, 28, 53–55]. X-ray patterns of two different superlattices are shown in Figure 3.2. First, notice that the two superlattices have different preferred orientations. The

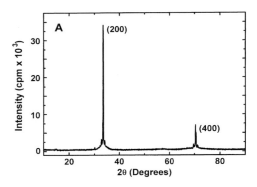

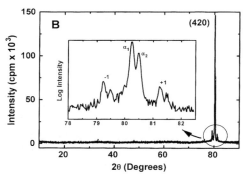

Figure 3.2. X-ray diffraction patterns of two electrodeposited Pb–Tl–O compositional superlattices. The top superlattice has a strong [100] orientation and a modulation wavelength of 13.3 nm, and the bottom superlattice has a strong [210] orientation and a modulation wavelength of 11.8 nm. The superlattices are epitaxial structures that follow the crystallographic orientation of the prelayer. The spacing of the satellites around the Bragg peaks is used to determine the modulation wavelengths.

top superlattice has a [100] orientation and the bottom superlattice has a [210] orientation. These preferred orientations were induced by first depositing oriented prelayers. Since the superlattices grow epitaxially, this orientation is followed throughout the entire growth process. In addition, the Bragg peaks are flanked by superlattice satellites. The satellites are caused by the superperiodicity in the system, because the X-ray pattern is the Fourier transform of the product of the lattice and modulation functions convoluted with the basis [31, 53]. The wavelength of the periodicity, Λ, can be calculated from the satellite spacing according to,

$$\Lambda = [(L_1 - L_2)\lambda]/[2(\sin\theta_1 - \sin\theta_2)] \tag{3.9}$$

where θ is the X-ray wavelength used, L is the order of the reflection, and θ is the diffraction angle [26, 47, 54]. In Eq. (3.9), $L = 0$ for the Bragg reflection, while the first satellite at lower angle has the value $L = -1$, and the first satellite at higher angle has the value $L = +1$. The superlattice at the top of Figure 3.2 has a modulation wavelength of 13.3 nm, and the superlattice at the bottom of the figure has a modulation wavelength of 11.8 nm. As the modulation wavelength increases, the satellites move closer to the Bragg peaks.

X-ray diffraction also provides a measure of the squareness of the composition profile in a superlattice. Since the X-ray pattern is the Fourier transform of the

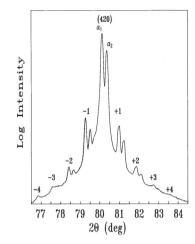

Figure 3.3. X-ray diffraction pattern of a Pb–Tl–O superlattice produced under potential control. Superlattice satellites out to the fourth order are seen around the (420) Bragg reflection. The modulation wavelength calculated from the satellite spacing is 13.4 nm. The X-ray radiation is CuK. The splitting of the peaks is caused by the presence of α_1 and α_2 radiation (wavelengths of 0.1540562 and 0.1544390 nm, respectively). (Figure adapted from [28]).

product of the lattice and modulation functions convoluted with the basis, the relative intensities of the various orders of X-ray satellites is a function of the modulation function. For example, if the composition is modulated sinusoidally in a superlattice, only first-order satellites are found to flank the Bragg reflections. Only one Fourier term is needed to describe a sinusoidal function. For square profiles, higher-order Fourier terms are necessary, and higher-order satellites are observed. The X-ray diffraction pattern of a Pb–Tl–O superlattice that was produced by modulating the electrode potential between 52 and 242 mV vs. SCE is shown in Figure 3.3 [28]. The superlattice, which has an X-ray modulation wavelength of 13.4 nm and an STM modulation wavelength of 12.7 nm, has superlattice satellites out to fourth order. The existence of higher-order satellites shows that the superlattice has a relatively square composition profile. Superlattices grown by pulsing the applied current rather than the applied potential had nearly sinusoidal composition profiles, and only first-order satellites were observed in the X-ray diffraction pattern [28].

3.4.2 Scanning Probe Microscopy

Another way to characterize superlattices is by scanning tunneling microscopy. Cleaved cross sections of both compositional and defect chemistry superlattices can be readily imaged in the STM [27–29, 31]. An STM image of a compositional superlattice is shown in the top of Figure 3.4. Since the layers of the superlattice are only 3 nm thick and the total thickness of the sample is 10 μm, this superlattice consists of over 3 000 layers. The modulation wavelength can be determined using Fourier analysis of the STM images [29]. A comparison of modulation wavelengths measured by STM and X-ray diffraction and calculated from Faraday's law for samples with modulation wavelengths ranging from 3 to 15 nm is shown in the

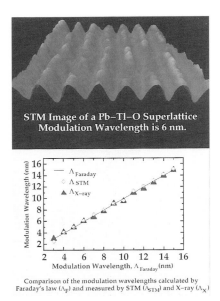

Figure 3.4. The top of the figure shows a scanning tunneling microscope (STM) image of a cleaved cross section of a Pb–Tl–O superlattice with a modulation wavelength of only 6 nm. The bottom of the figure shows a comparison of modulation wavelengths measured by STM and X-ray diffraction to those calculated from Faraday's law.

bottom of Figure 3.4. The STM is especially well suited to the measurement of modulation wavelengths that are too large to measure by X-ray diffraction but too small to measure by scanning electron microscopy. The STM can also give qualitative information about the composition profile in electrodeposited superlattices. For example, superlattices grown under current control have nearly sinusoidal profiles, while superlattices grown under potential control have much more abrupt profiles [28].

3.5 *In Situ* Studies of Epitaxial Growth

In addition to this qualitative information that is obtained *after* deposition of the superlattices, it is also possible to use the current-time transients that result during potentiostatic growth to both quantify and tune the composition profiles of superlattices *during* deposition [30]. Basically, when one of the precursors is in low concentration in the solution, and the layer is grown at high overpotential, the composition is graded throughout the layer with a $(time)^{-1/2}$ dependence. Superlattices grown at lower potentials in which both reactants are deposited under activation control have square composition profiles. The calculated profiles are shown in Figure 3.5 for three Pb–Tl–O superlattices that were grown by pulsing between 70–mV and 150, 230, or 260 mV vs. SCE. The composition of the 150 mV layer is relatively constant at 64% lead, while the lead content of the 260 mV layer varies from 39 to 76% through the layer. The graded composition profile may be desirable for some

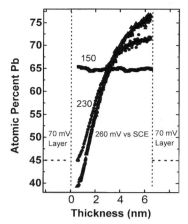

Figure 3.5. Composition profiles calculated by the Cottrell method for Pb–Tl–O superlattices grown by pulsing the potential between 70 mV and 150, 230, and 260 mV vs. SCE. This method can be used to both calculate and tailor composition profiles from graded to very abrupt in real time during the deposition process. (Figure adapted from Reference [30]).

applications. For example, grading the composition and lattice parameter may inhibit misfit dislocation formation in strained-layer superlattices. In semiconductor devices for optical or electronic applications, however, it may be desirable to have square composition profiles. The important point is that the electrochemical method is ideal for both measuring and tailoring the interface symmetry and composition profile in real time on a nanometer scale.

3.6 Electrodeposition of Nanocomposites

Nanophase materials are also of considerable interest because of their enhanced mechanical properties relative to bulk materials [56]. One reason for the different properties of these materials is that an increasing fraction of the atoms occupy sites at interfaces. It has been estimated that only about 3% of the atoms in a material are at boundaries when the grain size is 100 nm, but this increases to 25–50% when the grain size approaches 5–nm [57]. In metals there is typically an increase in the yield strength with decreased grain size as described by the well-known Hall–Petch relation, which describes the yield strength as a linear function of the inverse square root of the grain size. The behavior is due to the influence of grain boundaries on dislocation motion. As the grain size approaches the nanometer scale, large increases in strength can be obtained. In the nanoregime, however, the crystallite size becomes smaller than the characteristic length scales associated with dislocation generation and glide, which are the typical processes that determine mechanical behavior in metals. Deviations from the Hall–Petch relationship have been observed in nanocrystalline materials, even softening at the smallest grain sizes. This softening has been observed in nanocrystalline electrodeposits of Ni with grain sizes below approximately 12 nm [58].

Thick films of metals with nanocrystalline or amorphous structures for mechanical applications can be produced by codepositing a metalloid element such as phosphorous or boron with nickel and other iron group elements [59]. Another approach is to take advantage of the high supersaturation that is achievable during the very high peak current densities possible in pulse plating [60]. This is the electrochemical version of splat cooling of molten metals.

Electrochemical deposition is a very attractive processing route for the synthesis of composite materials. The low processing temperatures minimize the problems of chemical interaction and thermally induced stresses that are often serious problems in the conventional sintering, vapor phase, or liquid metal processes used to fabricate composites. A simple approach is to suspend particulate material in the plating electrolyte and codeposit this with the metallic matrix. This can be accomplished both by electroless deposition and by electroplating. Commercial applications of this approach include codeposition of alumina, silicon carbide, or diamond with a metal such as nickel. The particulates are normally in the micrometer range, but there have been a few studies involving nanoparticles [56]. A challenge in this work is to prevent agglomeration of the particles prior to codeposition.

Another electrochemical scheme for growing composites is electrochemical infiltration. Recently, the Sheppard group has used electrochemical infiltration to fill the 5 nm pores of a silica xerogel film with nickel [61]. In this manner, room temperature processing has been used to synthesize three-dimensionally interconnected nanoscale networks of metal and ceramic.

The Switzer group has recently shown that it is possible to electrodeposit nanocomposites of copper metal and cuprous oxide from alkaline aqueous solutions of copper(II) lactate [62, 63]. A fascinating feature of this system is that the electrode potential spontaneously oscillates when the films are deposited at a constant applied current density (galvanostatic deposition). Composites are deposited when the system oscillates. At constant pH the copper content in the films increases as the applied current density is increased. At constant current density the copper content increases as the pH is decreased. At pH 9 the phase composition varies from pure cuprous oxide at cathodic current densities below 0.1 mA cm^{-2} to 96 mole percent copper at 2.5 mA cm^{-2}. The cuprous oxide crystallites in the composites have the optical properties of quantum dots. For a film grown at a current density of 0.5 mA cm^{-2}, the optical absorption edge blueshifts from 2.1 eV for the bulk material to 2.5 eV for the nanophase crystallites [62]. In addition, discrete features develop in the spectrum, and there is approximately a twentyfold increase in the absorption coefficients for the quantum-confined semiconductor.

Potential oscillations at an applied current density of 0.5 mA cm^{-2} are shown in Figure 3.6 for copper lactate solutions in the pH range of 8.7–9.7. The oscillation period is a function of the solution pH, varying from 69 seconds at pH 8.7, to 11 seconds at pH 9.7 [63]. No oscillations were observed if the pH was below 8.5 or above 10. A map of the current densities and pH values for which oscillations are observed is shown in Figure 3.7. Nanocomposites are formed in the cross-hatched region of the figure. The oscillations are quasiperiodic, since the oscillation amplitude remains fairly constant, but the oscillation period increases slightly throughout the deposition. All the solutions used to observe the oscillations in Figure 3.6 were

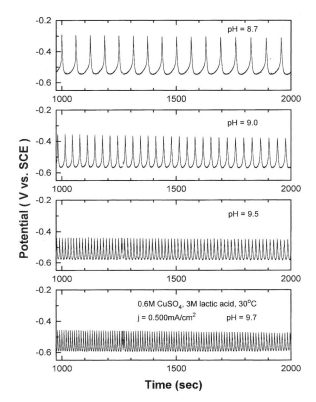

Figure 3.6. Potential oscillations as a function of pH observed during the growth of copper/cuprous oxide nanocomposites. The applied current density was 0.5 mA cm^{-2}. The solution was 0.6 M CuSO$_4$ and 3 M lactate ion at 30 °C. (Figure adapted from Reference [63]).

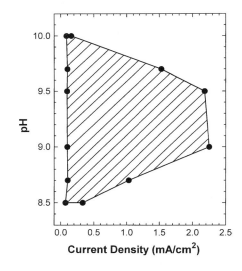

Figure 3.7. Map of the pH and current density over which potential oscialltions occur for the electrodeposition of copper/cuprous oxide nanocomposites. The solution was 0.6 M CuSO$_4$ and 3 M lactate ion at 30 °C. The potential oscillates and composites are deposited in the crosshatched region of the figure. (Figure adapted from Reference [63]).

Table 3.1. Phase composition, oscillation period, modulation wavelength, and cuprous oxide layer thickness as a function of solution pH for a series of copper/cuprous oxide nanocomposites. The applied current density was 0.5 mA cm^{-2}.

pH	Cu content (mol %)	Oscillation period (seconds)	Modulation wavelength (nm)	Cu$_2$O layer thickness (nm)
8.7	80	69	19	8
9.0	74	36	11	6
9.5	42	15	7	5
9.7	7	11	7	6

stirred. If the solutions are not stirred, the oscillations are not periodic and "ring out" in the time period shown in Figure 3.6. The potential oscillations observed during the deposition of all of the composites would suggest that these materials are modulated. Table 3.1 shows the phase composition, oscillation period, modulation wavelength, and cuprous oxide layer thickness as a function of pH for a series of copper/cuprous oxide nanocomposites grown at a current density of 0.5 mA cm^{-2} [63]. The oscillations for these composites are shown in Figure 3.6. The phase compositions in Table 3.1 were measured by X-ray diffraction, and the modulation wavelengths and layer thicknesses were calculated from the oscillation period using Faraday's law. Work is in progress to experimentally measure by TEM, STM, and Auger depth profiling the thickness of copper metal and cuprous oxide in these materials.

We chose the Cu/Cu$_2$O system to electrodeposit because of the dramatic differences between the two materials. Selected properties of cuprous oxide are summarized in Table 3.2 [63]. Cuprous oxide is a relatively nontoxic p-type semiconductor with a bandgap of 2.17 eV. Interest in the semiconducting properties of the material began as early as 1926, when Grondahl produced the Cu/Cu$_2$O rectifier [64]. Cuprous oxide is also a textbook example of a Wannier–Mott excitonic solid [65]. Up to ten hydrogenlike exciton lines can be seen in the absorption spectrum of Cu$_2$O at low temperatures [66–68]. Unbound excitons in Cu$_2$O have a binding energy of 150 meV, and a radius of 0.7 nm. The physics community is presently very excited about the excitonic properties of Cu$_2$O, since there is evidence that Bose–Einstein condensation of excitons can occur in the material [69]. Since the effective mass is close to that of a free electron, the de Broglie wavelength is large, and condensation of excitons can be observed at relatively high temperatures. Because excitons, like photons, are bosons, they can be made to propagate through a solid coherently [70–71]. Studies of Bose–Einstein condensation of excitons in Cu$_2$O have all been on single crystals. We are interested in producing nanostructures in which the dimensions of the Cu$_2$O are comparable to the exciton de Broglie wavelength. Nanoscale confinement in these materials should push the threshold for observation of Bose–Einstein condensation to even higher temperatures.

Table 3.2. Properties of cuprous oxide (from [68]).

Name	Value
Lattice parameter	0.427 nm
Bravais lattice	Primitive cubic
Space group	Pn $\bar{3}$
Formula units per unit cell	2
Copper special positions	(1/4, 1/4, 1/4); (3/4, 3/4, 1/4); (3/4, 1/4, 3/4); (1/4, 3/4, 3/4)
Oxygen special position	(0, 0, 0); (1/2, 1/2, 1/2)
Density	6.10 g cm^{-3}
Dielectric constant	7.11
Resistivity	3×10^6 ohm-cm
Majority carriers	Holes (p-type)
Bandgap (@ 4 K)	2.17 eV
Electron mass	$0.84 m_0$
Hole mass	$0.61 m_0$
Orthoexciton mass	$3.0 m_0$
Exciton radius	0.7 nm
Exciton binding energy	150 meV

3.7 The Future

What lies ahead in the area of nanoscale materials electrodeposition? There will certainly be increased emphasis on the electrodeposition of nanophase materials for magnetic applications, such as giant magnetoresistance materials [48–51], and for optical and electrical applications. Another big area of work should be the codeposition of very dissimilar materials, such as the copper/cuprous oxide nano-composites described in this chapter. The low processing temperatures of electro-deposition allow the codeposition of materials (ceramics, metals, polymers, semi-conductors) that would not tolerate each other at the high temperatures used for traditional thermal processing. There are really a lot of interesting possibilities.

Acknowlegements

The author would like to acknowledge the hard and enthusiastic work of a large number of graduate students and postdocs whose names are in the references. He would also like to acknowledge financial support of his work on nanoscale materials deposition by National Science Foundation Grants DMR-9020026 and DMR-9202872, Office of Naval Research Grants N00014-91-1-1499, N00014-94-1-0917, and N00014-96-0984, the University of Missouri Research Board, Mitsubishi Kasei, and Unocal Corporation.

Chapter 4

Size and Morphology Control of Nanoparticle Growth in Organized Surfactant Assemblies

M. P. Pileni

4.1 Introduction

The fabrication of assemblies of perfect nanometer-scale crystallites (quantum crystals) identically replicated in unlimited quantities in such a state that they can be manipulated and understood as pure macromolecular substances is the ultimate challenge of modern materials research with outstanding fundamental and potential technological consequences [1]. These potentialities are mainly due to the unusual dependence of the electronic properties on the size of the particles. Optical properties are one area where nanoparticles have markedly different properties from the bulk phase. These are described in terms of size quantization [2]. Since the size dependence of the band levels of semiconductor particles results in a shift of optical spectrum, quantum mechanical descriptions of the shift were carried out by several researchers [3–10]. At nanometer size, crystallites of semiconductors are influenced by the quantum confinement of the electronic states and modified from those of the bulk crystals. Colloidal dispersions of metals exhibit absorption bands or broad regions of absorption in the UV–visible range. These are due to the excitation of plasma resonances or interband transitions. They are characteristic properties of the metallic nature of the particles. The optical spectra of colloidal metallic particles have been described, including the effect of the size of the particle [11–13]. Indeed, if the particle dimensions are smaller than the mean free path of the conduction electrons, collisions of these electrons with the particle surface are noted. If the particle radii, R, are comparable with the mean free path of conduction electrons, L, the collisions of conduction electrons with the particle surface become important. Thus the effective mean free path is less than that in bulk materials. The electron energy bands are quantized and the number of discrete energy levels is of the order of magnitude of the number of atoms in the crystal.

In biomineralization, inorganic precipitates form under the full control of an organic tissue matrix. This control includes manipulation of the local concentration of the precipitants, the presence of nucleating surfaces, and the presence of inhibitors in solution. Particle size, shape, and orientation are regulated by the matrix.

Such *in situ* deposition processes also avoid the difficulties of handling nanosized particles while avoiding aggregation and without the need for large amounts of surfactant to keep them in suspension.

In terms of growth of particles, some analogies between surfactant self-assemblies and natural media can be proposed. In both cases, the growth of particles needs a supersaturated media where the nucleation takes place. Increasingly chemists are contributing to the synthesis of advanced materials with enhanced or novel properties by using colloidal assemblies as templates. In solution, surfactant molecules self-assemble to form aggregates [14]. At low concentration, the aggregates are generally globular micelles, but these micelles can grow upon an increase of surfactant concentration and/or upon addition of salt, alcohols, etc. In this case, micelles have been shown to grow to elongated, more or less flexible, rodlike micelles [15–22], in agreement with theoretical prediction on micellization [23, 24].

The preparation and characterization of these colloids have thus motivated a vast amount of work [25]. Various colloidal methods are used to control the size and/or polydispersity of the particles, using reverse [2] and normal [26] micelles, Langmuir–Blodgett films [27, 28], zeolites [29], or multilayer cast film [30, 31]. The achievement of an accurate control of the particle size and stability and a precisely controllable reactivity of the small particles are required to enable attachment of the particles to the surface of a substrate or to other particles without leading to coalescence and hence loss of the particles' size-induced electronic properties. Difficulties turn out to be manipulating nearly monodispersed nanometer-size crystallites of arbitrary diameter.

There are a number of reasons for forming films of inorganic particles' attached to or embedded just under the surface. Moreover, the ability to assemble particles into well-defined two- and three-dimensional spatial configurations should produce interesting properties as new collective physical behavior [32]. The development of a general procedure for the fabrication of "quantum" crystals is a major challenge of future research. One of the approaches to obtain 2D and 3D structure is to assemble nanoparticles themselves in ordered arrays. This requires a hard sphere repulsion, a controlled size distribution, and the inherent van der Waals attraction between particles and dispersion forces. The polydispersity in particle size prevents fabrication of such well-defined two- or three-dimensional structures. Recently, spontaneous arrangements with semiconductors such as Ag_2S [33] and CdSe [34] or metallic particles such as gold or silver [35, 36] have been reported. FCC arrangements of particles of different sizes have been characterized for microcrystals of silver sulfide nanoparticles [33]. Similar results were also reported in the case of CdSe nanoparticles [34] or gold metallic nanoparticles [35, 36].

In this chapter we demonstrate that colloidal assemblies are good candidates to be used as biomimetic routes for controlling the size and shape of nanoparticles. We present some examples in which the optical properties of nanoparticles vary with their size and shape. In the case of magnetic particles, we demonstrate a change in the magnetic properties with particle size. Some of these particles are able to arrange themselves and form monolayers in a hexagonal network or crystal with a centered cubic face.

4.2 Reverse Micelles

Reverse micelles are well known to be spherical water-in-oil droplets, stabilized by a monolayer of surfactant. The most used surfactant is sodium bis(2-ethylhexyl) sulfosuccinate, also known as Na(AOT). The phase diagram of Na(AOT)–water–isooctane shows a very large domain of water-in-oil droplets. This is why this surfactant is the most commonly used to form reverse micelles [2, 37]. The water pool diameter is related to the water content, $w = [H_2O]/[AOT]$, of the droplet as follows [2]:

$$D \text{ (nm)} = 0.3 \, w$$

If we take into account the existing domain of water-in-oil droplets in the phase diagram, the diameter of the droplet varies from 0.5 nm to 18 nm.

Reverse micelles are dynamic [38–41], and attractive interactions between droplets take place. The intermicellar potential decreases either by decreasing the number of carbon atoms of the bulk solvent or by increasing the number of droplets. This is because of the discrete nature of solvent molecules and is attributed to the appearance of depletion forces between two micelles (between the two droplets, the solvent is driven off). When the droplets are in contact, forming a dimmer, they exchange their water content. This exchange process is associated with the interface rigidity that corresponds to the binding elastic modulus of the interface. Hence, by collisions the droplets exchange their water content and again form two independent droplets. This process has been used to make nanosized material by either chemical reduction of metallic ions or coprecipitation reactions. These various factors (water content, intermicellar potentials) control the size of the particles.

When the syntheses are performed in reverse micelles, the nanoparticles formed are dispersed in the solution. This allows one to determine the absorption spectrum of the particles dispersed in the solvent. If we take a drop of this solution, it is possible, by TEM, to determine the average size of the particles. In this section, we present syntheses of metallic copper particles and semiconductor semimagnetic quantum dots. It is demonstrated that it is possible to make nanosized particles of a highly oxidable material and to control its size. Similarly, a solid solution of $Cd_{1-y}Mn_yS$ nanoparticles is made. As examples, syntheses and optical properties of semiconductor semimagnetic are presented.

4.2.1 Syntheses and Optical Properties of Metallic Copper Particles

Copper in its metallic form is well known to be oxidable. In the bulk phase, the surface is passivated. Copper nanosized particles have been extensively studied by ultra vacuum techniques. In solution, the copper nanosized particles are usually unstable. The metal is immediately transformed into its oxide form. Reverse micelles are good candidates for making nanosized metallic copper particles [42–44]. To

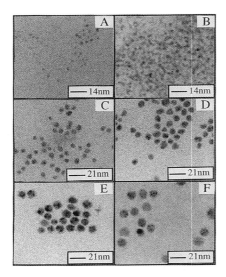

Figure 4.1. Electron microscopy patterns of copper metallic particles synthesized in $Cu(AOT)_2$–NaAOT– water–isooctane reverse micelles for various water contents: $[Na(AOT)] = 9 \times 10^{-2}$ M; $[Cu(AOT)_2] = 10^{-2}$ M; $[N_2H_4] = 3 \times 10^{-2}$ M: $w = 1$; 2; 3; 4; 5; 10; and 15.

form this material, a functionalized surfactant such as copper bis(2-ethylhexyl) sulfosuccinate, $Cu(AOT)_2$, is needed.

When $Cu(AOT)_2$ is replaced by copper sulfate (Cu^{2+}), metallic copper particles are not formed. They exist in their oxidized form. Furthermore, a few minutes after beginning the reaction, the particles flocculate. Mixed micelles made of sodium and copper bis(2-ethylhexyl) sulfosuccinate are prepared. The water content, $w = [H_2O]/[Na(AOT) + Cu(AOT)_2]$, is fixed at a given value. This solution is mixed to a Na(AOT) micellar solution having the same water content and filled with hydrazine. The chemical reduction of copper bis(2-ethylhexyl) sulfosuccinate takes place and metallic copper particles are formed. They are characterized by TEM, electron diffraction, and absorption spectroscopy. Figure 4.1 shows an increase in the particle size with increasing water content. The absorption spectrum of colloidal particles obtained for various water contents shows a progressive appearance of the plasmon peak (Figure 4.2). Hence, a direct relationship between the absorption spectrum and the size of the particles is obtained. This enables us to obtain a calibration curve relating the particle diameter to its absorption spectrum (Figure 4.3).

4.2.2 Syntheses and Optical Properties of Semiconductor Semimagnetic Quantum Dots

Syntheses of $Cd_{1-y}Mn_yS$ are performed in reverse micelles [45]. A coprecipitation takes place by mixing two micellar solutions having the same water content, $w = [H_2O]/[AOT]$: one is made of 0.1 M Na(AOT) containing S^{2-} ions; the other is a mixed micellar solution made of $Cd(AOT)_2$, $Mn(AOT)_2$, and Na(AOT). The

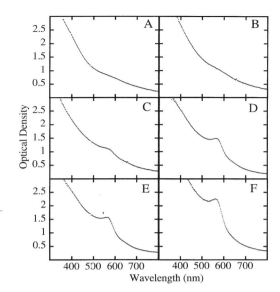

Figure 4.2. Absorption spectra of copper metallic particles synthesized in Cu(AOT)$_2$–NaAOT–water–isooctane reverse micelles at various water contents. [Na(AOT)] = 9×10^{-2} M; [Cu(AOT)$_2$] = 10^{-2} M; [N$_2$H$_4$] = 3×10^{-2} M: w = 1; 2; 3; 4; 5; 10; and 15.

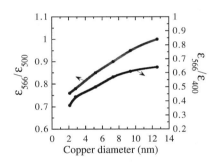

Figure 4.3. Variation of ratio of the extinction coefficients $\varepsilon_{566}/\varepsilon_{500}$ and $\varepsilon_{566}/\varepsilon_{400}$ of the colloidal copper solution with the size of the copper particles synthesized in Cu(AOT)$_2$–NaAOT–water–isooctane reverse micelles: [Na(AOT)] = 9×10^{-2} M; [Cu(AOT)$_2$] = 10^{-2} M; [N$_2$H$_4$] = 3×10^{-2} M.

syntheses are performed for various ratios of Mn(AOT)$_2$, Cd(AOT)$_2$, and water content. The composition of the particles is measured by energy dispersion spectroscopy (EDS). Only a small amount of manganese remains in the CdS matrix. A formation of a Cd$_{1-y}$Mn$_y$S solid solution is obtained.

When the syntheses are performed for various water contents and for a fixed composition, y, an increase in particle size is observed. Figure 4.4 shows a red shift in the absorption spectra when the particle size increases. This behavior is similar to that observed previously [2, 3–11, 46, 47]. Syntheses performed at fixed water content and for various compositions induce formation of particles having the same size and differing in their compositions. Figure 4.5 shows the change in the absorption spectrum with composition (the average diameter of the particles is 3 nm). The direct band gap of the nanocrystallites is deduced from the following equation

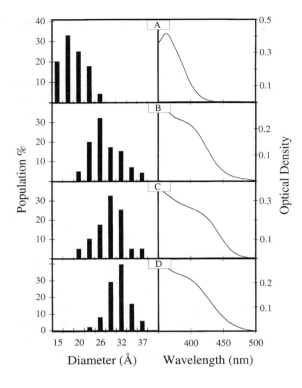

Figure 4.4. Histograms deduced from TEM patterns and absorption spectra of $Cd_{0.9}Mn_{0.1}S$ particles synthesized for various water contents $w = 5$ (a), 20 (b), 30 (c), and 40 (d).

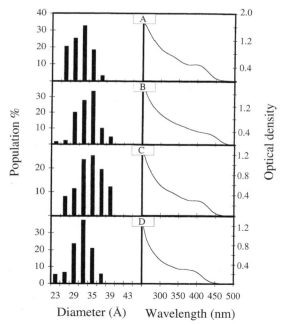

Figure 4.5. Histograms deduced from TEM patterns and absorption spectra of $Cd_{1-y}Mn_yS$ for various y: $y = 0$ (a), 0.08 (b), 0.12 (c), 0.23 (d). The size of the particles is kept equal to 3 nm.

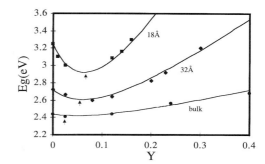

Figure 4.6. Bandgap (E_g) $Cd_{1-y}Mn_yS$ crystallites versus manganese composition y. ●: bulk[34]; ◆: diameter $= 32$ Å; ■: diameter $= 18$ Å.

[47]: $\sigma h\nu = (h\nu - Eg)^{1/2}$. From the absorption spectra recorded for various compositions, the energy band gap is deduced. It is compared to that given in the literature for the bulk phase [48]. The behavior of the energy band gap depends on the composition and particle size (Figure 4.6), as follows:

(i) For a given composition, y, the energy band gap increases with a decrease in particle size. This phenomenon is attributed to a quantum size effect.
(ii) With increasing composition, y, the energy band gap decreases and then increases (Figure 4.6). The variation of the energy band gap with composition is more pronounced when the particle size decreases. The position of the energy band gap minimum depends on the particle size. For particles having an average diameter equal to 2 nm and 3.2 nm, the minimum of the energy band gap is reached at $y = 0.05$ and 0.08 respectively. Similar behavior is obtained in the bulk phase. Opposite to what is observed with quantum dots, the variation of the energy band gap with composition, y, is not very pronounced. The minimum is not well defined (about $y = 0.02 - 0.05$). These strong changes in the behavior of the energy band gap with the size of nanoparticles and with the bulk phase could be attributed to perturbation induced by hybridation of magnetic cation orbitals (Mn^{2+}) with the band structure and to exchange interactions in a confined regime.

4.3 Oil-in-Water Micelles

Dissolved in water at a concentration below the critical micellar concentration, (or c.m.c.), the surfactant behaves as a strong electrolyte, entirely dissociated, whereas above the c.m.c., the monomers form spheroidal aggregates called normal micelles. The most often used surfactant is sodium dodecyl sulfate, Na(DS). The size, shape, and degree of ionization for micelles change with increasing surfactant concentration [49, 50].

Divalent dodecyl sulfate is made by mixing an aqueous solution of sodium dodecyl sulfate (0.1 M) with divalent ion derivatives (0.1 M), as described elsewhere

[51]. The solution is kept at 2 °C and a precipitate appears. It is washed several times with a 0.1 M solution of divalent ions and recrystallized in distilled water. Divalent dodecyl sulfate, $X(DS)_2$, forms micellar aggregates above the c.m.c. It does not differ greatly with various divalent counterions, staying in the region of 10^{-3} M. The shape and size of these aggregates have been determined by small-angle X-ray scattering (SAXS) and by light scattering [52]. Prolate ellipsoidal micelles with a hydrodynamic radius equal to 2.7 nm are found.

We used oil-in-water micelles either to make magnetic fluid [53] and control the size of the particles [54, 55] or to control the shape of metallic copper particles [26]. Two examples are presented in the following sections.

4.3.1 Magnetic Fluids: Syntheses and Properties

Aqueous methylamine, CH_3NH_3OH, is added to a mixed micellar solution formed by $Co(DS)_2$ and $Fe(DS)_2$ surfactants. The solution is stirred for two hours at room temperature. A magnetic precipitate appears. The supernatant is removed and replaced by pure bulk aqueous phase. The precipitate is redispersed and a brown magnetic suspension is obtained. This is usually called magnetic fluid. The percentage of surfactant remaining in solution is less than 0.1% in weight. The composition in metal of cobalt ferrite particles consists of Fe(III) and Co(II). Electron diffractogram patterns are in good agreement with the intense peaks listed for Co-Fe_2O_4 in standard reference tables. This indicates the formation of material having an inverted spinel crystalline structure as in the bulk phase. EDS confirms the relative ratio of cobalt and iron elements in the $CoFe_2O_4$ particles (the percentages of the iron and cobalt elements are found to be equal to 65.63% and 34.37% respectively). The Mössbauer spectrum of particles of 5 nm diameter recorded at 4.2 K is similar to that observed for larger $CoFe_2O_4$ particles (30–350 nm) [56] with the sixth B-site line apparently less intense than the sixth A-site line, as observed for bulk cobalt ferrite. By using the same procedure, Fe_3O_4 nanoparticles have been obtained from syntheses performed with $Fe(DS)_2$ as the reactant [57].

Several syntheses have been performed by increasing the $Fe(DS)_2$ concentration, from $6.5.10^{-3}$ M to $2.6.10^{-2}$ M, keeping the $[Co(DS)_2]/[Fe(DS)_2]$ and $[Fe(DS)_2]/[CH_3NH_3OH]$ ratios equal to 0.325 and 1.3×10^{-2} respectively. From TEM patterns and histograms, Figure 4.7 shows an increase in the particle size with an increase in $Fe(DS)_2$ concentration. Hence, control of the size is obtained by changing the $Fe(DS)_2$ concentration by a factor of two. This control of particle size is obtained without large changes in the experimental conditions. This allows us to assume, to a first approximation, that the surface of the particle keeps the same composition. This makes possible the study of the relationship between size and magnetic properties.

Figure 4.8 shows the magnetization curve, obtained at 200 K, for particles having an average diameter equal to 2, 3, and 5 nm respectively. The initial susceptibility shows no hysteresis; that is, both remanence and coercitivity are zero. This indicates superparamagnetic behavior, as expected for the nanoscale dimensions of the particles. The average magnetic size of the particles can be deduced from simulation

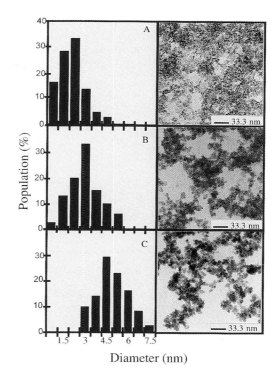

Figure 4.7. Electron microscopy patterns and histograms of magnetic fluid made at various surfatant concentrations, keeping $[Co(DS)_2]/[Fe(DS)_2] = 0.325$, $[Co(DS)_2]/[NH_2CH_3] = 1.3 \times 10^{-2}$. $[Fe(DS)2] = 6.5 \times 10^{-3}$ M (a), $[Fe(DS)] = 1.3 \times 10^{-2}$ M (b), $[Fe(DS)_2] = 2.6 \times 10^{-2}$ M (c).

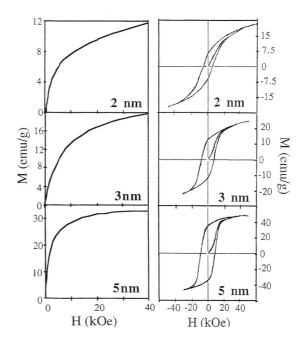

Figure 4.8. Left: Magnetization as a function of the applied field at 200 K for particles shown in Figure 4.1. Right: Magnetization as a function of the applied field at 10 K for for 2 nm (a), 3 nm (b), and 5 nm (c) cobalt ferrite fluid. Volume fraction = 1%.

Table 4.1. Variation of the saturation magnetization at 200 K (M_{s200}) and 10 K (M_{s10}); the average diameter of the particles deduced from TEM, D_{TEM} (nm), from simulation, D_M (nm), and from SAXS, D_{SAXS} (nm); the polydispersity in size determined by TEM, σ_{TEM}, and from simulation, σ_M, the ratio of the remanence and saturation magnetization.

D_{TEM} (nm)	2	3	5
σ_{TEM} (%)	37	36	23
D_{SAXS} (nm)	2.6	3.6	5.4
D_M (nm)	2	3	4.2
σ_M (%)	42	40	35
M_{s200} (emu g^{-1})	14	22	35
M_{s10} (emu g^{-1})	23	31	50
M_{s200}/M_{s10}	0.60	0.71	0.70
M_{r10}/M_{s10}	0.31	0.43	0.74
H_c (kOe)	5	7.5	9
$\chi_{200} \cdot 10^4$ (Oe)	2	2.4	4
$K_A \times 10^{-7}$ (erg cm^{-3})	7	3	1

of the Langevin relationship, assuming a log normal size distribution [58]. A good agreement between the size determined by TEM and SAXS [55] and the magnetization curve is observed (Table 4.1).

The magnetic field needed to reach the saturation magnetization depends on the size of the particles. For particles having an average size equal to 2 nm and 3 nm, the saturation is not reached even for a magnetic field equal to 40 kOe. On the other hand, it is obtained with 5 nm particles. For particles having an average size equal to 2 and 3 nm, the saturation magnetization is deduced from zero extrapolation of M vs. $1/H$. It decreases with a decrease in particle size. Even for the larger particles (5 nm diameter), the saturation magnetization is less than the bulk value (Table 4.1). This could be explained by an increase in noncollinear structure when the particle size decreases. Cobalt ferrite, known to have a relatively high magnetocrystalline anisotropy, has a noncollinear structure [59, 60]. This increases with coating by surfactant and with a decrease in particle size in the range of few hundred ångstroms.

The magnetic particles are frozen in zero field at 10 K. Figure 4.8 shows the presence of hysteresis with an increase in the coercitivity with particle size. The ratio of the remanence to saturation magnetizations, M_r/M_s, deduced from the magnetization curve, decreases with a decrease in particle size (Table 4.1). The reduced remanence, M_r/M_s, depends on the magnetocrystalline anisotropy constant, the median diameter and the standard deviation of the system [61]. For random distribution of easy magnetic axes of particles with cubic magnetocrystalline anisotropy, the reduced remanence is expected to be equal to 0.83 at 0 K. For particles having 5 nm as an average diameter, Table 4.1 shows a M_r/M_s ratio equal to 0.74 at 10 K. The large remanence and coercitivity values and the remanence-to-saturation-magnetization ratio indicate that 5 nm particles consist of randomly oriented equiaxial particles with cubic magnetocrystalline anisotropy [62]. Table 4.1 shows a progressive decrease of M_r/M_s ratio with a decrease in particle size. This is ex-

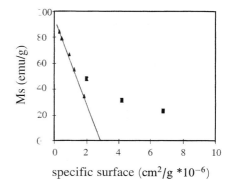

Figure 4.9. Variation of the saturation magnitization with specific surface at 10 (■) obtained in our experimental conditions and the results obtained by Mollar et al. at 4.2 K (▲)

plained as a progressive change of magnetocrystalline anisotropy from cubic (for 5 nm) to axial (for 2 nm and 3 nm) structure. This is confirmed from the simulation of the Mössbauer spectra recorded at various temperatures and various particle sizes [63].

Figure 4.9 shows a decrease in the saturation magnetization with specific surface per area. Mollar et al. [64] have established, for $CoFe_2O_4$ particles with diameters ranging from 6.2 to 33 nm, that the saturation magnetization (measured at 4.2 K) in a strong magnetic field decreases linearly with an increase in specific surface area. In the range 2–5 nm, the measurements performed at 10 K show similar behavior, as has been proposed by Mollar et al. [64] (Figure 4.9). The experiments performed by these two groups were not carried out at the same temperature (4.2 and 10 K respectively), but the change of the saturation magnetization with temperature is small enough (Figure 4.9) to allow comparison. For particles of a similar size (6.2 and 5 nm respectively) a relatively good agreement of the saturation magnetization given by the two groups is observed. Mollar et al. [64] found a linear relationship of the saturation magnetization with the specific area. The extrapolation of the slope to zero magnetization is obtained for particles having an average diameter equal to 4 nm. This has been explained, as has been demonstrated for films made of ferromagnetic transition metals [65, 66], as resulting from a formation of layers that remain nonmagnetic ("dead layers"). The thickness of the layer is evaluated to be twice the lattice parameter. For $CoFe_2O_4$, this corresponds to 1.6 nm. In our experimental conditions, this explanation can be excluded for the following reasons:

(i) The saturation magnetization for 4 nm particles is equal to 40 emu g^{-1}, whereas it is found equal to zero from extrapolation of the Mollar curve (Figure 4.9).
(ii) Magnetization is still obtained for particles with an average diameter equal to 2 nm. This is not compatible with a dead layer having a thickness equal to 1.6 nm. The decrease in the saturation magnetization with a decrease in size of the particles is attributed to an increase in the noncollinearity of the structure.

The samples are cooled in zero field to 10 K. The magnetization is measured as a function of temperature in a 100 Oe field. As expected, the blocking temperature

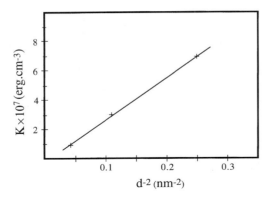

Figure 4.10. Variation of the anisotry energy, K_A, with d^{-2}, where d is the diameter of the particle.

increases with particle size (Table 4.1). From the blocking temperature, the anisotropy constant is deduced. It decreases with a decrease in particle size. It varies as d^{-2}, where d is the diameter of the particle (Figure 4.10). It is larger than the bulk value of $CoFe_2O_4$ material. Such variation of the anisotropy constant with the particle diameter is confirmed by Mössbauer spectroscopy [63].

4.3.2 Control of the Shape of Metallic Copper Particles

Syntheses of metallic copper particles are performed in an aqueous solution containing copper(II) dodecyl sulfate, $Cu(DS)_2$. The copper ions associated with the surfactant are reduced by sodium borohydride, $NaBH_4$. The ratio $NaBH_4/Cu(DS)_2$ is equal to 2. The syntheses are performed at and above the c.m.c. In all the cases, formation of metallic copper aggregates is observed. These are characterized by electron diffraction and from the absorption spectrum of the colloidal solution. At the c.m.c., $(1.2 \times 10^{-3}$ M), the colloidal solution is characterized by an absorption spectrum centered at 570 nm. A drop of this solution is deposited on a carbon grid and TEM measurements show the formation of large domains of aggregates arranged in an interconnected network (Figures 4.11(a) and 4.12). The expansion of the TEM pattern (Figure 4.12(b)) shows that the network corresponds to a change in the particle shape and not to aggregation of small particles in strong interaction.

Synthesis performed at 2×10^{-3} M $Cu(DS)_2$ induces the formation of elongated particles. The maximum of the absorption spectrum is characterized by a maximum centered at 564 nm. When the syntheses are performed at higher $Cu(DS)_2$ concentration, the size of the rods decreases with the formation of more spherical particles, (Figure 4.11).

The absorption spectrum of 10 nm spherical metallic copper particles is characterized by a plasmon peak centered at 558 nm (Figure 4.2(f)). Figure 4.13 shows a red shift in the plasmon peak when the shape of the metallic copper particles changes from spheres to rods. Such changes in the absorption spectra of metallic copper particles can be related to those predicted at various r values, where r is defined as the ratio of the length to the diameter of a cylinder [67–69]. The plasmon

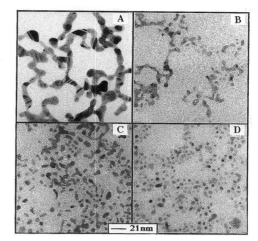

Figure 4.11. Electron microscopy of colloidal copper dispersion prepared in a pure copper dodecyl sulfate solution: $[Cu(DS)2] = 1.2 \times 10^{-3}$ M, 2×10^{-3} M, 3×10^{-3} M, 10^{-2} M, $[NaBH4] = 2[Cu(DS)2]$.

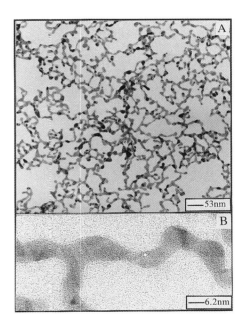

Figure 4.12. Electron microscopy (a) and an expanded picture (b) of the copper network prepared in a pure copper dodecyl sulfate solution: $[Cu(DS)2] = 1.2 \times 10^{-3}$ M, $[NaBH4] = 2.4 \times 10^{-3}$ M.

peak due to the rod particles is centered at 570 nm (Figure 4.13). According to simulated absorption spectra [26], the plasmon peak centered at 570 nm corresponds to an r value equal to 2.5. From the imagery analysis of the skeleton of the interconnected network (Figure 4.12(a)) corresponding to a plasmon peak centered at 570 nm, the average length of linear strands is found to be equal to 22 nm. The average minor diameter is equal to 6.5 nm. Hence, the r value is found to be equal to 3.3. From theoretical predictions, such an r value ($r = 3.5$) corresponds to a

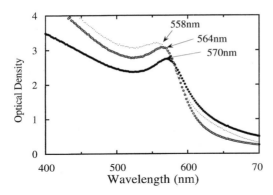

Figure 4.13. Absorption spectra of interconnected (●●●●), elongated (○○○○) and spherical (. . . .) colloidal particles [22].

higher shift in the plasmon peak (620 nm) than that obtained (570 nm). This difference between r and the maximum of the plasmon peak is explained by the fact that the simulation is related to one size of cylindrical particles. In the present experiment, the particles are interconnected and not isolated cylinders. Furthermore, the polydispersity has to be taken into account. A qualitative shift in the maximum of the plasmon peak with particle shape is in good agreement to what is predicted. Hence, the shift and the growth in the absorption band, compared to spherical and elongated particles, indicate that the interconnected network, observed by TEM, exists in solution. This is due to a change in the effective mean free path of the conduction electron and not caused by the evaporation process of the solutions.

4.4 Interconnected Systems

The 5×10^{-2} M Cu(AOT)$_2$ in isooctane solution is an isotropic phase. By water addition, the phase diagram evolves progressively. At $w = 5.5$ a phase transition takes place. The lower phase is optically clear and keeps the blue color characteristic of Cu(AOT)$_2$. The upper phase is pure isooctane. If the water content is increased from $w = 5.5$ to 11, the following are observed:

(i) An increase in the isooctane volume (upper phase). This induces an increase in the Cu(AOT)$_2$ concentration in the lower phase.
(ii) An increase in the conductivity of the lower phase, whereas that of the upper phase is similar to that obtained in pure isooctane.
(iii) A progressive increase in the viscosity of the lower phase and then a decrease. The maximum in viscosity is reached for a w value close to 7.
(iv) A characteristic scatter of cylinders, observed by SAXS [70]. The gyration radius of the normal section and persistence length of the cylinders, $\langle l \rangle$, are deduced, for various water contents. Table 4.2 shows no drastic changes in either the gyration radius or in the persistence length with an increase in water

Table 4.2. Gyration radius, R_g, and persistence length (nm), $\langle l \rangle$.

w	6.5	7.5	8.5	9.5	10	11
R_g (nm)	1	0.9	1	1	1	1.1
$\langle l \rangle$ (nm)	3	3	3.2	3.1	3	3.1

content. From this, the average size of the cylinders is deduced. This remains constant when the water content increases. The errors in the values given in Table 4.2 are evaluated to 10%.

(v) Freeze fracture replicas show a homogeneous system made up from only very small objects.

The increase in conductivity and viscosity of the lower phase with an increase in the water content can be related to the increase in $Cu(AOT)_2$ concentration: At low water content (below $w = 5.5$), the water-in-oil phase is rather diluted. The increase in water content from $w = 5.5$ to 11 induces a phase transition with an increase in the $Cu(AOT)_2$ concentration in the lower phase. This favors an increase in the number of connections between cylinders to form a bicontinuous network. Similar behavior has already been observed with other self-assembled surfactants. It is attributed to the formation of disordered open connected microemulsions [71–73]. At high water content, $w > 7$, the decrease in viscosity is explained in terms of branching of one cylinder into another with, locally, a saddle structure. Similar behavior has been observed in oil-in-water micelles [74].

Similar structure, with interconnected cylinders having similar persistence length, is observed in the range $w = 30$–35. Syntheses of metallic copper particles are performed in the two domains ($5.5 < w < 11$ and $30 < w < 35$) [75].

(i) Syntheses performed at $w = 6$ show (Figure 4.14) the formation of a relatively large amount of metallic copper cylinders (32%) in coexistence with 68% of spheres. The average diameter of the spherical particles is 9.5 ± 1.1 nm. The histogram shows the formation of large cylinders (Figure 4.14(c)). The average length-to-width ratio of the cylinder is found to be equal to 3.5 with 40% polydispersity. The length and width of the cylinders are equal to 22.6 ± 5.4 nm and 6.7 ± 1.4 nm respectively.

(ii) Figure 4.15 shows a TEM pattern obtained at $w = 34$. As at lower water content, cylindrical (42%) and spherical (58%) nanoparticles are observed. The average diameter of spherical particles is equal to 9.5 ± 0.9 nm. The length and width of the cylinders are equal to 19.8 ± 2.7 nm and 6.5 ± 0.8 nm respectively.

Syntheses performed in these two domains ($5.5 < w < 11$ and $30 < w < 35$) show very strong correlation and similar data. A very great similarity in the size and shape of nanoparticles is obtained from syntheses performed in the two parts of the phase diagram having the same structure (interconnected cylinders) and differing in their water content. No other particle shapes have been observed. In both regions:

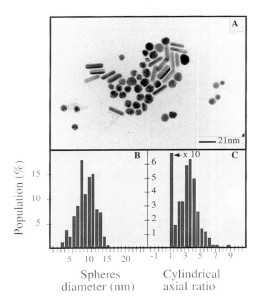

Figure 4.14. TEM pattern obtained after synthesis at $w = 6$, $[Cu(AOT)_2] = 5 \times 10^{-2}$ M (a), and histograms of the diameter of the spheres (b), and ratio of the length to the width of the cylinders (c).

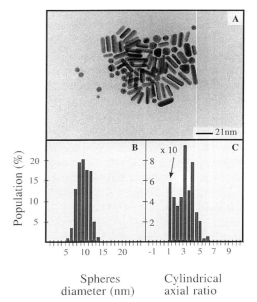

Figure 4.15. TEM pattern obtained after synthesis at $w = 34$, $[Cu(AOT)_2] = 5 \times 10^{-2}$ M (a), and histograms of the diameter of the spheres (b), and ratio of the length to the width of the cylinders (c).

(i) The average diameter of the spherical particles is the same (9.5 nm). The poly-dispersity is a little higher at low water content (Table 4.3).

(ii) The size of the cylinders remains identical in the two domains (Table 4.3). The same average diameter and same ratio of cylinder axes ($\neq 3.3$) are observed at low ($5.5 < w < 11$) and high ($30 < w < 35$) water content. Because of this great

Table 4.3. Variation with water content, w, of the average diameter of the spheres, $\langle d_s \rangle$, the polydispersity of the spheres, σ_s, the percentage of spheres, $\%_s$, the percentage of cylinders, $\%_c$, the percentage expressed in weight of copper metallic cylindrical particles, $\%_c$(weight), the average length, $\langle L_c \rangle$, and width, $\langle l_c \rangle$, of the cylinders, the polydispersity in the average length, σ_{Lc}, and width, σ_{lc}, of the cylinders, the average ratio of the cylinder axes, $\langle L_c/l_c \rangle$, and the polydispersity in this ratio, $\sigma_{Lc/lc}$.

w	$\langle d_s \rangle$ (nm)	σ_s (%)	$\%_s$	$\%_c$	$\%_c$ (weight)	$\langle L_c \rangle$ (nm)
6	9.5	27	68	32	45.5	22.6
34	9.5	19	58	42	51.4	19.8

w	σ_{Lc} (%)	$\langle l_c \rangle$ (nm)	σ_{lc} (%)	$\langle L_c/l_c \rangle$	$\sigma_{Lc/lc}$ (%)
6	24	6.7	21	3.5	40
34	27	6.5	24	3.2	33

similarity in various experimental conditions, this phenomenon is attributed to the structure of the colloid used as a template. This could be explained as in nature [76], where the key step in the control of mineralization is the initial isolation of a space.

4.5 Onion and Planar Lamellar Phases in Equilibrium

With the addition of water to a 5×10^{-2} M Cu(AOT)$_2$ in isooctane ($11 < w < 15$), a birefringent phase appears in equilibrium with the inverted phase and isooctane. By increasing the water content to $w = 15$, the inverted phase progressively disappears. From freeze-fracture images the coexistence of a rather well ordered planar lamellar phase (Figure 4.16(a)) with poorly ordered spherulites (Figure 4.16(b)) is observed.

Syntheses show the formation of rodlike metallic copper particles (Figure 4.17(a)). The diameter of the rods varies from 10 to 30 nm and the length from 300 to 1500 nm. High-resolution electron microscopy shown in Figure 4.17(b) indicates a high crystallinity with a very low number of defects. The formation of such large rods having a high crystallinity indicates that the phase structure governs a slow nucleation.

4.6 Spherulites

Water is added to 5×10^{-2} M Cu(AOT)$_2$ in isooctane solution. The overall water concentration varies from 1.5 to 2 M ($15 < w < 20$). Immediately after water addi-

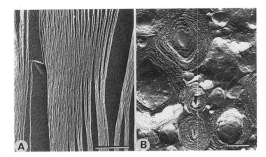

Figure 4.16. Freeze-fracture electron micrographs of $w = 14$ samples (a, b). Note the presence in both samples of planar lamellar phases and spherulites. The bars represents 500 nm.

Figure 4.17. (a) Electron micrographs of copper rods prepared in lamellar phases La. (b) High-resolution electron microscopy of a rod particle.

tion, a phase transition occurs. The lower phase contains $Cu(AOT)_2$, water, and isooctane, whereas the upper part is pure isooctane. The lower phase is birefringent and its conductivity remains unchanged with the various amounts of water added (12.8 µS). SAXS measurements reveal one Bragg peak with no second-order peaks and a strong increase in the scatter at low angles. The characteristic distance, d, is deduced. It increases linearly with increasing water concentration. This is explained in the following way. The equilibrium with almost pure isooctane forces the isooctane lamellae to be maximally swelled. Addition of water causes swelling of the aqueous lamellae and increases the characteristic distance, d. However, d is smaller than the value expected from the measured isooctane content. This may be due to onionlike lamellar structures containing the excess isooctane in between the spherulites and in their centers. This is confirmed by the freeze-fracture replicas, which

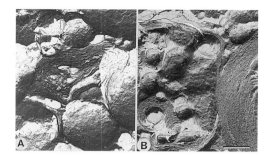

Figure 4.18. Freeze-fracture electron micrographs of $w = 13$ (a) and $w = 22$ (b) samples. The bar represents 1000 nm.

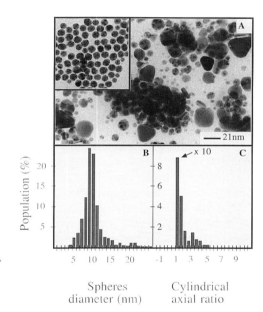

Figure 4.19. TEM pattern obtained after synthesis at $w = 18$, $[Cu(AOT)_2] = 5 \times 10^{-2}$ M (a), and histograms of the diameter of the spheres (b) and the ratio of the length to the width of the cylinders (c).

show almost exclusively the presence of spherulites (Figure 4.18). These are formed without any external forces, in contrast to those usually observed after shearing of lamellar phases [77, 78]. The size of the spherulites varies greatly (from 100 to 8000 nm).

Syntheses show the formation of particles having a higher polydispersity in size and shape (Figure 4.19) compared to what is observed in the other parts of the phase diagram. Figure 4.19 shows the formation of triangles, squares, cylinders, and spheres. However, in some regions of the carbon grid, self-assemblies made of spherical particles are observed (insert in Figure 4.19(a), 24% of particles are characterized by a spheroidal shape). Because of the strong change in the particle shape, it is difficult to produce histograms. The size distribution has been measured in the following way: triangular and tetrahedral particles have been assimilated to spheres. When the difference in length and width of particles was greater than 3 nm, it was assumed that the particles were cylinders.

4.7 Self-organization of Nanoparticles in 2D and 3D Superlattices

Several factors act simultaneously to produce a self-assembly made of nanoparticles and organized 2D and 3D superlattices. The low polydispersity in particle size and the interactions between particles are strong enough to induce such organization.

Metallic silver, $(Ag)_n$, and silver sulfide, $(Ag_2S)_n$, nanoparticles organize themselves in a very large network. In both cases, the syntheses are performed in mixed reverse micelles made of Ag(AOT) and Na(AOT). In both cases, control of particle size is obtained. However, some differences appear: with $(Ag)_n$ particles the average diameter varies from 2 to 6 nm [79]. With $(Ag_2S)_n$, it varies from 2 to 10 nm [80]. To stabilize the particles and to prevent their growth, 1 µl ml^{-1} of pure dodecanethiol is added to the reverse micellar system containing the particles. This induces a selective reaction at the interface, with covalent attachment, between thio derivatives and silver atoms [81, 82]. The micellar solution is evaporated at 60 °C and a solid mixture made of dodecanethiol coated nanoparticles and surfactant is obtained. To remove the AOT and excess dodecanethiol surfactant, a large amount of ethanol is added. The particles remaining are dried and dispersed in heptane. Surprisingly, the behavior differs for $(Ag_2S)_n$ and $(Ag)_n$ nanoparticles. With $(Ag_2S)_n$, only a small fraction of the particles are dispersed in heptane and size selection takes place. The polydispersity drops from 30% to 14%. With $(Ag)_n$ slight size selection occurs and the polydispersity varies from 43% to 37%.

In both cases, when the particles are dispersed either in reverse micellar solution or in heptane, the solutions are optically clear. This permits one to follow the colloidal particles by spectroscopy. As predicted by simulation, the absorption spectrum of $(Ag)_n$ nanoparticles varies with particle size: the plasmon peak decreases with a decrease in particle size [79]. The attachment of dodecanthiol on the silver particles can be monitored by UV–visible spectroscopy: Figure 4.20 (dotted line) shows a drastic decrease in the extinction coefficient of the plasmon band after dodecanthiol addition, whereas below 300 nm, no change in UV absorption is

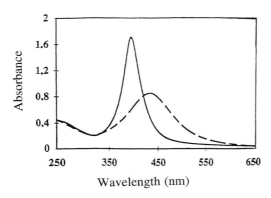

Figure 4.20. UV–visible absorption spectra of silver colloidal solution after synthesis in reverse micelles (—) and after the addition of dodecanthiol to the micellar solution (---).

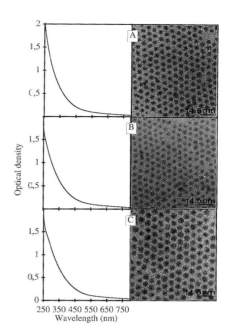

Figure 4.21. T.E.M patterns and absorption spectra of coated $(Ag_2S)_n$ particles differing in their size.

observed. This strong decreases in the intensity and the red shift of the maximum observed in the absorption spectrum for coated particles are due to a change in the free electron density. This induces changes in the surface plasmon band of silver particles [83] and leads to a variation of the width and maximum of the plasmon band absorption [83, 84]. Similar behavior has been observed by NaSH addition to silver colloids in aqueous solution [85]. Contrary to what is observed for other nanoparticles, the absorption spectrum of $(Ag_2S)_n$ particles is not structured and a long tail is observed. However, a blue shift compared to the optical band edge of bulk silver sulfide, which is well known to be at 1240 nm (1 eV), is obtained. Surprisingly, this does not change with the size of the particles (in the range 2–10 nm) (Figure 4.21). For nanocrystals with direct gap absorption, a sharp absorption onset with multiple discrete features consistent with quantum confinement is observed [86]. For indirect transition in nanocrystals, the electronic absorption shows no discrete features in the visible–IR region. However, for particles having an indirect transition, such as PbS [87] or CdSe, submitted to high pressure [88] a blue shift compared to the bulk phase with a decrease in particle size is observed. In the case of silver sulfide, a direct and an indirect transition take place. So, whatever the transition is, we would expect a large change in the energy band gap with a change in particle size. Because of the similarity of the Ag$_2$S absorption spectra of particles observed before and after extraction from reverse micelles, this phenomenon cannot be attributed either to polydispersity in size or to an enhancement of the excitonic absorption by change in the particle surface, as described by Wang [89]. In fact, the polydispersity in size decreases from 30% to 14% after extraction from micelles.

Because of the fact that size selection takes place during the extraction of $(Ag_2S)_n$ particles from the micelles, self-assemblies in 2D and 3D dimensions are obtained. On the other hand, for $(Ag)_n$ particles a size-selected precipitation has to be performed. In the following section the two self-organizations are presented.

4.7.1 Silver Sulfide, $(Ag_2S)_n$, Self-assemblies

By using a dilute solution of $(Ag_2S)_n$ nanocrystallites (particle volume fraction, ϕ, equal to 0.01%), monolayers of particles are formed. The particles are organized in a hexagonal network. However, the area covered by the monolayer of $(Ag_2S)_n$ nanocrystallites differs strongly with the preparation. When particles are deposited on the support, drop by drop while one waits for solvent evaporation before adding another droplet, monolayers are formed on a very small area [33]. The particles are arranged in a hexagonal network with an average interparticle distance equal to 2 nm. However, the micrographs are not totally covered by particles and large domains on the TEM grid are free of particles (Figure 4.22(a)). Instead of adding drop after drop of the solution, several drops are added immediately and then the solvent is evaporated. Large domains of monolayer are obtained (Figure 4.22(b)). The surface of the TEM grid, which is not covered by particles, decreases greatly. The monolayer domain is very large and it forms long monolayer ribbons [90]. The

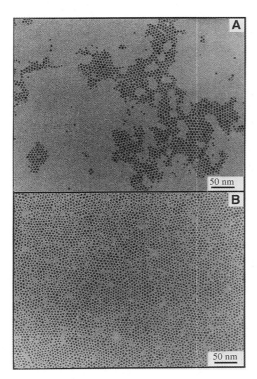

Figure 4.22. T.E.M. micrograph of a monolayer of 5.8 nm Ag$_2$S particles obtained by deposit: (a) a drop of the colloidal solution, the solvent is dried, and the procedure is repeated; (b) several drops of the solution, and then the solvent is removed.

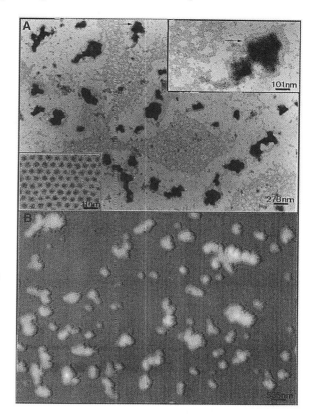

Figure 4.23. Islands on a large range made of 4 nm Ag$_2$S nanocrystallites: (a) TEM experiment, image scale 5 mm × 3.7 mm, insert: (a$_1$) magnification of monolayers in coexistence with an island, (a$_2$) high magnification of a monolayer; (b) TMAFM experiment, image scale 10 mm × 7 mm, z range: 100 mm from black to white.

length and width of the ribbons depend on the particle size. Whatever the size of particles, the ribbons are always very long. Direct observations using the TEM allowed an estimate to be made of the length of the ribbons: about 100 m. This feature cannot be reproduced on a micrograph since the size of the particles is too small compared to the length of the ribbons. On the other hand, the width of monolayer ribbons varies from 0.3 to 1 m when the particle size increases from 3 to 5.8 nm.

At high particle concentration (about 10^{-3} M), the solution remains optically clear. Supports used for TEM and TMAFM (tapping model atomic force microscopy) experiments are immersed in the solution for two hours and then dried at room temperature. TEM images reveal the formation of large aggregates (Figure 4.23(a)) over large areas [91]. Figure 4.23(b) shows that a similar pattern over a large area is obtained by TMAFM experiments as observed by TEM (Figure 4.23(a)). The size and the average distribution of the islands are similar. A cross section of one of these islands has been performed. Figure 4.24 shows the formation of a large aggregate with the form of a truncated pyramid. The height of the aggregate presented on Figure 4.24 is 116 nm and it is 1 m long. If we take into account the average diameter of the particles (4 nm) and the average distance between particles (1.77 nm), then there is an average of 20 layers.

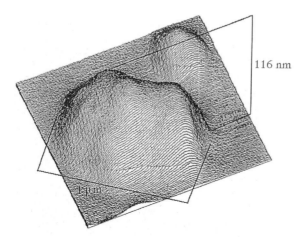

Figure 4.24. T.M.A.F.M. experiment: cross section of an island made of 4nm $(Ag_2S)_n$ nanocrystallites.

Magnification of one of these islands (Figure 4.25) shows that they are made of nanosized particles. Higher resolution of the island shows that the particles are highly oriented in a fourfold symmetry, attributed to the orientation of the particles in the {001} plane of a face-centered cubic (f.c.c.) structure, is observed. By tilting the sample, other orientations could be found, such as the {110} plane of face-centered cubic packing of nanocrystallites. This behavior is observed for various particle sizes. Hence, it is possible to make crystals of $(Ag_2S)_n$ particles that differ in size.

4.7.2 Self-assemblies Made with Silver Metallic Nanoparticles

In the case of $(Ag)_n$ particles, most of the coated particles with dodecanethiol are redispersed in heptane. The polydispersity in size is rather large (Figure 4.26(a, b)). To reduce the polydispersity, size-selected precipitation, SSP, is used. This method is based on the mixture of two miscible solvents that differ in their ability to dissolve the surfactant alkyl chains. The silver-coated particles are highly soluble in hexane and poorly in pyridine. Thus, a progressive addition of pyridine to hexane solution containing the silver-coated particles is performed. At a given volume of pyridine (which corresponds to roughly 50%), the solution becomes cloudy and a precipitate appears. This corresponds to the agglomeration of the largest particles as a result of their greater Van der Waals interactions [92–94]. The solution is centrifuged and an agglomerated fraction rich in large particles is collected, leaving the smallest particles in the supernatant. The agglomeration of the largest particles is reversible and the precipitate, redispersed in hexane, forms a homogeneous clear solution. An increase in the average diameter and a decrease in the polydispersity compared to what it was observed before the size selection is observed. This procedure is also performed with the supernatant, which contains the smallest particles. It is repeated

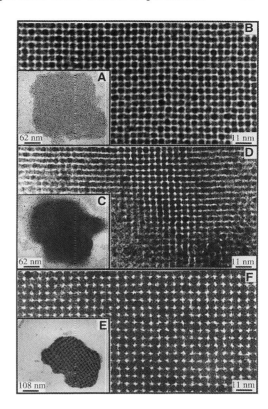

Figure 4.25. TEM and magnification of an island of Ag$_2$S in an {001} plane of a close-packed structure made of nanoparticles having an average size equal to 3 nm (a, b), 4 nm (c, d), and 6 nm (e, f).

several times and a strong decrease in the average particle size and its distribution is observed. When the polydispersity in size is small enough, self-organization is observed. The silver nanosized particles form a hexagonal network with an average distance between particles equal to 2 nm (Figure 4.26(c)). To build a 3D self-organization, we keep the particles obtained after one SSP. Particles having an average diameter equal to 4.1 nm are dispersed in hexane. By leaving a drop of the solution on a TEM carbon grid, an imperfect organization (Figure 4.27(a)) is obtained. By leaving the carbon grid in the solution for 3 hours, the TEM pattern becomes completely covered by a monolayer made of particles (Figure 4.27(b)). These are organized in a hexagonal close-packed network. In some region of the TEM pattern, a difference in the contrast can be observed. This can be attributed to the start of 3D self-organization. The increase of the immersion time until complete evaporation of the solvent leads to the formation of large aggregates. Figure 4.28(a) shows a rather high orientation of these aggregates around a large hole or ring. The average distance between the oriented aggregates varies from 20 to 60 nm. The magnification of the aggregates indicates that they are formed by (Ag)$_n$ nanoparticles. The average size of these aggregates ranges from 0.03 to 0.55 μm^2. High magnification of one of these aggregates shows that the particles are arranged in

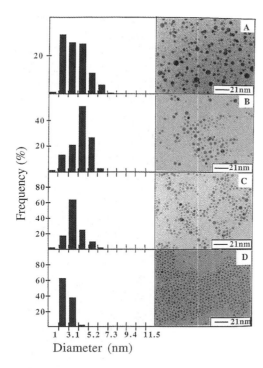

Figure 4.26. TEM micrograph and histograms of the size of silver particles: in a reverse micelle (a), after extraction (b), and at the end of the size-selected precipitation process (see text) (c).

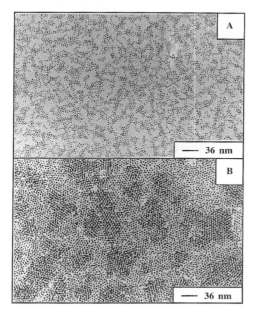

Figure 4.27. TEM micrographs of monolayers obtained by depositing a droplet of silver colloidal solution on a grid (a) or by immersing a TEM grid into the solution for 3 hours (b).

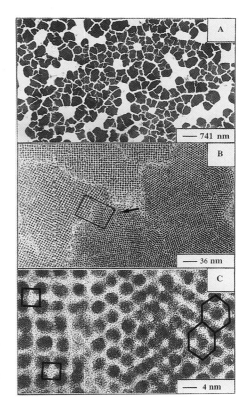

Figure 4.28. TEM micrograph of the grid at high magnification showing the different orientation of the silver particles: (a) general view, (b) magnification of one of these aggregates, and (C) magnification of the indicated region.

two different symmetries. Figure 4.28(b) shows the formation of a polycrystal. Magnification of Figure 4.28(b) (Figure 4.28(c)) shows either a hexagonal or a cubic arrangement of nanoparticles.

The transition from one structure to another is abrupt and there is a strong analogy with "atomic" polycrystals with a small grain known as nanocrystals. Each domain or grain has a different orientation. This clearly shows that the stacking of nanoparticles is periodic and not random. The "pseudo hexagonal" structure corresponds to the stacking of a {110} plane of the f.c.c. structure. On the same pattern there is observed (Figure 4.28(c)) a four fold symmetry, which is again characteristic of the stacking of {001} planes of the cubic structure. This cannot be found in the hexagonal structure. As matter of fact, no direction in a perfect hexagonal compact structure for which the projected positions of the particles can take this configuration. This is confirmed by TEM experiments performed at various tilt angles. It is always possible to find an orientation for which the stacking appears to be periodic. Hence, by tilting a sample having a pseudo hexagonal structure, a four fold symmetry is obtained. From this information, it is concluded that the large aggregates of silver particles are formed by the stacking of monolayers in a face-centered cubic arrangement.

4.8 Conclusions

The data presented in this paper show that control of the particle size is obtained by using reverse micelles as nanoreactors. By coating the particles, and after extraction, size selection takes place, inducing a 2D and 3D superlattice organized in a hexagonal and face-centered cubic structure respectively. This long-range organization of nanoparticles could open up a very large area of research. We can expect to obtain changes in the optical properties of nanoparticles when they are arranged in a network. When the particles are metallic, we can expect changes in the plasmon peak in position and shape.

In the case of semiconductors, the band gap could differ. The major goal would be to see if the nanoparticles have magnetic properties: long-range collective effects could occur.

If the syntheses are performed in other colloidal assemblies, the shape and the size of the particles differ markedly. Cylindrical nanoparticles mixed with spherical particles are obtained, whereas various structures can be formed when the colloidal assemblies are spherulites. One of the challenges is to manipulate the surface to be able to select the particles by shape. Another is the study of the optical properties of metallic cylinders differing in size. If these data are generalized to other materials, these systems could be very powerful for applications. If the particles have magnetic properties, these data would favor a better understanding of magnetism not only for basic physics but also for information storage, color imaging, and bioprocessing, where ferromagnets are of great technological importance.

Acknowledgements

I would like to thank my coworkers who participated in this work: T. Abdelhafed, F. Billoudet, J. Cizeron, S. Didierjean, N. Duxin, N. Feltin, A. Filankembo, L. Francois, J. F. Hochepied, D. Ingert, L. Levy, J. L. Lindor, Dr. I. Lisiecki, Dr. L. Motte, Dr. N. Moumen, P. Millard, A. Ngo, Dr. C. Petit, and J. Tanori. Special thanks are due to A. Filankembo, who was responsible for most of the figures and to Dr. I. Lisiecki who read this chapter.

References

[1] Special issue: Nanostructured Materials, *Chem. Mater.* **1996**, *8*, No. 5.
[2] M. P. Pileni, *J. Phys. Chem.* **1993**, *97*, 6961–6973.
[3] L. E. Brus, *J. Chem. Phys.* **1983**, *79*, 5566–5571.
[4] R. Rossetti, J. L. Ellison, J. M. Bigson, L. E. Brus, *J. Chem. Phys.* **1984**, *80*, 4464–4469.

[5] A. J. Nozik, F. Williams, M. T. Nenadocic, T. Rajh, C. I. Micic, *J. Phys. Chem.* **1985**, *89*, 397–399.

[6] C. Petit, M. P. Pileni, *J. Phys. Chem.* **1988**, *92*, 2282–2287.

[7] Y. Kayanuma, *Phys. Rev. B.* **1988**, *38*, 9797–9805.

[8] P. E. Lippens, M. Lannoo, *Phys. Rev. B.* **1989**, *39*, 10935–10942.

[9] Y. Z. Hu, M. Lindberg, S. W. Koch, *Phys. Rev. B* **1990**, *42*, 1713–1732.

[10] Y. Wang, N. Herron, *Phys. Rev. B* **1990**, *41*, 6079–6081.

[11] A. E. Hughes, S. C. Jain, *Adv. Phys.* **1979**, *28*, 717–828.

[12] A. Wokaun, J. P. Gordon, P. F. Liao, *Phys. Rev. Lett.* **1982**, *48*, 957–960.

[13] M. Meier, A. Wokaun, *Optics Lett.* **1983**, *8*, 581–583.

[14] R. G. Larson, *Rheol. Acta* **1992**, *31*, 497–500.

[15] S. J. Chen, D. F. Evans, B. W. Ninham, D. J. Mitchell, F. D. Blum, S. Pickup, *J. Phys. Chem.* **1986**, *90*, 842.

[16] D. F. Evans, D. J. Mitchell, B. W. Ninham, *J. Phys. Chem.* **1986**, *90*, 2817.

[17] I. S. Barnes, S. T. Hyde, B. W. Ninham, P. J. Deerian, M. Drifford, T. N. Zemb, T. N. *J. Phys. Chem.* **1988**, *92*, 2286.

[18] N. Mazer, G. Benedek, M. C. Carey, *J. Phys. Chem.* **1976**, *80*, 1075–1085.

[19] D. Blankschtein, G. M. Thurston, G. Benedek, *Phys. Rev. Lett.* **1986**, *85*, 7268–7270.

[20] G. Porte, J. Appell, Y. J. Poggi, *Phys. Chem.* **1980**, *84*, 3105.

[21] H. Hoffmann, G. Platz, W. Ulbricht, *J. Phys. Chem.* **1981**, *85*, 3160–3167.

[22] J. R. Mishic, M. R. Fisch, *J. Chem. Phys.* **1990**, *92*, 3222–3229.

[23] D. J. Mitchell, B. W. Ninham, *J. Chem. Soc. Faraday Trans. 2* **1981**, *77*, 601–629.

[24] S. A. Safran, L. A. Turkevich, P. A. Pincus, *J. Phys. Lett.* **1984**, *45*, L69–71.

[25] *Clusters and Colloids* (Ed: G. Schmid), VCH, Weinhem **1994**.

[26] I. Lisiecki, F. Billoudet, M. P. Pileni, *J. Phys. Chem.* **1996**, *100*, 4160–4166.

[27] J. H. Fendler, F. C. Meldrum, *Advanced Materials* **1995**, *7*, 607–632.

[28] J. H. Fendler, *Chem. Mater.* **1996**, *8*, 1616–1624.

[29] N. Herron, Y. Wang, M. Eddy, G. D. Stucky, D. E. Cox, K. Moller, T. Bein, *J. Amer. Chem. Soc.* **1989**, *111*, 530–540.

[30] N. Kimizuka, T. Kunitake, *Adv. Mater.* **1996**, *8*, 89–91.

[31] K. Okada, K. Sakata, T. Kunitake, *Chem. Mater.* **1990**, *2*, 89–91.

[32] D. Heitmam, J. P. Kotthaus, *Phys. Today* **1993**, 56–63.

[33] L. Motte, F. Billoudet, M. P. Pileni, *J. Phys. Chem.* **1995**, *99*, 16425–16429.

[34] C. B. Murray, C. R. Kagan, M. G. Bawendi, *Science* **1995**, *270*, 1335–1338.

[35] R. L. Whetten, J. T. Khoury, M. M. Alvarez, S. Murthy, I. Vezmar, Z. L. Wang, C. C. Cleveland, W. D. Luedtke, U. Landman, *Advanced Materials* **1996**, *8*, 429–433.

[36] M. Brust, D. Bethell, D. J. Schiffrin, C. J. Kiely, *Advanced Materials* **1995**, *7*, 771–797.

[37] *Reactivity in Reverse Micelles* (Ed: M. P. Pileni), Elsevier **1989**.

[38] T. F. Towey, A. Khan-Lodl, B. H. Robinson, *J. Chem. Soc. Faraday Trans. 2* **1990**, 86, 3757–3762.

[39] C. Robertus, J. G. H. Joosten, Y. K. Levine, *J. Chem. Phys.* **1990**, *93*, 10, 7293–7300.

[40] G. Cassin, J. P. Badiali, M. P. Pileni, *J. Phys. Chem.* **1995**, *99*, 12941–12946.

[41] T. K. Jain, G. Cassin, J. P. Badiali, M. P. Pileni, *Langmuir* **1996**, *12*, 2408–2411.

[42] I. Lisiecki, M. P. Pileni, *J. Am. Chem. Soc.* **1993**, *115*, 3887–3496.

[43] I. Lisiecki, M. P. Pileni, *J. Phys. Chem.* **1995**, *99*, 5077–5082.

[44] I. Lisiecki, M. Borjling, L. Motte, B. Ninham, M. P. Pileni, *Langmuir* **1995**, 2385–2392.

[45] L. Levy, J. F. Hochepied, M. P. Pileni, *J. Phys. Chem.* **1996**, *100*, 18322–18326.

[46] J. Nosaka, *J. Phys. Chem.* **1991**, *95*, 5054–5058.

[47] Y. Wang, N. Herron, *J. Phys. Chem.* **1991**, *95*, 525–532.

[48] M. Ikeda, K. Itoh, S. Hisano, *J. Phys. Soc. Jpn.* **1968**, *25*, 455–460.

[49] I. Satake, I. Iwamatsu, S. Hosokawa, R. Matuura, *Bull. Chem. Soc. Japan* **1962**, Vol. *36*, No. 2, 204.

[50] J. B. Hayter, J. Penfold, *Colloid & Polymer Sci.* **1983**, *261*, 1022–1025.

[51] Y. Moroi, K. Motomura, R. J. Matuura, *Colloid Interface Sci.* **1974**, *46*, 111–117.

[52] C. Petit, T. K. Jain, F. Billoudet, M. P. Pileni, *Langmuir* **1994**, *10*, 4446.

[53] N. Moumen, P. Veillet, M. P. Pileni, *J. Mag. Mag. Mat.* **1995**, *149*, 42.

[54] N. Moumen, M. P. Pileni, *J. Phys. Chem.* **1996**, *100*, 1867–1873.
[55] N. Moumen, M. P. Pileni, *Chem. Mat.* **1996**, *8*, 1128–1134.
[56] K. Haneda, A. H. Morish, *J. Appl. Phys.* **1988**, *63*, 4258–4260.
[57] N. Feltin, M. P. Pileni, Langmuir, **1997**, *13*, 3927.
[58] S. W. Charles, J. Popplewell, *Ferromagnetic Materials,* Vol. 2 (Ed: Wohlfarth), North Holland, Amsterdam **1982**.
[59] K. Haneda, *Can. J. Phys.* **1987**, *65*, 1233–1244.
[60] P. M. de Backer, E. De Grave, R. E. Vandenberghe, L. H. Bowen, *Hyperfine Int.* **1990**, *54*, 493–498.
[61] S. W. Charles, R. Chandrasekhar, K. O'Grady, M. J. Walker, *Appl. Phys.* **1988**, *64*, 5840–5842.
[62] *Magnetism and Mettallurgy* (Eds: A. E. Berkowtiz, E. Kneller), Academic Press **1969**, Vol. *1,* Chap. 8.
[63] N. Moumen, P. Bonville, M. P. Pileni, *J. Phys. Chem.* **1996**, *100*, 14410–14416.
[64] P. Mollard, P. Germi, A. Rousset, *Physica B + C* **1977**, *86–88*, 1393–1394.
[65] L. Liebermann, D. R. Fredkin, H. B. Shore, *Phys. Rev. Letters* **1969**, *22*, 539–541.
[66] L. Liebermann, J. Clinton, D. M. Edwards, J. Mathon, *Phys. Rev. Letters* **1970**, *25*, 232–235.
[67] E. J. Zeman, G. C. Schatz, *J. Phys. Chem.* **1987**, *91*, 634–643.
[68] M. P. Cline, P. W. Barber, R. K. Chang, *J. Opt. Soc. Am.* **1986**, *B3*, 15–21.
[69] D. S. Wang, M. Kerber, *Phys. Rev. B* **1981**, *24*, 1777. J. I. Gersten, A. Nitzan, in *Surface Enhanced Raman Scattering* (Eds: R. K. Chang, T. E. Furtak), Plenum, New York **1982**, p. 89.
[70] J. Tanori, T. Gulik-Krzywicki, M. P. Pileni, *Langmuir* **1997**, *13*, 633.
[71] S. J. Chen, D. F. Evans, B. W. Ninham, D. J. Mitchell, F. D. Blum, S. Pickup, *J. Phys. Chem.* **1986**, *90*, 842–847.
[72] D. F. Evans, D. J. Mitchell, B. W. Ninham, *J. Phys. Chem.* **1986**, *90*, 2817–2825.
[73] I. S. Barnes, S. T. Hyde, B. W. Ninham, P. J. Deerian, M. Drifford, T. N. Zemb, *J. Phys. Chem.* **1988**, *92*, 2286–2293.
[74] J. Appell, G. Porte, J. F. Berret, D. C. Roux, *Progr. Colloid Polym. Sci.* **1994**, *97*, 233–237.
[75] J. Tanori, M. P. Pileni, *Langmuir* **1997**, *13*, 639.
[76] P. Calvert, P. Rieke, *Chem. Mater.* **1996**, *8*, 1715–1727.
[77] O. Diat, D. Roux, *J. Phys. II, Fr.* **1993**, *3*, 9.
[78] J. F. Berret, D. C. Roux, G. Porte, P. Lindner, *Europhys. lett.* **1994**, *25*, 521–526.
[79] C. Petit, P. Lixon, M. P. Pileni, *J. Phys. Chem.* **1993**, *97*, 12974–12983.
[80] L. Motte, F. Billoudet, M. P. Pileni, *J. Materials Science* **1996**, *31*, 38–42.
[81] C. D. Bain, E. B. Troughton, Y. T. Tao, J. Evall, G. M. Whitesides, R. Nuzzo, *J. Am. Chem. Soc.* **1989**, *111*, 7155–7164.
[82] C. D. Bain, J. Evall, G. M. Whitesides, *J. Am. Chem. Soc.* **1989**, *111*, 7164–7165.
[83] K. P. Charles, F. Frank, W. Schulze, *Ber. Bunsenges. Phys. Chem.* **1984**, *88*, 354–359.
[84] U. Kreibig, *J. Phys. F: Metal. Phys.* **1974**, *4*, 999–1001.
[85] P. Mulvaney, T. Linnert, A. Henglein, *J. Phys. Chem.* **1991**, *95*, 7843.
[86] N. Herron, Y. Wang, M. M. Eddy, G. D. Stucky, D. E. Cox, K. Moller, T. Bein, *J. Am. Chem. Soc.* **1989**, *111*, 530.
[87] Y. Wang, N. Herron, *J. Phys. Chem.* **1987**, *91*, 257.
[88] S. H. Tolbert, A. M. Herhold, C. S. Johnson, A. P. Alivisatos, *Phys. Rev. Lett.* **1994**, *73*, 3266.
[89] Y. Wang, A. Suna, J. McHugh, E. F. Hilinski, P. A. Lucas, R. D. Johnson, *J. Chem. Phys.* **1990**, *92*, 6927.
[90] L. Motte, F. Billoudet, E. Lacaze, J. Douin, M. P. Pileni, *J. Phys. Chem.* **1997**, *107*, 138.
[91] L. Motte, F. Billoudet, E. Lacaze, M. P. Pileni, *Advanced Mat.* **1996**, *8*, 1018.
[92] A. Taleb, C. Petit, M. P. Pileni, *Chem. Mat.* **1997**, *9*, 950.
[93] R. L. Whetten, J. T. Khoury, M. M. Alvarez, S. Murthy, I. Vezmar, Z. L. Wang, C. C. Cleveland, W. D. Luedtke, U. Landman, *Advanced Materials* **1996**, *8*, 429.
[94] M. Brust, D. Bethell, D. J. Schiffrin, C. J. Kiely, *Advanced Materials* **1995**, *7*, 9071.

Chapter 5

Synthesis of Silicon Nanoclusters

R. A. Bley and S. M. Kauzlarich

5.1 Introduction

Semiconductor nanoclusters have received much attention in recent years because of their potential for use in the fabrication of optoelectronic devices [1–4]. Flat panel displays, optoelectronic sensors, solid state lasers, and electronic devices having nanoscale electronic circuitry made up of "single-electron transistors" are just some of the outcomes envisioned from the work currently going on in semiconductor nanocluster research [5]. Most of this work has focused on the II–VI and III–V binary semiconductors owing to the direct gap band structure these materials possess and the optoelectronic properties that result from their direct gap [1, 2]. Because of their band structure these semiconductors can readily be made to photoluminesce and electroluminesce and are therefore well suited for use in devices requiring these unique optoelectronic properties. Silicon has only recently been added to the list of potential candidates for use in such devices [4, 6]. Its addition has occurred because of the surprising discovery that nanoparticles of silicon can be made to luminesce efficiently in the visible region of the optical spectrum in spite of bulk silicon's indirect band gap.

5.2 Quantum Confinement

The unusual electronic and optical properties that make these semiconductor particles so useful result from the finite number of quantum states available to valence electrons in the clusters [7]. These clusters no longer possess a true conduction band consisting of a plethora of energy levels but have specific energy states at explicit levels. A major consequence of this is that the semiconductor's bandgap will increase by an amount that is inversely related to the size of the cluster. This is experimentally observed as a blueshift in absorption onset as the size of the particles decreases. The electrons require a greater amount of energy to enter the lowest available energy level because of the fewer available states in the material's valence band. The subsequent emission spectrum also displays a blueshift as the excited electrons relax back to their ground state. The shift in the emission spectrum results from the same change in available energy states that produces the shift in the absorption spectrum.

This shift to higher energy in the absorption and emission spectrum manifests itself as the size of the semiconductor particles becomes comparable to the diameter of the Wannier exciton in the bulk semiconductor. At such small sizes, excited electrons and their coincident holes are confined in all three dimensions to form what is commonly referred to as a "quantum dot". The word "quantum" is applied here to emphasize that the particles' unusual optical and electronic properties result from confinement of excited electrons to the finite number of quantum energy states available. The word "dot" refers to the fact that the confinement is in three dimensions as opposed to a quantum wire, which is confined in two dimensions, or a quantum well, which is confined along one dimension. Why these discrete states should exist rather than the continuum of states in a band can best be understood by following what happens to the electronic structure of a semiconductor cluster as its size increases from just a few atoms to a few thousand or more atoms, a size that generally no longer exhibits quantum size effects.

A small semiconductor cluster made up of only several atoms will have only a few energy levels (antibonding molecular orbitals) available to any excited electrons. This situation is similar to what is found in molecules. Consequently, the clusters' optical and electronic properties will be very similar to those of a molecule. As the number of atoms in the cluster increases, more quantum states are added. This is illustrated in Figure 5.1 by a widening of both the band representing the

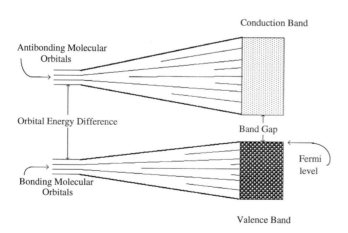

Figure 5.1. This figure shows the gradual changes that take place in the bonding and antibonding orbitals as atoms are added to a several atom semiconductor cluster. When enough atoms have been added to the cluster, a band structure is formed that has a continuum of possible states rather than the discrete states found in molecules.

occupied bonding molecular orbitals of the valence band and the unoccupied anti-bonding molecular orbitals of the conduction band.

The result is an increase in the total number of possible energy levels for excited electrons to occupy in the cluster, in addition to an overall decrease in the energy difference between the HOMO and LUMO (in molecular terminology) or between the Fermi level and conduction band (in solid state terminology). The quantum mechanical description of this phenomenon is that the electrons (quantum mechanical waves) confined in these small regions will only have allowed energies that correspond to the standing wave patterns available to them in the different sized clusters. This process of increasing available states while decreasing the energy difference between the HOMO and LUMO continues as more atoms are added to the cluster until there is essentially a continuum of available energy levels for excited electrons and the energy difference between the HOMO and LUMO is equivalent to the bandgap of the bulk semiconductor.

At this point the electronic and optical properties of the cluster are essentially the same as that of the bulk material. Adding more atoms only increases the size of the cluster and no longer effects the cluster's optoelectronic properties. Within the quantum confinement regime materials having properties in between those of the molecular and bulk material can be made. This allows for the design of specific electrical and optical properties in the semiconductors and provides enormous versatility with regard to their potential applications.

5.3 Development of Semiconductor Nanoclusters

To date, most of the work investigating semiconductor nanoclusters has centered around direct gap II–VI and III–V materials [2, 8, 9]. This is because of the greater ease in their synthesis and because of the advantages of utilizing a material that inherently possesses the necessary optoelectronic properties that give these clusters their usefulness: that is, because of their direct bandgap, these semiconductors will photoluminesce efficiently. The most thoroughly developed chemical syntheses for the nanoclusters are those of the II–VI binary semiconductors [2, 8]. These compounds have small solubility products and are therefore well suited for solution synthesis. In addition, it is relatively easy to obtain crystalline particles having a single structure type [10]. By varying the concentration and/or temperature, the size of the crystallites can be manipulated to a fair degree. Improvement over size control can be achieved through the use of several reaction media such as zeolites, porous glass, gels, and micelle media. The greatest success has been achieved with CdSe [11]. Nanoparticles of CdSe were originally produced in a pure stable form by growing the clusters in inverse micelles followed by passivation of the reactive surfaces of the clusters using organic constituents. More recently, a greatly improved method has been developed [8].

In addition to the II–VI semiconductors, progress has also been made in the synthesis of III–V quantum dots, although not to so great an extent [12, 13]. Although the chemical synthesis of II–VI quantum dots is the most highly advanced, to date, it has actually been the III–V semiconductors such as gallium arsenide that have been the most successfully exploited for use. However, one of the most promising potential applications for semiconductor optoelectronics would be to provide a direct link between electronic data processing and optical telecommunications. This means that high-performance optoelectronics needs to be successfully integrated with silicon electronics, a task proving difficult for the III–V semiconductors. Placing silicon nanoparticles on silicon integrated circuits would prove much easier than is the case for other semiconductors, since it is the lack of a common lattice that has been responsible for the difficulty in integrating silicon with the III–V semiconductors. Indirect bandgap semiconductors, such as silicon and germanium, however, have remained relatively uninvestigated as materials for optical applications until recently [4, 6]. A notable exception to this has been silver halides, which have been extensively used in the photography industry [14, 15].

5.3.1 Development of Silicon Nanoclusters

The suggestion that the luminescence seen in porous silicon was the result of quantum confinement [16] provided the impetus for synthesizing Si nanoclusters. The paper in which this was reported captured the imagination of scientists and initiated a new wave of research directed at exploring the optoelectronic properties of indirect bandgap semiconductor nanoclusters, particularly those of silicon [4, 6]. The realization that silicon nanoclusters could not only luminesce efficiently but could do so in the red region of the visible spectrum, rather than the near IR, where bulk silicon's bandgap lies, provided incentive for research aimed at synthesizing silicon quantum dots. Nanometer-sized silicon particles also are found to luminesce in the blue region of the visible spectrum when illuminated with ultraviolet light or when an electric current is passed through them, providing even greater incentive for an efficient chemical synthesis. The red luminescence has generally been attributed to quantum confinement, but the origin of the blue luminescence remains in dispute. The blue luminescence is of particular interest because this wavelength is not often seen in other semiconductor photoluminescence systems and could provide a new medium for display devices and indicator lights.

How nanosized particles of silicon luminesce despite silicon's indirect bandgap is not yet fully understood, but it is generally agreed that quantum confinement is involved. This is the most convincing explanation for the blueshifts observed in the absorption and emission spectra in these materials [4]. The fact that silicon has been well studied and is widely available should help make the integration of any new technologies based on silicon with those of the already-existing technologies that much easier.

5.3.2 Crystalline Structure of Silicon

The most common crystal lattice of silicon is the diamond structure, where all of the silicon atoms are tetrahedrally coordinated with other silicon atoms in a continuous

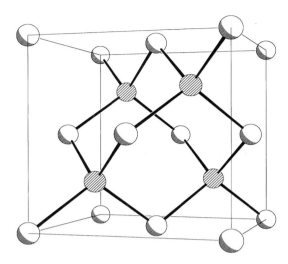

Figure 5.2. The diamond structure is made from a single element forming a face-centered cubic (fcc) lattice that has half of its tet rahedral holes filled by additional atoms of that same element. The distribution of the filled tetrahedral holes follows the pattern of that of the zincblende structure type, where every other hole for both types of tetrahedral site is occupied. In this figure, the shaded atoms represent the silicon atoms that occupy the tetrahedral sites. The unshaded atoms are the silicon atoms forming the fcc lattice.

three-dimensional array. This structure can best be viewed as a face-centered cubic lattice of silicon atoms that has half of its tetrahedral holes occupied by additional silicon atoms, as shown in Figure 5.2.

Another way of describing this structure is that it consists of a three-dimensional continuous array of puckered six-member rings of silicon, similar in form to the chair conformation of cyclohexane, which is illustrated in Figure 5.3. Other interesting forms of this element include several high-pressure phases [17] and amorphous silicon, which has the best photovoltaic properties suitable for the conversion of sunlight into electricity found to date.

5.3.3 Band Structure of Silicon

The bandgap of silicon at 25 °C is 1.12 eV and so resides in the infrared region of the spectrum. The band structure of this semiconductor is such that the lowest energy level available to an electron going from the valence band to the conduction band or vice versa has a so-called "forbidden transition" associated with it where the electron must undergo a change in its momentum for the transition to occur. This is referred to as an indirect gap semiconductor and ordinarily requires phonon assistance to make these lower-energy transitions that require a change in momentum. Figure 5.4 shows the energies of the conduction and valence bands for both a direct gap (GaAs) and indirect gap (Si) semiconductor plotted against the wave-vector, k.

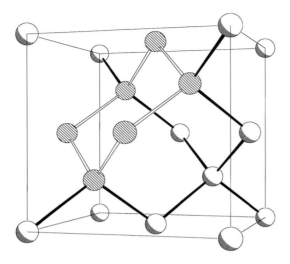

Figure 5.3. The six shaded silicon atoms illustrate that the structure formed by these atoms follows a motif that is similar in configuration to that of the chair conformation of cyclohexane. This motif is attached to other similarly formed motifs directed along different axis so as to form a continuos three-dimensional array of six-membered rings.

Direct Band Gap
Semiconductors:

III-V (GaAs, InP, etc)
II-VI (CdS, CdSe, etc)

Indirect Band Gap
Semiconductors:

IV (Si, Ge)
I-VII (AgBr)

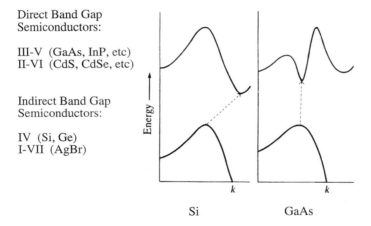

Figure 5.4. The band structure of the semiconductors, Si and GaAs. The indirect gap of Si requires a change in momentum for an electron going from one band into the other. The direct gap of GaAs requires no change in momentum for the transition to occur.

This plot shows how the energy associated with an electron in these materials will vary with its momentum as it travels through the solid. It shows that in the direct gap semiconductor the lowest energy level in the conduction band is directly over the highest energy level in the valence band and so is associated with the same value of momentum. Therefore, no change in momentum occurs when an electron is excited to the valence band or when it relaxes back down to the conduction band. In the indirect gap semiconductor the lowest valence level is at a different wave-vector value. As such it generally requires the assistance of a phonon before the electron can become excited or before the electron relaxes back down to the valence band. This results in the electron giving up its energy in the form of heat rather than through the emission of a photon upon relaxation. The reason that nanosized particles of silicon are able to overcome this restriction has been attributed to kinetic factors [18]. An early hypothesis suggested that the silicon nanoparticles developed a quasi-direct bandgap between the valence and conduction bands that accompanied the development of quantum size effects [7]. However, this is more likely in the nanocluster, where silicon retains its indirect bandgap but the rate of nonradiative relaxation decreases to such an extent that the radiative relaxation of the clusters becomes a competing mechanism for relaxation [19]. In bulk diamond-structured silicon, electron–hole recombination is governed by nonradiative three-body Auger processes in addition to their recombination via defects and impurities, which are also nonradiative. Both of these processes are reduced to a significant degree in quantum dots because the electron–hole pairs are confined within the nanoparticle's structure and so are not free to move about. For some applications, especially those involving optoelectronics, this may present difficulties if relaxation times are of importance. This is indeed the case with regard to using silicon nanoparticles for the switching of electronic signals, such as those used in data processing, into optical signals, like those used in telecommunications, and vice versa. Silicon nanoparticles would seem to be the ideal material for this type of device if it were not for the fact that the rate of conversion between an optical and an electronic signal in these particles is currently too slow to make this a viable process. This could change, however, as our understanding of these systems increases. Even if these indirect bandgap semiconductors never prove viable as an efficient means to interconvert electronic and optical signals, the reasons previously given (their use in display panels, optical sensors, and solid state lasers and for the fabrication of electronic circuits having single-electron transistors) are more than adequate to justify searching for an efficient means to produce these particles.

5.4 Synthetic Methods of Silicon Nanocluster Production

5.4.1 Decomposition of Silanes

The methods that have been used to synthesize silicon nanoclusters that can be made into a colloidal solution are primarily of four different types. The most successful method has used the gas phase decomposition of silanes [20–22]. In this method, a series of higher silanes, disilanes, and silylene polymeric species are formed at high temperatures (850–1050 °C) in an enclosed vessel. The many rate constants that are known for this system allow it to be modeled. However, the mechanisms responsible for producing diamond-structured silicon are not yet understood. Most single crystals are octahedral, displaying bulk silicon's lowest energy ⟨111⟩ facets on the octahedron's surfaces. This suggests that thermodynamic equilibrium is achieved during formation of the particles. It is also thought that hydrogen termination may occur on the lattice surfaces as the crystals cool. It is conceivable that nucleation, growth, annealing, and possibly hydrogen surface termination are all part of the mechanism taking place in this synthesis. There is little size control during the synthesis step of this high-temperature process other than what can be achieved by changes in the initial reactant concentrations and the length of time over which agglomeration is allowed to take place. This results in a comparatively wide size distribution, which can be greatly narrowed through use of size-selective precipitation.

5.4.2 Silicon Nanoparticles from Porous Silicon

A second method used to synthesize silicon nanoparticles involves the ultrasonic dispersion of porous silicon in different solvents [23–26]. The porous silicon is first produced by anodic electrochemical etching of silicon wafers in an HF solution. This produces a highly porous layer on the side of the wafer in contact with the HF solution. This porous layer is made up of columns of silicon left after the HF etching of the silicon wafer during anodization. Tiny crystallites, having hydrogen-terminated surfaces, are attached to small branches coming off of the silicon pillars. It is generally agreed that these particles are responsible for the red photo- or electroluminescence observed for porous silicon. Changing the composition of the anodization solution, the HF concentration, the current density used during anodization, and the length of time of anodization can all be used to some degree to change the morphology and size of the crystallites. However, a wide range still results regardless of the set of parameters used. The method we have used to collect these crystallites involves mechanically removing the porous silicon layers from the silicon wafer substrate followed by placing the material in a Schlenk flask with an appropriate solvent. This is then placed in an ultrasonic bath for up to seven days. The result is a colloid having particles that average between 1 and 10 nm in size in addition to agglomerates which can be as large as 50 nm. This is a relatively inexpensive method for producing silicon nanoparticles that luminesce, but the effort necessary to control size distribution and for quantitative characterization is prohibitive for large-scale syntheses.

5.4.3 Solution Synthesis of Silicon Nanoparticles

To date, there are two solution phase syntheses that have been reported for silicon nanocluster production. One is based on the reduction of $SiCl_4$ and $RSiCl_3$ by sodium metal in nonpolar organic solvents [27]. The synthesis is carried out in a bomb at relatively high temperatures (385 °C) and pressures ($\times 100$ atm) with rapid stirring for 3 to 7 days. Upon cooling, the product is filtered and washed with hexane, methanol, ether, and water to remove Na, NaCl, and hydrocarbon residue. The resulting material is then dried under vacuum. Analysis using transmission electron microscopy of the product from the reaction where R = hydrogen in $RSiCl_3$ reveals a size distribution of 5 to 3000 nm for silicon crystallites, which are mostly hexagonally shaped. Most are single crystals but a few of the smaller crystallites are aggregates made up of two or three individual crystallites. The R group in the

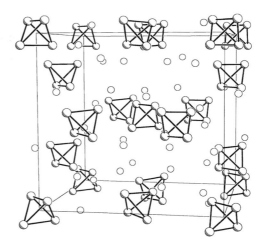

Figure 5.5. The unit cell of KSi. This has a cubic structure made up of four-atom silicon tetrahedral anions interspersed between the potassium cations.

$RSiCl_3$ species is used to cap the surface of the clusters. This allows for a fair degree of control over size distribution in the clusters where R = octyl. The diameter of the various crystallites in this reaction only varies from 2 to 9 nm. The yield from this reaction is low at less than 10%.

The other method, which produces silicon nanoclusters at much lower pressures and temperatures than that described above, has been recently achieved [28] by use of the intermetallic Zintl salt KSi [29, 30] as starting material (see Figure 5.5).

The success observed for the solution syntheses of group II–VI and III–V nanoclusters is in large part due to their precursors' ability to solubilize. This puts group IV semiconductors at somewhat of a disadvantage because they generally exhibit a more covalent nature in their chemistry. Zintl compounds have only recently been explored as potential precursors in the synthesis of new compounds [31, 32]. This is surprising, given that Zintl had suggested such a use himself because of the unusual bonding, structural configuration, and oxidation states that these compounds possessed. Relatively recent work has shown that these compounds are useful for the synthesis of novel clusters [33–35] in addition to new materials. KSi consists of covalently bonded Si_4^{4-} anionic clusters that are separated from the other anion clusters by four K^+ cations that cap the four faces of the tetrahedral Si_4^{4-} (Figure 5.6) [29, 30]. The anionic cluster is isostructural and isoelectronic with that of white phosphorous.

Each of the group IV atoms in the anionic cluster formally possesses a charge of -1, for a total charge of -4 on each cluster. These clusters are suspended in an appropriate coordinating solvent and are reacted with the (formally) cationic species of Si^{4+} from $SiCl_4$ to produce the silicon nanoclusters.

The addition of KSi to dried and degassed glyme or diglyme results in a faint greenish color appearing in the liquid, suggesting that there is some degree of solvation taking place. Most of the KSi remains as solid. Heating and stirring the sample results in a deepening of the green color, although even under these conditions most of the material remains undissolved. Excess $SiCl_4$ is added to the

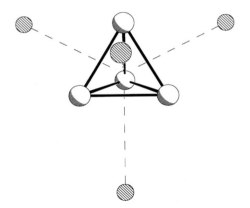

Figure 5.6. The silicon tetrahedral cluster of KSi showing how the nearest neighbors of potassium are situated with respect too the cluster.

sample and the resulting solution/suspension is refluxed for 48 to 96 hours under argon at atmospheric pressure.

$$4n\text{KSi} + n\text{SiCl}_4 \rightarrow \text{Si nanoparticles} + 4n\text{KCl} \tag{5.1}$$

If the solvent and excess SiCl_4 are removed at this stage and X-ray powder diffraction is used to probe the resulting dark-gray sample, only KCl is observed. This confirms the expected formation of KCl salt but does not establish what state the silicon is in, since it may be amorphous or in the form of crystallites too small to diffract X-rays. Microprobe elemental analysis confirms that there are two distinct phases in the powder. KCl is observed along with a silicon–oxygen phase having a silicon-to-oxygen ratio of 7 to 1. Because of the nature of the microprobe method used, slight oxygen exposure is unavoidable. The expected chlorine termination on the silicon particles is apparently too reactive with oxygen to survive this exposure. If a sample at this stage in the synthesis is washed with water several times to remove the KCl and then examined using a microprobe, only a silicon–oxygen phase is observed. Here, the silicon-to-oxygen ratio is close to 2 to 3, almost that of SiO_2. In this case the sample has been exposed to both water and oxygen for long periods and has undergone extensive oxidation.

In order to stabilize the surface of the silicon particles, after the initial Reaction shown in reaction (5.1) has gone to completion, the excess SiCl_4 and solvent are removed and then more dried and degassed solvent and methanol are added. This is stirred for 1 to 12 hours.

$$-\text{SiCl} + \text{HOCH}_3 \rightarrow -\text{SiOCH}_3 + \text{HCl} \tag{5.2}$$

This results in a much more stable product that is hydrophobic and is therefore easily isolated from the KCl salt. The clusters have also been successfully terminated using methyl lithium instead of methanol to produce a methyl-terminated surface rather than the methoxy termination. From this final product a colloidal suspension can be made using hexane as the solvent. Six to eight percent of the product goes into solution, leaving the remainder as a flocculent undissolved solid.

5.5 Characterization

The largest amount of information on colloidal silicon nanoparticles is for nano-particles capped with SiO_2 [4, 18, 19, 21, 36, 37]. Detailed characterization will be presented for the low-temperature-solution method and compared with silicon nanoclusters produced by other methods.

5.5.1 Infrared Spectroscopy

FTIR spectroscopy, high-resolution transmission electron microscopy (HRTEM), along with UV–visible and photoluminescence spectroscopies have been used to characterize the resulting product. FTIR data was collected for the different colloid samples to determine whether the expected silicon–oxygen bonds and hydrocarbon groups were present on the nanoparticles' surfaces. Figure 5.7 shows the IR spectrum of a diglyme sample that had been refluxed for 84 hours and then terminated with methanol to produce a methoxy-terminated sample.

This spectrum is typical of all of the colloid samples having methoxy termination and clearly shows the silicon–oxygen peak near 1100 cm^{-1} and the saturated hydrocarbon peak just below 3000 cm^{-1}.

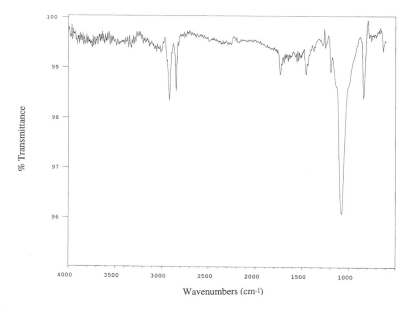

Figure 5.7. FTIR spectrum of silicon methoxy-terminated nanoparticles after evaporation of the colloid solvent. The two prominent features are the silicon–oxygen bonds, indicated by their characteristic peak around 1100 cm^{-1}, and the carbon–hydrogen bonds of saturated hydrocarbons just below 3000 cm^{-1}.

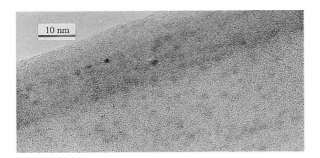

Figure 5.8. Bright-field HRTEM micrographs of silicon nanoparticles. Many small (1.5–2 nm) silicon particles are visible here that have been deposited on an amorphous carbon substrate by evaporation of solvent from the colloid suspension. Lattice fringes of 3.1 Å that correspond to the {111} lattice planes of silicon are observable on large particles found with these smaller particles. The smaller particles do not have observable lattice fringes but are of the same material as the larger particles, as is demonstrated by their diffraction overlap.

5.5.2 Electron Microscopy

HRTEM shows the flocculent precipitate to consist mostly of amorphous material. This is also true of the colloidal suspension obtained from the reaction done in THF. This is not true, however, of the colloidal suspensions obtained from the glyme and diglyme reactions. Figure 5.8 shows silicon particles obtained from a sample prepared using diglyme as the reaction solvent and having a reflux time of 48 hours.

Most particles are between 1.5 and 2.0 nm in diameter and are too small for resolution of their lattice fringes. However, larger agglomerates have also been found to form that can be around 30 nm or more in diameter. These agglomerates clearly show lattice fringes that correspond to the {111} planes (≈ 3.1 Å) in silicon having the diamond structure. Lattice fringes are discernible in the larger agglomerate because interference from the amorphous carbon substrate on which the particle rests does not manifest itself as strongly in the larger particles as it does in the smaller particles. The electron beam diffracted off the agglomerate is much more intense yet only contends with the same substrate interference as that of the beam diffracted from the smaller particles. The selected area electron diffraction pattern of the area shown in Figure 5.8 was taken and has rings that correspond to the {111} and {220} lattice planes of diamond-structure silicon. Rings are seen because the diffraction originates from a large number of very small particles rather than from a single crystal. When electron diffraction patterns are obtained for areas containing both the small particles and larger agglomerates, the diffuse ring pattern with overlapping intense intermittent spots is obtained. This results from the larger silicon particles producing spots that precisely overlap the ring pattern formed by the many smaller silicon particles.

The time the reaction solution was allowed to reflux was increased in order to determine the effect on the crystallites' morphology. Figure 5.9 shows a micrograph

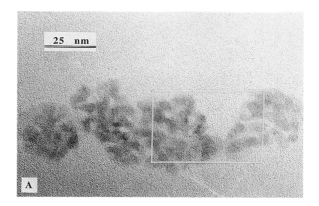

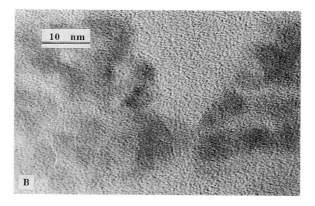

Figure 5.9. Bright-field HRTEM micrograph of silicon nanoclusters having a reflux time of 84 hours. The reaction time has been increased from 48 to 84 hours to determine what effects this will have on the particles' morphology. Most of these particles are between 3 and 6 nm in diameter. The longer reflux time has resulted in an increased average diameter of the clusters.

obtained from a reaction similar to that described previously, but here the reaction time was increased from 48 to 84 hours. This micrograph shows particles between 3 and 6 nm in diameter.

Most of these single crystallites have begun to agglomerate. In spite of their small size, resolution of their lattice fringes is possible for these clusters. The mechanism for the formation of these agglomerates is uncertain as yet, but the fact that they form is consistent with many nanocluster systems, both nonsemiconductor and semiconductor [38, 39]. Figure 5.9(b) shows an enlargement of the area outlined in Figure 5.9(a). Here the lattice fringes are more obvious than in the smaller particles of Figure 5.8.

5.5.3 Absorption Spectrum

Figure 5.10 shows the UV–Vis spectrum of a sample from a diglyme reaction that had been refluxed for 48 hours. This spectrum is fairly typical in its shape with regard to the other silicon colloid samples in that individual transition peaks are not observed.

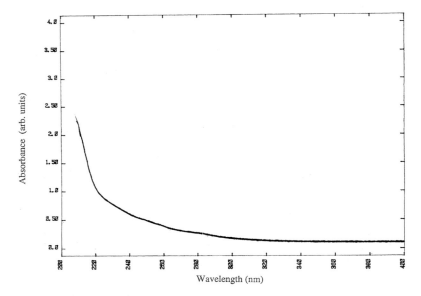

Figure 5.10. UV–visible spectrum of colloid sample having a reflux reaction time of 48 hours. The rather featureless shape of the spectrum is typically found for nanoparticles of silicon. The large blueshift in absorption is due to the very small size of the largest crystallites.

Absorption begins around 320 nm and gradually rises until around 220 nm where it begins to rise more rapidly. While the shape of this spectrum is consistent with other colloid systems of nanosized silicon, the absolute magnitude of the absorption is not the same [21]. This is probably because the spectrum reflects most closely the characteristics of the largest-sized particles in the colloid. The particles made using this method have a very narrow size distribution with the largest of the particles still quite small. The fact that the absorption onset is so strongly blueshifted supports the HRTEM data regarding particle size. The quantum mechanical explanation of this is that the small-sized particles obtained in this synthesis have a wider bandgap than particles using other synthetic methods. A greater energy is therefore required before excitation will take place. The energy required for absorption onset of silicon particles of this size has been calculated and is in good agreement with what is observed here [40].

5.5.4 Photoluminescence Spectroscopy

These particles, like other silicon nanoclusters, will luminesce when excited through absorption of a photon. The luminescence has generally been in two different areas of the visible spectrum: a blue peak centered around 400 nm and a red peak centered around 650 nm. To date, most silicon nanoclusters have had either hydrogen or SiO_2 on the surfaces of the particles. Different synthetic methods have produced different optical results that, because of the differences in the surface morphologies and size distributions of the particles, cannot be compared in any meaningful quantitative way. How different terminating species affect the photoluminescence properties of the clusters will be a great help in probing surface and quantum confinement effects.

The red peak observed in the photoluminescence spectra of these clusters is believed by most investigators to result from quantum confinement, while the origin of the blue peak is much less certain. Many models have been proposed to explain the blue peak but no single model has been generally accepted. This is because most of the models explain the properties from a particular synthetic method [41]. Most of these models propose that oxides of silicon or surface states coming from silicon–oxygen interfaces are the recombination centers responsible for the luminescence. However, if the recombination of electron–hole pairs results from interband recombination, it is highly unlikely that the recombination centers reside in SiO_2 because its bandgap of approximately 8 eV is far too energetic to account for the blue photoluminescence of only 3 eV. On the other hand, nonstoichiometric silicon oxides, SiO_x, where x is between 1.4 and 1.6, have bandgaps of the appropriate energy differences of around 3–4 eV and so could account for the photoemision observed [42, 43]. Additionally, defects in SiO_2 are known to emit in both the blue and red regions of the visible spectrum [44]. Hydroxyl groups absorbed on a SiO_2 surface have also been suggested as possible sites for recombination.

For both the methoxy- and methyl-terminated clusters produced by this new method, the spectra obtained using size-selective photoluminescence spectroscopy are consistent with a quantum confinement model for the luminescence. Figure 5.11

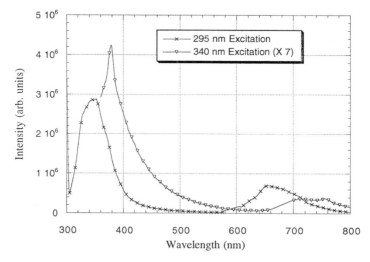

Figure 5.11. The size-selective photoluminescence spectra of the colloid containing methyl-terminated silicon clusters. The emission spectra obtained from the higher-energy excitation exhibits a blueshift relative to the lower-energy excitation.

shows the shift in emission spectra of the particles terminated with methyl groups and having a size of around 2.5 nm in diameter.

There is a shift in the emission spectra that coincides with the shift in the excitation energy because only a specific size of cluster becomes excited with any particular excitation wavelength. This shifting is to be expected if quantum confinement is responsible for the luminescence. Since both the red and blue signals shift their energies, this would seem to indicate that both are originating from quantum confinement.

A new interpretation of the photoluminescence from silicon nanoclusters is possible from the data obtained from experiments done on these clusters and silicon nanoclusters produced from porous silicon [42]. The explanation for both the blue and red luminescence shifts comes from quantum size effects. The blue luminescence results from direct interband recombination of the electron–hole pairs, and not the red emission, as is generally thought to be the case. This is reasonable, not only because the blue luminescence approaches more nearly the energy value where absorption takes place, but also because this agrees much better with theoretical calculations of where the bandgap should lie in silicon clusters having these sizes [40]. This suggests that the red luminescence originates from a trapped electron that had been residing at an intermediate energy level between the conduction and valence bands. If this electron relaxes radiatively, this could explain why the emission still exhibits quantum confinement effects even though it does not occur from electrons going directly from the conduction band to the valence band.

5.6 Summary

The development of research aimed at producing silicon nanoclusters has been presented. Although several methods for producing colloidal solutions have been presented, this chapter has focused on a new synthetic method. This method produces particles that lend themselves to the easy manipulation of their surfaces in addition to their being of uniform size. The fact that this is a solution route at normal atmospheric pressure also makes it an attractive method because of the ease with which they are made and because of the much lower costs associated with the manufacture of the silicon particles. Characterization of the nanoclusters produced by this method have been presented and a discussion of the origin of the photoluminescence provided.

Acknowledgments

The authors have benefited from discussions with Philip P. Power, Howard W. H. Lee, and Gilardo Delgado. Financial support from the National Science Foundation (DMR-9505565) and the Campus Laboratory Collaboration Program of the University of California is gratefully acknowledged.

References

[1] A. Henglein, *Chem. Rev.* **1989**, *89*, 1861–1873.
[2] M. L. Steigerwald, L. B. Brus, *Annu. Rev. Mater. Sci.* **1989**, *19*, 471–495.
[3] Y. Wang, N. Herron, *J. Phys. Chem.* **1991**, *95*, 525–532.
[4] L. Brus, *J. Phys. Chem.* **1994**, *98*, 3575–3581.
[5] R. W. Siegel, *Scientific American* **1996**, December, 74–79.
[6] Ö. Dag, A. Kuperman, G. A. Ozin, *Adv. Mater.* **1995**, *7*, 72–78.
[7] L. Brus, *J. Phys. Chem.* **1986**, *90*, 2555–2560.
[8] C. B. Murray, D. J. Norris, M. G. Bawendi, *J. Am. Chem. Soc.* **1993**, *115*, 8706–8715.
[9] D. Heitmann, J. P. Kotthaus, *Physics Today* **1993**, 56–63.
[10] M. L. Steigerwald, L. E. Brus, *Acc. Chem. Res.* **1990**, *23*, 183–188.
[11] M. L. Steigerwald, A. P. Alivisatos, J. M. Gibson, J. M. Harris, R. Kortan, A. J. Muller, A. M. Thayer, T. M. Duncan, D. C. Douglass, L. E. Brus, *J. Am. Chem. Soc.* **1988**, *110*, 3046–3050.
[12] S. S. Kher, R. L. Wells, *Chem. Mater.* **1994**, *6*, 2056–2062.
[13] S. S. Kher, R. L. Wells, *Mater. Res. Soc. Symp.* **1994**, *351*, 293–298.
[14] T. Takagaharar, K. Takeda, *Phys. Rev. B* **1992**, *46*, 15578–15581.
[15] H. Weller, Angew. Chem. *Int. Ed. Engl.* **1993**, *32*, 41–53.
[16] L. T. Canham, *Appl. Phys. Lett.* **1990**, *57*, 1046–1048.
[17] M. Imai, K. Yaoita, Y. Katayama, J. Chem, K. Tsuji, *J. Non-crystalline Solids* **1992**, *150*, 49–52.
[18] L. E. Brus, *Semiconductor Nanocrystals, Microelectronics, and Solar Cells*, Houston, Texas **1995**, pp. 1–21.
[19] L. E. Brus, P. F. Szajowski, W. L. Wilson, T. D. Harris, S. Schuppler, P. H. Citrin, *J. Am. Chem. Soc.* **1995**, *117*, 2915–2922.
[20] A. Fojtik, A. Henglein, *Chem. Phys. Lett.* **1994**, *221*, 363–367.
[21] K. A. Littau, P. J. Szajowshki, A. J. Muller, A. R. Kortan, L. E. Brus, *J. Phys. Chem.* **1993**, *97*, 1224–1230.
[22] D. Zhang, R. M. Kolbas, P. Mehta, A. K. Singh, D. J. Lichtenwalner, K. Y. Hsieh, A. I. Kingon, *Mat. Res. Soc. Symp. Proc.* **1992**, *256*, 35–40.
[23] J. L. Heinrich, C. L. Curtis, G. M. Credo, K. L. Kavanagh, M. J. Sailor, *Science* **1992**, *255*, 66–68.
[24] S. Berhane, S. M. Kauzlarich, K. Nishimura, R. L. Smith, J. E. Davis, H. W. H. Lee, M. L. S. Olson, L. L. Chase, *Mat. Res. Soc. Symp. Proc.* **1993**, *298*, 99–102.
[25] R. A. Bley, S. M. Kauzlarich, J. E. Davis, H. W. H. Lee, *Chem. Mater.* **1996**, *8*, 1881–1888.
[26] R. A. Bley, S. M. Kauzlarich, H. W. H. Lee, J. E. Davis, *Mat. Res. Soc. Symp. Proc.* **1994**, *351*, 275–280.
[27] J. R. Heath, *Science* **1992**, *258*, 1131–1133.
[28] R. A. Bley, S. M. Kauzlarich, *J. Am. Chem. Soc.* **1996**, *118*, 12461–12462.
[29] R. Schäfer, W. Klemm, *Z. Anorg. Allg. Chem.* **1961**, *312*, 214–220.
[30] E. Busmann, *Z. Anorg. Allg. Chem.* **1961**, *313*, 90–106.
[31] R. C. Haushalter, C. J. O'Connor, J. P. Haushalter, A. M. Umarji, G. K. Shenoy, *Angew. Chem., Int. Ed. Eng.* **1984**, *23*, 169–170.
[32] C. J. O'Connor, J.-S. Jung, J. H. Zhang, in: *Chemistry, Structure, and Bonding of Zintl Phases and Ions* (Ed: S. M. Kauzlarich), VCH, New York **1996**, p. 275–299.
[33] S. Charles, J. C. Fettinger, B. W. Eichhorn, *J. Am. Chem. Soc.* **1995**, *117*, 5303–5311.
[34] R. Ahlrichs, D. Fenske, K. Fromm, H. Krautscheid, U. Krautscheid, O. Treutler, *Chem. Eur. J.* **1996**, *2*, 238–244.

[35] J. D. Corbett, *Chem. Rev.* **1985**, *85*, 383–397.
[36] W. L. Wilson, P. F. Szajowski, L. E. Brus, *Science* **1993**, *262*, 1242–1244.
[37] S. Schuppler, S. L. Friedman, M. A. Marcus, D. L. Adler, Y.-H. Xie, F. M. Ross, Y. J. Chabal, T. D. Harris, L. E. Brus, W. L. Brown, E. E. Chaban, P. F. Szajowshki, S. B. Christman, P. H. Citrin, *Phys. Rev. B* **1995**, *52*, 4910–4925.
[38] R. G. Freeman, M. B. Hommer, K. C. Grabar, M. A. Jackson, M. J. Natan, *J. Phys. Chem.* **1996**, *100*, 718–724.
[39] C. Allain, M. Cloitre, M. Wafra, *Phys. Rev. Lett.* **1995**, *70*, 1478–1481.
[40] C. Dellerue, G. Allan, M. Lannoo, *Phys. Rev. B* **1993**, *48*, 11024–11036.
[41] D. L. Griscom, *J. Ceram, Soc. Jpn.* **1991**, *99*, 923–942.
[42] G. R. Delgato, H. W. H. Lee, R. A. Bley, S. M. Kauzlarich, in preparation.
[43] G. R. Delgado, H. W. H. Lee, S. M. Kauzlarich, R. A. Bley, **1997**, unpublished research.
[44] L. N. Skuja, A. N. Streletsky, A. B. Pakovich, *Sol. State Commun.* **1984**, *150*, 1069.

Chapter 6

Two-Dimensional Crystal Growth of Fullerenes and Nanoparticles

D. M. Guldi

6.1 Introduction

The feasibility of large-scale production of buckminsterfullerene (C_{60}) and its higher analogues (C_{70}, C_{76}, C_{78}, C_{84}, etc.) stimulated broad and interdisciplinary interest in the chemical and physical properties of these pure carbon allotropes [1, 2]. Besides the relative ease of its chromatographic separation [3] and the lack of any coexisting isomers, the intriguing spherical symmetry of icosahedral C_{60} makes this, the most abundant fullerene, a central topic in the chemistry of fullerenes [4–8]. The unique symmetrical shape, the large size of the π system and characteristic physicochemical properties, such as facile reduction and photosensitization, of buckminsterfullerene raised the expectation that C_{60} might play an active role in biological relevant processes [8–13]. As a consequence, new materials with a wide range of unique and spectacular physicochemical properties have been discovered that prompted the exploration of potential applications ranging from drug delivery to advanced nano-structured devices [14–16].

The first part of this chapter will review fullerene-based two- and three-dimensional crystals as fabricated by ultrahigh vacuum deposition techniques. The quality and morphology of deposited fullerene films and the resulting properties will be summarized as a function of the substrate and sample preparation. This part will be concluded by a selective overview of the formation of Langmuir and Langmuir–Blodgett films at the air–water interface and on solid substrates, respectively. The second part will focus on the systematic variation of parameters in fullerene derivatives based on covalent functionalization with hydrophilic addends, which govern the unambiguous formation of truly two-dimensional fullerene crystals. Various versatile methodologies for the formation of monolayers, such as Langmuir–Blodgett films or self-assembly, will be described, with emphasis on their potential to construct organic films on the order of the molecular level and their important contributions to the field of fullerenes. Finally, in the last part, some potential future applications of this technologically interesting material will be discussed.

Ground state C_{60} has remarkable electron acceptor properties, capable of accommodating as many as six electrons, yet displays a surprisingly moderate reduction potential in dichloromethane of -0.44 V vs. SCE for the formation of C_{60}^{-} [3, 17, 18]. Photoexcitation of C_{60}, on the other hand, facilitates the reduction of singlet-excited $^1C_{60}$ and triplet-excited $^3C_{60}$, which have redox potentials of 1.3 V and 1.14 V vs. SCE, respectively [19]. Hence, C_{60} can be expected to be a potential and powerful electron acceptor moiety in artificial photosynthesis.

In the crystalline form, C_{60} molecules occupy the sites of a face-centered cubic (fcc) lattice with a large rotation disorder at these lattice sites. The molecular orbitals h_u and t_{1u} broaden into respectively the valence and conduction bands of the solid, with a bandgap of about 2 eV [20]. The optical and electronic properties of solid C_{60} film and of crystals [21–61], which are insulators, have been studied with increasing interest, particularly since the discovery of superconductivity [15, 62, 63] upon doping them with alkali metals or other materials. C_{60} has become a potential building block for new materials that may possess reproducible electronic switching and memory properties.

Well-ordered three-dimensional monolayered films are of great interest because of the valuable insights they provide regarding molecule interactions and their potential application to important technologies related to coatings and surface modifications. The optical properties of C_{60} films are profoundly controlled by the deposition conditions and by impurities or disordered structures. Thus, an absolutely essential requirement for the exploration of these properties is the incorporation of fullerenes in well-defined two-dimensional arrays and three-dimensional networks. Despite extensive efforts to form stable and well-ordered monolayered fullerene films, the strong π–π interactions and the resulting tendency to form aggregates precludes the formation of stable monolayers and Langmuir–Blodgett films at the air–water interface. Currently available data suggest three promising approaches: (i) amphiphilic functionalization of pristine C_{60} via covalent attachment of hydrophilic groups, (ii) reduction of the hydrophobic surface via controlled multiple functionalization, or (iii) self-assembly via electrostatic attractions of oppositely charged species.

6.2 Pristine Fullerenes

6.2.1 Films of Pristine Fullerenes, C_{60} and C_{70}

Ultrahigh vacuum deposition techniques, traditionally developed for the fabrication of silicon and semiconductor thin films, have been applied to make high-quality molecular thin films. The structure and quality of fullerene-based films have been determined by transmission electron microscopy (TEM), reflection high-energy electron diffractometry (RHEED), X-ray diffraction, infrared and ultraviolet–visible spectroscopy, Raman spectroscopy, and atomic force microscopy (AFM) [21–60].

The quality of molecular thin films depends strongly on the interaction between the molecules and between the substrates employed for their growth. Since fullerenes display much stronger molecular interactions and higher stability than conventional organic molecules, their utilization for molecular thin films, by means of organic molecular beam epitaxy, has been vigorously investigated. The character and strength of the interaction between C_{60} and the substrate essentially determines the morphology of the film growth [54, 64]. In the case of strong interactions, which have been regarded as chemisorption, the mobility of the absorbed molecule is limited, and successive deposition leads to polycrystalline grains of small diameter [58, 65]. In contrast, for substrates, such as {001} KBr, [49, 66] {0001} MoS_2, [67] {0001} GaSe, [67] {111} CaF, [68] {111} Si, [36, 68] GeS, {111} GaAs, [35, 68] Sb, [69] freshly cleaved mica, [26, 27, 33, 39], and layered materials, [64] sufficiently strong Van der Waals interaction between the fullerene molecules subdues the interaction between the substrate and individual C_{60} molecules. This leads, in turn, to a highly effective surface mobility of the physisorbed fullerene molecules and to a crystalline film growth characterized by fairly large grains. Predominantly a face-centered cubic (fcc) structure with a series of closed-packed planes {111} oriented with respect to the substrate plane, or a hexagonal close-packed (hcp) structure, with {0001} close-packed planes, is found. Charge transfer into the lowest un-occupied molecular orbital band (LUMO) of C_{60} leads to a strong interaction with the substrate and reduces the effective surface mobility, as has been observed for Cu {111}, Au {110}, and a variety of metals including Ag, Mg, Cr, and Bi [22, 23, 30, 38, 54, 55, 65, 70]. This has been demonstrated by luminescence studies of C_{60} adsorbed onto a {111} Ag surface.

Consequently, deposition of the first monolayer is a crucial factor, determining the growth of the subsequent layers and consequently the crystallinity of films consisting of many monolayers. This has been impressively documented by strikingly different crystallinity of C_{60} films on hydrophobic (passivated) and hydrophilic (nonpassivated) Si. For example, films on a passivated substrate were crystalline with an fcc structure and a noticeable {111} texture, while films grown under similar conditions on a nonpassivated Si substrate were amorphous. The amorphous character has been ascribed to the fullerene's interaction with the hydrophilic substrate [54].

It should be noted that fabrication of fullerene films on gold surfaces gives the opportunity for recording real-time surface-enhanced Raman (SER) spectra. Vibrational spectra of the fullerene's π radical anion indicate significant perturbations in the bonding and symmetry characteristics of C_{60} [71].

For electrochemical studies, thin films of fullerenes were initially formed by the drop-coating methodology, which is based on evaporation on an electrode surface of a known volume of the fullerene solution [61, 71–76]. Films were discontinuous and contained entrapped solvent molecules. Their electrochemical behavior displayed different degrees of reversibility and stability with respect to exposure to consecutive redox cycles. Alternatively, Langmuir–Blodgett films of pristine C_{60} were subjected to a large number of electrochemical studies. Electrochemical reduction was found to form insoluble films, owing to the incorporation of charge-compensating countercations into the film, or resulted in dissolution, since the reduced forms of C_{60} are more soluble than the nonreduced form. The large separation between cathodic and anodic waves, indicative of a high degree of irreversibility, has been attributed to structural rearrangements upon the reduction and reoxidation process.

Besides very unusual properties, such as conductivity and superconductivity, thin solid films of C_{60} and C_{70}, as prepared by thermal evaporation, exhibit high optical nonlinearities. As a consequence of the presence of 60 carbon atoms with 60 delocalized π-electrons, large third-order nonlinear optical and second-harmonic responses of C_{60} in solution (benzene) and in thin films have been reported [77–79].

Several studies have demonstrated the successful incorporation of C_{60} into polymeric structures by following two general routes: (i) in-chain addition (pearl necklace) or (ii) on-chain addition (pendant) polymers [80, 81]. Reports on pendant C_{60} copolymers focus on the functionalization with different amine-, azide-, ethylene propylene terpolymer, polystyrene, poly(oxyethylene), and poly(oxypropylene) precursors [23, 82–86]. On the other hand, pearl necklace polymers were reported for $(-C_{60}Pd-)_n$, which was formed by the periodic linkage of C_{60} and Pd monomers [87] and upon reaction with the diradical species p-xylylene [88]. An alternative approach envisages the fabrication of an all-carbon polymer consisting exclusively of fullerenes in which adjacent fullerenes are linked by covalent bonds. Thin solid films of C_{60} and C_{70} are sensitive to UV–visible illumination and, in the absence of a triplet quencher (oxygen), phototransformation via $2 + 2$ cycloaddition into a polymeric solid that is insoluble in common solvents occurs. While in the case of pristine C_{60} the phototransformation into a polymeric solid is reversible, illumination of C_{70} leads to a random and irreversible photodimerization [89–92].

Finally, the intriguing design of supramolecular composites of conjugated polymers as electron donors and C_{60} as electron acceptor should be mentioned. Polymeric semiconductors have been shown to be effective electron donors upon photoexcitation of the valence band electrons across the bandgap into the conduction band. Photoexcited states of C_{60}, on the other hand, act as strong electron acceptors. Thus, these composites exhibit ultrafast, reversible, metastable photoinduced electron transfer and charge separation. This area has been reviewed elsewhere [7].

6.2.2 Langmuir–Blodgett Films of Pristine Fullerenes, C_{60} and C_{70}

The intriguing electronic, spectroscopic, and structural properties make C_{60} an interesting building block for Langmuir–Blodgett and self-assembled monolayers. These techniques are also considered to have the potential to construct organic films controlled on the order of the molecular level. The large van de Waals radius of 10.0 Å facilitates surface imaging of fullerene-adsorbed layers by scanning-probe microscopies. Furthermore, based on the remarkable redox features, electrochemical studies on monolayered fullerene films can be employed as a powerful tool for an unambiguous surface characterization.

Spreading an organic solution of a surfactant or fullerene on an aqueous solution in a Langmuir trough equipped with a movable barrier that controls the surface pressure leads to monolayer formation at the water–air interface. Subsequent to the evaporation of the solvent, the surfactant molecules align relatively far from each other in a two-dimensional gaseous state with low surface pressures (Π). An increase of the surface pressure by moving the barrier results through an intermediate gaseous-to-liquid state in a liquid phase in which the surfactant molecules begin to assemble. Further compression leads to the transformation of the surfactant to their two-dimensional closed-packed solid state.

When a Langmuir film is transferred to a solid substrate, it is then referred to as a Langmuir–Blodgett (LB) film. Typically, amphiphilic compounds with a hydrophilic tail that interacts with or immerses into the aqueous subphase and a hydrophobic head group are employed for the production of LB films. Although pristine C_{60} is insoluble in water, the physical and chemical properties of this spherical carbon allotrope are strikingly different from those of classical self-assembling amphiphiles, whose structures have both hydrophobic and hydrophilic domains, usually employed to form Langmuir–Blodgett films. Despite the fact that pristine fullerenes do not possess any amphiphilic character, initial reports demonstrate the unambiguous formation of monolayer C_{60} films at the air–water interface with a dynamic molecular area of 98 $Å^2$ molecule^{-1} and a molecular radius of 5.6 Å (from X-ray diffraction data on C_{60} powders, the van der Waals diameter of pristine C_{60} is 10.0 Å; the van der Waals area is then 78.5 $Å^2$, which, for a planar close packing of spherical molecules, becomes 86.6 $Å^2$) [93]. This suggests that the observed data are in line with the molecular fingerprint of C_{60}. However, reproducible formation of high-quality monolayered Langmuir–Blodgett films renders these experiments difficult [94–111]. As a result of the predominant hydrophobic nature of the fullerene core, strong three-dimensional interactions among the hydrophobic fullerene moieties play a key role in governing the stability of true fullerene monolayers and clean transformation into a solid two-dimensional film. Visualizing the compression of pristine fullerene films on the air–water interface by either surface pressure (Π) vs. surface area (A) isotherms or complementary Brewster angle microscopy [112] reveals unambiguous evidence for the formation of multilayers, even if one employs dilute solutions ($\sim 1 \times 10^{-5}$ M in various solvents) and low compression rates. In particular, spreading C_{60} at the gas–water interface gave rise to multilayer films with limiting molecular areas between 20 and 30 $Å^2$ molecule^{-1}. These values, obtained by extrapolation of the surface pressure (Π) vs. molecular area (A) slope to zero surface pressure, are much smaller than those estimated by a hexagonal space-filling model for the molecular surface area of C_{60}. These data indicate that, at zero film pressure, when the fullerene domains are floating on the water surface, there are, on the average, four fullerene molecules stacked on top of each other, and the data demonstrate that aggregation occurs immediately upon the spreading of the fullerene solution. In spite of numerous efforts, monolayer formation has not been unequivocally confirmed on spreading pristine fullerenes on water surfaces. It is, however, remarkable that C_{60} and C_{70} form stable films of any type at the air–water interface with high surface pressure and sustain large attractive forces between the fullerene molecules with the formation of rigid films.

Brewster angle micrographs taken immediately after spreading of pristine C_{60} on the water surface show randomly shaped, relatively bright islands (Figure 6.1), in-

Figure 6.1. Brewster angle microscopic image of a pristine C_{60} layer on water; domains observed at $\Pi = 0$ mN m^{-1} (the length of the image shown is 200 µm).

dicating the formation of thicker films before and at the early stages of the compression, which is in accord with the observed Π vs. A isotherm for pristine C_{60}.

Scanning tunneling microscopy (STM), high-resolution transmission electron microscopy (HRTEM) images, and low-angle X-ray diffraction patterns showed that films of pristine C_{60} have two-dimensional and three-dimensional crystalline regions. The C_{60} molecules, however, pack in a face-centered cubic (fcc) pattern (cell constant of 14.2 Å), rather than in hexagonal close-packed structures [105].

Relatively homogenous C_{60} Langmuir films were prepared by employing a subphase of phenol aqueous solution. The homogeneity of the Langmuir films was attributed to an effectively promoted spread of the benzene–C_{60} solution on the water surface. High-resolution transmission microscopy of the manufactured films revealed, however, polydomain polymorphic structure, such as hexagonal and distorted hexagonal forms, and an amorphous-like disordered form [113].

The superconducting transition temperature T_c of alkali-metal-doped C_{60} ranges from 18 K (K_3C_{60}) [62] to 33 K ($RbCs_2C_{60}$) [42] and even K-doped thin C_{60} films have a superconducting onset around 16 K [62]. This enabled the fabrication of multilayered C_{60} LB films and the subsequent potassium doping. The resulting intercalated film on a poly(ethylene terephthalate) substrate displays superconductivity with T_c(onset) at 12.9 K, as measured by microwave low-field signal and electron spin resonance [94, 107]. A parallel approach, by means of doping with RbN_3, gave a T_c(onset) of 23 K [114].

6.2.3 Langmuir–Blodgett Films of Pristine C_{60}/Amphiphilic Matrix Molecules

The most common way to prevent fullerenes from forming aggregates via strong π–π interactions is to employ surfactants that contain appropriately balanced hydrophobic and hydrophilic moieties. Thus, addition of amphiphilic matrix molecules, such as arachidic acid [93, 95, 97, 99, 111], aza-crown molecules [115], or long-chain alcohols [98] has been carried out. This should lead to a dilution of fullerene cores at the air–water interface, and subsequently to a separation of individual molecules in the resulting film. Two-component monolayered films (C_{60}/AA) displayed dynamic molecular areas identical to those of the matrix molecule (AA) alone. This suggests the formation of heterogeneous films in which the AA molecules align at the air–water interface while the fullerene cores are cushioned on the AA layer, rather than being embedded within the matrix molecule.

The host–guest chemistry of C_{60} comprises two supramolecular approaches: (i) functionalization of the fullerene core, for example with crown ethers, or (ii) formation of inclusion complexes with suitably sized host molecules (cyclodextrins [116] and calixarenes [117]), which leaves the C_{60} intact. Particularly innovative studies focus on the true monolayer formation of C_{60} mixtures with amphiphilic compounds, such as acylated aza-crown ethers [115] or calix[8]arenes [112], containing lipophilic cavities to accommodate the fullerene molecule. On the basis of the experimental observations, namely surface pressure vs. surface area isotherms, Brewster angle microscope images, small-angle X-ray scattering, and atomic force microscopy, it has been proposed that the fullerene molecule is located inside the cavity of the host complexes.

A successful approach for the design of pure C_{60} monolayered films involves the formation of a mixed C_{60}/matrix monolayer, followed by transfer of the monolayer film to a solid substrate and subsequent extraction of the matrix molecule with an appropriate solvent that selectively dissolves the matrix molecule. Surface pressure vs. surface area isotherms of a fullerene/hexadecyl-bis(ethyenedithio)-tetrathiafulvalene mixture exhibit a more homogeneous distribution of the C_{60} molecules in this matrix, although the morphology remains poor. The dissolution of the fullerene was significantly improved by employing SURF. Subsequent removal of the surfactant from the deposited film leads to the formation of a very uniform amorphous C_{60} film [118].

6.3 Langmuir–Blodgett Films of Functionalized Fullerene Derivatives

The unique reactivity of buckminsterfullerene (C_{60}) stimulated broad and interdisciplinary interest in modifying its polyfunctional structure, which contains 30 reactive double bonds located at the junctions of two hexagons via an extended number of addition reactions [4, 8, 81, 119]. In principle, covalent functionalization allows the fusing of the properties of the fullerene core with those of the functionalizing addends. The various types of functionalized fullerene derivatives synthesized so far include (i) cyclic adducts formed via cycloaddition [120–123], (ii) C_{60}–R_x-type derivatives resulting from radical addition or by thereaction of $C_{60}{}^{n-}$ anions with various alkyl halides [124, 125], (iii) adducts involving triangular bridging of a C–C bond [126–129], and (iv) organometallic derivatives [130–133]. Thus, organic functionalization of fullerenes has received interdisciplinary attention as a powerful methodology for the fabrication of new derivatives with promising two-dimensional Langmuir–Blodgett properties.

6.3.1 Monofunctionalized Fullerene Derivatives

Successful spreading of monolayers requires the film-forming material to float on the subphase surface and, at the same time, be insoluble in it. Surface pressure (Π) vs. surface area (A) isotherms of various monofunctionalized fullerenes with either hydrophobic or polar addends exhibiting high collapse pressures (ca. 70 mN m^{-1}), which is indicative of the formation of stable films on the water surface [96, 134–136]. The cross sectional areas per molecule are, however, appreciably smaller than estimated by the hexagonal close-packed model for fullerenes in monolayers. This difference suggests the formation of multilayer structures on the water surface, a hypothesis that is being substantiated in different studies by *in situ* UV–visible spectroscopy, Brewster angle microscopy, optical light microscopy, and atomic force microscopic measurements of transferred multilayered films [134, 136–138]. Thus, it has been shown that functionalization of C_{60} with hydrophobic addends is

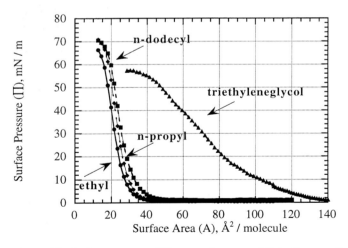

Figure 6.2. Surface pressure (Π) vs. surface area (A) isotherms of monofunctionalized fullerenes (C_{60}[C(COO(CH$_2$CH$_2$O)$_3$CH$_3$)$_2$], C_{60}[C(COOEt)$_2$], C_{60}[C(COOpropyl)$_2$], and C_{60}[C(COOdodecyl)$_2$]).

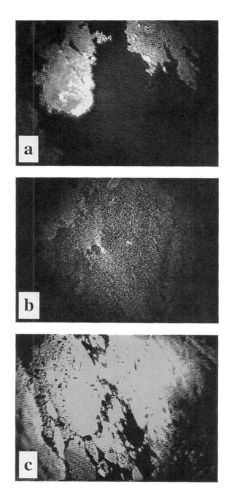

Figure 6.3. Brewster angle microscopic images of a $C_{60}[C(COOEt)_2]$ layer on water. (a) Domains observed at $\Pi = 0$ mN m^{-1}; (b) condensed film recorded at $\Pi = 40$ mN m^{-1}; (c) domain structures observed after expansion (the length of the image shown is 200 μm).

not sufficient to prevent the strong hydrophobic three-dimensional interactions among the fullerene moieties and to stabilize fullerene monolayers at the air–water interface. Furthermore, it has been shown that the partially polar character of ester or methoxy groups does not compensate the strong π–π interaction among the fullerene moieties.

Valuable information with respect to the stacking properties of functionalized fullerenes towards the formation of monolayered assemblies was obtained by means of Brewster angle microscopy (BAM). Thus, images taken immediately after spreading of a toluene solution of $C_{60}[C(COOEt)_2]$ on the water surface show randomly shaped, relatively bright islands (Figure 6.3), indicating the formation of thicker films before and at the early stages of the compression. The bright islands could be pushed together by lateral compression without an apparent change in their brightness. These results indicate uneven packing and the absence of appreciable phase transition, in accord with the observed Π vs. A isotherm for $C_{60}[C(COOEt)_2]$.

A systematic variation of structural parameters was investigated to relate between the fullerene core–addend structure and the ability to form two-dimensional close-packed monolayers at the air–water interface [138, 139]. Factors such as the length of alkyl chains (ethyl, *n*-propyl, and dodecyl), different degree of hydrophilicity (ester derivative vs. free acid), the size of the addend (cyclopropyl vs. tetrahydronaphtalene), and the presence of aromatic cores were carefully altered. Synthesis of fullerene derivatives with covalently attached polar and hydrophilic addends affords structures that resemble the configuration of conventional amphiphiles and, in turn, lead to molecular areas similar to pristine C_{60}. A different parameter, which accounts for the Langmuir–Blodgett behavior, is the size of the addend's head group, e.g. the larger the head group, the greater the dynamic molecular area. The conclusion of these studies is that utilization of fullerenes, functionalized with strongly hydrophilic addends and large head groups, seems to be the most promising approach for the fabrication of stable and compressable two-dimensional fullerene Langmuir–Blodgett films.

6.3.2 Monofunctionalized Fullerene Derivatives Bearing Hydrophilic Groups

The importance of having the right hydrophobic–hydrophilic balance is demonstrated by stable monolayer formation from C_{60} derivatives with highly hydrophilic head groups. Consequently, functionalization with hydrophilic addends, [140, 141] such as cryptate molecules [138], triethyleneglycol monomethyl ether [137, 142], benzocrowns [143–145], N-acetyl pyrrolidine derivatives [143–147], carboxylic acid groups [148, 149], or $C_{60}O$ [150] increases the amphiphilic character of the fullerene core significantly. In turn, the hydrophilic head groups enhance the interaction with the aqueous subphase and allow a two-dimensional fixation of the C_{60} at the air–water interface.

In line with the above, Π–A isotherms of these derivatives display liquidlike regions at low surface pressure and condensed regions at higher pressures, indicating a phase transition. For example, $C_{60}[C(COO(CH_2CH_2O)_3CH_3)_2]$ extrapolation from the condensed phase to $\Pi = 0$ shows a limiting molecular area (A), which is in satisfactory agreement with that reported for fullerene (93 Å^2 molecule^{-1}; see Figure 6.2). This indicates the formation of a true monolayer in which the polar side chain of the fullerene is directed to and hydrated by water (see Figure 6.4).

BAM images, recorded during the compressing of this fullerene derivative (Figure 6.5), revealed the presence of domains even at a very low surface pressure. They enlarge gradually with increasing surface pressure and their brightness increases until the film collapses. These effects are in agreement with the observed phase transition and indicate formation of a two-dimensional fullerene monolayer. Formation of a truly monolayered $C_{60}[C(COO(CH_2CH_2O)_3CH_3)_2]$ film was further substantiated by complementary atomic force microscopic measurements. AFM

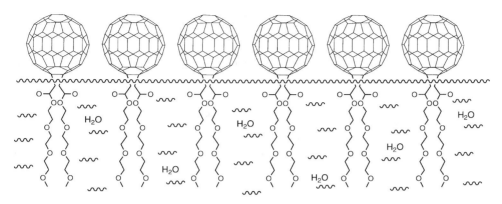

Figure 6.4. Schematic representation of a monolayer of a functionalized C_{60} derivative functionalized with a long hydrophilic chain ($C_{60}[C(COO(CH_2CH_2O)_3CH_3)_2]$).

images of films prepared from $C_{60}[C(COO(CH_2CH_2O)_3CH_3)_2]$, at 20 mN m^{-1} surface pressure, revealed a thickness of the LB films of 7 $\pm$ 3 Å, which corresponds well to the diameter of C_{60} ($\sim$10.0 Å).

Mono- and multilayers were successfully transferred to solid substrates by the LB technique. Their transfer was monitored by absorption spectrophotometry. The absorption spectra of Langmuir–Blodgett films, prepared from $C_{60}[C(COO(CH_2CH_2O)_3CH_3)_2]$, are characterized by maxima at 338 nm, 267 nm, and 222 nm. These maxima are reminiscent of those found for LB films of other amphiphilic fullerene derivatives [138].

Several important properties have been found for fullerene monolayered films: (i) excellent transferability to solid substrates with ratios close to unity [137–139, 151–153], (ii) Z-type or Y-type deposition [137–139, 141, 151, 152], (iii) close packing of the C_{60} moiety [135, 149], (iv) oblique orientation of the molecular axis with respect to the film surface [148], (v) a repeat distance similar to the molecular length [148], (vi) an enhanced second-order nonlinear optical property relative to vapor-deposited thin films of C_{60} [144, 153], (vii) second- and third-order nonlinear susceptibility [134, 135, 141, 153].

A comprehensive study on different fulleropyrrolidines has attempted to tune the fullerene's ability to form true monolayers by modifying the amphiphilic structure of the investigated derivatives [146, 147]. While spreading N-methylfulleropyrrolidine on the air–water interface afforded multilayered structures, introduction of a polar amide group improved the two-dimensional structure of the resulting fullerene film. Only a perfluoroalkyl chain prevented sufficient fullerene aggregation, and true monolayer films were observed that were formed independently of the spreading conditions.

Successfully transferred monolayers of amphiphilic N-methylfulleropyrrolidine onto SnO_2 electrodes exhibited photocurrents upon illumination. Parameters, such as bias voltage or electron donor/acceptor in the electrolyte or electronic effects of the functionalizing addend, that govern the magnitude of the photocurrent were

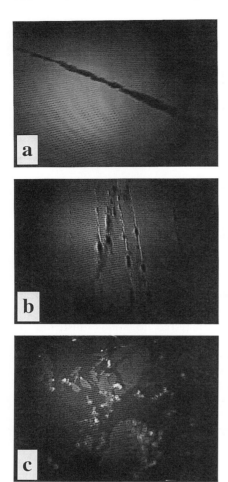

Figure 6.5. Brewster angle microscopic images of a functionalized C_{60} derivative functionalized with a long hydrophilic chain ($C_{60}[C(COO(CH_2CH_2O)_3CH_3)_2]$) layer on water. (a) Condensed film recorded at $\Pi = 40$ mN m^{-1}; (b) at collapse; and (c) taken after expansion (the length of the image shown is 200 μm).

systematically altered. The photocurrent response for the investigated mono-functionalized derivatives was found to be higher relative to pristine C_{60} and displayed a further enhancement upon functionalization with electron-donating groups [154].

Interestingly, ionic fullerene derivatives, such as $C_{60}[C(COO^-)_2]$, which display an appropriately adjusted hydrophobic–hydrophilic balance, form stable Langmuir monolayers on pure water and in solutions containing divalent Ca^{2+} and Cd^{2+} cations. Furthermore, cation head group interactions resulted in the intercalation of Ca^{2+} cations between carboxylates of two adjacent malonate head groups. This is shown by an increased dynamic molecular area compared with the monolayer on pure water. Expanded films of Ca^{2+}-fullerene intercalated monolayers were compressed to solid phases and transferred onto quartz substrates [149].

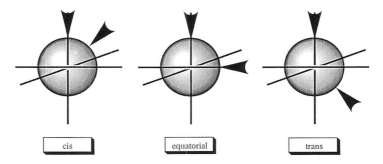

Figure 6.6. Configuration of cis-, equatorial-, and trans-isomers.

6.3.3 Multiply Functionalized Fullerene Derivatives

As an alternative to the amphiphilic concept of employing a hydrophilic tail and hydrophobic head group, the possibility of reducing the strong cohesive interactions between individual fullerene molecules by addition of two functionalizing groups, namely bis(ethoxycarbonyl)methylene groups, to pristine C_{60} was investigated. At the same time the presence of four ester functionalities should introduce regions with partially amphiphilic character to the adduct. The controlled variation of the stereochemical positioning of two bis(ethoxycarbonyl)methylene groups on C_{60} had significant consequences regarding the coexistence of mono- and multilayered structures (see Figure 6.6) [139].

Thus, placement of a second bis(ethoxycarbonyl) methylene group on nearly the opposite site of the fullerene core (trans-2-C_{60}[C(COOEt)$_2$]$_2$) is sufficient to prevent the strong hydrophobic interactions among the fullerene moieties and to stabilize fullerene monolayers at the air–water interface. Placing the second addend closer to the position of the first one obviated the possibility of hydrogen bond formation and thus multilayer formation has been observed upon spreading of the equatorial-C_{60}[C(COOEt)$_2$]$_2$ and trans-3-C_{60}[C(COOEt)$_2$]$_2$ isomer on water surfaces. The true monolayered assembly of the trans-2-C_{60}[C(COOEt)$_2$]$_2$ derivatives can be schematically envisaged by interaction of these molecules with the subphase via hydrogen bonding through their carboxylic groups. Conceivably, hydrogen bonding constrains the fullerene derivatives on the water surface. The unique configuration of the two addends, at nearly opposite poles of the C_{60} sphere, prevents the hydrophobic core from the formation of multilayers, at least along one dimension; and the strong interaction with the aqueous subphase prevents stacking along the second dimension.

The surface pressure vs. surface area isotherm, Brewster angle microscopic, and atomic force microscopic measurements of a trisfunctionalized derivative, namely e,e,e-C_{60}[C(COOEt)$_2$]$_3$, provided evidence for the formation of a stable and high-quality monolayer upon the compression, up to 35 mN m^{-1}, on a water surface in a Langmuir film balance. Compression to higher pressures resulted in the irreversible

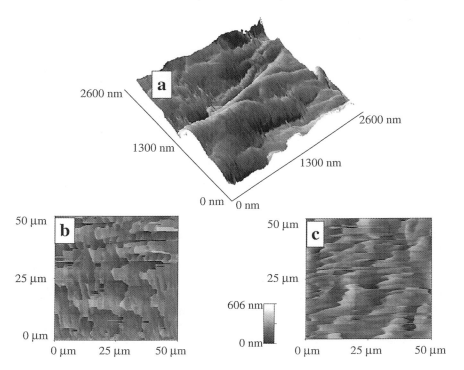

Figure 6.7. (a) Three-dimensional AFM image of *e,e,e*-$C_{60}[C(COOEt)_2]_3$ transferred from the water surface to freshly cleaved mica at $\Pi = 70$ mN m^{-1}. (b, c) Two-dimensional AFM image of *e,e,e*-$C_{60}[C(COOEt)_2]_3$ transferred from the water surface to freshly cleaved mica at $\Pi = 70$ mN m^{-1} and scanned perpendicular to (b) and along (c) the rods.

transformation to rods (1 μm diameter and up to 100 μm long) with porous and oriented structures (see Figure 6.7) [151].

Despite the lack of a definite amphiphilic character, polyamine adducts, namely $C_{60}[NH_2(CH_2)_2CH_3]_{12}$, form stable and homogeneously ordered Langmuir monolayers on water [155, 156]. The enlarged limiting molecular areas (A) obtained by extrapolation from the condensed phase to $\Pi = 0$ relative to pristine C_{60} were rationalized in terms of interdigitation of hydrocarbon chains with neighboring fullerene adducts. In line with the concept that short hydrocarbon chains should favor the formation of homogeneous monolayers, neutron and X-ray scattering of a monolayered $C_{60}[NH_2(CH_2)_{11}CH_3]_x$ (with $x = 5 \pm 3$) film uncovered its inhomogeneous structure.

6.3.4 Transfer to Solid Substrates

Attempts to transfer rigid multilayered films of pristine C_{60} to solid substrates was found to be extremely difficult and did not result in defined multilayer struc-

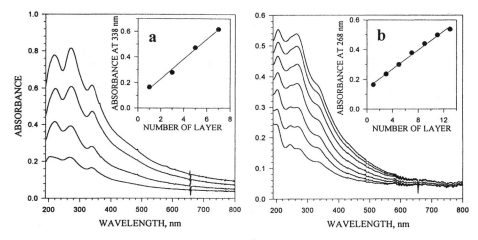

Figure 6.8. UV–visible absorption spectra of Langmuir–Blodgett films: (a) $C_{60}[C(COOEt)_2]$ on optical quartz plate with 1, 3, 5, and 7 monolayer (from bottom to top); absorbance at 338 nm is plotted as a function of number of layers in the insert. (b) $C_{60}[C(COO(CH_2CH_2O)_3CH_3)_2]$ on optical quartz plate with 1, 3, 5, 7, 9, 11, and 13 monolayers (from bottom to top); absorbance at 268 nm is plotted, in the insert, as a function of number of layers.

tures. In contrast, mono- and multilayers prepared from mono-($C_{60}[C(COOEt)_2]$, $C_{60}[C(COOpropyl)_2]$, $C_{60}[C(COOdodecyl)_2]$), bis-(equatorial-$C_{60}[C(COOEt)_2]_2$, trans-2-$C_{60}[C(COOEt)_2]_2$, trans-3-$C_{60}[C(COOEt)_2]_2$), and trisfunctionalized fullerene derivatives (*e,e,e*-$C_{60}[C(COOEt)_2]_3$) have been successfully transferred to solid substrates by the LB technique [137–139, 151–153]. The transfer was monitored by absorption spectrophotometry. The absorption spectra of Langmuir–Blodgett films, prepared from $C_{60}[C(COO(CH_2CH_2O)_3CH_3)_2]$, are characterized by maxima at 338 nm, 267 nm, and 222 nm (see Figure 6.8(a)), as reported for monofunctionalized C_{60} derivatives [157–159]. Similar results were observed for $C_{60}[C(COOEt)_2]$ (see Figure 6.8(b)), $C_{60}[C(COOpropyl)_2]$, and $C_{60}[C(COOdodecyl)_2]$. A linear dependence of the absorbance at 338 nm vs. the number of layers indicates satisfactory stacking of these monofunctionalized fullerene derivatives in the LB films, but the higher slopes observed for $C_{60}[C(COOEt)_2]$, $C_{60}[C(COOpropyl)_2]$, and $C_{60}[C(COOdodecyl)_2]$ are indicative of multilayer rather than monolayer stacking.

Perfect transfer ratios are found in the linearity of absorbance vs. number-of-layers plots for the equatorial-$C_{60}[C(COOEt)_2]_2$, trans-3-$C_{60}[C(COOEt)_2]_2$, trans-2-$C_{60}[C(COOEt)_2]_2$, and *e,e,e*-$C_{60}[C(COOEt)_2]_3$. The slopes of these plots can be categorized into three groups: (i) 0.062 for the LB film of $C_{60}[C(COOEt)_2]$; (ii) 0.04 for equatorial-$C_{60}[C(COOEt)_2]_2$ and 0.029 for trans-3-$C_{60}[C(COOEt)_2]_2$; and (iii) 0.023 for trans-2-$C_{60}[C(COOEt)_2]_2$ and 0.02 for $C_{60}[C(COO(CH_2CH_2O)_3CH_3)_2]$. Once again, the markedly large slopes for $C_{60}[C(COOEt)_2]$ compared to those true monolayers indicate higher two-dimensional concentration, implying the formation of multilayers.

6.4 Fullerenes Covalently Attached to Self-assembled Monolayers and Self-assembled Monolayers of Functionalized Fullerene Derivatives

Self-assembled monolayers develop upon immersion of a substrate into an organic solution of a suitable surfactant. This methodology is attractive since it avoids the complex requirements for the production of Langmuir–Blodgett films. Organosulfur compounds such as alkyl and aromatic thiols are known to form close-packed and well-ordered monolayers, so-called self-assembled monolayers (SAMs), on gold or silver surfaces via chemisorptive S–Au or S–Ag bonds.

SAMs of alkanethiol derivatives are interesting because of their structural analogy to biomembranes, their ease of preparation, their apparent stability, and their potentiality to functionalized C_{60}. Thus, an alternative approach for the preparation of stable fullerene two-dimensional films involves the attachment of C_{60} to self-assembled monolayers by a terminal amine (see Figure 6.9) or pyridine groups. This is based, for example, on the known ability of primary and secondary amines to undergo addition reaction with the fullerene's C=C double bonds and could be demonstrated by covalent linkage of pristine fullerenes to aminopropylsilanized indium–tin-oxide (ITO) substrates, [160] gold surfaces modified with 2-amino-ethanethiol (cysteamine), [161] and pyridly terminated silicon oxide surface in the presence of OsO_4 [162]. Contact angle measurements for H_2O of the resulting C_{60}-modified ITO surface $(\Theta = 72°)$ revealed its hydrophobicity relative to pure ITO $(\Theta = 20°)$ [160].

Monolayers were characterized by contact angle measurements, high-resolution X-ray diffraction, X-ray photoelectron spectroscopy, electrochemistry, and quartz crystal microbalance measurements. Cathodic shifts of the fullerene's first two reduction steps relative to pristine C_{60} are consistent with a covalent functionalization of the self-assembled fullerenes to the terminal amine groups of the modified sub-

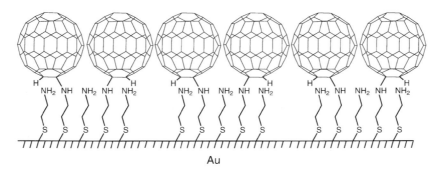

Figure 6.9. Schematic representation of a covalently attached monolayer of C_{60} on a self-assembled layer of 2-aminoethanethiol (cysteamine) on gold.

strate. Applying a very negative potential to the C_{60}-modified electrode over a long time period resulted in a significant fullerene desorption from the surface. This permits control of the surface coverage by chemical-induced absorption or electrochemical-induced desorption. In contrast to C_{60} films on aminopropylsilanized oxide substrates, the electrochemistry of fullerene SAMs formed on pure cystamine indicates a surprising resistance toward reduction. This has been ascribed to dense fullerene packing and ion transport inhibition and suggests films of higher quality on Au/thiol substrates compared to indium–tin-oxide (ITO)/siloxane analogues [160, 161].

A different approach focuses on the self-assembly of prefunctionalized fullerene derivatives. A monofunctionalized fullerene derivative bearing a thiol terminus spontaneously adsorbs onto an Au {111}/mica substrate to form highly ordered monolayer films [163]. The fullerene cores are found to dominate the adsorbate packing and atomic force microscopic images of the self-assembled monolayer reveal spherical features regularly spaced in a distorted hexagonal arrangement. The fullerene–fullerene nearest neighbor distance of 10.9 Å is reminiscent of the distance for the {111} face of crystalline C_{60}.

Molecular films were observed for fullerenes covalently bonded to the functional surface of self-assembled monolayers (azide-terminated) on a silicon substrate. Atomic force microscopy (AFM) measurements revealed mechanically robust, molecularly smooth, and homogeneous areas in which fullerene molecules are packed in unordered lattices corresponding to the {h00} faces of a face-centered cubic unit cell with edge length $a = 14$ Å. This type of two-dimensional self-assembled fullerene film has attracted attention as a potential boundary lubricant, since it exhibits no wear and material transfer between sliding surfaces. These films were subjected to friction force microscopy and showed very high wear stability with friction coefficients similar to sublimed fullerene films [164, 165].

Another intriguing report shows the adsorption of fullerenes onto the surfaces of etched, single-crystal n-CdS and n-CdSe semiconductors. As a consequence of electron-withdrawing effects from the semiconductor bulk to surface electronic states, the semiconductors band-edge photoluminescence is noticeably quenched, with a dead-layer thickness as large as ~ 300 Å. The photoluminescence quenching by fullerene solutions was found to be concentration dependent and could be detected at submicromolar concentrations. Large binding constants in the range 10^5–10^6 M^{-1} were determined for these surfaces. In conclusion, bulk photoluminescence from CdS was shown to act as a sensitive probe of surface adduct formation with fullerenes [166].

Molecular recognition principles in conjunction with self-assembled monolayers have been employed to form a monolayer film of a fullerene derivative that is functionalized with a crown ether and an ammonium-terminated alkanethiolate–gold surface. The determination of the surface coverage, as derived from Osteryoung square wave voltammetry (OSVW), yielded a value $(1.4 \times 10^{-10}$ mol cm$^{-2})$ well in accordance with an fcc closed-packed packing of C_{60} ($\sim 1.9 \times 10^{-10}$ mol cm^{-2}). Desorption experiments confirmed the fully reversible nature of this process, i.e. there is no covalent and irreversible linkage of the fullerene moiety to the modified surface, and substantiated the presence of specific interactions between the crown ether functionality and the ammonium terminal group [152].

6.4.1 Self-assembled Monolayers of Fullerene Containing Supramolecular Dyads

Typically, photoactive donor–bridge–acceptor dyads are based on composites containing chromophoric acceptor moieties such as porphyrins, phthalocyanines, or ruthenium-(II)-tris(bipyridine) and suitable electron donors (ferrocene). In line with the "photosensitive donor–bridge–acceptor dyads" concept, various covalently linked fullerene/chromophore assemblies have been synthesized recently and were subjected to photoinduced electron and energy transfer studies [167–183]. In solution, picosecond-resolved photolysis of covalently linked fullerene/ferrocene-based donor–bridge–acceptor dyads shows intramolecular quenching of the fullerene's excited singlet state, resulting in the formation of charge-separated ion radical pairs [170]. Similarly, photoinduced charge separation and subsequent charge recombination were observed in a series of porphyrin/fullerene dyads [172, 174].

A very intriguing report shows the spontaneous self-assembly of a fullerene/porphyrin-based donor–bridge–acceptor dyad, bearing a methylthio group at the porphyrin moiety, on an Au surface. Upon illumination of this photosensitive dyad, the authors detected a photocurrent, which, it has been proposed, evolved from an electron relay via the excited singlet or triplet states of the porphyrin and the C_{60} core [184].

6.4.2 Self-assembly of Functionalized Fullerene Derivatives via Electrostatic Interaction

Self-assembly of individual molecules by electrostatic interactions leads to the preparation of films composed of self-assembled layers of a water-soluble, negatively charged fullerene derivative (F), and a polydiallyldimethylammonium polycation (P). This leads to the fabrication of densely packed, highly ordered monolayer films. Preparation of self-assembled n layers of $(P/F)_n$ films consisted of the alternating immersion of a substrate into an aqueous solution of polydiallyldimethylammonium and into a fullerene solution. Subsequent P/F sandwich units were self-assembled by repeating these steps n times to produce films consisting of n sandwich units, $(P/F)_n$ (see Figure 6.10) [185].

Successful self-assembly of successive layers of $(P/F)_n$ was confirmed by monitoring the characteristic fullerene absorption peaks at 212, 230, and 295 nm using spectrophotometry after each step. The linearity of the plot of absorbencies vs. number of layers indicates the uniformity of the assembled P/F sandwich units.

Surface plasmon spectroscopy provided a convenient means for monitoring the thickness regularities of the buildup of the $(P/F)_n$ film. The self-assembly of each

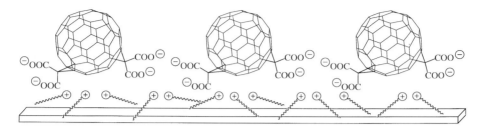

Figure 6.10. Schematic representation of a self-assembled layer of negatively charged fullerene derivative (F) on a polydiallyldimethylammonium polycation (P).

additional P and F layer is seen to result in the gradual shift of the surface plasmon curve to a higher degree and hence to a greater thickness.

6.5 Outlook and Application

The development of versatile methodologies focusing on the covalent functionalization of fullerenes provides an outstanding assortment of tailor-made building blocks for technologically interesting materials with new and perhaps even more spectacular properties than pristine fullerenes. However, their high instability at elevated temperatures, which are employed for vapor deposition of C_{60} in vacuum, limits the investigations of vaporized thin films based on functionalized fullerene derivatives. A first report on a fullerene derivative bearing a cyclic siloxane addend $C_{60}(CH_2Si(CH_3)_2)_2O$ is a breakthrough in this direction. It demonstrates the successful fabrication of its vacuum deposited thin film with a thermal stability below 300 °C [186].

Balancing the hydrophobic and hydrophilic parts in fullerene derivatives has been demonstrated to be of paramount importance for forming high-quality monolayers and LB films from functionalized fullerenes. Preparation of Langmuir–Blodgett films from these fascinating molecules, deposited on the surface of appropriate electrodes, permits the viable construction of ultrathin modified electrodes with the high density of electroactive sites for their application as sensors, photoelectrochemical devices, or photoelectrochemical information storage devices. In line with a potential application falls the intriguing report on a complex film of 2,6-bis(2,2-bicyanovinyl)pyridine (BDCP) and C_{60}. A sandwich-like device thereof, Ag/C_{60}-BDCP/Ag, showed stable and reproducible bistable electronic switching and memory phenomena [187].

Results on various fullerene-containing donor–bridge–acceptor dyads demonstrate that C_{60} is a new promising building block as an acceptor unit in artificial photosynthetic models. Thus, incorporation of fullerene dyads into well-defined two-dimensional arrays and three-dimensional networks is a very challenging goal with a great potential.

Finally, developments like the recently reported design and self-organization of fullerene-based lipid bilayer membranes have a great potential for their application as molecular lipid films [188].

Acknowledgment

This work was supported by the Office of Basic Energy Sciences of the Department of Energy and is contribution No. NDRL-3983 from the Notre Dame Radiation Laboratory.

References

[1] W. Krätschmer, L. D. Lamb, K. Fostiropoulos, D. R. Huffman, *Nature* **1990**, *347*, 354–358.
[2] H. W. Kroto, J. R. Heath, S. C. O'Brien, R. F. Curl, R. E. Smalley, *Nature* **1985**, *318*, 162–164.
[3] P.-M. Allemand, A. Koch, F. Wudl, Y. Rubin, F. Diederich, M. M. Alvarez, S. J. Anz, R. L. Whetten, *J. Am. Chem. Soc.* **1991**, *113*, 1050–1051.
[4] A. Hirsch, *Synthesis* **1995**, 895–913.
[5] R. C. Haddon, *J. Am. Chem. Soc.* **1996**, *118*, 3041–3042.
[6] C. A. Mirkin, W. B. Caldwell, *Tetrahedron* **1996**, *52*, 5113–5130.
[7] N. S. Sariciftci, *Prog. Quant. Electr.* **1995**, *19*, 131–159.
[8] A. W. Jensen, S. R. Wilson, D. I. Schuster, *Bioorg. Med. Chem.* **1996**, *4*, 767–779.
[9] A. S. Boutorine, H. Tokuyama, M. Takasugi, H. Isobe, E. Nakamura, C. Helene, *Angew. Chem.* **1994**, *106*, 2526– 2529.
[10] Y. N. Yamakoshi, T. Yagami, S. Sueyoshi, N. Miyata, *J. Org. Chem.* **1996**, *61*, 7236–7237.
[11] K. Irie, Y. Nakamura, H. Ohigashi, H. Tokuyama, S. Yamago, E. Nakamura, *Biosci. Biotech. Biochem.* **1996**, *60*, 1359–1361.
[12] H. Tokuyama, S. Yamago, E. Nakamura, T. Shiraki, Y. Sugiura, *J. Am. Chem. Soc.* **1993**, *115*, 7918–7919.
[13] R. Sijbesma, G. Srdanov, F. Wudl, J. A. Castoro, C. Wilkins, S. H. Friedman, D. L. DeCamp, G. L. Kenyon, *J. Am. Chem. Soc.* **1993**, *115*, 6510–6512.
[14] R. C. Haddon, T. Siegrist, R. M. Fleming, P. M. Bridenbaugh, R. A. Laudise, *J. Mater. Chem.* **1995**, *5*, 1719–1724.
[15] R. C. Haddon, A. F. Hebard, M. J. Rosseinsky, D. W. Murphy, S. J. Duclos, K. B. Lyons, B. Miller, J. M. Rosamilia, R. M. Fleming, A. R. Kortan, S. H. Glarum, A. V. Makhija, A. J. Muller, R. H. Eick, S. M. Zahurak, R. Tycko, G. Dabbagh, F. A. Thiel, *Nature* **1991**, *350*, 320–322.
[16] R. C. Haddon, A. S. Perel, R. C. Morris, T. T. M. Plastra, A. F. Hebrad, R. M. Fleming, *Appl. Phys. Lett.* **1995**, *67*, 121–123.
[17] D. Dubois, K. M. Kadish, S. Flanagan, L. J. Wilson, *J. Am. Chem. Soc.* **1991**, *113*, 7773–7774.
[18] D. Dubois, K. M. Kadish, S. Flanagan, R. E. Haufler, L. P. F. Chibante, L. J. Wilson, *J. Am. Chem. Soc.* **1991**, *113*, 4364–4366.

[19] C. S. Foote, *Top. in Curr. Chem.* **1994**, *169*, 348–363.
[20] M. K. Kelly, P. Etchegoin, D. Fuch, W. Krätschmer, K. Fostiropulos, *Phys. Rev. B* **1992**, *46*, 4963–4968.
[21] E. I. Altman, R. J. Colton, *Surf. Sci.* **1992**, *279*, 49–67.
[22] E. I. Altman, R. J. Colton, *J. Phys. Rev. B* **1993**, *48*, 18244–18249.
[23] E. I. Altman, R. J. Colton, *Surf. Sci.* **1993**, *295*, 13–33.
[24] B. Bhushan, J. Ruan, B. K. Gupta, *J. Phys. D.* **1993**, *26*, 1319–1322.
[25] H.-G. Busmann, R. Hiss, H. Gaber, I. V. Hertel, *Surf. Sci.* **1993**, *289*, 381–388.
[26] A. Fartash, *Appl. Phys. Lett.* **1994**, *64*, 1877–1879.
[27] J. E. Fischer, E. Werwa, P. A. Heiney, *Appl. Phys. A* **1993**, *56*, 193–196.
[28] A. V. Hamza, M. Balooch, *Chem. Phys. Lett* **1992**, *198*, 603–608.
[29] A. V. Hamza, M. Balooch, *Chem. Phys. Lett.* **1993**, *201*, 404–408.
[30] T. Hashizume, K. Motai, X. D. Wang, H. Shinohara, Y. Saito, Y. Maruyama, K. Ohno, Y. Kawazoe, Y. Nishina, H. W. Pickering, Y. Kuk, T. Sakurai, *Phys. Rev. Lett.* **1993**, *71*, 2959–2962.
[31] S. Henke, K. H. Thürer, J. K. N. Lindner, B. Rauschenbach, B. Stritzker, *J. Appl. Phys.* **1994**, *76*, 3337–3340.
[32] S. Howells, T. Chen, M. Gallagher, D. Sarid, D. L. Lichtenberger, L. L. Wright, C. D. Ray, D. R. Huffman, L. D. Lamb, *Surf. Sci* **1992**, *274*, 141–146.
[33] W. Krakow, N. M. Rivera, R. A. Roy, R. S. Ruoff, J. J. Cuomo, *J. Mater. Res.* **1992**, *7*, 784–787.
[34] Y. Z. Li, J. C. Patrin, M. Chander, J. H. Weaver, L. P. F. Chibante, R. E. Smalley, *Science* **1991**, *252*, 547–548.
[35] Y. Z. Li, M. Chander, J. C. Patrin, J. H. Weaver, L. P. F. Chibante, R. E. Smalley, *Science* **1991**, *253*, 429–433.
[36] Y. Z. Li, M. Chander, J. C. Patrin, J. H. Weaver, L. P. F. Chibante, R. E. Smalley, *Phys. Rev. B* **1992**, *45*, 13837– 13840.
[37] A. J. Maxwell, P. A. Brühwiler, A. Nilsson, N. Martensson, P. Rudolf, *Phys. Rev. B* **1994**, *49*, 10717–10725.
[38] K. Motai, T. Hashizume, H. Shinohara, Y. Saito, H. W. Pickering, Y. Nishina, T. Sakurai, *Jpn. J. Appl. Phys.* **1993**, *32*, L450–L453.
[39] D. Schmicker, S. Schmidt, J. G. Skofronick, J. P. Toennies, R. Vollmer, *Phys. Rev. B* **1991**, *44*, 10995–10997.
[40] A. Sellidj, B. E. Koel, *J. Phys. Chem.* **1993**, *97*, 10076–10082.
[41] S. Suto, A. Kasuya, O. Ikeno, N. Horiguchi, Y. Achiba, T. Goto, Y. Nishina, *J. Electron Spectrosc. Relat. Phenom.* **1993**, *64*, 877–882.
[42] K. Tanigaki, T. W. Ebbesen, S. Saito, J. Mizuki, J. S. Tsai, Y. Kubo, S. Kuroshima, *Nature* **1991**, *352*, 222–223.
[43] A. Tokmakoff, D. R. Haynes, S. M. George, *Chem. Phys. Lett.* **1991**, *186*, 450–455.
[44] W. M. Tong, D. A. A. Ohlberg, H. K. You, R. S. Williams, S. J. Anz, M. M. Alvarez, R. L. Whetten, F. Rubin, F. Diederich, *J. Phys. Chem.* **1991**, *95*, 4709–4712.
[45] X.-D. Wang, H. Hashizume, T. Shinohara, Y. Saito, Y. Nishina, T. Sakurai, *Phys. Rev. B* **1993**, *47*, 15923–15930.
[46] R. J. Wilson, G. Meijer, D. S. Bethune, R. D. Johnson, D. D. Chambliss, M. S. deVries, H. E. Hunziker, H. R. Wendt, *Nature* **1990**, *348*, 621–622.
[47] J. Wragg, J. E. Chamberlain, H. W. White, W. Krätschmer, D. R. Huffman, *Nature* **1990**, *348*, 623–624.
[48] Y. Zhang, X. Gao, M. J. Weaver, *J. Phys. Chem.* **1992**, *96*, 510–513.
[49] W.-B. Zhao, X.-D. Zhang, K.-J. Luo, J. Chen, Z.-Y. Ye, J.-L. Zhang, C.-Y. Li, D.-L. Yin, Z.-N. Gu, X.-H. Zhou, Z.-X. Jin, *Thin Solid Films* **1993**, *232*, 149–153.
[50] W. Zhao, W. L. Zhou, L.-Q. Chen, Y.-Z. Huang, Z.-B. Zhang, K. K. Fung, Z.-X. Zhao, *J. Solid State Chem.* **1994**, *112*, 412–417.
[51] W. L. Zhou, W. Zhao, K. K. Fung, L. Q. Chen, Z. B. Zhang, *Physica C* **1993**, *214*, 19–24.
[52] A. Dereux, C. Girard, O. J. F. Martin, P. Lambin, H. Richter, *J. Chem. Phys.* **1994**, *101*, 10975–10979.
[53] P. Dietz, K. Fostiropoulos, W. Krätschmer, P. K. Hansma, *Appl. Phys. Lett.* **1992**, *60*, 62–64.

[54] A. F. Hebard, O. Zhou, Q. Zhong, R. M. Fleming, R. C. Haddon, *Thin Solid Films* **1995**, *257*, 147–153.
[55] C. Joachim, J. K. Gimzewski, R. R. Schlittler, C. Chavy, *Phys. Rev. Lett.* **1995**, *74*, 2102–2105.
[56] S.-L. Ren, K. A. Wang, P. Zhou, Y. Wang, A. M. Rao, M. S. Meier, J. P. Selegue, P. C. Eklund, *Appl. Phys. Lett.* **1992**, *61*, 124–126.
[57] E. J. Snyder, M. S. Anderson, W. M. Tong, R. S. Williams, S. J. Anz, M. M. Alvarez, Y. Rubin, F. N. Diederich, R. L. Whetten, *Science* **1991**, *253*, 171–173.
[58] Y. Takahashi, K. Hayashi, *J. Electron Microsc.* **1994**, *43*, 378–385.
[59] J. H. Weaver, *Acc. Chem. Res.* **1992**, *25*, 143–149.
[60] R. J. Wilson, G. Meijer, D. S. Bethune, R. D. Johnson, D. D. Chambliss, M. S. deVries, H. E. Hunziker, H. R. Wendt, *Nature* **1990**, *348*, 621–622.
[61] B. Miller, J. M. Rosamilia, G. Dabbagh, R. Tycko, R. C. Haddon, A. J. Muller, W. Wilson, D. W. Murphy, A. F. Hebard, *J. Am. Chem. Soc.* **1991**, *113*, 6291–6293.
[62] A. F. Hebard, M. J. Rosseinsky, R. C. Haddon, D. W. Murphy, S. H. Glarum, T. T. M. Palstra, A. P. Ramirez, A. R. Kortan, *Nature* **1991**, *350*, 600–601.
[63] C. M. Varma, J. Zaanen, K. Raghavachari, *Science* **1991**, *254*, 989–992.
[64] K. Tanigaki, S. Kuroshima, T. W. Ebbesen, *Thin Solid Films* **1995**, *257*, 154–165.
[65] J. K. Gimzewski, S. Modesti, R. R. Schlittler, *Phys. Rev. Lett.* **1994**, *72*, 1036–1039.
[66] T. Ichihashi, K. Tanigaki, T. W. Ebbesen, S. Kuroshima, S. Iijima, *Chem. Phys. Lett.* **1992**, *190*, 179–183.
[67] M. Sakurai, H. Tada, K. Saiki, A. Koma, H. Funasaka, Y. Kishimoto, *Chem. Phys. Lett.* **1993**, *208*, 425–430.
[68] A. Koma, *Thin Solid Films* **1992**, *216*, 72–76.
[69] J. A. Dura, P. M. Pippenger, N. J. Halas, X. Z. Xiong, P. C. Chow, S. C. Moss, *Appl. Phys. Lett.* **1993**, *63*, 3443–3445.
[70] J. E. Rowe, P. Rudolf, L. H. Tjeng, R. A. Malic, G. Meigs, C. T. Chen, *Int. J. Mod. Phys. B* **1992**, *6*, 3909–3913.
[71] Y. Zhang, G. Edens, M. J. Weaver, *J. Am. Chem. Soc.* **1991**, *113*, 9395–9397.
[72] L. M. Goldenberg, *J. Electroanal. Chem.* **1994**, *379*, 3–19.
[73] J. Chlistunoff, D. Cliffel, A. J. Bard, *Thin Solid Films* **1995**, *257*, 166–184.
[74] C. Jehoulet, A. J. Bard, F. Wudl, *J. Am. Chem. Soc.* **1991**, *113*, 5456–5457.
[75] L. Seger, L.-Q. Wen, J. B. Schlenoff, *J. Electrochem. Soc.* **1991**, *138*, L81–L83.
[76] R. G. Compton, R. A. Spackman, R. G. Wellington, M. L. H. Green, J. Turner, *J. Electroanal. Chem. Interfacial Electrochem.* **1992**, *327*, 337–341.
[77] X. K. Wang, T. G. Zhang, P. M. Lundquist, W. P. Lin, Z. Y. Xu, G. K. Wong, J. B. Ketterson, R. P. H. Chang, *Thin Solid Films* **1995**, *257*, 244–247.
[78] Z. H. Kafafi, J. R. Lindle, R. G. S. Pong, F. J. Bratoli, L. J. Lingg, J. Milliken, *Chem. Phys. Lett.* **1992**, *188*, 492–496.
[79] F. Kajzar, C. Taliani, R. Danieli, S. Rossini, R. Zamboni, *Chem. Phys. Lett.* **1994**, *217*, 418–422.
[80] I. Amato, *Science* **1991**, *254*, 30–31.
[81] R. Taylor, D. R. M. Walton, *Nature* **1993**, *363*, 685–693.
[82] L. Y. Chiang, R. B. Upasani, J. W. Swirczewski, *J. Am. Chem. Soc.* **1992**, *114*, 10154–10157.
[83] T. Suzuki, Q. Li, K. C. Khemani, F. Wudl, Ö. Almarsson, *J. Am. Chem. Soc.* **1992**, *114*, 7300–7301.
[84] T. Suzuki, Q. Li, K. C. Khemani, F. Wudl, *J. Am. Chem. Soc.* **1992**, *114*, 7301–7302.
[85] K. E. Geckeler, A. Hirsch, *J. Am. Chem. Soc.* **1993**, *115*, 3850–3851.
[86] T. Benincori, E. Brenna, F. Sannicolo, L. Trimarco, G. Zotti, P. Sozzani, *Angew. Chem. Int. Ed. Engl.* **1996**, *35*, 648–651.
[87] H. Nagashima, A. Nakaoka, Y. Saito, M. Kato, T. Kawanishi, K. Itoh, *J. Chem. Soc., Chem. Commun.* **1992**, 377–379.
[88] D. A. Loy, R. A. Assink, *J. Am. Chem. Soc.* **1992**, *114*, 3977–3978.
[89] P. Zhou, Z.-H. Dong, A. M. Rao, P. C. Eklund, *Chem. Phys. Lett.* **1993**, *211*, 337–340.
[90] Y. Wang, J. M. Holden, Z.-H. Dong, X.-X. Bi, P. C. Eklund, *Chem. Phys. Lett.* **1993**, *211*, 341–345.

[91] Y. Wang, J. M. Holden, X.-X. Bi, P. C. Eklund, *Chem. Phys. Lett.* **1994**, *217*, 413–417.

[92] P. C. Eklund, A. M. Rao, P. Zhou, Y. Wang, J. M. Holden, *Thin Solid Films* **1995**, *257*, 185–203.

[93] Y. S. Obeng, A. J. Bard, *J. Am. Chem. Soc.* **1991**, *113*, 6279–6280.

[94] P. Wang, R. M. Metzger, S. Bandow, Y. Maruyama, *J. Phys. Chem.* **1993**, *97*, 2926–2927.

[95] G. Williams, C. Pearson, M. R. Bryce, M. C. Petty, *Thin Solid Films* **1992**, *209*, 150–152.

[96] G. Williams, A. Soi, A. Hirsch, M. R. Bryce, M. C. Petty, *Thin Solid Films* **1993**, *230*, 73–77.

[97] Y. Xu, J. Guo, C. Long, Y. Li, Y. Liu, Y. Yao, D. Zhu, *Thin Solid Films* **1994**, *242*, 45–49.

[98] J. Milliken, D. D. Dominguez, H. H. Nelson, W. R. Barger, *Chem. Mater.* **1992**, *4*, 252–254.

[99] T. Nakamura, H. Tachibana, M. Yumara, M. Matsumoto, R. Azumi, M. Tanaka, Y. Kawabata, *Langmuir* **1992**, *8*, 4–6.

[100] A. A. Kharlamov, L. A. Chernozatonskii, A. A. Dityat'ev, *Chem. Phys. Lett.* **1994**, *219*, 457–461.

[101] C.-F. Long, Y. Xu, F.-X. Guo, Y.-L. Li, D.-F. Xu, Y.-X. Yao, D.-B. Zhu, *Solid State Commun.* **1992**, *82*, 381–383.

[102] M. Iwahashi, K. Kikuchi, Y. Achiba, I. Ikemoto, T. Araki, T. Mochida, S.-I. Yokoi, A. Tanaka, K. Iriyama, *Langmuir* **1992**, *8*, 2980–2984.

[103] R. Castillo, S. Ramos, J. Ruiz-Garcia, *J. Phys. Chem.* **1996**, *100*, 15235–15241.

[104] R. Back, R. B. Lennox, *J. Phys. Chem.* **1992**, *96*, 8149–8152.

[105] P. Wang, M. Shamsuzzoha, X.-L. Wu, W.-J. Lee, R. M. Metzger, *J. Phys. Chem.* **1992**, *96*, 9025–9028.

[106] P. Wang, M. Shamsuzzoha, W.-J. Lee, X.-L. Wu, R. M. Metzger, *Synth. Met.* **1993**, *55*, 3104–3109.

[107] P. Wang, Y. Maruyama, R. M. Metzger, *Langmuir* **1996**, *12*, 3932–3937.

[108] T. Nakamura, H. Tachibana, M. Yumara, M. Matsumoto, W. Tagaki, *Synth. Met.* **1993**, *55*, 3131–3136.

[109] L. O. S. Bulhoes, Y. S. Obeng, A. J. Bard, *Chem. Matr.* **1993**, *5*, 110–114.

[110] C. Jehoulet, Y. S. Obeng, Y.-T. Kim, F. Zhou, A. J. Bard, *J. Am. Chem. Soc.* **1992**, *114*, 4237–4247.

[111] Y. Xiao, Z. Yao, D. Jin, F. Yan, Q. Xue, *J. Phys. Chem.* **1993**, *97*, 7072–7074.

[112] R. Castillo, S. Ramos, R. Cruz, M. Martinez, F. Lara, J. Ruiz-Garcia, *J. Phys. Chem.* **1996**, *100*, 709–713.

[113] Y. Tomioka, M. Ishibashi, H. Kajiyama, Y. Taniguchi, *Langmuir* **1993**, *9*, 32–35.

[114] K. Ikegami, S.-I. Kuroda, M. Matsumoto, T. Nakamura, *J. Appl. Phys. Lett.* **1995**, *34*, L1227–L1229.

[115] F. Diederich, J. Effing, U. Jonas, L. Jullien, T. Plesnivy, H. Ringsdorf, C. Thilgen, D. Weinstein, *Angew. Chem. Int. Ed. Engl.* **1992**, *31*, 1599–1601.

[116] T. Andersson, K. Nilsson, M. Sundahl, G. Westman, O. Wennerström, *J. Chem. Soc., Chem. Commun.* **1992**, 604–605.

[117] M. Takeshita, T. Suzuki, S. Shinkai, *J. Chem. Soc., Chem. Commun.* **1994**, 2587–2588.

[118] T. S. Berzina, V. I. Troitsky, O. Y. Neilands, I. V. Sudmale, C. Nicolini, *Thin Solid Films* **1995**, *256*, 186–191.

[119] F. Diederich, C. Thilgen, *Science* **1996**, *271*, 317–323.

[120] P. Belik, A. Gügel, J. Spickerman, K. Müllen, *Angew. Chem. Int. Ed. Engl.* **1993**, *32*, 78–81.

[121] M. Maggini, G. Scorrano, M. Prato, *J. Am. Chem. Soc.* **1993**, *115*, 9798–9799.

[122] M. Prato, M. Maggini, C. Giacometti, G. Scorrano, G. Sandona, G. Farnia, *Tetrahedron* **1996**, *52*, 5221–5234.

[123] S. R. Wilson, N. Kaprindis, Y. Wu, D. I. Schuster, *J. Am. Chem. Soc.* **1993**, *115*, 8495–8496.

[124] J. W. Bausch, G. K. S. Prakash, G. A. Olah, D. S. Tse, D. C. Lorents, Y. K. Bae, R. Malhotra, *J. Am. Chem. Soc.* **1991**, *113*, 3205–3206.

[125] C. Caron, R. Subramanian, F. D'Souza, J. Kim, W. Kutner, M. T. Jones, K. M. Kadish, *J. Am. Chem. Soc.* **1993**, *115*, 8505–8506.

[126] C. Bingel, *Chem. Ber.* **1993**, *126*, 1957–1959.

[127] A. Hirsch, I. Lamparth, H. R. Karfunkel, *Angew. Chem. Int. Ed. Engl.* **1994**, *33*, 437–440.

[128] I. Lamparth, C. Maichle-Mössmer, A. Hirsch, *Angew. Chem. Int. Ed. Engl.* **1995**, *34*, 1607–1609.

[129] G. Schick, A. Hirsch, H. Mauser, T. Clark, *Chem. Eur. J.* **1996**, *2*, 935–943.
[130] S. Schreiner, T. N. Gallagher, H. K. Parsons, *Inorg. Chem.* **1994**, *33*, 3021–3022.
[131] S. A. Lerke, D. H. Evans, P. J. Fagan, *J. Electroanalytical Chem.* **1995**, *383*, 127–132.
[132] P. J. Fagan, P. J. Krusic, D. H. Evans, S. A. Lerke, E. Johnston, *J. Am. Chem. Soc.* **1992**, *114*, 9697–9699.
[133] S. A. Lerke, B. A. Parkinson, D. H. Evans, P. J. Fagan, *J. Am. Chem. Soc.* **1992**, *114*, 7807–7813.
[134] D. Zhou, L. Gan, C. Luo, H. Tan, C. Huang, Z. Liu, Z. Wu, X. Zhao, X. Xia, S. Zhang, F. Sun, Z. Xia, Y. Zou, *Chem. Phys. Lett.* **1995**, *235*, 548–551.
[135] S. Ma, X. Lu, J. Chen, K. Han, L. Liu, Z. Huang, R. Cai, G. Wang, W. Wang, Y. Li, *J. Phys. Chem.* **1996**, *100*, 16629–16632.
[136] S. Ravine, F. LePecq, C. Mingotaud, P. Delhaes, J. C. Hummelen, F. Wudl, L. K. Patterson, *J. Phys. Chem.* **1995**, *99*, 9551–9557.
[137] D. M. Guldi, Y. Tian, J. H. Fendler, H. Hungerbühler, K.-D. Asmus, *J. Phys. Chem.* **1995**, *99*, 17673–17676.
[138] U. Jonas, F. Cardullo, P. Belik, F. Diederich, A. Gügel, E. Harth, A. Herrmann, L. Isaacs, K. Müllen, K. H. Ringsdorf, C. Thilgen, P. Uhlmann, A. Vasella, C. A. A. Waldraff, M. Walter, *Chem. Eur. J.* **1995**, *1*, 243–251.
[139] Y. Tian, J. H. Fendler, H. Hungerbühler, D. M. Guldi, K.-D. Asmus, *Supramolecular Science* **1997**, in press.
[140] L. M. Goldenberg, G. Williams, M. R. Bryce, A. P. Monkman, M. C. Petty, A. Hirsch, A. Soi, *J. Chem. Soc., Chem. Commun.* **1993**, 1310–1312.
[141] D. Zhou, L. Gan, C. Luo, H. Tan, C. Huang, G. Yao, X. Zhao, Z. Liu, X. Xia, B. Zhang, *J. Phys. Chem.* **1996**, *100*, 3150–3156.
[142] C. J. Hawker, P. M. Saville, J. W. White, *J. Org. Chem.* **1994**, *59*, 3503–3505.
[143] L. Isaacs, A. Ehrsig, F. Diederich, *Helv. Chim. Acta* **1993**, *76*, 1231–1250.
[144] D. A. Leigh, A. E. Moody, F. A. Wade, T. A. King, D. West, G. S. Bahra, *Langmuir* **1995**, *11*, 2334–2336.
[145] F. Diederich, U. Jonas, V. Gramlich, A. Herrmann, H. Ringsdorf, C. Thilgen, *Helv. Chim. Acta* **1993**, *76*, 2445–2453.
[146] M. Maggini, A. Karlsson, L. Pasimeni, G. Scorrano, M. Prato, L. Valli, *Tetrahedron Lett.* **1994**, *35*, 2985–2988.
[147] M. Maggini, L. Pasimeni, M. Prato, G. Scorrano, L. Valli, *Langmuir* **1994**, *10*, 4164–4166.
[148] M. Matsumoto, H. Tachibana, R. Azumi, M. Tanaka, T. Nakamura, G. Yunome, M. Abe, S. Yamago, E. Nakamura, *Langmuir* **1995**, *11*, 660–665.
[149] H. M. Patel, J. M. Didymus, K. K. W. Wong, A. Hirsch, A. Skiebe, I. Lamparth, S. Mann, *J. Chem. Soc., Chem. Commun.* **1996**, 611–612.
[150] C. N. C. Maliszewskyj, P. A. Heiney, D. R. Jones, R. M. Strongin, M. A. Cichy, A. B. Smith, *Langmuir* **1993**, *9*, 1439–1441.
[151] D. M. Guldi, Y. Tian, J. H. Fendler, H. Hungerbühler, K.-D. Asmus, *J. Phys. Chem.* **1996**, *100*, 2753–2758.
[152] F. Arias, L. A. Godinez, S. R. Wilson, A. E. Kaifer, L. Echegoyen, *J. Am. Chem. Soc.* **1996**, *118*, 6086–6087.
[153] L. B. Gan, D. J. Zhou, C. P. Luo, C. H. Huang, T. K. Li, J. Bai, X. S. Zhao, X. H. Xia, *J. Phys. Chem.* **1994**, *98*, 12459–12461.
[154] C. Luo, C. Huang, L. Gan, D. Zhou, W. Xia, Q. Zhuang, Y. Zhao, Y. Huang, *J. Phys. Chem.* **1996**, *100*, 16685–16689.
[155] D. Vaknin, J. Y. Wang, R. A. Uphaus, *Langmuir* **1995**, *11*, 1435–1438.
[156] J. Y. Wang, D. Vaknin, R. A. Uphaus, K. Kjaer, M. Lösche, *Thin Solid Films* **1994**, *242*, 40–44.
[157] J. L. Anderson, Y.-Z. An, Y. Rubin, C. S. Foote, *J. Am. Chem. Soc.* **1994**, *116*, 9763–9764.
[158] D. M. Guldi, K.-D. Asmus, *J. Phys. Chem A* **1997**, *101*, 1472–1481.
[159] D. M. Guldi, H. Hungerbühler, K.-D. Asmus, *J. Phys. Chem.* **1995**, *99*, 13487–13493.
[160] K. Chen, W. B. Caldwell, C. A. Mirkin, *J. Am. Chem. Soc.* **1993**, *115*, 1193–1194.
[161] W. B. Caldwell, K. Chen, C. A. Mirkin, S. J. Babinec, *Langmuir* **1993**, *9*, 1945–1947.

[162] J. A. Chupa, S. Xu, R. F. Fischetti, R. M. Strongin, J. P. McCauley, A. B. Smith, J. K. Blasie, L. J. Peticolas, J. C. Bean, *J. Am. Chem. Soc.* **1993**, *115*, 4383–4384.

[163] X. Shi, W. B. Caldwell, K. Chen, C. A. Mirkin, *J. Am. Chem. Soc.* **1994**, *116*, 11598–11599.

[164] V. V. Tsukruk, L. M. Lander, W. J. Brittain, *Langmuir* **1994**, *10*, 996–999.

[165] V. V. Tsukruk, M. P. Everson, L. M. Lander, W. J. Brittain, *Langmuir* **1996**, *12*, 3905–3911.

[166] J. Z. Zhang, M. J. Geselbracht, A. B. Ellis, *J. Am. Chem. Soc.* **1993**, *115*, 7789–7793.

[167] D. Armspach, E. C. Constable, F. Diederich, C. E. Housecroft, J.-F. Nierengarten, *J. Chem. Soc., Chem. Commun.* **1996**, 2009–2010.

[168] F. Diederich, C. Dietrich-Buchecker, J.-F. Nierengarten, J.-P. Sauvage, *J. Chem. Soc., Chem. Commun.* **1995**, 781–782.

[169] T. Drovetskaya, C. A. Reed, P. Boyd, *Tetrahedron Lett.* **1995**, *36*, 7971–7975.

[170] D. M. Guldi, M. Maggini, G. Scoranno, M. Prato, *J. Am. Chem. Soc.* **1997**, *119*, 974–980.

[171] H. Imahori, K. Hagiwara, T. Akiyama, S. Taniguchi, T. Okada, Y. Sakata, *Chem. Lett.* **1995**, 265–266.

[172] H. Imahori, K. Hagiwara, M. Aoki, T. Akiyama, S. Taniguchi, T. Okada, M. Shirakawa, Y. Sakata, *J. Am. Chem. Soc.* **1996**, *118*, 11771–11782.

[173] S. I. Khan, A. M. Oliver, M. N. Paddon-Row, Y. Rubin, *J. Am. Chem. Soc.* **1993**, *115*, 4919–4920.

[174] D. Kuciauskas, S. Lin, G. R. Seely, A. L. Moore, T. A. Moore, D. Gust, T. Drovetskaya, C. A. Reed, P. D. W. Boyd, *J. Phys. Chem.* **1996**, *100*, 15926–15932.

[175] P. A. Liddell, J. P. Sumida, A. N. M. Pherson, L. Noss, G. R. Seely, K. N. Clark, A. L. Moore, T. A. Moore, D. Gust, *Photochem. and Photobio.* **1994**, *60*, 537–541.

[176] T. G. Linssen, K. Dürr, M. Hanack, A. Hirsch, *J. Chem. Soc., Chem. Commun.* **1995**, 103–104.

[177] M. Maggini, A. Donò, G. Scorrano, M. Prato, *J. Chem. Soc., Chem. Commun.* **1995**, 845–846.

[178] Y. Nakamura, T. Minowa, Y. Hayashida, S. Tobita, H. Shizuka, J. Nishimura, *J. Chem. Soc., Faraday Trans.* **1996**, *92*, 377–382.

[179] M. Rasinkangas, T. T. Pakkanen, T. A. Pakkanen, *J. Organomet. Chem.* **1994**, *476*, C6–C8.

[180] N. S. Sariciftci, F. Wudl, A. J. Heeger, M. Maggini, G. Scorrano, M. Prato, J. Bourassa, P. C. Ford, *Chem. Phys. Lett.* **1995**, *247*, 510–514.

[181] M. J. Shephard, M. N. Paddon-Row, *Aust. J. Chem.* **1996**, *49*, 395–403.

[182] R. M. Williams, J. M. Zwier, J. W. Verhoeven, *J. Am. Chem. Soc.* **1995**, *117*, 4093–4099.

[183] R. M. Williams, M. Koeberg, J. M. Lawson, Y.-Z. An, Y. Rubin, M. N. Paddon-Row, J. W. Verhoeven, *J. Org. Chem.* **1996**, *61*, 5055–5062.

[184] T. Akiyama, H. Imahori, A. Ajawakom, Y. Sakata, *Chem. Lett.* **1996**, 907–908.

[185] N. A. Kotov, D. M. Guldi, J. H. Fendler, to be published.

[186] H. Nagashima, Y. Kato, H. Satoh, N. Kamegashima, K. Itoh, K. Oi, Y. Saito, *Chem. Lett.* **1996**, 519–520.

[187] M. Ouyang, K. Z. Wang, H. X. Zhang, Z. Q. Xue, C. H. Huang, D. Qiang, *Appl. Phys. Lett.* **1996**, *68*, 2441–2443.

[188] H. Murakami, Y. Watanabe, N. Nakashima, *J. Am. Chem. Soc.* **1996**, *118*, 4484–4485.

Chapter 7

Metal Colloids in Block Copolymer Micelles: Formation and Material Properties

L. Bronstein, M. Antonietti, and P. Valetsky

7.1 Introduction

Over the last few years, interest in the synthesis and properties of colloidal metal particles and metal clusters has steadily grown because of their unique properties, and great expectations have been raised for the application of metal colloids as catalysts, semiconductors, or magnetic fluids with a switchable rheology [1–3]. For the noble metal colloids, the interest mainly focuses on catalytic applications [3]. The advantages of colloidally dispersed noble metals are the advantages of colloidal particles in general: they have huge surfaces and they exhibit unique activities and spectroscopic features owing to the size quantization of most electronic properties. For colloids that are designed for special magnetic properties the size and shape of particles strongly control the type of magnetism, i.e. the whole transition from para- through superparamagnetic to ferromagnetic behavior can be adjusted by particle size.

The formation of metal colloids is, however, still far from being sufficiently well handled, and two of the main problems remain in the synthesis of metal colloids. On the one hand, it is still desired to gain a better control of the particle growth in terms of particle size, particle size distribution, and structure of the particles. On the other hand, the stabilization of the colloids is also far from being perfect, and quantities like durability in catalytic reactions are directly related to a more efficient particle stabilization.

Consequently, a variety of methods for the preparation and stabilization of metal colloids exists, the advantages and disadvantages of which have been discussed in detail elsewhere [4]. Examples are the preparation of metal particles in a polymeric environment (in both solids and liquids [5–7]), in vesicles [8], and in microemulsions [9–12]. To solve problems both of size control and of stabilization, recently another approach to the preparation of metal nanoparticles was proposed almost simultaneously by at least five different working groups [13–25].

The main concept of this approach is the formation of metal nanoparticles in some microheterogeneities created in polymers, which in this case can be regarded as a restricted reaction environment or a "microreactor". Usually, the micelle cores of amphiphilic block copolymers are taken as such a well-defined microheterogeneity, but mesophases of solid block copolymer films [13, 14] or the microphase-separated structures of polyelectrolyte gels [26] have been used too.

The block copolymer micelles, especially, formed by well-defined block copolymers with well-chosen block length and chemical composition in selective solvents turned out to be an excellent model system, and it is easy to delineate their advantages in colloid synthesis and properties, compared to the classical stabilization systems by surfactants or in microemulsions:

- the structural relaxation times of this nanostructured environment can be adjusted, i.e. they can be long compared to the metal colloid growth process;
- polymers as colloidal stabilizers are very effective steric stabilizers or electrostatic stabilizers;
- corrosion of metal particles or metal leaking can be avoided by binding the metal salts, the metal particles, but also possible side products to the polymeric microenvironment.

This chapter is mainly focused on the discussion of these possibilities, which are obtained with amphiphilic block copolymers as colloidal stabilizers. Its organization is as follows. In the first part, a literature review of current activities related to the application of block copolymers as stabilizers for metal colloids and semiconductor particles is given. In the second part, we discuss the chemical variety of block copolymers being applied in context with colloid synthesis and stabilization as well as some of the rules that govern micelle formation as well as metal salt uptake. We then summarize data on the synthesis of the metal colloids in an inverse phase situation or in inverse block copolymer micelles, i.e. the polymers are usually dissolved in organic solvents, whereas the colloid is embedded in the micelle core. Depending on the reaction pathway, three cases can be distinguished: the metal colloid is generated in the continuous (organic) phase, in an interface reaction, or in the micelle core. The last scenario is also called the "true nanoreactor" approach, since both size and properties are predominantly controlled by the architecture of the micelle core. In the last part, we focus on the synthesis of metal colloids in polar or aqueous media, which is presumably the technologically most relevant case, but which has also turned out to be the most complicated case. Here, the metal salts, being the source for both metal and semiconductor colloids, are usually present in both the continuous and the dispersed phase, which makes reaction handling rather complicated. This problem can, however, be attacked by the design and synthesis of appropriate new, so-called "double hydrophilic" block copolymers.

7.2 Overview of Current Activities on Amphiphilic Block Copolymers as Tailored Protecting Systems for Colloids

In 1992, Saito et al. [17] prepared silver colloids in a poly-2-vinylpyridine block of polystyrene-b-poly-2-vinylpyridine (PS-b-P2VP) by soaking partly cross linked film (to preserve the initial morphology) with Ag compounds and consequent reduction of silver from AgI.

In the same year, one of the present authors demonstrated the possibility of synthesizing narrowly distributed gold colloids in the micelles of polystyrene-b-poly-4-vinylpyridine (PS-b-P4VP) [18]. In this paper, the nanoreactor concept resulting in one colloid particle per micelle was demonstrated for the first time.

Cohen and Schrock and coworkers have elaborated the synthesis of Pt and Pd nanoclusters within microphase-separated diblock copolymers by ring-opening metathesis polymerization of nonbornene-derived organometallic complexes and η^3-1-phenylallyl, then followed by static casting of films and subsequent reduction of the organometallic complexes using the molecular hydrogen [15]. Au and Ag nanoclusters were prepared through chemical modification of precursor diblock copolymer by gold and silver compounds [13]. In this case, nanoclusters in microphase-separated blocks of different morphologies were produced by heating at 150 °C on the grid of an electron microscope.

In 1995, A. Eisenberg et al. [21, 22] used the block copolymer approach for the preparation of semiconductor particles (CdS, PbS). Changing the parameters of ionomers and inert blocks, they seemed to be able to control the size of growing CdS and PbS particles.

Recently [27], we took up the work on these systems by describing the synthesis of PS-b-P4VP block copolymers with varying chain lengths via anionic polymerization and the structure analysis of the block copolymer micelles in various selective solvents (solvents for PS). Such strongly segregated diblock copolymer systems practically do not exhibit both a critical micelle concentration and unimers in solution, which is typical for traditional surfactants [28].

To provide the microreactor concept, the block forming the micelle core should have groups that could coordinate to metal compounds. Since P4VP is a strong metal-chelating agent, we have suggested using this polymer to fix a large number of metal ions in the micelle core, the size of which only depends on the aggregation number [4, 23].

Nearly at the same time, M. Möller et al. [19, 20, 29] employed polystyrene-poly-2-vinylpyridines (PS-b-P2VP) and polystyrene-polyethylene oxides (PS-b-PEO) as media for the preparation of gold colloids. First, PS-b-P2VP were used only as an extracting system for metal colloids that were previously prepared in aqueous solutions [20]; here, the advantages of block copolymers were not completely used. Recently, the same authors have described the formation of metal colloids mainly in films and on the grid for electron microscope examination, observing the formation of both one colloid per micelle core and many colloids in micelles depending on the colloid preparation [19, 29].

In 1996, we extended the chemical basis of this approach by reporting a tailor-made synthesis of amphiphilic block copolymers with one metal binding block via polymer analogous conversion of simple polystyrene-b-polybutadiene (PS-b-PB) block copolymers. In this case, the chemical nature of the metal binding group was varied over a wider range, and substitution patterns that can act as donor or acceptor groups as well as multidentate ligands were employed [30, 31].

Another variation of the application of amphiphilic block copolymers was recently developed by A. Mayer and J. Mark [24, 25]. Some palladium, platinum, silver, and gold colloids were prepared by reducing the corresponding metal precursor, presumably in the continuous phase, in the presence of protective amphiphilic block copolymers. For this purpose, polystyrene-b-polyethyleneoxide (PS-b-PEO) and polystyrene-b-polymethacrylic acid (PS-b-PMAA) were applied in alcoholic solutions where these polymers form micelles with a PS core. Since the PS micelle core does not like metal salts, the metal colloids end up in the shell and partly on the surface of micelles. In this case, neither the size control of the microreactor concept nor the stabilization power of block copolymers are used, but from the viewpoint of catalytic applications, the formed particles in the corona might be accessible for substrates and can display high catalytic activity [25].

7.3 Chemistry of Amphiphilic Block Copolymers and Their Aggregation Behavior; Loading of the Micelles and Binding inside the Micelles

7.3.1 Amphiphilic Block Copolymers

To prepare the metal colloids in block copolymer micelles one can employ two approaches to the synthesis of amphiphilic copolymers with a block-type structure. One is the direct living polymerization of two monomers (for instance, styrene and 4-vinylpyridine or styrene and methyl methacrylate), resulting in the formation of well-characterized block copolymers. In case one component cannot be polymerized according to a living mechanism (anionic, cationic, or group transfer polymerization), macromonomer synthesis, or the capping with special end groups for restarting, chain transfer or termination is also possible.

Table 7.1 illustrates the list of monomers used by us in anionic copolymerization

Table 7.1. Structures of block copolymers synthesized by anionic polymerization.

Monomer	Polymer
4-vinylpyridine	
butadiene	
methacrylic acid	

with styrene and structures of block copolymers. Generally, living polymerization requires high purity during the reaction, tedious isolation procedures or/and the use of the protecting group chemistry. This makes such amphiphilic block copolymers, with a few exceptions, very expensive and prohibits many technological applications.

An alternative approach, chosen by us, is the chemical modification of a trivial block copolymer such as PS-b-PB. Besides price and simplicity, this approach also allows one to broaden the variety of amphiphilic block copolymers to some species that are not accessible by direct polymerization (Table 7.2) [30, 31]. The described amphiphilic block copolymers contained several heteroatoms per monomer that

Table 7.2. Structures of polybutadiene blocks after modification.

Structure of modified block

provided a good binding of metal salts in micelle cores. Metal colloids of gold, silver, palladium, and rhodium were prepared from these block copolymers.

7.3.2 Aggregation Behavior of Amphiphilic Block Copolymer Micelles

Since the structure and dynamics of the block copolymer aggregates, the micelles, play a predominant role in the size and morphology control of the metal colloids formed inside, such as in the ideal case – the use of the micelle cores as "nanoreactors", a better understanding of the underlying mechanisms of micelle formation is required to visualize the fundamental possibilities of this approach as well as the basis of an advanced morphology control.

Micelle formation is generally obtained in selective solvents; these are solvents for one of the polymer blocks. Most of the block copolymers described above form micelles either with the more polar or with the more unpolar block pointing outwards, i.e. regular or inverse micelles, depending on the polarity of the solvent. Since we are interested in obtaining stable assemblies with the colloid, a micelle morphology with the functional groups located in the micelle core is desired.

Such dissolved, swollen micelles of amphiphilic block copolymers usually exhibit diameters D_h such that 50 nm $< D_h <$ 200 nm, with polydispersity between 5% and 30% Gaussian width. Since block copolymer micelles are much more stable than surfactant assemblies, the micelle formation can be visualized by electron microscopy. Figure 7.1 shows a typical illustration characterizing these micelles; owing to the lack of contrast, the sample was shadowed with Pt/C.

The spherical shape of the micelles, as well as their rather narrow size distribution, is easily recognized; the different shades of gray are due to the formation of stable multilayers. For more quantitative considerations, similar experiments and static light scattering reveal that the size of these micelles is perfectly controlled by the length of the outer, dissolved block (N_A) and of the core-forming block (N_B) as well as the interface energy between core and solvent. The relation between aggre-

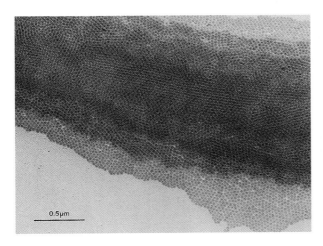

0.5μm

Figure 7.1. TEM micrograph of PS-b-P4VP micelles (reproduced by permission of *Nachr. Chem. Tech.* Lab.).

gation number Z as a function of these quantities for the limit of a high interface was experimentally found to be [30, 31]:

$$Z = Z_0 N_A^2 N_B^{-0.8} \tag{7.1}$$

$$Z_0 = 36\pi \frac{V_M^2}{A_M^3} \tag{7.2}$$

The quantity Z_0 contains all geometric characteristics (the monomer volume V_M and the stabilized interface area per molecule A_M) and is defined analogously to the surfactant ratio of low-molecular-weight surfactants. It was shown that these relations hold for chemically very different systems, such as nonionic surfactants and charged block copolymers [31].

7.3.3 Metal Salt Incorporation

The formation of colloidal particles in PS-b-P4VP micelles was studied very thoroughly and enabled the elaboration of the main principles of controlling the particle size. So the results on metal salt uptake obtained with these block copolymers as a model will be discussed in more detail in this chapter.

Already applied in the last section, a number of PS-b-P4VP block copolymers were synthesized that allowed us to change the micelle size and hydrodynamic characteristics over a rather wide range. Table 7.3 summarizes the molecular and micellar parameters of the applied set of PS-b-P4VP.

The dissolution of metal salts in the solution of PS-b-P4VP micelles that are otherwise insoluble in the solvent means the incorporation of these salts only in micelle cores due to coordination with corresponding groups of polymer, as described in previous papers [4, 23]. For example, with the introduction of $HAuCl_4 \cdot 3H_2O$ it was found that the micelle size stays the same after the incorporation of a metal compound (Table 7.4) or even after the reduction of $HAuCl_4$ with sodium borohydride in toluene solution (Table 7.5). We have observed such a stability of the micelles for all series of N_A and N_B (even when N_A is nearly equal to N_B) in PS-b-P4VP. This is easily understood in the quantitative description of the rules of micelle formation, where going beyond a critical value of the interface

Table 7.3. The characteristics of micelles derived from PS-b-P4VP in toluene.

Polymer	N_A*	N_B**	R_g (nm)	R_h (nm)	Z
PS-1,2	262	960	74.7	69.6	199
PS-3,2	35	122	10.2	12.3	34
PS-3,3	45	122	8.3	16.5	54
PS-3,4	63	122	22.9	21.1	123
PS-6,2	65	118	16.0	18.5	147
PS-6,3	123	118	12.5	24.5	310

Table 7.4. The characteristics of PS-b-P4VP micelles filled with HAuCl₄.

Polymer	N_A	Z	R_h (nm)	R_h (+ HAuCl₄) (nm)
PS-1,2	262	199	69.6	68.5
PS-3,4	63	123	21.1	20.5
PS-3,3	45	54	16.0	16.2
PS-3,2	35	34	12.5	13.7

Table 7.5. Hydrodynamic radius of PS-b-P4VP micelles before and after reduction with NaBH₄.

Polymer	Au:N	R_h(HAuCl₄)	R_h(after reduction)
PS-1,2	1:3	68.5	70.2
PS-3,4	1:3	20.5	20.9
PS-3,4	1:9	20.7	21.0
PS-6,3	1:3	24.7	24.9

energy between micelle core does not change the micellar characteristics; there is a saturation phenomenon in the "superstrong segregation limit" [32, 33]. For PS-b-P4VP, a deviating behavior where micelle formation is enhanced by metal salts is only seen for very small polymer chains [33].

This is not true for PS-b-P2VP (with $N_A = N_B = 190$), where the authors observed the change of micellar characteristics after incorporation of different amounts of HAuCl₄, i.e. the micelle size increases with the amount of gold compound [29]. Later, it will even be shown that micelle formation actually requires the presence of salt. In this "weak segregation limit," the micelle size delicately depends on changes of the polarity. Since such changes also occur during colloid formation, direct morphology control is much harder to obtain in this limit.

The interaction of metal ions with 4VP units was studied by FTIR spectroscopy. For all metal salts, except of strong acids (e.g. HAuCl₄), similar changes proceed. The FTIR spectrum of the polymer prepared by interaction of PS-1,2 with AuCl₃ at a feeding ratio Au:N = 1:3 shows the weakening of bands characteristic of pyridine units (1555 and 1415 cm^{-1}) and appearance of new bands at 1635 and 1616 cm^{-1} that are responsible for vibrations of pyridine complexes with metal salts. The ratio 4VP:Au = 1:3 is the experimentally observed solubilization limit for AuCl₃ in these blocks; a further excess stays insoluble in reaction solution, even after some days of stirring.

Pd colloids in block copolymer micelles have been studied via complex formation of 4VP units with Na₂PdCl₄ and Pd(CH₃COO)₂. The former is strongly insoluble in toluene, and the penetration into micelles happens very slowly (for 3–7 days) depending on the feeding molar ratio Pd:N. The solubility of Pd(CH₃COO)₂ in toluene provides its fast dissolution in reaction media. FTIR spectra of both palladium-containing block copolymers show the weakened bands (1555 and 1415 cm^{-1}) of 4VP and the appearance of new bands at 1616 and 1431 cm^{-1} assigned to complexes of 4VP units with metal salts.

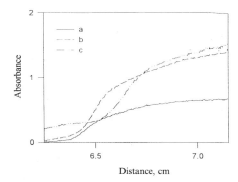

Figure 7.2. Sedimentation of PS-3,4-Pd(II) micellar solutions (toluene, 4 g l^{-1}) in ultracentrifuge with a rate of 10 000 U min^{-1} (UV detection: l = 400 nm) for samples Pd:N equal to: (a) 1:5, (b) 1:3, (c) 1:2.

Because the Pd(CH$_3$COO)$_2$ is soluble in toluene, it can be present both inside and outside micelles. To check whether the soluble Pd salt is consumed by the micelle core, ultracentrifugation was employed. Since the block copolymer/metal salt hybrid solution exhibits a yellow color, the simple sedimentation run where the optical absorption is measured at 400 nm permits one to judge the location of the salt. Ultracentrifugation was performed for three solutions with molar ratios Pd/4Vp:1/2, 1/3, 1/5. In just 4 hours complete uptake of the salt by micelles was achieved if the molar ratio Pd/4VP did not exceed 1/3 (Figure 7.2).

For HAuCl$_4$ · 3H$_2$O, besides complexation with metal salt, the protonation of vinylpyridine units by HCl must be considered. Indeed the interaction of PS-1,2 with HCl (water solution) induces a significant change in FTIR spectra. It leads to vanishing bands at 1555 and 1415 cm^{-1} and the appearance of some new strong bands at 1637, 1618, and 1506 cm^{-1}. The FTIR spectrum of a polymer prepared by the interaction of PS-1,2 and HAuCl$_4$ contains a set of bands that can be assigned both to complexation (as with AuCl$_3$) and to protonation reaction. The relative importance of these two binding mechanisms depends on metal load and time; within some hours and using protonation only, a limiting molar ratio Au:N = 1:1 can be obtained. The FTIR spectrum of Au polymer with molar ratio 1:1 does not consist of bands that are characteristic of pyridine units but contains bands that can be assigned to protonated 4VP units and their complexes with the gold salt.

7.4 Synthesis of Metal Colloids in the Presence of Amphiphilic Block Copolymers in Organic Solvents

7.4.1 Synthesis of Metal Colloids Inside the Micellar Cores: The Nanoreactor Concept

The formation of metal colloids from metal-salt-loaded micelles depends strongly on the reducing agent because the relative rate of nucleation to growth of colloids

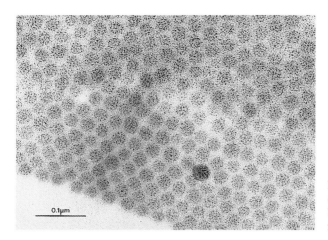

0.1μm

Figure 7.3. TEM micrograph of Au colloids produced with NaBH$_4$ reduction in PS-b-P4VP micelles.

manages the particle size. If we can manage to differentiate the growth of different crystal surfaces by interaction with the block copolymer, the particle morphology can also be controlled.

In the following sections, we have to distinguish between the reduction with sodium borohydride (NaBH$_4$) or superhydride (LiBEt$_3$H), which provides fast, homogeneous nucleation, and the reducing agent triethylsilane, (Et)$_3$SiH, which is a homogeneous reducing agent and acts very slowly.

7.4.2 Fast Homogeneous Reduction

Reduction of metal salts in block copolymers micelles with sodium borohydride and superhydride can be illustrated by the example of gold particles formation.

Using TEM in all cases of sodium borohydride (or superhydride) reduction we observe the micelles to be filled with a lot of metal particles that are strongly stabilized by micelle cores (Figure 7.3). It should be underlined that such a morphology, which we have called the "raspberry", is very important from the point of view of catalysis, and it can be one of the main advantages of preparing metal colloids in block copolymer micelles cores because in this case very small active particles can be stabilized.

In the SAXS curve of the NaBH$_4$-reduced gold colloids (Figure 7.4) below $s = 0.1$ nm^{-1}, clusterlike scattering is observed that lies beyond the resolution of the diffractometer. At higher s values, we observe a less structured fading of the scattering curve that we know from the aggregation of clusters. The proposed raspberrylike arrangement of single colloids in one well-defined micelle core results in a similar scattering curve.

Using wide-angle X-ray scattering of the gold colloids prepared in PS-6,3 with a molar ratio N:Au = 3:1 and reduced by superhydride, it was found that even small colloids have a crystalline structure.

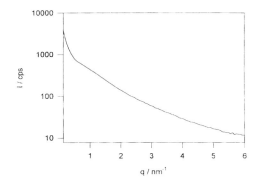

Figure 7.4. SAXS curve of the NaBH$_4$-reduced gold colloids.

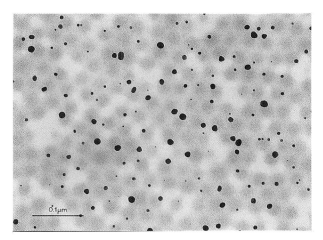

Figure 7.5. TEM micrograph of gold colloids reduced with triethylsilane in PS-6,3-Au (N:Au = 6:1).

It must be mentioned that the relative rates of nucleation and growth depend on metal type. Pd colloids prepared by NaBH$_4$ reduction are significantly smaller compared to gold particles.

7.4.3 Slow Homogeneous Reduction

Obviously, fast reduction by superhydride and NaBH$_4$ cannot provide the subtle size control of the metal colloids via the size of the micellar core because of the many particles formed per micelle core, and the size of the particles being mainly limited by the rate of nucleation. For direct control of the particle size (the "nanoreactor") we have used triethylsilane as the reducing agent. It is homogenous and works fairly slowly (2 hours for gold colloids).

The morphology of the particles in the case of silane reduction was studied by means of PS-6,3 forming very stable micelles at molar ratios N:Au = 3:1 and 6:1. Figure 7.5 shows an ELMI picture where it is seen that triethylsilane reduction re-

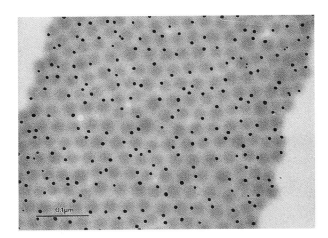

Figure 7.6. TEM micrograph of gold colloids reduced with triethylsilane in PS-6,3-Au (N:Au = 6:1) with a cross linked micelle core.

sults mainly in the formation of one colloid per micelle. Just a few micelles are empty, and others contain two colloids. The last observation might be caused by a collision-induced exchange of the colloidal core (the situation is even more complicated and presumably works over an intermediate micelle fusion: see Reference [34]). To prevent such events and to keep the micelle cores as molecular entities, the micelle cores were slightly cross linked with *p*-xylylenedibromide (5% per 4VP). To provide a higher mobility inside micelle cores a small percentage of methanol was additionally added to toluene. Figure 7.6 shows the electron micrograph of PS-6,3-Au (N:Au = 6:1) with a cross linked micelle core, which was reduced by triethylsilane.

With these two experimental tricks, the formation of one particle per micelle is obtained. The size of these colloids was determined with SAXS to be $d = 4.0$ nm with a Gaussian width $s = 0.19$, which exactly corresponds to the metal load in each micelle core prior to reduction. In this case, the colloid size control was really provided by the micelle architecture, i.e. we observe fulfillment of the "true nanoreactor."

7.4.4 Generation of the Metal Colloids with Heterogeneous Interface Reactions

Because hydrazine as a reduction agent forms a second droplet phase in the toluene solution of the block copolymer micelles, a complicated course of reduction has to be expected. Indeed for the hydrazine reduction of $HAuCl_4$ in PS-b-P4VP micelles a significant change in UV spectra was observed throughout the course of reduction (Figure 7.7). In the beginning of the reaction, immediately after addition of the reducing agent, the spectrum is characterized by a split plasmon resonance and violet color, which is usually related to unstable gold colloids. The violet gold colloids of the present experiments are, however, far from being unstable: once separated from the hydrazine, they can be stored for months without any change of the spectral or colloidal properties [35].

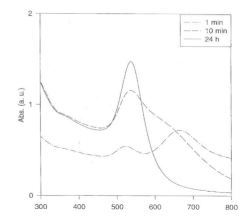

Figure 7.7. UV spectra of PS-3,2-Au (N:Au = 9:1) reduced with hydrazine at different stages of the reduction process (reproduced by permission of *Coll. & Polym. Sci.*).

After 10 minutes of reduction, the noble metal colloids are characterized by a broad, single plasmon resonance, having a tail toward longer wavelengths. This band continuously narrows with further progress of the reaction and is slightly shifted towards shorter wavelengths. In just 30 minutes the tail becomes less pronounced, and the 24 h measurement shown in Figure 7.7 is typical for the fully reacted particles. It must be underlined that the observed phenomena seem to be rather general: similar spectra and a related time dependence were already described for other synthetic approaches to well-defined and stable gold colloids [36, 37]. In Reference [37], the occurrence of the second, long-wavelength absorption was explained by the existence of "large, fluffy aggregates," which enable electronic coupling of the plasmon resonance between different particles. This follows the classical interpretation, where these experiments are to be described as a deaggregation and decreasing particle size with passage of time, but it is really difficult to imagine that colloids shrink with time.

To clarify this phenomenon, high-resolution transmission electron microscopy was performed with the same samples where the UV spectra were recorded. The corresponding electron micrographs are presented in Figure 7.8. It turned out that gold colloids formed after 1 minutes reduction period are not spherical: all of them have about the same shape, which is clearly anisometric and sometimes slightly bent. Owing to the typical shape and the violet color of their solutions, we have called this type of colloidal architecture "aubergine" morphology.

A more careful look at the micrograph reveals that each "aubergine" consists of two, sometimes three, primary, close-to-spherical particles that seem to be aggregated and glued together. The longish shape might be a memory that this colloid nucleation and aggregation took place on the two-dimensional surface of the heterophasic reducing agent, the hydrazine droplets. Since all particles have about the same size and shape, this aggregation does not occur at random but is obviously heavily controlled by the block copolymers in the hydrophobic half-room near the colloids.

After 10 minutes, the aubergine morphology begins to vanish, and the steady progress of reduction leads to the reappearance of spherical particles. Finally all

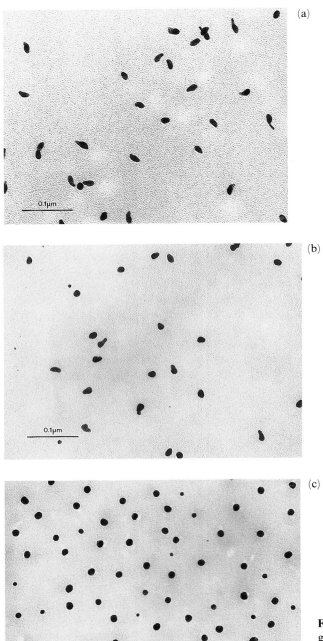

Figure 7.8. TEM micrographs of gold colloids reduced with hydrazine after different reduction times: (a) 1 min; (b) 10 min; (c) 24 h (reproduced by permission of *Coll. & Polym. Sci.*).

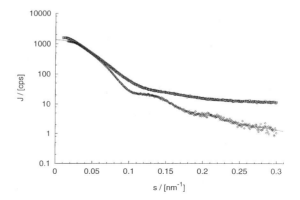

Figure 7.9. SAXS diffractograms of PS-3,2-Au (N:Au = 9:1) after 1 min (upper curve) and 24 h (lower curve) reduction with hydrazine. The straight line represents a fit with the form factors of a sphere (reproduced by permission of *Coll. & Polym. Sci.*).

particles are nearly spherical and are significantly smaller than the aubergines, but slightly larger than aubergine-constituting particles. A statistical evaluation of the electron micrographs leads to the conclusion that aubergine aggregates exhibit a mean size of about 24×10 nm, including nuclei with anestimated diameter of roughly 8 nm, while the 24 h colloids show a sphere diameter of 12 nm.

Our model of this process is that the nonreduced salt in the early stages of reaction acts as a glue for the formation of aubergines; during the progress of reduction the salt is consumed and leads to the increase in size of the primary particles, coupled with a deaggregation process. Further progress in reaction only results in a ripening of the gold colloids: they become slightly bigger and more monodisperse. From electron microscopy, we find no evidence that the observed blueshift of the plasmon band and the end of the reaction is still due to continued deaggregation. Thus, we can conclude that the wavelength change has to be due to a change in the dielectric environment of each particle.

The colloidal particles formed at different stages of the reduction process were also examined by X-ray analysis [35]. The small-angle diffractograms of the auberginelike colloids and the spherical colloids present at the end of the reduction are compared in Figure 7.9.

The scattering from the particles produced in the late stages of reaction are perfectly described with a form factor of a sphere with sharp, well-defined interface and a homogeneous density. The fit of the experimental curve with this structure model (straight line) works almost perfectly and results in a mean diameter of $d = 12.5$ nm and a Gaussian width of the size distribution of 13%.

The scattering of the auberginelike particles is remarkably different and can be described by the form factor of prolate ellipsoids. Since the polydispersity and axial ratio have a similar influence on the scattering curve of such particles, a quantitative fit of these data only reproduces the input information of electron microscopy.

It must be underlined that the metal colloids are far too large to result solely from the metal load of one micelle, and the proposed heterogeneous reduction scheme employing opened micelles offers an explanation for this difference. On the other hand, overstoichiometric load of some micelles requires the coexistence of empty micelles too. First experimental indication for those free micelles was already seen in electron microscopy experiments, where background textures typical for thin block copolymer films were observed.

The most elegant and simple proof of coexistence of free micelles was received from ultracentrifugation with continuous measurement of the optical absorption at two different wavelengths [35]. The profile of the optical density at the wavelength of the polymer absorption indicates that the colloidal solution contains two different, but well-defined, species with different sedimentation coefficients. The mass of block copolymer bound in these structures is about balanced, as taken from the absolute values of the step heights. On the other hand, the absorption at the wavelength of the plasmon band was observed from more dense species with the higher sedimentation coefficient only. So both free and filled micelles exist.

7.4.5 Homogeneous Colloid Production and Heteroaggregation with Amphiphilic Block Copolymer Micelles

Here we would like to depict the case observed for Co nanoparticles prepared from dicobalt octacarbonyl, $Co_2(CO)_8$. $Co_2(CO)_8$ is soluble in toluene, and it is straightforward to assume that it is present both inside and outside the micelles. We would not expect that $Co_2(CO)_8$ is consumed by micelle cores because zero-valent Co cannot strongly coordinate with 4VP units. However, FTIR spectra of PS-b-P4VP after the addition of $Co_2(CO)_8$ in the carbonyl region does not correspond to $Co_2(CO)_8$ [38] and shows mainly one wide band at 1884 cm^{-1}, which is characteristic of the anionic species $Co(CO)_{4-}$ of cationic–anionic complexes that can form in DMF [6] and pyridine [39]. Therefore, on the basis of the FTIR data, we suggest that dissolution of $Co_2(CO)_8$ in block copolymer solution leads to the formation of a similar cationic–anionic complex with VP units with the structure $[Co(VP)_6]^{2+}[Co(CO)_4]_{-2}$, i.e. three Co atoms per six 4VP units can really be fixed to the micelle core, the rest acting as a homogeneous Co source.

To clarify the shape and size of particles formed TEM was employed. Figure 7.10 shows TEM images of Co particles prepared in PS-1,2 micelles after $Co_2(CO)_8$ incorporation at three feeding ratios N:Co = 1:1 (a), 1:2 (b), and 1:3 (c). For a feeding ratio of N:Co = 1:1 mainly spherical particles form (a slight asymmetry is seen) with a mean diameter of 10 nm with a standard deviation of 1.5 nm. The feeding ratio N:Co = 1:2 results in the formation of fluffy, starlike aggregates (about 20–23 nm in diameter and having an irregular shape), which seem to consist of primary, anisometric clusters. The situation changes again for the feeding ratio N:Co = 1:3, which produces two kinds of particle: smaller spheres and bigger cubes. Here, the statistical evaluation of the micrographs gave a mean diameter of spheres of 10 nm, (the same size and polydispersity as the spherical particles with the 1:1 feed), and cubes with mean side of 21 nm (with a standard deviation of 1.5 nm). The size distribution for both kinds of particle is quite narrow. In some pictures it is seen that the cubic particles have a strong tendency to chain or trail formation, with a lining up of a number of cubes.

From TEM micrographs it is also seen that the spherical particles (N:Co = 1:1)

(a)

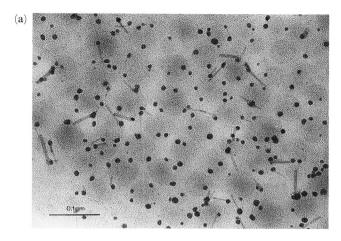

(b)

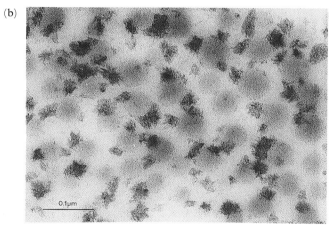

(c)

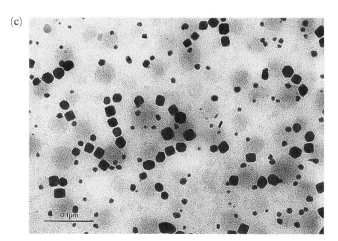

Figure 7.10. TEM micrographs of Co nanoparticles prepared in PS-1,2 at molar ratios N:Co = 1:1 (a), 1:2 (b), 1:3 (c).

are mainly located in the micelle cores. Even the larger stars (N:Co = 1:2) seem to start their growth from the cores. At the molar ratio N:Co = 1:3, only the cubic particles are placed outside of micelle cores, while the spherical particles still form in the cores.

Obviously, the feeding ratio heavily influences the particle size and shape, but in all cases the polymeric amphiphiles take part in the particle formation, and the following reaction scenario can be given. $Co_2(CO)_8$ is present both in solution and in the micelle core (complex). After rapid heating to 110 °C (decomposition starts at 60 °C), elementary Co is formed that quickly aggregates to Co clusters and spherical particles. The formation of such particles is strongly controlled by interaction with 4VP units. For higher Co loads, the $Co_2(CO)_8$ present in the continuous phase acts as a Co source that feeds further growth of the particles, and either spherical colloids 10 nm in size or starlike clusters are obtained. Only in the case of a larger Co excess do colloids also nucleate in the continuous solvent phase, and cubic colloids are formed.

A similar phenomenon, i.e. the influence of the molar ratio of a polymer/metal compound on the shape of particles formed was described in Reference [40] for the preparation of Pt colloids in an aqueous medium. Changing the loading of K_2PtCl_4 towards polyacrylic acid, the authors changed the shape from spheres to cubes and triangles. The related shape control opens the great opportunity for regulating both magnetic and catalytic properties.

7.5 Synthesis of Metal Colloids in the Presence of Amphiphilic Block Copolymers in Water or Related Polar Solvents

7.5.1 Micelle Formation due to Hydrophobic/Hydrophilic Block Copolymers and Interaction of Metal Salts with the Hydrophilic Shell

This approach was recently developed by A. Mayer and J. Mark [24, 25]. It was mentioned in Section 7.2 that two block copolymers, PS-b-PEO and PS-b-PMAA, were used for the preparation of platinum, silver, and gold colloids in alcoholic solutions where these polymers form micelles with a PS core. The chosen PS-PEO contained a long PS block ($M_n = 29\,800$) that forms a large glassy micelle core, while the PEO block was shorter ($M_n = 8\,400$). This situation allows the formation of metal colloids in the shell only, and the block copolymer micelle is not a closed "nanoreactor." The formed colloids seem to precipitate onto the glassy core, and from the viewpoint of a potential catalytic application, such particles might be better accessible for the substrates and might display a high catalytic activity [25].

We applied the similar approach for the stabilization of various metal colloids

(Au, Pd, Pt, Rh, Cu, Ni) in water in the presence of a short, nonionic PS-PEO block copolymer (SE1030 from Goldschmidt AG with $M_N^{PS} = 1\,000$ and $M_N^{PEO} = 3\,000$). The choice of water is obviously the best from the viewpoint of environmentally friendly processes for catalyst preparation. Because the PS block in this case is short, the micelle core is still in the liquid state, and we expect that it could step in stabilization and the shape control owing to hydrophobic interactions between the metal colloid and the hydrophobic core. The reduction of a number of metal salts in the presence of SE1030 leads to the formation of metal colloids that were characterized by UV–vis spectroscopy (for Au, Rh, and Cu) or TEM (other types). Depending on metal type and because of the rather small polyethyleneoxide chains employed for steric stabilization, the stability of the colloid is mediocre but usually extends over some days. This is sufficient for handling and deposition on inorganic supports. Ultracentrifugation measurements carried out for the most stable system of this series, gold colloids, revealed that the sedimentation of the block copolymer micelles and metal colloids proceeds separately. This means that the binding between the amphiphilic blocks and the colloids and the related stabilization in such systems is rather weak, and more effective binding requires additional functional groups carried by the metal binding block.

7.5.2 Colloid Synthesis In "Double-Hydrophilic" Block Copolymers

To avoid this stabilization failure of amphiphilic block copolymers and the formation of particles in the shell only (in aqueous media) we have proposed the use of "double-hydrophilic" block copolymers for metal colloid formation where one of the water soluble blocks is able to coordinate the metal while the other, noninteracting block provides a remaining good solubility in water. For this, polyethyleneoxide-b-polyethyleneimine (PEO-b-PEI) was synthesized, and its interaction with metal salts was studied [41]. Using DLS it was found that PEO-b-PEI in water forms no micelles but exhibits at higher concentrations composition fluctuations that are rather dynamic structures. The correlation length of these fluctuations depends on the polymer concentration, is sensitive to ultrasonification (i.e. a bicontinuous structure), and varies in the range 300–500 nm. The low absolute scattering intensity of the PEO-b-PEI solutions underlines the low modulation depth of the fluctuations. Since the PEI block contains NH groups that are able to coordinate with metal ions [42], a sharpening of these fluctuations or micelle formation in water has to be expected.

It was found that the introduction of $AuCl_3$, H_2PtCl_6, and $PdCl_2$ indeed induces micelle formation. The characteristics of micelles (Table 7.6) were found to depend on the metal type and molar ratio of the polymer/salt. Very big and broadly distributed aggregates were detected for $PdCl_2$, while loading with H_2PtCl_6 results in narrowly distributed micelles. The micelle size for this system increases with increasing amounts of H_2PtCl_6. The reduction of H_2PtCl_6 with hydrazine leads to a decrease in the size of the compound system; H_2 reduction results in practically no changes.

Table 7.6. The micellar characteristics of PEO-b-PEI micelles in water after addition of metal salts and their reduction.

Polymer system	Molar ratio NH:metal	D_h (nm)	σ (%)
PEO-b-PEI + H$_2$PtCl$_6$ (salt)	8:1	137.9	32.1
	4:1	154.3	28.9
	2:1	176.2	31.8
	1:1	200.8	29.9
PEO-b-PEI + H$_2$PtCl$_6$[a]	4:1	110.0	23.6
PEO-b-PEI + H$_2$PtCl$_6$[b]	4:1	148.9	26.6
PEO-b-PEI + PdCl$_2$	4:1	232.1	56.7
PEO-b-PEI + PdCl$_2$[a]	4:1	188.1	61.2
PEO-b-PEI + AuCl$_3$	24:1	60.0	53.9
(self-reduction, 5 hours)	18:1	61.8	60.9
	12:1	62.5	54.4

[a] Reduced with hydrazine.
[b] Reduced with H$_2$.

For this well-defined model system, some additional dependencies have been tested, for instance the dependence of the size of metal colloids formed on the type of reducing agent. It is clearly seen with transmission electron microscopy that Pt particles in PEO-b-PEI reduced by H$_2$ (Figure 7.11(a)) and hydrazine (Figure 7.11(b)) exhibit a very different morphology. Hydrogen reduction, which proceeds rather slowly, provides the formation of small, well-defined particles (with a mean diameter of about 3–5 nm), whereas the hydrazine reduction lead to big aggregates (30–50 nm in size), which, however, are constructed from very tiny particles (1–2 nm). Obviously, the faster hydrazine reduction leads to smaller particles the large surfaces of which are imperfectly stabilized and are inclined to aggregate, while particles formed during hydrogen reduction seem to be large enough for the amount of stabilizer present in the reaction medium.

Using TEM it is also found that in this special system the particle size of Pt colloids reduced with H$_2$ depends on metal loading: the smaller the amount of metal salt added, the smaller the particles. In addition, a change in Pt loading was found to control not only the particle size but also the particle shape. For a molar ratio NH/Pt = 4/1, only nearly spherical particles were observed (Figure 7.11(a)). Opposite to this, the TEM micrograph (PEO-b-PEI-Pt; NH/Pt = 8/1) presented in Figure 7.11(c) displays the appearance of cubes and triangles along with rather spherical Pt particles. Such nonspherical morphologies are rather interesting, since they disclose higher amounts of a specific crystal plane, which might be coupled to special catalytic activities. A similar influencing of the morphology of Pt colloids by polymer templates was also reported in Reference [40], however, this was active only at much lower concentrations. Our system allows the preparation of such particles at higher concentrations of reagents (by at least one order of magnitude), which is a necessity for technical catalytic applications.

(a)

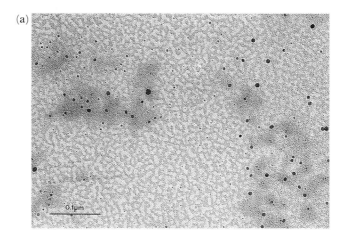

(b)

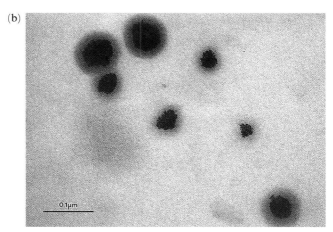

(c)

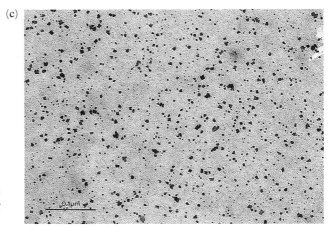

Figure 7.11. TEM micrographs of Pt colloids prepared in PEO-b-PEI at a molar ratio NH/Pt = 4/1 and reduced with H$_2$ (a) and hydrazine (b) and at a molar ratio NH/Pt = 8/1 after H$_2$ reduction (c).

7.6 Catalytic Properties of Metal Colloids Stabilized by Amphiphilic Block Copolymers

Because one main application for noble metal colloids lies in catalysis, the metal colloids prepared in the micelles of amphiphilic block copolymers have been studied in hydrogenation. Such hybrid systems turned out to be very active, and their activity was found to depend strongly on a number of parameters that are closely related both to colloid morphology and to the chemical surrounding of the colloids.

As a model reaction, the hydrogenation of cyclohexene to cyclohexane was chosen. Because the dependence of cyclohexane formation on time in the hydrogenation of cyclohexene with all systems prepared by us was linear for up to 75% of the conversion and no induction period was observed, the catalytic activity of metal colloids was compared with the amount of cyclohexane produced after 30 minutes of hydrogenation only. As a standard, an optimized technical catalyst, Pd on activated carbon (Aldrich), was used in all catalytic reactions.

The influence of the type of reducing agent on the activities of Pd colloids derived from PS-3,4 and $Pd(CH_3COO)_2$ is presented in Table 7.7. All systems behave according to our expectations, which relate the activity to the colloid size, i.e. the smaller the colloid size, the higher the reactivity. Table 7.7 shows that reduction by hydrazine-hydrate leads to the formation of practically inactive colloids, while reduction by $NaBH_4$ and superhydride results in Pd nanoparticles that exhibit a conversion of about 28–33%. Reduction by $(C_2H_5)_3SiH$ produces an intermediate result. In the context of variation of the reducing agent, it is worth noting that its remainders or reaction products after reduction might remain in the micelle core and could influence the catalytic properties very strongly [43].

The influence of the micelle size on catalytic activity was studied by means of the block copolymers presented in Table 7.8. One can note that though the subtle influence of micelle parameters was not observed, the transition from comparatively weakly segregated micelles (PS-1,2 and PS-3,4) to strongly segregated ones (PS-7,1 and PS-5,2) decreases the catalytic activity. The cross linking of the micelle core,

Table 7.7. Influence of reducing agent type on conversion of cyclohexene into cyclohexane for Pd colloids derived from PS-3,4 and $Pd(CH_3COO)_2$.

Reducing agent	Cyclohexane (%)
Without reducing agent	4.2
Superhydride	32.7
$NaBH_4$	28.3
$N_2H_4H_2O$	1.0
$(C_2H_5)_3SiH$	10.8
Superhydride (deposited on Al_2O_3)	45.5
Pd on activated carbon (Aldrich) (1 wt. % Pd)	40.0

Table 7.8. The influence of characteristics of block copolymers on catalytic activity of Pd colloids derived from $Pd(CH_3COO)_2$.

Polymer	N_A*	N_B**	R_h (nm)	Z	Cyclohexane (%)
PS-1,2	262	960	69.6	199	32.8
PS-3,4	35	122	12.3	34	32.7
PS-7,1	102	97	44.0	302	19.3
PS-5,2	123	145	66.5	551	16.1
PS-5,2*	123	145	66.5	551	1.0

* Cross linked.

which was carried out for better size control of the colloid growth (see Section 7.4), even leads to practically inactive species. Obviously, a minimal polymer mobility in the micelle cores surrounding the metal colloids is required.

The reactivity, activity, and selectivity of the catalysts is also given by the type of metal and the colloid composition (monometallic or bimetallic colloids). As shown in the literature [44–47], the formation of a bimetallic Pd–Au increases the activity of the Pd in such colloids in hydrogenation. To check this influence for the present systems, bimetallic colloids on the basis of Pd/Pt and Pd/Au were prepared too. The activities of these colloids in cyclohexene hydrogenation compared to four different pure noble metal colloids are given in Table 7.9. This table also contains the data of metal colloids stabilized in block copolymers deposited on activated Al_2O_3, which significantly increases the activity and stability in catalysis. The most active and stable catalyst within all described variations is a bimetallic colloid, $Au/Pd = 1/5$, produced in a raspberry morphology and deposited on Al_2O_3. The same catalyst turned out to be the best in the selective hydrogenation of 1,3-cyclohexadiene to cyclohexene. For hydrogenation of 1,3-cyclooctadiene, the examined Pd colloids in their block copolymer shell generally exhibit a 100% selectivity.

A number of our colloids compete well with the activity of the commercial cata-

Table 7.9. The catalytic activity of mono- and bimetallic colloids in the hydrogenation of cyclohexene.

Catalyst	Cyclohexane (%)
PS-3,4; $[Rh(CO)_2Cl]_2$	26.4
PS-3,4; $Pd(CH_3COO)_2$	32.7
PS-3,4; $K[Pt(C_2H_4)Cl_3]$	60.7
PS-3,4; $HAuCl_4 \cdot 3H_2O$	0.0
PS-3,4; $Pd(CH_3COO)_2$, $K[Pt(C_2H_4)Cl_3]$ (Pd:Pt = 4:1)	55.5
PS-3,4; $Pd(CH_3COO)_2$, $HAuCl_4 \cdot 3H_2O$ (Pd:Au = 4:1)	65.7
PS-3,4; $Pd(CH_3COO)_2$, $HAuCl_4 \cdot 3H_2O$ (Pd:Au = 5:1)	71.3
PS-3,4; $Pd(CH_3COO)_2$, $HAuCl_4 \cdot 3H_2O$ (Pd:Au = 5:1) deposited on Al_2O_3	72.5

lysts but show at the same time higher stability as well as selectivity. Considering these facts, the further development of similarly complex-structured noble metal colloids seems to be very promising.

7.7 Magnetic Properties of Co-colloids Stabilized by Amphiphilic Block Copolymers

Besides catalytic properties, two other applications are particularly attractive for metal colloids. On the one hand, these are semiconductor properties, which cover a lot of possible applications for colloidal particles [1]. Although semiconductors can also be prepared in block copolymer micelles [21, 22], we omit their discussion, since their synthesis and properties would require a separate chapter.

On the other hand, metal colloids are also interesting for magnetic materials the properties of which can be switched by changing the size and shape of particles. In this chapter we shall discuss only the magnetic properties of Co colloids studied, the synthesis of which was described above.

It is known that the magnetic properties of Co particles are strongly influenced by their size, which enables one to use ferromagnetic resonance (FMR) for the examination of particle size, especially as a correlation between FMR linewidth and particle size exists [48, 49]. Particles smaller than 1 nm, for instance, exhibit magnetic disorder [50] and produce no FMR signal.

The data of elementary analysis and FMR for the Co nanocrystals derived from PS-1,2 and CoCl$_2$ are summarized in Table 7.10. The very small Co colloids prepared by the superhydride reduction of CoCl$_2$ in block copolymer micelles indeed give no FMR signal. These samples were dried and heated in vacuum-sealed ampoules for 2 hours at 200 °C, which exceeds the glass transition temperature T_g of both blocks. After this heating cycle, the samples were reexamined by FMR. In

Table 7.10. The data of elemental analysis and FMR for fcc Co species derived from PS-1,2 and CoCl$_2$.

Sample	Co (wt. %)[a]	H_0 (Oe)	ΔH (Oe)	H_c (Oe)
PS-1,2-Co-1				
liquid				
toluene	2.34	–	–	0
solid	2.34	2975	650	425
PS-1,2-Co-2				
liquid, THF	8.26	2550[b]	1250[b]	0
solid	8.26	2950	500	675

[a] Very weak signal, which is responsible for particles formed outside of micelles.
[b] In the solid polymer film.

both cases such heating leads to the formation of spherical particles about 3–5 nm in size for PS-1,2-Co-1 and PS-1,2-Co-2 ($H_0 = 2950$ Oe; a value of 2975 Oe is typical for spherical particles [48]). Table 7.10 shows the very high values of coercive force for these samples. Similar values of H_c have been described for Co particles prepared in stabilizing polymers [51] at a tenfold-higher Co content (75 wt. %). The polymer materials derived from block copolymer micelles therefore possess very high values of the coercive force (despit the very low Co content), which is evidence of a very high specific magnetization density.

In the case of the thermolysis of $Co_2(CO)_8$ at 110 °C as a source of elementary Co, the magnetic properties as well as the shape and size of the formed Co nanoparticles can be controlled by the balance of 4VP/Co.

For a feeding ratio N:Co = 2:1 (for PS-3,4) the Co particles formed are neither ferromagnetic nor superparamagnetic. This corresponds to a particle size of less than 1 nm. At feeding ratios of PVP:Co of 1.5:1 and 1.75:1, spherical particles are formed: the position of the absorption maximum H_0 exactly corresponds to one of the spheres. Moreover, such molar ratios also lead to the preparation of superparamagnetic particles, since $H_c = 0$.

A further increase of the Co content in the reaction medium to and above the balance of N:Co leads to a change in the position of the absorption maximum H_0, which indicates the formation of anisometric particles (see Section 7.4.3). These samples exhibit ferromagnetic behavior: their coercive force varies from 250 to 475 Oe, depending on the Co content and polymer sample.

The high values of coercive force and magnetization, the ability to control particle size, shape, and the related type of magnetism, and the possibility of producing both very stable magnetic fluids and magnetic polymeric films and coatings adds another facet that makes block copolymer/metal colloid hybrids promising materials with high technological relevance.

7.8 Conclusion and Outlook

In this chapter we have described a novel approach for the preparation of nanoparticles in block copolymer micelles and those opportunities which such a means of stabilization provides, i.e. control of particle size and particle size distribution as well as particle shape and morphology. Nanoparticles prepared in block copolymers can fulfill the demands of any application that can be expected for nanoparticles depending on their nature. Noble metal monometallic and bimetallic colloids are of interest in many catalytic reactions; nanoparticles of magnetic materials formed in block copolymers are promising from the viewpoint of preparation of magnetic polymeric materials with switchable magnetic properties depending on the shape and size of the particles; plenty of semiconductor nanoparticles will meet many of the needs of microelectronics.

The use of different kinds of block copolymer enables applications both in organic and aqueous media to be catered for. For amphiphilic block copolymers in an organic medium this approach allows one to prepare hybrid systems containing one particle per block copolymer micelle whose size exactly matches the metal loading or micelles filled with many colloids that can be desirable for catalytic applications.

In water or another polar medium, amphiphilic block copolymers are not desirable for the introduction of metal salts in the micelle core. For this, double-hydrophilic block copolymers are recommended when the micelle core forms owing to interaction with metal compounds.

Because nanoparticles are stabilized in block copolymers where one block serves for the interaction with metal nanoparticles and another one provides the solubility, such stabilization has great endurance that also makes such nanodispersed polymeric systems highly promising. Unlike many other kinds of stabilizers such as vesicles, zeolites, and surfactant micelles except for polymeric ones, block copolymers form films and coatings, so nanoparticle-containing block copolymer hybrid systems can be prepared and used as solutions, powders, films, and coatings that provide a variety of possible applications. In our opinion, the further study of metal, metal oxide, and other metal compound nanoparticle formation in block copolymer micelles can develop strongly and modify some technologies based on nanoparticles.

References

[1] A. Henglein, *Chem. Rev.* **1989**, *89*, 1861–1873.
[2] S. Oggawa, Y. Hayashi, N. Kobayashi, T. Tokizaki, A. Nakamura, *Jpn. J. Appl. Phys.* **1994**, *33*, L331–L333.
[3] L. N. Lewis, *Chem. Rev.* **1993**, *93*, 2693–2730.
[4] M. Antonietti, E. Wenz, L. Bronstein, M. Seregina, *Adv. Mater.* **1995**, *7*, 1000–1005.
[5] L. M. Bronstein, P. M. Valetsky, *Inorg. & Organometal. Polym.* **1994**, *4*, 415–423.
[6] L. M. Bronstein, E. Sh. Mirzoeva, P. M. Valetsky, S. P. Solodovnikov, R. A. Register, *J. Mater. Chem.* **1995**, 5, 1197–1201.
[7] M. Harada, K. Asakura, N. Toshima, *J. Phys. Chem.* **1994**, 98, 2653–2662.
[8] K. Kurihara, J. H. Fendler, *Abstr. Am. Chem. Soc.* **1983**, *186*, 41.
[9] M. Boutonnet, J. Kizling, P. Stenius, G. Maire, *Colloids Surf.* **1982**, *5*, 209–225.
[10] R. Touroude, P. Girard, G. Maire, J. Kizling, M. Boutonet-Kisling, P. Stenius, *Colloids & Surf.* **1992**, *67*, 9–12.
[11] C. Petit, P. Lixon, M. J. Pileni, *Phys. Chem.* **1993**, 97, 12974–12983.
[12] P. Barnickel, A. Wokaun, W. Sager, H. F. Eicke, *J. Colloid Interface Sci.* **1992**, *148*, 80–90.
[13] Y. N. C. Chan, R. R. Schrock, R. E. Cohen, *Chem. Mater.* **1992**, *4*, 24–27.
[14] V. Sankaran, J. Yue, R. E. Cohen, R. R. Schrock, R. J. Silbey, *Chem. Mater.* **1993**, *5*, 1133–1142.
[15] Y. N. C. Chan, G. S. W. Craig, R. R. Schrock, R. E. Cohen, *Chem. Mater.* **1992**, *4*, 885–894.
[16] J. Yue, V. Sankaran, R. E. Cohen, R. R. Schrock, *J. Amer. Chem. Soc.* **1993**, *115*, 4409–4410.
[17] H. Saito, S. Okamura, K. Ishizu, *Polymer* **1992**, *33*, 1099–1101.
[18] M. Antonietti, S. Heinz, *Nachr. Chem. Lab. Tech.* **1992**, *40*, 308–314.
[19] J. P. Spatz, A. Roescher, M. Möller, *Adv. Mater.* **1996**, *8*, 337–340.
[20] A. Roescher, M. Möller, *Adv. Mater.* **1995**, *7*, 151–154.
[21] M. Moffit, L. McMahon, V. Pessel, A. Eisenberg, *Chem. Mater.* **1995**, *7*, 1185–1192.
[22] M. Moffit, A. Eisenberg, *Chem. Mater.* **1995**, *7*, 1178–1184.

[23] M. Antonietti, E. Wenz, L. Bronstein, M. Seregina, *PMSE Preprints*, Chicago **1995**, Vol. *73*, 283–284.

[24] A. B. R. Mayer, J. E. Mark, *Polymer Preprints*, New Orleans **1996**, Vol. *74*, 459–460.

[25] A. B. R. Mayer, J. E. Mark, *J. Coll. Polym. Sci.*, in press.

[26] L. M. Bronstein, O. A. Platonova, A. N. Yakunin, I. M. Yanovskaya, P. M. Valetsky, A. T. Dembo, E. E. Makhaeva, A. V. Mironov, A. R. Khokhlov, *Macromolecules*, in press.

[27] M. Antonietti, S. Heinz, C. Rosenauer, M. Schmidt, *Macromolecules* **1994**, *27*, 3276–3285.

[28] S. Förster, M. Zisenis, E. Wenz, M. Antonetti, *J. Chem. Phys.* **1996**, *104*, 9956–9968.

[29] J. P. Spatz, S. Sheiko, M. Möller, *Macromolecules* **1996**, *29*, 3220–3226.

[30] M. Antonietti, S. Förster, S. Oestreich, J. Hartmann, *Macromolecules* **1996**, *29*, 3800–3806.

[31] M. Antonietti, S. Förster, S. Oestreich, J. Hartmann, E. Wenz, *Nachr. Chem. Lab. Tech.* **1996**, *44*, 579–586.

[32] E. Wenz, doctoral thesis, Berlin **1996**.

[33] M. Antonietti, E. Wenz, S. Förster, to be published.

[34] S. Oestreich, M. Antonietti, to be published.

[35] M. Antonietti, A. Thünemann, E. Wenz, *Coll. & Polym. Sci.* **1996**, *274*, 795–801.

[36] P. Wilcoxon, R. L. Williamson, R. Baughman, *J. Chem. Phys.* **1993**, *98*, 9933–9950.

[37] M. K. Chow, C. F. Zukoski, *J. Coll. Inerfac. Sci.* **1994**, *165*, 97–109.

[38] K. Nakamoto, *Infrared and Raman Spectra of Inorganic and Coordination Compounds*, Wiley, New York **1986**, p. 168.

[39] R. A. Friedel, I. Wender, S. L. Shufler, H. W. Sternberg, *J. Amer. Chem. Soc.* **1955,** 77, 3951–3958.

[40] T. S. Ahmandi, Z. L. Wang, T. C. Green, A. Henglein, M. A. El-Sayed, *Science* **1996**, 272, 1924–1926.

[41] L. Bronstein, M. Sedlak, J. Hartmann, M. Breulmann, H. Cölfen, M. Antonietti, *PMSE Preprints*, San Francisco, **1997**, in press.

[42] M. Michaelis, A. Henglein, *J. Phys. Chem.* **1992**, *96*, 4719–4724.

[43] R. Hughes, *Deactivation of Catalysis*, Academic Press, London **1984**.

[44] N. Toshima, T. Yonezawa, K. Kushihashi, *J. Chem. Soc.-Far. Trans.* **1993**, *89*, 2537–2543.

[45] T. Toshima, M. Harada, Y. Yamazaki, K. Asakura, *J. Phys. Chem.* **1992**, *96*, 9927–9933.

[46] D. Mandler, I. Willner, *J. Phys. Chem.* **1987**, *91*, 3600–3605.

[47] R. Touroude, P. Girard, G. Maire, J. Kizling, M. Boutonnet-Kizling, P. Stenius, *Colloid. & Surf.* **1992**, *67*, 9–19.

[48] C. P. Bean, J. D. Livingston, D. S. Rodbell, *Acta Metall.* **1957**, *5*, 682–684.

[49] V. K. Sharma, A. Baiker, *J. Chem. Phys.* **1981,** 75, 5596–5601.

[50] S. V. Vonsovskii, *Magnetism*, Nauka, Moscow **1971**, pp. 800–812.

[51] P. H. Hess, P. H. Parker, *J. Appl. Pol. Sci.* **1966**, *10*, 1915–1927.

Chapter 8

Plasma-Produced Silicon Nanoparticle Growth and Crystallization Processes

J. Dutta, H. Hofmann, C. Hollenstein, and H. Hofmeister

8.1 Introduction

Nanophase or cluster-assembled materials consist of small atomic clusters fused into a bulk material. Clusters and cluster-assembled materials have gained increasing interest owing to the possibilities of fabricating nanostructured materials showing novel properties [1–3]. These materials are being considered for next-generation use in structural and functional applications [1, 2]. Clusters are either produced by wet chemical synthesis methods, self-organization mechanisms, or gas phase condensation methods [6]. Gas phase condensation techniques are considered to be most appropriate for the preparation of nonagglomerated powders [6, 7].

Low-temperature powder synthesis processes have several advantages. Thermal plasma such as an electric arc or a plasma jet is used for powder synthesis with possible high yields of pure powders. Nonthermal plasma, on the other hand, works under small input power and flow rates. The low pressure, low temperature, and longer residence time of the powder precursors produced in these plasmas render this process interesting for the fabrication of nanoscaled powders [7].

Silicon and its alloys are extensively used in the electronics industries and also in the fabrication of high-temperature structural ceramics. The semiconducting properties of silicon have been used in the development of modern electronics and information technology. In recent times increasing interest has been shown for studies on the properties and structure of silicon clusters, partly because of the observation of visible light emission from porous silicon, obtained from electrochemically anodized silicon wafers [8]. Nanocrystalline silicon has been developed by a variety of methods, including standard wet chemical synthesis and more sophisticated cluster-assembling techniques [9, 10].

We have recently shown the possibility of synthesizing well-controlled silicon nanoparticles from gas phase reactions in a radio frequency (RF) discharge in silane [11]. The size of the particles varies between <10 and 200 nm, depending on the plasma characteristics, composition, and gas flow.

Powder formation in reactive plasmas has been observed and reported in the literature since the early 1970s [12, 13]. To apply suitable plasma deposition processes for the fabrication of thin films, plasma conditions that in particular led to powder contamination were carefully avoided in order to minimize reactor and film contamination. The knowledge gained in the control of powder contamination in films can now be profitably utilized for the controlled synthesis of nanoscaled powders.

In this review we will discuss the growth of these particles in the gas phase and the subsequent agglomeration processes that lead to the formation of powder. We will limit ourselves to the formation mechanism, the microstructure, and some essential properties of nanoscaled silicon powders obtained in a radio frequency silane plasma (in some cases diluted with noble gases). Powder formation in these plasmas has been extensively investigated and can be considered a model for future research and development efforts for the production of nanoscaled materials. We will also discuss the crystallization phenomena of these particles, as it is of interest for the understanding of confinement effects and also for controlling the microstructure of advanced ceramics.

8.2 Experimental Methods

8.2.1 Powder Preparation and Annealing

A conventional parallel-plate RF capacitive reactor has been used for the synthesis of nanoscaled silicon powders and for the *in situ* diagnostics [11, 14]. Plasma power of typically a few watts at 13–80 MHz was utilized [15]. RF power modulation, when used, was obtained by mixing a low-frequency square wave (10 Hz–20 kHz) into the RF generator signal using a balanced mixer before amplification by the RF wideband power amplifier (10 kHz–200 MHz) (Figure 8.1). The gas inlet was placed in the side wall of the reactor with a gas diffuser in front to reduce powder perturbation by gas drag. For certain experiments a gas shower incorporated in the RF electrode was used for distributing the gas uniformly in the whole discharge volume. Typically 30 sccm of gas (silane) was used at operating pressures of 0.1 mbar. The temperature of the grounded electrode was varied from room tem-

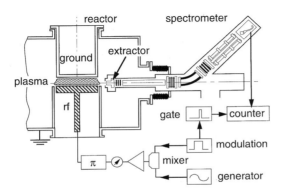

Figure 8.1. Schematic diagram of the reactor, including a block diagram of the RF power modulation and time-resolved mass spectrometry.

perature to 200 °C. The powders were collected from the electrodes or the sampling probe adjacent to the electrode gap.

Annealing of the samples was carried out at various temperatures and for different periods of time in a reducing atmosphere (forming gas, 92% N_2 + 8% H_2) in a SiC furnace with excess carbon, in order to reduce oxidation of the particles during the annealing process.

8.2.2 Transmission Electron Microscopy

The powders were dispersed by ultrasonic agitation in isopropanol and then transferred to carbon-coated copper grids for electron microscopy inspection. A holey carbon film covering the copper microgrids was used to allow high-resolution electron microscopy (HREM) imaging without disturbance from the support. Conventional transmission electron microscopy (TEM) of the powders was done in a JEM 100C operated at 100 kV while the HREM was carried out in a JEM 4000EX operated at 400 kV. Electron micrographs were taken with the objective lens appropriately defocused so as to achieve optimum contrast (i.e. near Scherzer defocus) of the amorphous particles and to allow imaging of the channels characteristic of the lattice of (110)-oriented crystallites as bright dots. Real and Fourier space image processing of the micrographs, recorded at 5×10^5 magnification and digitized with 8 bit depth of gray level, by means of the NIH "image" [16] and GATAN "digital micrograph" programs aimed at noise reduction, contrast enhancement, and reciprocal space characterization. Computer simulation of the image contrast was carried out according to the multislice algorithm ("MacTempas") applied to models of Si clusters created by means of the "Cerius" molecular dynamics package. The calculations were carried out for characteristic imaging parameters of 400 kV acceleration voltage, 1 mm spherical aberration coefficient, 8 nm focus spread, and 0.6 mrad beam divergence, respectively.

8.2.3 Vibrational Spectroscopy

Infrared spectra of pellets made by mixing the powders with potassium bromide (KBr) were recorded using a Nicolet 510 FTIR spectrometer [17]. Raman spectra were recorded with a standard double monochromator (Dilor) using the 514.5 nm line of an Ar^+ ion laser. The samples were kept at room temperature in nitrogen for the IR measurements and in air during the Raman measurements. Laser power for the Raman measurements were chosen as low as 0.2 mW on a focus diameter of 250 μm to avoid heating the samples [18].

8.2.4 Plasma and *In Situ* Powder Diagnostics

To diagnose in detail the powder formation in these RF plasmas, mass spectrometry (to characterize the plasma composition) and *in situ* Raleigh/Mie scattering

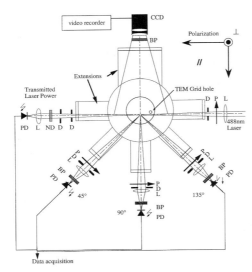

Figure 8.2. Schematic diagram of the top view of multiangle polarization-sensitive light-scattering setup showing the laser beam transmission measurement (extinction) and the three-angle detection of the laser beam scattering. P: polarizer, L: lens, D: diaphragm, BP: 488 nm bandpass filter, ND: neutral density filter, PD: photodetector.

(for the detection of the particles) were utilized. Other diagnostic tools such as optical-emission spectroscopy and microwave interferometry were used to investigate the behavior and correlation with powder-forming parameters of dusty plasmas. Two different types of differentially pumped mass spectrometer, namely a Hidden Analytical Limited plasma monitor type EQP 500 with a mass limit of 512 amu [14, 19, 20] and a Balzers plasma process monitor type PPM 421 with a mass range option up to 2048 amu [21] were independently employed, both equipped for neutral analysis, positive and negative ion extraction, and mass and energy measurements. Both instruments also allowed time-resolved measurements with a time resolution of about 2 µs [14].

The multiangle polarization-sensitive light-scattering system used to determine the particle size (r) and particle number density (n_{part}) is shown in Figure 8.2 [22]. An argon–ion laser operating at 488 nm was used as the light source. The detection system consists of three identical independent detectors composed of a polarizer, a diaphragm to define the solid angle, a collection lens, a bandpass filter, and a calibrated silicon photodiode. The scattering volume used in these experiments was 1 mm^3. Three scattered signals (45°, 90°, 135°) and the transmitted intensity were simultaneously digitized by an acquisition system. Four photodiode signals were measured, starting from plasma ignition in four different laser and detector polarization combinations. The present design of the light scattering allows particle size determination down to radii of 2 nm and particle number densities in the range 10^{14}–10^{17} m^{-3}. In contrast to other laser scattering methods [23], multiangle polarization-sensitive light scattering also provides information about the sphericity of the dust particles, which is important for a proper understanding of the associated agglomeration processes. The detection of any cross polarization signal gives additional valuable information about the particle shape.

Elastic light scattering by spherical homogeneous particles can be described by the Rayleigh–Gans theory for particles smaller than $\sim 0.1\ \lambda$ and by the Lorenz–Mie theory for larger particles [24, 25]. An iterative procedure was used to determine the particle radius and the particle number density [26]. The degree of linear polarization, which depends on the complex refractive index (m), is independent of the particle number density, but is a multivalued function of the size parameter. To solve the indeterminate problem of four unknown quantities (m_r, m_i, r, n_{part}) from three independent measurements, the scattering angular disymmetry method has to be used.

Depending on the type of information being looked for with respect to powder formation, other light-scattering systems have been used. In particular, for the study of the powder dynamics in these plasmas either white or laser light (Ar–ion and He–Ne laser) has been used as a light source. For powder dynamics investigation the beam was widened through a four-prism beam expander to illuminate the whole electrode gap, and the scattered light at 90° was detected by means of a CCD camera, which was then treated using an image-processing program [27, 28].

A system similar to the one described above has been used to measure the visible photoluminescence of the particles suspended in the plasma where the detection system of the light-scattering arrangement was replaced by a collection lens and a monochromator [29]. This arrangement allowed the measurement of the emission spectra.

Finally, *in situ* infrared absorption spectroscopy was used to obtain information about the chemical nature of the growing particles, its transition from cluster to particle, and the depletion of the feed gas [30]. The spectra were collected using a Bruker IFS-66 Fourier spectrometer (spectral range 600–5000 cm^{-1}, spectral resolution 1 cm^{-1}). The exiting infrared light beam from this instrument was sent through the plasma via ZnSe windows and detected by means of an external nitrogen-cooled HgCdTe detector.

8.3 Structure of the Silicon Nanoparticles

8.3.1 Morphology

The as-prepared powder varied from yellowish to reddish-brown to black in color and consisted of very small particles with sizes ranging between 8 and about 200 nm, as observed by transmission electron microscopy. As illustrated by the conventional electron micrograph shown in Figure 8.3, the particles are agglomerated in black-berrylike or even cauliflowerlike shapes, forming a highly porous powder with a large specific surface area (up to 162 m^2 g^{-1}) [31]. HREM generally shows an amorphous contrast appearance throughout the material with no indications of crystalline ordering, as is shown in the HREM micrographs captured at low magnification in Figure 8.4 [32] and also confirmed by the diffuse rings observed in the selected area electron diffraction (SAED) patterns in Figure 8.5. Individual particle surfaces are rough, as shown in Figure 8.6, with nearly spherical entities of 1 to 2 nm

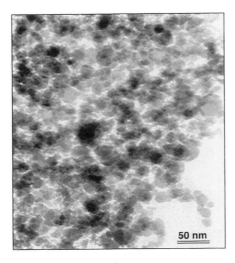

Figure 8.3. TEM image of as prepared Si particles showing the agglomeration.

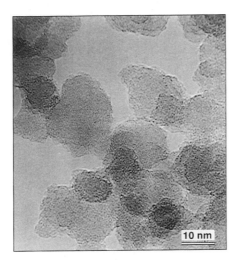

Figure 8.4. HREM image of as-prepared Si particles.

showing annular contrast features that protrude from the surface, suggesting that the particle formation process takes place by the agglomeration of clusters. Careful HREM imaging reveals annular (encircled) as well as fringelike (arrow) contrast features, about 1.5 nm in size, within individual amorphous particles, as shown in Figure 8.7. These observations suggest the presence of partially ordered regions, the dimensions of which are below the size of the smallest crystallites observed upon annealing, as will be discussed in Section 8.5 [33]. This ordering in the nanometric scale is between the short-range order of the amorphous network and the long-range translational order of crystals and has been attributed to arising as a result of the presence of clusters that are formed during the particle synthesis process, as will be discussed in Section 8.4. Fivefold twinned structures formed during crystal-

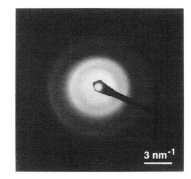

Figure 8.5. SAED pattern of the amorphous Si powder.

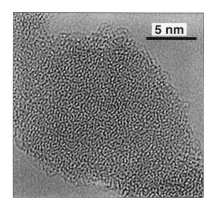

Figure 8.6. HREM image of an individual particle revealing its high surface roughness.

lization at elevated temperatures are believed to nucleate at such preexisting seeds [34–36].

The stability of clusters of carbon, silicon, and germanium having a cage or a cage–core configuration, and their possible role in the crystallization of the diamond cubic (dc) materials, has been discussed recently in the literature [34, 35, 37–40]. A common feature of all these models is the presence of inherent noncrystallographic symmetries (pentagonal rings), which, owing to a possible enhanced nucleation probability, may play an essential role in the formation of crystalline phases. The Ge_{15} cluster is proposed to serve as the nucleus of fivefold twinned structures in thin-film formation processes by physical vapor deposition [35]. Attachment in crystallographically favorable positions of atoms from the surrounding amorphous phase on the 15-atom seed has been proposed to lead to a fivefold twinned crystallite. Such fivefold twins are among the first ordered structures formed with the onset of crystallization in these materials [33].

Image contrast calculations according to appropriate structural models may help to substantiate the interpretation of contrast features discussed above. A first calculation of HREM image contrasts was attempted utilizing a relatively simple cluster model consisting of 45 and 75 atoms [41]. The image contrast calculations for two characteristic defocus settings for a 100-atom cluster in different highly

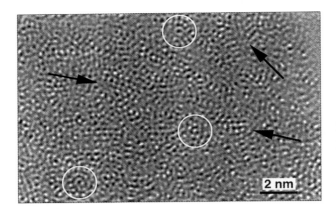

Figure 8.7. Part of an amorphous Si particle exhibiting respectively annular (marked by circles) and fringelike (marked by arrows) contrast features.

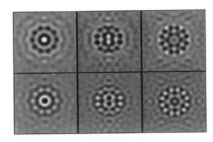

Figure 8.8. Image contrast calculations for a 100-Si-atom cluster model in three highly symmetric orientations with respect to the electron beam. Top and bottom of the image pairs correspond to two characteristic defocus settings.

symmetric orientations of the model with respect to the electron beam are shown in Figure 8.8. The left-hand pair corresponds to a fivefold orientation, the middle one to a pseudotwofold orientation, and the right-hand pair to a threefold orientation. The Si_{100} cluster is based on a 20-atom core in the shape of a pentagonal dodecahedron, 12 pentagonal faces of which are decorated by truncated pentagonal bipyramids [42]. A direct comparison of experimental and calculated image contrasts has to take into consideration that (i) the size and structure of the clusters may not necessarily correspond completely to the model used, (ii) arbitrary orientations of such clusters may coexist, and (iii) the clusters are not isolated but embedded in a matrix consisting of a random network that acts as "glue" in coexistence with other clusters. Dedicated methods of image processing that allow one to discriminate corresponding image contrast by image evaluation, pattern recognition, and template matching are not yet available for amorphous structures.

8.3.2 Vibrational Spectroscopy

8.3.3 Infrared Spectra

Figure 8.9 shows the infrared spectrum of a typical powder sample measured in the transmission mode. Strong bands at 640, 840, and 880 cm^{-1} can be assigned to the

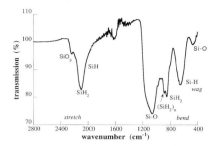

Figure 8.9. Infrared spectrum in transmission mode of a typical powder sample.

silicon–hydrogen [Si–H and SiH$_2$] rocking, bending, and scissors mode vibrations, while the bands between 2000 and 2100 cm^{-1} arise from the silicon–hydrogen [Si–H, SiH$_2$, and (SiH$_2$)$_n$] stretching vibrations. The Si–H, SiH$_2$, and (SiH$_2$)$_n$ stretching modes are usually observed in highly porous a-Si:H films with large void fractions [43]. The Si–H stretching bond is predominant in the amorphous silicon films. The SiH$_2$ stretching (at 2080 cm^{-1}) and for some powders the (SiH$_2$)$_n$ stretching (at 2100 cm^{-1}) are relatively more dominant modes compared to the normally dominant Si–H modes in a-Si:H thin films. Hydrogen content in the powders, as estimated from the integration of the wagging mode of Si–H, was about 30%, which agrees very well with the estimation made from hydrogen effusion experiments [17]. Exposure to the atmosphere oxidizes the powders, as can be noted from the prominent silicon–oxygen absorption band centered around 1050 cm^{-1}, oxygen–hydrogen bending centered at 1630 cm^{-1}, and the O–Si–O wagging mode vibrations at 460 cm^{-1}. In addition there is also the presence of the Si–O$_3$ mode at 2250 cm^{-1}, signifying the presence of oxidized surfaces. Oxidation occurs immediately after exposure to the atmosphere, as has also been reported for silicon powders prepared in high vacuum that oxidized rapidly even when they were left under vacuum [44].

8.3.4 Raman Spectra

Phonon confinement effects in silicon particles become predominant for particles with diameters less than 10 nm. The Raman spectra of some silicon nanopowders show a broad structure between 430 and 530 cm^{-1}, arising from amorphous silicon with several distinct superimposed peaks at 480 cm^{-1} or higher, which were assigned to molecularlike or localized modes arising from small ordered regions (Figure 8.10) [45]. Typical Raman spectra of silicon particles show three major features: vibrations due to silicon oxide and silicon acoustical phonons between 250 and 450 cm^{-1}, silicon optical phonon vibrations between 450 and 550 cm^{-1}, and the 550 and 700 cm^{-1} spectral region arising from silicon–hydrogen rocking vibrations (Fig 8.11) [46]. As we observe in the infrared spectra, a nonnegligible quantity of oxygen is present in the as-prepared powder upon exposure to the atmosphere. The changes in the oxide stoichiometry during the annealing procedures lead to the disappearance of crystalline order symmetry-forbidden contributions to the Raman

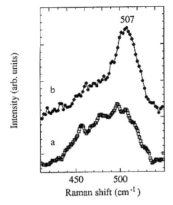

Figure 8.10. Raman spectrum of very fine silicon nanopowder showing vibrational modes arising from localized states.

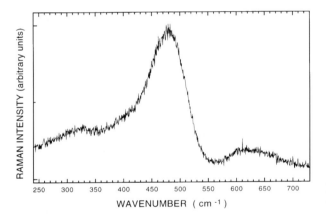

Figure 8.11. Raman spectrum of typical silicon nanopowder.

spectra from acoustical phonons. The Raman intensity in this region is considerably reduced upon annealing the samples at temperatures above 700 °C (Figure 8.12). The broad structure around 620 cm^{-1} was assigned to Si–H rocking modes [47], but peaks at similar frequencies have also been suggested to arise owing to combinations of optical and acoustical phonons in Si structures of reduced dimensionality such as porous Si or from accidental critical points [48, 49]. When the samples were annealed to 500 °C, the Raman intensity of this peak reduced, although the samples remained amorphous (as will be discussed in Section 8.5) and at temperatures above 650 °C it disappeared completely. The structural changes that occur at the annealing temperatures under consideration may be related to Si–H vibrational bands and agree pretty well with the hydrogen evolution characteristics [17]. Since hydrogen can also be expected on the surface of porous silicon, as has been reported in References [48] and [49], it was argued that the 580 to 680 cm^{-1} band involves Si–H vibrations [18].

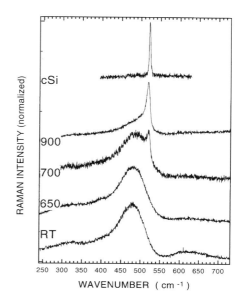

Figure 8.12. Raman spectrum of samples annealed for one hour at different temperatures (the crystalline silicon vibrational band is shown for reference).

8.4 Silicon Nanoparticle Synthesis and Related Properties

8.4.1 Powder Precursors

Extensive investigations have been carried out to identify the precursors and the associated gas phase reactions for powder formation in silane plasmas. Some researchers believe that the neutral radicals lead to particle formation by the insertion of lower silane radicals into higher saturated molecules [50, 51]. Positive ions have also been considered as a potential powder precursor, although it has been shown that activation barriers prevent the formation of high-mass cations [52]. Negative ions, first reported to exist in silane plasmas by Perrin et al., are trapped in these discharges by the sheath potentials, and it is generally accepted that plasma polymerization proceeds along a negative ion path [53, 54].

No high neutral masses or positive ions were observed in a silane plasma, as has been reported in the literature (Figure 8.13(a)). However, high-mass anions of 1600 amu containing up to 60 Si atoms could be detected in modulated silane plasmas [21]. The negative ion intensity and the presence of particulates in plasma depends strongly on the power modulation frequency. A clear anticorrelation between the powder appearance and negative ion intensity was found that has been explained to arise from the detrapping of negative ions during the plasma off phase.

For low modulation frequencies all the negative ions formed leave the discharge volume, whereas at high frequencies the plasma off time is too small to empty the discharge of all the negative ions. In this case the negative ions remain trapped, accumulate, and polymerize to higher masses, which in turn leads to powder for-

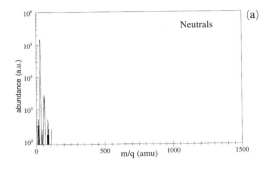

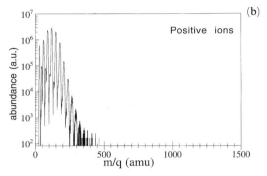

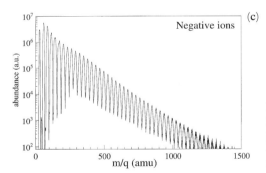

Figure 8.13. Mass spectra of neutral (a), positive (b), and negative (c) ion groups in a pure silane RF plasma. This raw data, with low mass resolution, is uncorrected for any mass-dependent falloff in sensitivity.

mation. This has been confirmed by observations made in time-resolved mass spectroscopic studies [14].

The ratio of H to Si atoms within each cluster of the spectrum shown in Figure 8.13(c) is plotted in Figure 8.14. A transition occurs in the maximum-abundance H:Si ratio at $n \sim 5$, passing from a (Si_nH_{2n+2}) dominance to a size-independent ratio H:Si = 4/3 for cluster sizes above $n \sim 10$. If the (Si_nH_{2n+2}) sequence were followed indefinitely, the ratio would have been 2 in the limit of large n [21], which demonstrates that silanions are highly cross linked, three-dimensional structures. Given the absence of double bonds in silicon clusters and neglecting dangling

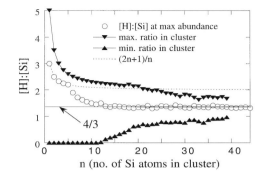

Figure 8.14. Ratio of hydrogen to silicon atoms of anions in spectra as presented in Figure 8.13.

bonds, H:Si = 4/3 corresponds to equal numbers of Si–H and Si–Si bonds within the most abundant clusters. The constant ratio also shows that pure Si_n cores with surface H atoms can be excluded as they cannot be regarded as cluster ions in the strict sense of the term $(SiH_m)–(SiH_4)_n$ for which H:Si → 4.

The regular nature of the mass distribution in Figure 8.13(c) suggests that a simple statistical approach might be sufficient to explain its form. A simulation of the mass spectrum by means of random bond theory [55] (Figure 8.15(a)) shows a good agreement with the measured mass spectrum (Figure 8.15(b)) [56]. This similarity

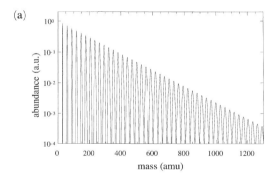

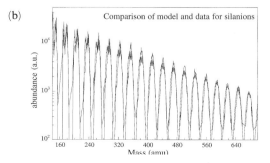

Figure 8.15. Comparison between measured and calculated anion abundances assuming a statistical distribution of hydrogen within the clusters (the exponential decay in amplitude is arbitrarily chosen to resemble the spectra presented in Figure 8.13(a) for comparison).

supports the hypotheses of the statistical nature of the hydrogen distribution within the Si anion clusters.

Silane is dissociated by electron impact, producing mainly neutral species like SiH, SiH_2, and SiH_3, as well as positively and negatively charged species. Whereas neutral precursors are argued predominantly to contribute to the deposition of a-Si:H thin films, the initial gas phase polymerization occurs through negative ion clustering like the condensation reaction (8.1):

$$Si_nH_x^- + SiH_4 \rightarrow (Si_{n+1}H_{x'}^-)^* + (H, H_2 \text{ products}) \tag{8.1}$$

where the initiating step is the formation of monosilicon hydride anions by dissociative electron attachment to silane [14].

Ion–ion recombination is a second possible pathway for polymerization:

$$Si_nH_x + Si_pH_q^+ \rightarrow Si_{n'} H_{x'} \rightarrow Si_nH_x^- + (Si, H, H_2 \text{ products}) \tag{8.2}$$

This reaction (8.2) results in a neutral cluster, and since heavy neutrals are not observed, it must be supposed that they are attached via

$$Si_nH_x + e^- (Si_nH_x^-)^* \rightarrow Si_nH_x^- \tag{8.3}$$

Low-energy electron attachment in which the parent negative ion survives against dissociation or auto attachment has been shown to be the dominant channel of large clusters. Therefore, the ion–molecule (Eq. (8.1)) and ion–ion reactions both eventually lead to stable, higher-mass negative ions.

8.4.2 Powder Formation and Agglomeration

Starting from silane plasma (SiH_4) under appropriate discharge conditions, a few tens of nanometer-sized particulates are formed shortly after plasma ignition. Figure 8.16 shows a typical time development of the particle radius and particle number

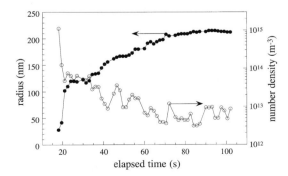

Figure 8.16. Time development of the particle radius and particle number density in a pure RF silane plasma.

density as a function of the discharge time, as obtained by the light-scattering method described above [22, 26, 29, 58]. Rayleigh–Mie scattering measurements show a strong agglomeration starting from particles around the present detection limit for the Rayleigh-scattering system. The agglomeration phase under these conditions always leads to particles about 100–200 nm in size. The agglomeration phase is accompanied by a decrease over several orders of magnitude of the particle number density. It should be mentioned here that at the beginning of the agglomeration phase, particle number densities in the order of 10^{15} m^{-3} are measured, which is just about comparable to the electron density. The particle size continues to slowly develop further by aggregation and processes analogous to film growth, leading to particles with a-Si:H properties as found by *in situ* IR absorption spectroscopy (see also Figure 8.9) [30].

The presence of cross polarization intensities and *ex situ* TEM measurements showed that during a time interval of the agglomeration phase, nonspherical particles do exist [22]. The presence of nonspherical particles restricts the use of the classical Mie scattering theory. Interestingly, later in the time development of clusters, nearly spherical particles can be found again. The structure observed by TEM analysis reveals densely packed spheres (with a "blackberry" structure) composed of spherical particles around 20 nm in size.

The particle size and particle number density could be modeled by a neutral agglomeration scheme, as is often employed in aerosol physics, namely the Brownian free molecular coagulation model [59, 60]. Figure 8.17 shows that, despite the simplicity of the model, excellent qualitative agreement between the measured particle size and particle number density is achieved. The particle size distribution as observed by TEM analysis can be well described by a log–normal distribution. The time development of the particle distribution could also be described by the Brownian free molecular coagulation model extended by including a self-preserving log–normal distribution during time development [61].

A model of the particle charging in the plasma based on the equality of the electron and ion current impinging on the particle together with the plasma neutrality shows that at low positive ion densities, particles as large as 40 nm in radius can be considered neutral [22]. Therefore neutral coagulation schemes can be applied to

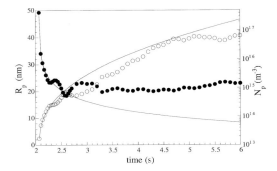

Figure 8.17. Time development of the particle radius (R_p) and number density (N_p) for early discharge times. Solid lines shows the best fit of the Brownian free molecule coagulation model.

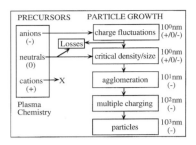

Figure 8.18. Tentative scheme for particle growth from molecules in low-pressure silane plasmas.

describe at least the initial step of the coagulation phase. The particles at the beginning of the agglomeration phase behave as neutrals and therefore their dwell time within the plasma is strongly reduced. Thus the gas drag, in particular in the shower head arrangement, may lead to large losses of these neutral or charge-fluctuating particle precursors, therefore inhibiting further agglomeration [62]. This fact may also explain the evidence that small-size nanoparticles (a few nanometers) can leave even a CW plasma and may therefore contribute to the growth of thin films [3]. These findings may lead to advanced techniques for incorporating controlled-size nanoparticles in a matrix for various applications.

The above investigation leads to a tentative scheme, as shown in Figure 8.18, for particle growth from molecules in low-pressure silane plasmas [21]. Powder precursors other than negative ions might not be excluded under different plasma operation conditions or using different monomers. For instance if a neutral cluster [64] can reach a critical size or become large enough to be negatively charged by the RF plasma, there is no need to invoke a slower, parallel anion pathway.

In situ light-scattering experiments showed that the particle formation process passes through three distinct phases: the initial clustering phase (plasma chemistry), a second phase consisting of the formation of larger primary particles, and finally the aggregation of primary particles into agglomerates. The scattering signal could not be detected initially after the ignition of the plasma (for the first 15–20 seconds). Around this period the negative ion signal, which is monitored by ion-mass spectroscopy, was found to increase as a function of time. After this stage of formation of the negative ions, Rayleigh scattering from the particles could be observed, which was used to determine the size of the primary particles (<30 nm). The number density of the particles was found to decrease rapidly concurrently with a rapid increase in the particle size, suggesting an agglomeration process. To sum up, the negative ions lose their charge as soon as they attain a critical size (of a few nm), whereupon these neutral clusters agglomerate into the primary particles by van der Waals interaction and form the spherical primary particles by further agglomeration and/or sintering processes. This agglomeration continues until the formation of powders so obtained. Shiratani et al. have also shown that the particle growth in low-pressure silane plasma includes nucleation, rapid growth, and growth saturation, which supports our observation and interpretation [65]. A schematic representation of the growth process is shown in Figure 8.19.

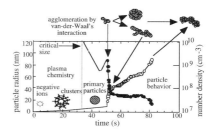

Figure 8.19. Schematic representation of the growth process of silicon nanopowders in low-pressure silane plasma.

8.4.3 Powder Dynamics

For future applications of plasma processes for the production of nanoscaled powders, it is of prime interest to understand the powder dynamics, since powder handling, including powder collection and powder introduction into the plasma, is essential for appropriate processing. Powder dynamics in these plasmas is governed by several forces acting on the particles [66]. Besides gravity and electrostatic forces, thermal gradients in the reactor lead to thermophoresis [67] and the gas flow induces important gas drag forces on the particles. Also the plasma itself induces forces such as the ion drag [67] that are supposed to play an important role in the plasma dynamics.

The time development of the powder dynamics has been studied by illuminating a cross section of the discharge by a polarized expanded laser beam, and the global spatiotemporal scattered light and extinction have been recorded by charge-coupled-device cameras [26, 27]. In pure silane discharges a confinement of the powder in two layers is usually observed [58], which also depends strongly on the electrode temperature [68]. The scattered intensities show alternate bright/dark regions in time and space that reverse according to the polarization. The temporal variations show uniform particle growth over large regions of the powder layers. High-contrast spatial intensities demonstrated the existence of particle size gradients for the steady state pure silane plasma and also for powder trapped in argon plasma [26]. Large differences in the trapping topography in silane and argon–silane plasmas and trapped powder in an argon plasma revealed the importance of the plasma chemistry in determining powder location.

8.4.4 *In Situ* Diagnostics of Powder Properties

Appropriate powder synthesis requires *in situ* powder diagnostics to monitor different properties such as chemical composition, degree of crystallinity, and crystallite size during the powder formation processes. Two-particle diagnostics has been used to probe *in situ* the particles, namely measurement of visible photoluminescence [29] and IR absorption spectroscopy [30, 69].

Visible photoluminescence at room temperature has been observed in amorphous hydrogenated silicon particles during their formation in the silane RF plasma [29].

The appearance of visible photoluminescence coincides with the particle agglomeration phase, as shown by light-scattering experiments.

The fact that the negatively charged particles are trapped in the RF discharge opens new ways for plasma treatment of these particles. Processes such as coating of small particles with functional layers needs *in situ* diagnostics sensitive to their chemical composition. IR absorption spectroscopy has been applied to determine *in situ* these properties of the growing particles. In particular this technique has been applied to hydrogenated silicon particles to investigate the oxidation mechanism and to determine the composition of plasma-produced SiO_x powders.

Hydrogenated silicon particles have been produced in an argon–silane plasma followed by a trapping phase in pure argon. The IR absorption spectra of these trapped particles show no "spontaneous" oxidation. However, if small amounts of oxygen are added to these Ar discharges, typical features associated with silicon–oxygen vibrational bands appeared in the IR spectra. These experiments clearly show that despite the high reactivity of the silicon particles, multiple plasma treatment is possible without apparent oxidation between successive steps. These experiments demonstrated for the first time that successive plasma post-treatments of plasma-produced particles were feasible.

Furthermore IR absorption spectroscopy has been employed to monitor the particle composition *in situ* during SiO_x particle synthesis. Figure 8.20 shows IR absorption spectra for plasmas with different silane/oxygen content and for compar-

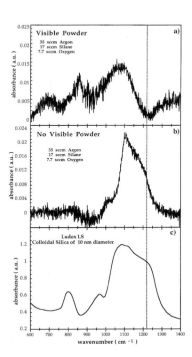

Figure 8.20. IR absorption spectra of plasma-produced SiO_x particles in a silane–oxygen plasma for (a) high and (b) low silane flow; (c) shows for comparison the IR absorption spectra of the commercial SiO_2 "Ludox LS" powder.

ison an absorption spectrum from commercial SiO_2 (Ludox) powder of comparable size. At low silane concentration nearly stochiometric SiO_2 particles, comparable to the commercial powder, are obtained. Increasing silane concentration leads to substoichiometric oxide powders.

Besides valuable information on the solid state nature of the particles, IR absorption spectroscopy gives information on the gas phase components such as silane depletion. In the case of the silane–oxygen plasma, it has been clearly seen that at low silane concentration, complete depletion occurs. The experiments described above clearly demonstrate that IR absorption spectroscopy is an important *in situ* diagnostic for powder synthesis and powder treatment by plasmas.

Particle formation [70–72], particle dynamics [26], and particle morphology [73] in noble-gas-diluted plasma was found to differ from the case in pure silane. In particular, argon- [70, 71, 73] and helium-diluted silane plasmas have been investigated [74, 75]. Adding hydrogen and helium to a silane plasma (low silane concentration) retards the powder formation process and reduces the powder density. At elevated silane concentration, powder disappears completely. On the other hand, argon dilution experiments reveal an acceleration of the powder formation compared to the pure silane case. Depletion might be responsible for many effects observed in noble-gas-diluted dusty plasmas, such as the appearance of different particle generations [76]. Differences in material properties and composition might also be found between powder produced in pure silane and silane noble gas mixture plasmas [73].

Besides processing aspects, dusty plasmas are of great interest for the foundations of plasma physics. Dust in the plasmas is known to change fundamental plasma properties such as electron density and temperature and they are also supected to be the reason for discharge transition [77], as frequently observed in RF discharges. To understand the recent experimental results on powder formation in plasmas, considerable efforts in the modeling and theory of dusty plasmas has been begun [54, 78, 79].

8.5 Silicon Nanoparticle Processing

8.5.1 Crystallization

The annealing of the powders in a furnace as well as during *in situ* heating in the transmission electron microscope leads to the crystallization of the amorphous powder. If they were annealed for 1 h between 300 and 600 °C, no apparent structural changes were observed. The nonhomogeneous atomic distribution with partially ordered regions, as revealed by dark-field imaging and HREM, which are observed in the as-prepared powders, become more pronounced, showing an increasing extent of annular contrast features. This structural rearrangement slightly below the threshold of long-range translational order is induced by H_2 removal [17]

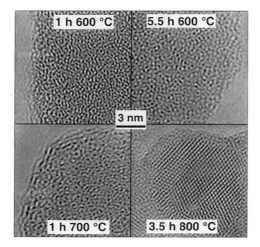

Figure 8.21. Individual powder particles after various stages of annealing, showing different degrees of oxide surface layer formation.

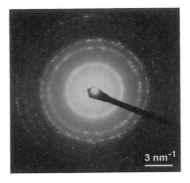

Figure 8.22. SAED pattern of the powder after 1 h annealing at 700 °C.

and the disintegration of SiO_x phases [80]. When the critical size, which is estimated to be about 3 nm, is exceeded [81], stable crystallites may nucleate around these localized ordered regions and the growth could proceed by rearrangement of atoms at the amorphous-to-crystalline interface.

Crystallization distinctly sets in upon annealing between 700 and 800 °C (Figure 8.21). The SAED pattern of the annealed powder is shown in Figure 8.22, which exhibits dotted rings superimposed on diffuse broad rings, which is characteristic of a low-dimensional crystalline phase in an amorphous matrix. Frequent events of small-scale crystalline ordering are observed, as shown in Figure 8.23, where the diffractogram obtained by fast Fourier transformation (FFT) of the image exhibits faint {111} spots. In addition, fast-grown fivefold twinned crystallites can be observed, an example of which is shown in Figure 8.24. The corresponding diffractogram (FFT) clearly exhibits 10 spots of {111} type of the five mutually twinned units azimuthally rotated to each other by about 72°. Even at this stage of crys-

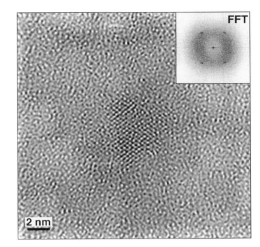

Figure 8.23. Small-scale crystalline ordering after 1 h at 700 °C with the diffractogram (FFT) shown in the inset.

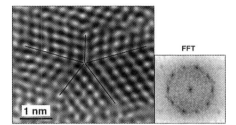

Figure 8.24. Fivefold twinned crystallite formed by 1 h annealing at 700 °C with the diffractogram (FFT) exhibiting a pseudopentagonal symmetry.

tallization, growth twinning is observed [33]. Since the excess twin energy is rather small, growth twinning is very common in dc semiconductors [34, 35, 82, 83]. Typical structures formed by growth twinning are shown in detail in Figure 8.25. Repeated twinning on alternate twin planes leads to the formation of parallel and azimuthally rotated twin lamellae and of additional multiple twin junctions (encircled, with the enlarged circle showing the original fivefold twin junction). The progress of crystallization upon annealing at 800–900 °C was monitored in detail by HREM. Up to 800 °C, crystallization is not observed uniformly throughout the particles and a certain proportion of amorphous portion remains in each particle. At 900 °C, in almost all particles an extended crystal lattice with characteristics of dc Si appears. Because of extensive growth twinning, a heavily faulted structure (an example is shown in Figure 8.26) is observed in many particles. The thermal treatment results in almost completely crystallized nanoparticles characterized by a high density of twin boundaries of various order. No considerable change in particle size or in the degree of agglomeration was observed. Besides amorphous remnants in the interior of particles, an amorphous shell covering the crystalline cores is distinctly visible.

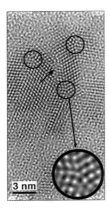

Figure 8.25. Fivefold twin junction (circle with arrow) surrounded by parallel and azimuthally rotated twin lamellae with additional multiple twin junctions (encircled) formed by growth twinning.

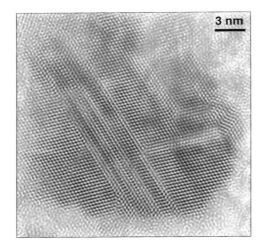

Figure 8.26. Heavily twinned structure in a crystallized region after 3.5 h at 800 °C.

As we have noticed from vibrational spectroscopy (Section 8.3) even in the as-prepared powder a nonnegligible quantity of oxygen contaminates the powder from exposure to the atmosphere [17]. Rapid formation of silicon oxide, which is nonstoichiometric (as interpreted from the infrared spectroscopy), begins with annealing at temperatures above 350 °C. The formation of an oxide shell was clearly visible by HREM only in samples annealed at 700 °C or higher, as can be recognized by comparison of the images shown in Figure 8.21. At 900 °C the oxide surface layer evolves to a thickness of about 2 nm. Because of this oxide layer, sintering of particles is effectively prevented [82].

It is interesting to note that the crystallite sizes are much smaller compared to the actual particle size, as the crystallization process is limited to the primary particle. Quantitative metallography as well as the analysis of the X-ray peaks result in a mean crystallite size of only 4.5 nm (700 °C) and 6 nm (900 °C) respectively in the polycrystalline particles. The kinetics of the crystallization as followed during *in situ*

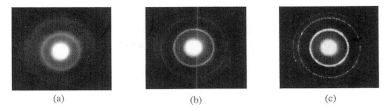

Figure 8.27. Electron diffraction pattern of the silicon powder: (a) as received; (b) 10 min at 650 °C; (c) 1 h at 650 °C.

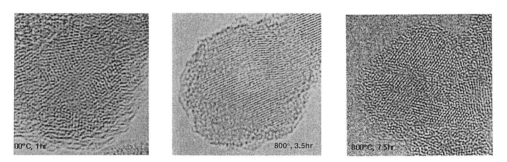

Figure 8.28. Grain growth of silicon particles at 800 °C.

annealing in TEM is shown in Figure 8.27. After 10 min annealing at 650 °C, the first signs of crystallization could be noticed. Complete crystallization (when the diffraction pattern did not change anymore) is reached after 1 h. During annealing, the grains grow from a mean diameter of 4.5 nm up to 10 nm after 7.5 h at 800 °C (Figure 8.28). In addition to the crystalline grains in the particles, an amorphous surface layer of silicon oxide of 1.5–2 nm is observed. The results of the *in situ* observation of the sintering behavior of the silicon powder mounted on a GATAN hot stage at 900 °C are shown in Figure 8.29(a), which is schematically represented in Figure 8.29(b). Figure 8.29(a) shows the difference in the microstructure after 10 min sintering. The shrinkage range is still very small, but we can observe some particle movement (rearrangement). This particle movement is more clearly shown in Figure 8.29(b), the arrows indicating the direction of the movement. A neck formation or a neck growth were, however, not observed.

8.5.2 Sintering

8.5.3 Thermodynamics

During heating up to the sintering temperature, the amorphous powder crystallizes. Because the thermodynamically stable minimum crystallite size of Si grains has

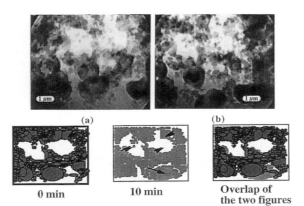

(a)   (b)

0 min   10 min   Overlap of
                              the two figures

Figure 8.29. Sintering behavior of Si powder. (a) *In situ* observation in the TEM. (b) Schematic view of the particle movement.

been reported to be between 4 [81] and 7–8 nm [86], we can expect to obtain monocrystalline powder particles upon annealing. The Si clusters in the powder particles act as seeds for the crystallization process [36]. A thermodynamic calculation carried out by Veprek et al. suggests a stable size of the crystalline seeds of 3 nm (or more than 1000 Si atoms per cluster) [51, 81]. The relatively large size of the nucleus also explains why no grains smaller than 3 nm could be observed. Additionally, the amorphous oxide layer around each grain inhibits the grain growth across the particle–particle interface, which limits the maximum grain size.

The nanosized a-Si particles show a very complex behavior during heating. The crystallization takes place at relatively high temperatures (>923 K), which is 0.55 times the melting temperature (typical crystallization temperatures of Si are between 710 and 870 K) [85]. Additionally, we cannot observe monocrystalline particles, the grain size after crystallization being only 3–4 nm, which implies that each particle consist of 4–6 grains and therefore also grain boundaries. Each particle is coated with an amorphous layer of an oxygen-rich Si compound (SiO_x). Figure 8.30 describes schematically the crystallization behavior of the amorphous Si particles.

From a thermodynamic point of view, the transformation from the amorphous phase to the nanocomposite consisting of c-Si, a-Si, and SiO_x may be regarded as a decomposition of the amorphous phase to the nanosized crystallites, an oxygen-rich surface layer, and the interfaces. In nanosized materials, and especially in the present example, this interface can be regarded as a separated phase in the Si particle, i.e.

Inhomogeneous solid + $(T > 700 \,°C) \rightarrow n$ crystallites + Interface
amorphous/gas   interface crystallite/amorphous +
                                                     interface amorphous/
                                                     amorphous +
                                                     interface amorphous/gas

or with reference to the notation in Figure 8.30,

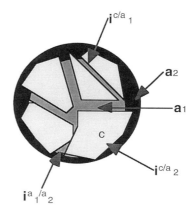

Figure 8.30. Schematic description of the crystallization amorphous Si.

$$a_1 + i_{a/g} \rightarrow c + i_{c/a} - i_{a_1/a_2} + a_2 + i_{a_2/g}$$

The change in the Gibbs free energy (G) for the overall transformation may be expressed by considering the atomic fraction of amorphous material (only Si) as x and the Gibbs free energy of formation of the different phases and interfaces as DG_f^i, $G(T)$ can be expressed as:

$$\Delta G(T) = (1 - x)\Delta G_{f^c}(T) + x\Delta G_{f^{ic/a_1}}(T) + k\Delta G_{f^{ic/a_2}}(T) + \Delta G_{f^{ia_1/a_2}}(T)$$
$$- \Delta G_{f^{a_1}}(T) \tag{8.4}$$

For geometrical reasons, we can say that the geometrical factor k in Eq. (8.4) will be equal to $(1 - x)$. Additionally $\Delta G_{f^{a_1/a_2}}(T)$ can be assumed to be negligible.

It is well known that nanocrystalline structure is not in equilibrium as a perfect crystal. Therefore, we cannot use the crystalline bulk state as the thermodynamic standard state, but we can define $\Delta G^i(T) = \Delta G^{c/a_1}(T) - \Delta G_{f^c}(T)$ and $\Delta G^a(T) = \Delta G_{f^a}T - \Delta G_{f^c}(T)$ as the excess Gibbs free energy for the interface and the amorphous phases respectively. Therefore, (8.4) can be simplified as:

$$\Delta G(T) = x\Delta G^i(T) - \Delta G^a(T) + (1 - x)\Delta G_{f^{ic/a_2}}(T) \tag{8.5}$$

From Eq. (8.5) we can conclude that the amorphous phase will be stabilized by the formation of a SiO_x layer at the surface of the particle.

The thermodynamic properties of an amorphous solid can be approximated by those of a supercooled liquid state for $T < T_m$. On the basis of classical thermodynamic theory, the excess enthalpic (ΔH^a), excess entropic (ΔS^a), and excess free (ΔG^a) energy for the amorphous phase related to the crystalline phase were calculated using for $\Delta Cp(T) = Cp^a(T) - Cp^c(T)$ the estimation of Battezatti and Garrone [87].

Figure 8.31 shows the calculated excess Gibbs free energy ΔG^a. For the values of $\Delta G^i(T)$, only an estimation based on the results for metals was done [86]. The ex-

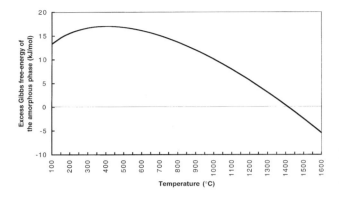

Figure 8.31. Calculated excess Gibbs free energy ΔG^a of the amorphous phase.

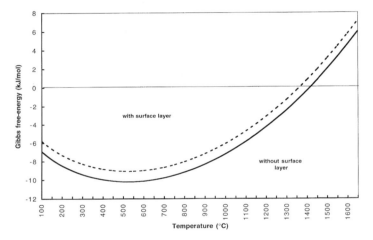

Figure 8.32. Calculated excess Gibbs free energy for the crystallization of a-Si.

cess volume of amorphous Si ($V_a - V_c$) is 5.4% [88]. This value of the excess volume is in a good agreement with the expected value for a microstructure with crystallites of 3–5 nm. At this relatively low excess volume, we can also expect a low $\Delta G^i(T)/\Delta G^a(T)$ ratio. From the results of Lu [86], we can find a $\Delta G^i(T)/\Delta G^a(T)$ ratio of 5 (at $T \approx 460$ °C) up to 2 (at $T \approx 1000$ °C). The value of $\Delta G^{ic/a_2}(T)$ was estimated to be 1.26 kJ mol^{-1} using an interface energy of 0.3 J m^{-2} and a specific surface area of 150 m^2 g^{-1}. Figure 8.32 clearly shows the stabilizing influence of the SiO$_x$ layer on nanosized particles. At 0.5 T_m, the Gibbs free energy is 1.4 kJ mol^{-1}, which was estimated to be 13% lower for particles with an oxide layer. With these values, we estimated the maximum atomic fraction of amorphous phase in the Si particle at 700 °C to be between 0.25 and 0.30.

The results of this estimation of the thermodynamic behavior of the relatively complex system is in good agreement with other observations. The results from the HRTEM investigation as well as Raman spectroscopy show the presence of a non-negligible amount of amorphous phases even after the crystallization [18, 33].

8.5.4 Kinetics

The sintering behavior of Si nanoparticles is not very well understood. Si is a co-valent material and the sintering temperatures of coarse Si grains is between 0.75 T_m (beginning of the densification) and 0.98 T_m (maximum density). It is well known that the melting temperature (T_m) is reduced with decreasing particle size; particles with a size >10 nm show a linear decrease in T_m [89]. For a particle size of 17 nm the melting temperature is 95% of the melting temperature of the bulk material. This small difference in T_m cannot really influence the sintering behavior of nanosized powders. On the other hand, the thermodynamic description of the crystallization behavior shows that, in this special case, the nanoparticles are stabilized by the oxide layer. Shi [89] showed that only free nanoparticles have a lower T_m, while nanoparticles embedded in a matrix have T_ms that depend upon the surrounding matrix. For example, it was observed that the melting point of tin (Sn) in an amorphous oxide matrix is much lower compared to the bulk melting point of Sn [90], whereas in a carbon matrix T_m increases. Unfortunately, no adequate interpretation has been attempted in the literature to explain the observations of super-heating as well as melting point suppression for the same nanocrystals in different matrices [89]. Hence the melting point of the Si crystals in the particles cannot be defined, but owing to similarities between Si and Sn we can assume a reduction in T_m and also a suppression of the sintering temperature.

Starting from the description of the microstructure of this system (Figures 8.28 and 8.30) we have to discuss the sintering behavior in more detail with a model of particles with a hard, not "sinterable", core and a soft sinterable surface layer. Following Jagota [91], for equal-sized, spherical, coated particles with a diameter d forming a packing with solid fraction r, the minimum coating thickness, c, to achieve full density is

$$c/d = (1/r)^{1/3} - 1 \tag{8.6}$$

In agglomerated powders with a value of r of 50%, a coating thickness of 2 nm is sufficient for complete densification, whereas in areas between the agglomerates a coating of 5.5 nm is necessary. Here we will concentrate on the sintering of the agglomerates where $c = 0.25d$, because only for this area is the observed coating of 2 nm enough for a full densification.

The relative density after 10 min sintering at 900 °C as estimated from Figure 8.29, using the normalized time t', can be defined as:

$$\tau' = [(3/4)p]1/3t\gamma/\eta(d+c) \tag{8.7}$$

where τ' = sintering time; γ = surface energy; η = viscosity of the coating at the sintering temperature (for SiO_2 at 900 °C: $\eta = 2 \times 10^{13}$ Pa s).

The surface energy of amorphous SiO_2 nanoparticles was obtained from the following equation [92]:

$$\gamma' = \gamma(1 - \theta_3(a/r) + \theta_4(a/r)^2 + \cdots) \tag{8.8}$$

where $\gamma' = $ surface tension of nanoparticles; $\theta = $ numerical coefficient θ_3, $\theta_4 = 1$, $r = $ radius of the particles; and $a = $ lattice constant, or here the distance between the next neighbors.

Using 0.2 J m^{-2} for g and 0.06 nm for a typical distance between tetrahedral SiO$_4$ [93], the calculated γ' for 17 nm amorphous SiO$_2$ particles is 0.19 J m^{-2}. With this value and a sintering time of 10 min, the change in the relative density is estimated to be 1%. These results show clearly that for this mechanism (viscous flow with hard core) in the system Si/SiO$_2$ even in the nanosized domain at temperatures $>0.5\ T_m$, very limited sintering can take place (sintering time 1 month). This model is therefore inadequate to describe the observed rearrangement of the nanoparticles upon annealing (Figure 8.29).

Therefore a model based on the work of Gryaznov and Trusov (see, for example, Reference [92]) was used for the explanation of the observed rearrangement. We can assume that internal plastic deformation is inhibited in Si nanoparticles, especially in the present case of polycrystalline particles, because the critical length for dislocation will be larger than the diameter of the crystallites or the particles. Therefore, the contribution of interparticle sliding to nanopowder shrinkage becomes substantial. In the agglomerates the typical pore sizes are of the same order as the particle sizes (as can be deduced from the green density of the nanocomposites) [94]. Such an ensemble of nanoparticles allows interparticle sliding where nanoparticles as a whole slip into pores. The driving force for such a process is the surface tension and the shrinkage rate, which can be approximately described by:

$$\mathrm{d}\rho/\mathrm{d}t = D_p\gamma r/kT \tag{8.9}$$

where $D_p = $ effective diffusion coefficient of nanoparticles in the agglomerate. The characteristic sintering time can thus be calculated as $\tau' = kT(d/a)^{1.5}/D_p$. D_p depends upon the particle size and the surface diffusion:

$$D_p = (a/d)^{3/2}D_s \tag{8.10}$$

where $D_s = 1.3 \times 10^{-12}$ m^2 s^{-1}, for Si nanoparticles, $D_p = 8.6 \times 10^{-16}$ m^2 s^{-1}, and therefore the characteristic sintering time is <1 s. This value is comparable with the characteristic sintering time of Ni nanoparticles at 600 K. These results show that the sintering of agglomerated Si nanoparticles coated with SiO$_2$ is fast at relatively high temperatures. This leads to a densification of the agglomerates, and a complete densification of the sample is impossible or possible only at very high temperatures, as observed for submicron or micron-sized Si powder. This observation is in agreement with the results presented by Kruis et al. [95], who reported the stabilization of phases and the microstructure by a second phase even in a completely different nanocomposite system. A thermodynamic explanation of this phenomenon is still not possible at the present time. It is interesting to note that the use of the classical approach of sintering theory to explain the sintering of Si nanoparticles gives much lower characteristic sintering times ($\tau' \ll 1$ s). The reason for this difference needs to be studied in greater details.

8.6 Conclusions and Prospects

Here we have discussed the formation of 20–30 nm-sized silicon particles from 2–3 nm clusters in a glow discharge of silane. Dusty plasmas and powder production in plasmas is a rapidly expanding topic and in particular the formation of nanometer-sized silicon-based powders has been extensively studied [96]. Most of these investigations have been performed in low-pressure RF silane or silane–noble-gas discharges. In these discharges, often operated near to conditions giving device-quality amorphous silicon, the negative ions are supposed to lead to powder formation. The plasma-produced particles show interesting properties such as visible photoluminescence, and new diagnostic methods allow the *in situ* determination of chemical and morphological properties of the particles formed. It has been shown that plasma processes are suitable not only for powder production but also for particle modification by adding, for instance, thin functional layers onto the particles or by the treatment of the surfaces of the particles.

High-resolution transmission electron microscopic studies have revealed the presence of 1.5–2 nm clusters showing medium-range order with noncrystallographic symmetry that were earlier predicted from Raman spectroscopic studies. This agrees well with the observations made during the Mie scattering experiments carried out during the particle synthesis and these clusters are believed to be the building blocks of the primary particles. Crystallization processes are governed by the clusters, which act as seeds. Besides small-scale crystalline ordering, from the very beginning of crystallization, fast-grown fivefold twinned crystallites are formed. The crystallization proceeds mainly by growth twinning, leading to a heavily faulted structure. The sintering behavior of nanosized Si powder was studied and it was observed that the powder morphology had a very important influence on the sintering behavior. Particles showed a distinct crystalline lattice at half the melting temperature of the bulk material. Additionally an amorphous silica layer on the surface of each particle, which forms during annealing experiments, influences the sintering behavior.

These observations will be useful for the study of the properties of sintered ceramics and the optoelectronic properties of the tiny crystallites, which show quantum confinement effects [97–99]. This study will also be helpful for the preparation of reaction-bonded silicon nitride and silicon carbide ceramics with tailored microstructure. Further understanding of the growth of these clusters in the plasma will enable one, in effect, to control the primary particle sizes as well as the crystallite sizes in the final sintered bodies.

Future efforts will go into the development of advanced powder handling, such as powder collection, powder feeding into the plasma, control of powder formation (size dispersion, agglomeration), and control of the composition of the particulates (chemical composition, crystallinity). Efforts also need to be made to understand the dispersion behavior of these particles so that appropriate surface treatment would allow the fabrication of less agglomerated powders. In order to utilize these powder particles in functional applications, studies on the disagglomeration of the particles, attaching appropriate polymeric chains, and the arrangement of these in two- or three-dimensional matrices will be interesting. The effects of different matrices on the thermodynamic and optoelectronic properties of these or other nanoparticles need to be studied in further detail in order to profitably utilize the novel properties in nanocomposites for suitable applications [100].

Acknowledgements

The authors would like to thank the members of their respective groups. Work at LTP was partially supported by Swiss Federal Grants FN 2100-039361.93/1 and at CRPP by Swiss Federal Research Grants BBW 93.0136 for the Brite-Euram project and BEW 9400051.

References

[1] *Crystals* (Eds.: P. Jena, S. N. Kanna, B. K. Rao), Kluwer Dodrecht **1992**.
[2] K. J. Klabunde, *Free Atoms, Clusters, and Nanoscale Particles*, Academic Press, San Diego, **1994**.
[3] D. M. Tanenbaum, A. L. Laracuente, A. Gallagher, *Appl. Phys. Lett.* **1996**, *68*, 1705–1707.
[4] A. J. Burggraaf, A. Winnubust, H. Verweij, *Proc. Third Euro-Ceramics* **1993**, *3*, pp. 561–576.
[5] R. Roy, *M. R. S. Symp. Proc*; *Materials Research Society* (Eds.: S. Pittsburgh, J. C. Parker, G. J. Thomas), Pittsburgh, USA **1993**, *286*, 241–250.
[6] R. W. Siegel, *NanoStructured Materials* **1994**, *4*, 121–138.
[7] H. Drost, H.-D. Klotz, R. Mach, K. Szulzewsky, in: *Advanced Materials '93*, (Eds.: N. Mizutani et al.); *Trans. Mat. Res. Soc. Jpn.*, Elsevier, Amsterdam **1994**, *14A*: I/A, 53 –57.
[8] L. T. Canham, *Appl. Phys. Lett.* **1990**, *57*, 1046–1048.
[9] D. H. Lowndes, D. B. Geohegan, A. A. Puretzky, D. P. Norton, C. M. Rouleau, *Science* **1996**, *273*, 898–903.
[10] A. Perez, P. Melinon, V. Depuis, P. Jensen, B. Prevel, J. Tuaillon, L. Bardotti, C. Martet, M. Treilleux, M. Broyer, M. Pellarin, J. L. Vialle, B. Palpant, J. Lerme, *J. Phys. D*, **1997**, *30*, 709–721.
[11] J. Dutta, R. Houriet, H. Hofmann, J. L. Dorier, A. A. Howling, C. Hollenstein, in: *Proc. of 6th European Conference on Applications of Surface & Interface Analysis* (ECASIA 95) (Eds.: H. J. Mathieu, B. Reihl, D. Briggs), Wiley, Chichester **1996**, pp. 483–486.
[12] H. Kobayashi, A. T. Bell, M. Shen, *Macromolecules* **1974**, *7*, 277–285.
[13] H. Yasuda, *Plasma Polymerization*, Academic Press, Orlando **1985**.
[14] A. A. Howling, L. Sansonnens, J.-L. Dorier, C. Hollenstein, *J. Appl. Phys.* **1994**, *75*, 1340–1353.
[15] A. A. Howling, C. Hollenstein, P.-J. Paris, F. Finger, U. Kroll, in: *XX International Conference on Phenomena in Ion ized Gases*, Pisa, Italy **1991**, *5*, 1089–1090.
[16] W. Rasband, *Public domain software*, US National Institute of Health.
[17] R. Houriet, J. Dutta, H. Hofmann, in: *Proc. of 6th European Conference on Applications of Surface & Interface Analysis* (ECASIA 95) (Eds.: H. J. Mathieu, B. Reihl, D. Briggs), Wiley, Chichester **1996**, pp. 491–495.
[18] Scholz, S. M; Dutta, J; Hofmann, H; Hofmeister, H. J. Mat. Sci. & Technol. **1997**, *13*, 327–333.

[19] A. A. Howling, J.-L. Dorier, C. Hollenstein, *Appl. Phys. Lett.* **1993**, *62*, 1341–1343.
[20] A. A. Howling, L. Sansonnens, J.-L. Dorier, C. Hollenstein, *J. Phys.* D **1993**, *26*, 1003–1006.
[21] A. A. Howling, C. Courteille, J.-L. Dorier, L. Sansonnens, C. Hollenstein, *Pure & Appl. Chem.* **1996**, *68*, 1017–1022.
[22] C. Courteille, C. Hollenstein, J.-L. Dorier, P. Gay, W. Schwarzenbach, A. A. Howling, E. Bertran, G. Viera, R. Martins, A. Macarico, *J. Appl. Phys.* **1996**, *80*, 2069–2078.
[23] M. Shiratani, H. Kawasaki, T. Fukuzawa, T. Yoshioka, Y. Ueda, S. Singh, Y. Watanabe, *J. Appl. Phys.* **1996**, *79*, 104–109.
[24] C. F. Bohren D. R. Huffmann, *Absorption and Scattering of Light by Small Particles*, Wiley, New York **1983**.
[25] H. C. van de Hulst, *Light Scattering by Small Particles*, Wiley, New York **1975**.
[26] J.-L. Dorier, C. Hollenstein, A. A. Howling, *J. Vac. Sci. Technol.* **1995**, A *13*, 918–926.
[27] J.-L. Dorier, C. Hollenstein, A. A. Howling, C. Courteille, W. Schwarzenbach, A. Merad, J. P. Boeuf, *IEEE Trans. Plasma Sci.* **1995**, *24*, 101–102.
[28] A. A. Howling, C. Hollenstein, P. J. Paris, *Appl. Phys. Lett.* **1991**, *59*, 1409–1411.
[29] C. Courteille, J.-L. Dorier, J. Dutta, C. Hollenstein, A. A. Howling, *J. Appl. Phys.* **1995**, *78*, 61–66.
[30] G. M. W. Kroesen, J. H. W. G. de Boer, L. Boufendi, F. Vivet, M. Khouli, A. Bouchoule, F. J. de Hoog, *J. Vac. Sci. Technol.* **1996**, A *14*, 546–549.
[31] J. Dutta, I. M. Reaney, C. Bossel, R. Houriet, H. Hofmann, *Nanostructured Materials* **1995**, *6*, 493–496.
[32] J. Dutta, H. Hofmann, R. Houriet, H. Hofmeister, C. Hollenstein, *Colloids & Surfaces* **1997**, *127*, 263–272.
[33] H. Hofmeister, J. Dutta H. Hofmann, *Phys. Rev.* **1996**, B *54*, 2856–2862.
[34] S. Matsumoto, Y. Matsui, *J. Mater. Sci.* **1983**, *18*, 1785–1793.
[35] T. Okabe, Y. Kagawa, S. Takai, *Phil. Mag. Lett.* **1991**, *63*, 233–239.
[36] J. Dutta, R. Houriet, H. Hofmann, H. Hofmeister, *Nanostructured Materials* **1997**, *9*, 359–362.
[37] P. R. Taylor E. Bylaska, J. H. Weare, R. Kawai, *J. Chem. Phys. Lett.* **1995**, *235*, 558–563.
[38] M. V. Ramakrishna, J. Pan, *J. Chem. Phys.* **1994**, *101*, 8108–8118.
[39] J. Pan, M. V. Ramakrishna, *Phys. Rev.* **1994**, B *50*, 15431–15434.
[40] U. R. Räthlisberger, W. Andreoni, M. Parinello, *Phys. Rev. Lett.* **1994**, *72*, 665–668.
[41] H. Hofmeister, J. Dutta, H. Hofmann, *Materials Science Forum* **1997**, *235–238*, 595–600.
[42] C. Gerstengarbe, *Publications of the 12. Electron Microscopy Conference* **1988**, Dresden, pp. 481–482 & A174.
[43] For example, in: "Materials Issues in Applications of Amorphous Silicon Technology," *Materials Research Society Symposium Proceedings, 49; Materials Research Society* (Eds.: D. Adler, A. Madan, M. J. Thompson), Pittsburgh **1985**.
[44] S. Hayashi, S. Kawata, H. M. Kim, K. Yamamoto, *Jpn. J. Appl. Phys.* **1993**, *32*, 4870–4877.
[45] J. Dutta, W. Bacsa, C. Hollenstein, *J. Appl. Phys.* **1995**, *77*, 3729–3733.
[46] S. Minomura, in: *Semiconductors and Semimetal, Vol. 21: Hydrogenated Amorphous Silicon*, Part A (Ed.: J. Pankove), Academic Press, Orlando **1984**, pp. 273–290.
[47] Y. Uchida, in: *Semiconductors and Semimetals, Vol. 21: Hydrogenated Amorphous Silicon*, Part A (Ed.: J. Pankove), Academic Press, Orlando **1984**, pp. 41–54.
[48] I. Gregora, B. Champagnon, A. Halimaoui, *J. Appl. Phys.* **1994**, *75*, 3034–3039.
[49] H. Tanino, A. Kuprin, H. Deai, N. Koshida, *Phys. Rev.* **1996**, B *53*, 1937–1947.
[50] M. J. Kushner, *J. Appl. Phys.* **1988**, *63*, 2532–2551.
[51] S. Veprek, K. Schopper, O. Ambacher, W. Rieger, M. G. J. Veprek-Heijman, *J. Electrochem. Soc.* **1993**, *140*, 1935–1942.
[52] M. L. Mandich, W. D. Reents, K. D. Kolenbrander, *Pure & Appl. Chem.* **1990**, *62*, 1653–1660.
[53] J. Perrin, A. Lloret, G. D. Rosny, J. P. M. Schmitt, *Int. J. Mass Spectrom. Ion Processes* **1984**, *57*, 249–281.

[54] Choi, S. J; Kushner, M. J. *J. Appl. Phys.* **1993**, *74*, 853–861.
[55] E. Bertran, J. Costa, G. Sardin, J. Campmany, J. L. And jar, A. Canillas, *Plasma Sources Sci. Technol.* **1994**, *3*, 348–354.
[56] C. Hollenstein, W. Schwarzenbach, A. A. Howling, C. Courteille, J.-L. Dorier, L. Sansonnens, *J. Vac. Sci. Technol.* **1996**, *A 14*, 535–539.
[57] I. Suers, L. G. Christophorou, J. G. Carter, *J. Chem. Phys.* **1979**, *71*, 3016–3024.
[58] Y. Watanabe, M. Shiratani, *Jpn. J. Appl. Phys.* **1993**, *32*, 3074–3080.
[59] G. Prado, J. Lahaye, in: *Proceedings of the GMR Symposium* **1980**, 143–175.
[60] S. K. Friedlander, *Smoke, Dust and Haze, Fundamentals of Aerosol Behaviour*, Wiley, New York **1977**.
[61] K. W. Lee, H. Chen, J. A. Gieseke, *Aerosol Science and Technol.* **1984**, *3*, 53–62.
[62] J. Goree, *Plasma Sources Sci. Technol.* **1994**, *3*, 400–406.
[63] P. R. I. Cabarrocas, P. Gay, A. Hadjadj, *J. Vac. Sci. Technol.* **1996**, *A 14*, 655–659.
[64] Y. Watanabe, M. Shiratani, T. Fukuzawa, H. Kawasaki, Y. Ueda, S. Singh, H. Ohkura, *J. Vac. Sci. Technol.* **1996**, *A 14*, 995–1001.
[65] M. Shiratani, H. Kawasaki, T. Fukuzawa, T. Watanabe, *J. Vac. Sci. & Technol. A.* **1996**, *14*, 603–607.
[66] J. E. Daugherty, D. B. Graves, *J. Appl. Phys.* **1995**, *78*, 2279–2287.
[67] J. Perrin, P. Molinas-Mata, P. Belenguer, *J. Phys. D: Appl. Phys.* **1994**, *27*, 2499–2507.
[68] J.-L. Dorier, C. Hollenstein, A. A. Howling, U. Kroll, *J. Vac. Sci. Technol.* **1992**, *A 10*, 1048–1052.
[69] W. W. Stoeffel, E. Stoeffel, G. M. W. Kroesen, M. Haverlag, J. H. W. G. de Boer, F. J. de Hoog, *Plasma Sources Sci. Technol.* **1994**, *3*, 321–324.
[70] A. Bouchoule, A. Plain, L. Boufendi, J.-P. Blondeau, C. Laure, *J. Appl. Phys.* **1991**, *70*, 1991–2000.
[71] A. Bouchoule, L. Boufendi, J. Hermann, A. Plain, T. Hbid, G. Kroesen, W. W. Stoeffel, *Pure & Appl. Chem* **1996**, *68*, 1121–1126.
[72] C. Courteille, L. Sansonnens, J. Dutta, J.-L. Dorier, C. Hollenstein, A. A. Howling, U. Kroll, in: *12th EC Photovoltaic Solar Energy Conference* (Eds.: H. S. Stephens & Associates), Amsterdam **1994**, Vol. 1, pp. 319–323.
[73] A. Bouchoule, L. Boufendi, *Plasma Sources Sci. Technol.* **1993**, *2*, 204–213.
[74] M. Shiratani, H. Kawasaki, T. Fukuzawa, H. Tsuruoka, T. Yoshioka, Y. Watanabe, *Appl. Phys. Lett.* **1994**, *65*, 1900–1902.
[75] Y. Watanabe, M. Shiratani, M. Yamashita, *Plasma Sources Sci. Technol.* **1993**, *2*, 35–39.
[76] L. Boufendi, A. Bouchoule, *Plasma Sources Sci. Technol.* **1994**, *3*, 262–267.
[77] J.-P. Boeuf, P. Belenguer, *J. Appl. Phys.* **1992**, *71*, 4751–4754.
[78] P. Belenguer, J.-P. Blondeau, L. Boufendi, M. Toogood, A. Plain, A. Bouchoule, C. Laure, J.-P. Boeuf, *Phys. Rev.* **1992**, *A 46*, 7923–7933.
[79] A. A. Fridman, L. Boufendi, T. Hbid, B. V. Potapkin, A. Bouchoule, *J. Appl. Phys.* **1996**, *79*, 1303–1314.
[80] W. Calleja, C. Falcony, A. Torres, M. Aceves R. Osorio, *Thin Solid Films* **1995**, *270*, 114–117.
[81] S. Veprek, Z. Iqbal, F. A. Sarott, *Phil. Mag.* B **1995**, *45*, 137–145.
[82] A. Nakamura, F. Emoto, E. Fujii, A. Yamaoto, Y. Uemoto, K. Senda, G. Kano, *J. Appl. Phys.* **1989**, *66*, 4248–4251.
[83] H. Hofmeister, T. Junghanns, *J. Non-Crystalline Solids* **1995**, *192/193*, 550–555.
[84] H. Hofmann, J. Dutta, *Sintering Technology* (Eds.: R. M. German, G. L. Messing, R. G. Cornwall), Mercel Dekker, New York **1996**, 101–108.
[85] T. Kretz, D. Pribat, P. Legagneux, F. Plais, *Jpn. J. Appl. Phys.* **1995**, *34*, L660–L663.
[86] K. Lu, *Phys. Rev. B* **1995**, *51*, 18–27.
[87] L. Battezatti, E. Garrone, *Z. Metallkde* **1984**, *75*, 305–310.
[88] J. S. Caster, M. O. Thompson, et al., *Appl. Phys. Lett.* **1994**, *64*, 437–439.
[89] F. G. Shi, *J. Mater. Res.* **1994**, *9*, 1307–1313.
[90] K. M. Unruh, B. M. Patterson, S. I. Shah, *J. Mater. Res.* **1992**, *7*, 214–218.
[91] A. Jagota, *J. Am. Ceram. Soc.* **1994**, *77*, 2237–2239.

[92] V. G. Gryaznov, L. I. Trusov, *Progress in Materials Science* **1993**, *37*, 289–401.

[93] R. K. Iler, *The Chemistry of Silica*, Wiley, New York **1979**.

[94] C. Bossel, J. Dutta, R. Houriet, J. Hilborn, H. Hofmann, *Mat. Sci. & Eng.* **A 1996**, *204*, 107–112.

[95] F. E. Kruis, K. A. Kusters, S. E. Pratsinis, *Aerosol Science & Technol.* **1993**, *19*, 514–526.

[96] For a review, see *Plasma Sources Sci. Technol.* **1994**, *3* and *J. Vac. Sci. Technol.* **1996**, *A14*.

[97] S. Furukawa, T. Miyasato, *Phys. Rev. B* **1988**, *38*, 5726–5731.

[98] W. L. Wilson, P. F. Szajowski, L. E. Brus, *Science* **1993**, *262*, 1242–1244.

[99] L. E. Brus, *J. Phys. Chem.* **1994**, *98*, 3575–3581.

[100] J.-C. Valmalette, L. Lemaire, G. L. Hornyak, J. Dutta, H. Hofmann, *Analusis* **1996**, *24*, M23–M25.

Chapter 9

Electron Transfer Processes in Nanostructured Semiconductor Thin Films

P. V. Kamat

9.1 Introduction

In recent years, researchers from various disciplines have been exploring novel and interesting properties of semiconductor nanoclusters (see, for example, recent review articles [1–11]). These ultrafine semiconductor particles have potential applications in the area of microelectronics, photovoltaics, imaging and display technologies, sensing devices, and thin film coatings. For example, by making use of the principles of photoelectrochemistry, semiconductor nanoclusters have been successfully employed in the conversion of light energy [10, 12, 13] and photocatalytic detoxification of air and water [14–19]. The unique properties and possible applications of semiconductor nanoclusters are summarized in Figure 9.1.

Under bandgap excitation, semiconductor nanoclusters act as short-circuited microelectrodes and directly oxidize and reduce the adsorbed substrates (Figure 9.2(a)). Alternatively they can promote a photocatalytic reaction by acting as me-

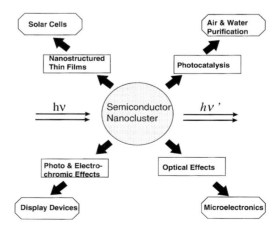

Figure 9.1. Applications of semiconductor nanoclusters.

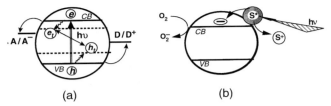

Figure 9.2. Photoinduced charge transfer processes in semiconductor nanoclusters: (a) under bandgap excitation, and (b) sensitized charge injection by exciting adsorbed sensitizer (S). *CB* and *VB* refer to conduction and valence bands of the semiconductor and e_t and h_t refer to trapped electrons and holes respectively.

diators for the charge transfer between two adsorbed molecules (Figure 9.2(b)). This process, which is commonly referred as photosensitization, is extensively used in photoelectrochemistry and imaging science. In the first case, the bandgap excitation of a semiconductor particle is followed by the charge transfer at the semiconductor–electrolyte interface. However in the second case, the semiconductor particle quenches the excited state by accepting an electron and then transfers the charge to another substrate or generates photocurrent. The energy of the conduction and valence bands of the semiconductor and redox potential of the adsorbed molecule control the reaction course of the photochemical reaction.

In earlier reviews we have focused on the aspects of charge separation in semiconductor nanocrystallites [3, 5, 10]. The efficiency of charge separation can be greatly improved by employing surface modifiers, composite systems, and sacrificial donors/acceptors. One collective way to utilize these photoinduced charges in semiconductor nanoclusters is to assemble them on a conducting surface in the form of thin films. The photogenerated charges in these semiconductor nanoclusters can then be utilized collectively to generate photocurrent or carry out selective oxidative and reductive processes. It should be noted that this area of material science is still in its infancy and has great potential for further technological advances.

9.2 Preparation and Characterization of Nanostructured Semiconductor Thin Films

Several efforts have been made in recent years to synthesize thin semiconductor films by chemical, electrochemical, and organized assembly methods [10, 12, 20–22]. A simple approach involves casting of thin films directly from colloidal semiconductor suspensions [22]. This method of preparation is relatively simple and inexpensive compared with the other existing methods such as chemical vapor deposition or molecular beam epitaxy. Preparation of semiconductor nanoclusters in polymer films [23–34] and LB films [35–39] has also been considered. The sol–gel technique has been found to be useful in developing nanostructured semiconductor membranes with either a 2D [40] or 3D configuration [41–46]. Organic-template-mediated synthesis has been employed to develop nanoporous tin(IV) sulfide materials [47]. The nanostructured films are highly porous and can easily be surface modified with sensitizers, redox couples, and other nanostructured semiconductors [20, 39, 48–65].

9.2.1 From Colloidal Suspensions

The nanostructured semiconductor films of different metal oxides, SnO_2 [55], ZnO [50, 51, 66–73], TiO_2 [37, 52, 54, 59, 61, 74–81], WO_3 [82, 83] and Fe_2O_3 [60] have been prepared from colloidal suspensions. By controlling the preparative conditions of semiconductor precursor colloids it is possible to tailor the properties of these semiconductor films. These thin metal oxide films exhibit interesting photochromic, electrochromic, photocatalytic, and photoelectrochemical properties that are inherited from the native colloids.

The synthetic procedure involves preparation of ultrasmall semiconductor particles (particle diameter 2–10 nm) in aqueous or ethanolic solutions by controlled hydrolysis. Colloidal suspension of SnO_2 (particle diameter ~ 10 nm) is also commercially available (Johnson Matthey). The colloidal suspension of the metal oxide semiconductor ($\sim 1\%$) is coated onto a conducting glass plate (referred to as an optically transparent electrode, OTE) and dried on a warm plate. The film is then annealed at 200–400 °C in air for about 1–2 hours. The conducting surface facilitates direct electrical contact to the nanostructured semiconductor thin films. The schematic diagram in Figure 9.3 illustrates the methodology of preparing thin film from colloidal suspensions.

This simple approach of coating preformed colloids on a desired surface and annealing produces a thin semiconductor film that is robust with excellent stability in

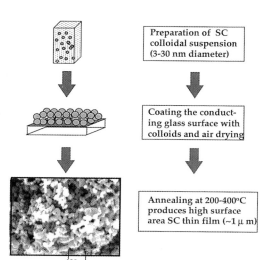

Figure 9.3. Methodology adapted for preparing nanostructured semiconductor thin films. A scanning electron micrograph of nanostructured TiO_2 film is also shown.

Preparation of SC colloidal suspension (3-30 nm diameter)

Coating the conducting glass surface with colloids and air drying

Annealing at 200-400°C produces high surface area SC thin film (~1 μ m)

100 nm

both acidic and alkaline media (pH range 1–13). Usually one can achieve a thickness in the range 0.1–1 mm. It may be necessary to optimize the concentration of precursor colloid for a particular application since higher colloid concentrations lead to cracking of the film. The above-mentioned procedure can be repeated several times to cast thicker films. (For example, see corresponding references for the methodology of preparation of SnO_2 [55], ZnO [51], TiO_2 [84], and WO_3 [83] films).

Transmission electron micrographs of nanostructured films prepared from colloidal suspensions show a three-dimensional network of metal oxide nanocrystallites of particle diameter < 5 nm [55]. No significant aggregation or sintering effects could be seen during the annealing process. XRD analysis has also confirmed the crystallinity of these nanostructured films. A similar approach that consists in mixing two or more components prior to casting films has been considered to prepare composite films. For example, thin films of SnO_2/TiO_2 composite semiconductors [85] and CdS in polymer (polyvinyl carbazole or polymethylmethacrylate) [86] have been synthesized. These composite films exhibit improved charge separation properties compared to single-component semiconductor films.

9.2.2 Chemical Precipitation

It is also possible to carry out chemical precipitation of the desired semiconductor directly on the surface of another semiconductor or a conducting surface. This technique is especially convenient to grow thin films of II–VI compound semiconductor nanocrystallites, as illustrated in Figure 9.4 [51, 87, 88]. For example, by successively dipping the ZnO film in Cd^{2+} and S^{2-} solutions one can cast a

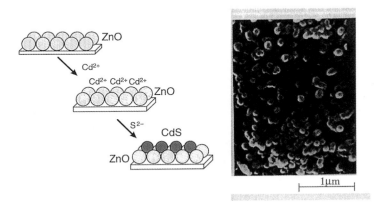

Figure 9.4. Designing semiconductor heterostructures with selective adsorption of ions from solution (right). Scanning electron micrograph of CdS nanocrystallites deposited on a ZnO thin film (left). The chemical precipitation was carried out by dipping ZnO film in $Cd(ClO_4)_2$ and Na_2S solutions sucessively. (From Reference [22]. Reprinted with permission from the American Chemical Society.)

thin film of CdS nanocrystallites. The yellow coloration of the film indicates formation of CdS on the ZnO surface. Such a chemical precipitation technique is especially useful for designing composite-type semiconductor films. The formation of uniform-size CdS clusters of diameter ~20 nm is evident from the scanning electron micrograph shown in Figure 9.4.

9.2.3 Electrochemical Deposition

Electrochemical deposition is also a convenient technique for casting thin films of II–VI compound semiconductors, viz., CdS, CdSe, CdTe, and several mixed semiconductor films. Quantum size TiO_2 crystallites have also been deposited on conductive surfaces by anodic oxidative hydrolysis of $TiCl_3$ [75]. The synthetic details of the electrochemically deposited semiconductor films are discussed elsewhere [20, 48, 63, 89]. Electrochemically deposited semiconductor films are strongly adherent to the substrate and are composed of aggregated nanocrystallites [20, 90, 91]. The nanocrystal size distribution (isolated or aggregated) can be controlled by the deposition current and temperature.

9.2.4 Self-assembled Layers

The surface-assembled layers of functionalized molecules that interact with the solid surface (e.g., thio compounds on gold surfaces) have been successively used to cast thin semiconductor films. Fendler and his coworkers [92, 93] have constructed ordered nanostructured films of layer-by-layer self-assembly of cationic poly(diallylmethylammonium chloride) and negatively charged solid particles on a variety of substrates. More details on this procedure can be found elsewhere [93].

9.2.5 Surface Modification

The porous metal oxide semiconductor films prepared from colloidal suspensions have a great affinity for interaction with organic dyes, redox couples, and organometallic complex molecules. This technique is especially convenient for extending the photoresponse of large-bandgap semiconductors with sensitizing dye molecules or for making the nanostructured films electrochemically active [55, 76, 94]. Because of the high porosity, large amounts of the sensitizing molecules (up to 0.1 M) can be incorporated in a nanocrystalline film of thickness of ~1 μm. Figure 9.5 shows deposition of cresyl violet on nanocrystalline SnO_2 film by the adsorption technique.

The adsorbed dye has a blueshifted absorption band compared to the absorption band of the monomer. Significant interaction between the adsorbed molecules often leads to the aggregation effects. The deposition of monomeric and aggregate forms of chlorophyll and chlorophyllin on ZnO, TiO_2 and SnO_2 films has also been carried out [49, 57, 96].

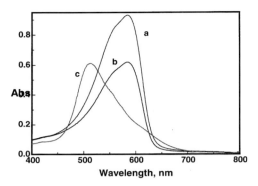

Figure 9.5. Absorption spectra of dye-modified SnO_2 film. The absorption spectra of the dye in aqueous solution before and after immersing the OTE/SnO_2 film shows the decrease in dye concentration as it is adsorbed on the SnO_2 nanocrystallites. (From Reference [95]. Reprinted with permission from the American Chemical Society.)

9.3 Optical Properties

9.3.1 Electron Storage and Photochromic Effects

Reversible electrochromic and photochromic effects can be observed with thin films made from metal oxide colloids such as WO_3, SnO_2, and TiO_2 [74, 80, 82, 83, 97–99]. A blue coloration quickly develops as a result of electron storage at the trap sites when these films are subjected to UV irradiation or to an electrochemical (cathodic) bias. This effect can also be induced by radiolysis [100]. A sonochromic effect has also been observed during the sonolysis of WO_3 colloidal suspension [101].

The changes in the absorption observed at a nanostructured WO_3 film at different applied potentials are shown in Figure 9.6. The film turns blue as we irradiate these films with UV light. This is indicated by the increase in the absorption at wavelengths greater than 500 nm.

Similar blue coloration was also observed when the film was subjected to a negative bias. The onset potential at which the electrochromic effect is observed corresponds to the flat band potential of the corresponding semiconductor and is dependent on the pH of the medium. By monitoring the optical absorbance, Fitzmaurice and his coworkers [74–105] have determined the flatband potentials of nanostructured films of TiO_2 and GaAs and the extinction coefficient of the trapped electrons. On the other hand, ZnO [70], SnO_2 [55], and CdS [106] films exhibit bleaching at potentials more negative than the flat band potential. The principle behind these chromic effects is shown in Figure 9.7.

The bandgap excitation of WO_3 particulate film leads to charge separation followed by trapping of charge carriers (Eq. 9.1a),

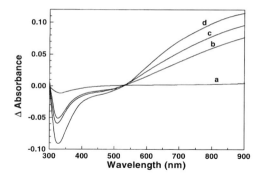

Figure 9.6. Effect of bandgap excitation on the absorption of WO$_3$ particulate film cast on an OTE glass plate. Difference absorption spectra were recorded, (a) before photolysis, (b) 5, (c) 10, and (d) 30 seconds after UV photolysis. The film was preannealed at a temperature of 423 K. (From Reference [83]. Reprinted with permission from the American Chemical Society.)

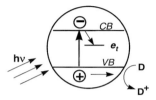

Figure 9.7. Photoinduced charge separation and electron trapping processes leading to photo-chromic effect in metal oxide semiconductor nanocrystallites. *CB* and *VB* refer to conduction and valence bands of the semiconductor, and e_t and h_t refer to trapped electrons and holes respectively.

$$WO_3 \; h\nu > E_g \rightarrow WO_3 \; (e_{CB} \ldots h_{VB}) \rightarrow WO_3(e_t + h_t) \tag{9.1a}$$

$$WO_3 \; (e_t + h_t) + D \rightarrow WO_3 \; (e_t) + D^+ \tag{9.1b}$$

where e_{CB} and e_{VB} refer to free charge carriers in the conduction and valence bands, e_t and h_t represent trapped electrons and holes, and D is a hole scavenger respectively. In the presence of a sacrificial electron donor such as oxalic acid the photo-generated holes are quickly scavenged (Reaction (9.1b)) thus reducing the possibility of charge recombination. Similar stabilization of trapped electrons has been observed for other metal oxide colloids such as TiO$_2$, ZnO, and WO$_3$. These electrons trapped at the metal ion sites (e.g., Ti^{4+} in TiO$_2$ or W^{6+} in WO$_3$) are known to exhibit characteristic broad absorption in the red–IR region. The photochromic effect observed with WO$_3$ particulate film was reversible. When the UV irradiation was stopped, the blue color slowly disappeared in air and the original color of the film was restored. The recovery in air was rather slow (10–20 min) since the reduction potential of O$_2$ is slightly more negative than the flat band potential of WO$_3$ ($E_{fb} \sim -0.2$ V vs. NHE).

Spectroelectrochemical and microwave absorption experiments suggest that trapped electrons are the major species responsible for the blue coloration of the film [58, 83]. The biphotonic dependence of microwave conductivity indicated that the free carriers in the conduction band of the semiconductor can only be achieved by reexciting the trapped charge carriers with a second photon. Since the trapped electrons have a long lifetime, this provides a convenient method of storing electrons. Investigations of the trapping process by picosecond laser flash photolysis indicate that this trapping process occurs in a subnanosecond timescale [82]. The electron storage effects in such semiconductor nanostructures have potential applications not only in electrochromic devices, but also in building electronic devices such as microcapacitors. Such a concept has also been demonstrated recently by employing a WO_3 electrode as a counterelectrode in a photochemical cell [107].

9.3.2 Photocurrent Generation

The semiconductor particles immobilized on a conducting glass behave as an interconnected array of microelectrodes. When subjected to bandgap excitation they are capable of generating photocurrent collectively. This makes them especially suitable as photosensitive electrodes in photoelectrochemical cells, the principle of which is illustrated in Figure 9.8.

The photoelectrochemical characteristics of these nanostructured semiconductor films are similar to those of a polycrystalline semiconductor material. Under illumination with UV light these metal oxide films undergo charge separation to form electron–hole pairs. The differing rates of electron or hole transfer into the electrolyte results in the accumulation of one of the charge carriers within the film [48, 55, 61, 63, 84, 108, 109]. For example, one can accumulate electrons in a nanocrystalline TiO_2 film if photogenerated holes are scavenged by hydroxide ions. The varying degree of electron accumulation changes the pseudo-Fermi level and creates a potential gradient to drive away the electrons towards the collecting surface. The

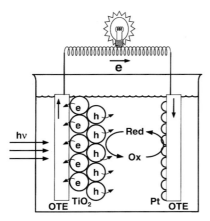

Figure 9.8. Mechanism of photocurrent generation in a nanostructured semiconductor film. The electrons accumulated in TiO_2 nanocrystallites are collected at the optically transparent electrode (OTE) surface and are transferred to the counter electrode via the external circuit.

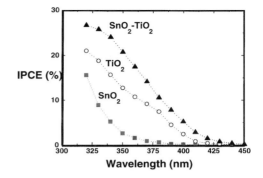

Figure 9.9. Photocurrent action spectrum of (a) OTE/SnO$_2$ (■), (b) OTE/TiO$_2$ (○), and (c) OTE/SnO$_2$/TiO$_2$, (▲) electrode in deaerated 0.02 M NaOH. (Excitation source: monochromatic light from the xenon lamp. The electrodes were maintained at a bias potential of 0.8 V. The composition of composite semiconductor film was 0.18 mg cm^{-2} of SnO$_2$ and 0.18 mg cm^{-2} of TiO$_2$.) See Equation (9.2) for the analysis of IPCE. (From Reference [85]. Reprinted with permission from the American Chemical Society.)

photoelectrochemical properties of these films are susceptible to surface-adsorbed species such as oxygen [84], charge recombination at the grain boundaries [55, 61], and leakage of electrons into the solution instead of generating photocurrent [63, 110]. However, a properly designed semiconductor particulate system and a suitably matched redox couple can greatly improve the efficiency of charge separation and thus the photoelectrochemical performance of the nanostructured semiconductor film.

The photoelectrochemical response and absorption spectra of the SnO$_2$, TiO$_2$, and SnO$_2$/TiO$_2$ nanoclusters immobilized on a conducting glass surface are shown in Figure 9.9. The incident photon-to-photocurrent efficiency (IPCE) was determined by measuring the photocurrent of the OTE/TiO$_2$ electrode at various excitation wavelengths and using the expression

$$\text{IPCE } (\%) = 100(1240 i_{\text{sc}})/(\lambda I_{\text{inc}}) \tag{9.2}$$

where i_{sc} is the short-circuit current photocurrent (A cm^{-2}), I_{inc} is the incident light intensity (W cm^{-2}) and λ is the excitation wavelength (nm). The onset of photocurrent is seen at the wavelengths ~ 360 and ~ 400 nm for SnO$_2$ and TiO$_2$ films, respectively. The increase in the photocurrent at excitation wavelengths below this onset wavelength closely matches the absorption characteristics of SnO$_2$ ($E_{\text{g}} = 3.5$ eV) and anatase TiO$_2$ ($E_{\text{g}} = 3.2$ eV). This indicates that the observed photocurrent is initiated by the excitation of semiconductor nanoclusters in the thin film. The composite film, OTE/SnO$_2$/TiO$_2$, exhibited higher IPCE than OTE/SnO$_2$ or OTE/TiO$_2$. A maximum IPCE of $\sim 25\%$ was obtained at 325 nm. The higher IPCE of the OTE/SnO$_2$/TiO$_2$ electrode is suggestive of increased charge separation in the composite film. The generation of anodic current is also indicative of the fact that the direction of flow of electrons is towards the OTE surface.

One can observe novel effects from the thin films consisting of semiconductor structures in the form of multiple quantum wells (MQW) or superlattices (SL). Nozik and his coworkers have carried out an photoelectrochemical study on 250 Å-thick $GaAs/GaAs_{0.5}P_{0.5}$ superlattice film [111–114]. The photocurrent action spectra of lattice-matched superlattice electrodes in photoelectrochemical cells show structure with stepped waves that correspond to the quantum states in the quantum wells of superlattice. The question regarding the hot electron transfer from upper quantum states into solution has also been addressed both with a kinetic model and experimental results [114–116]. The flat band potential and photocurrent spectra of a single quantum well was also investigated in these studies.

9.3.3 Sensitization of Large-Bandgap Semiconductors

The process of utilizing subbandgap excitations with dyes is referred to as photosensitization and is conveniently employed in color photography and other imaging science applications. This approach of light energy conversion is similar to plant photosynthesis, in which chlorophyll molecules act as light-absorbing antenna molecules. The dye-modified semiconductor films provide an efficient method for mimicking the photosynthetic process. Bignozzi et al. have presented a supramolecular approach for designing photosensitizers [117]. By optimizing the design of light-harvesting molecules (sensitizers) it should be possible to suppress the interfacial charge recombination and improve the cross section for light absorption.

The strong surface-bonding property of the nanostructured semiconductor films facilitates surface modification with electrostatic or charge transfer interactions. The high porosity of these films enables incorporation of sensitizing dyes in large concentrations. The nanostructured TiO_2 films modified with a ruthenium complex exhibit photoconversion efficiencies in the range of 10% [76], which is comparable to that of amorphous-silicon-based photovoltaic cells.

The photocurrent response evaluated in terms of photon-to-photocurrent efficiency (IPCE) of SnO_2 is shown in Figure 9.10 [55]. The IPCE maximum of the

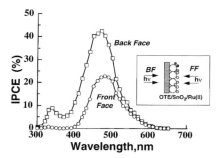

Figure 9.10. Sensitized photocurrent generation at an SnO_2 nanocrystalline semiconductor film modified with a ruthenium(II) polypyridyl complex $(Ru(bpy)_2(dcbpy)^{2+})$ in a photoelectrochemical cell containing 0.04 M I_2 and 0.5 M LiI in acetonitrile as electrolyte. (Reprinted with permission of the American Chemical Society. From Reference [55].)

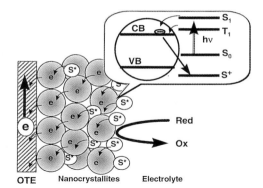

Figure 9.11. Mechanism of sensitized photo-current generation in a nanocrystalline semiconductor film. Charge injection from excited sensitizer (S*) into semiconductor nanocrystallites and regeneration of the sensitizer by the Red/Ox couple are two important steps in this process.

surface-modified SnO_2 film closely matches the absorption maximum of the sensitizer. The SnO_2 film, which is sensitive only to UV excitation (see Figure 9.9) prior to surface modification, responds to the visible light (wavelengths greater than 400 nm) as a result of surface modification. This shows that a photosensitization mechanism (Figure 9.11) is operative in extending the photocurrent response of the SnO_2 film.

When the electrode is illuminated with visible light, the sensitizer molecules absorb light and inject electrons into the SnO_2 particles. These electrons are then collected at the conducting glass surface to generate anodic photocurrent. The redox couple (e.g. I_3^-/I^-) present in the electrolyte quickly regenerates the sensitizer. By choosing an appropriate sensitizer it is possible to tune the photoresponse of these nanostructured semiconductor films. For example, sensitizing dyes such as chlorophyll *a* [118] and chlorophyll *b*, [49, 57], squaraines [62, 119], rhodamine [120], and oxazines [94] can extend the photoresponse of SnO_2 films to the red–infrared region. The maximum IPCE in the example discussed in Figure 9.10 (around 50%) shows that nearly half of the injected charge from the excited sensitizer is lost as a result of recombination with the oxidized sensitizer. By optimizing the operating conditions it is possible to improve the performance of the IPCE of the sensitizer-based cells. Ru-complex modified TiO_2 nanostructured films exhibit an IPCE of nearly 90% under optimized light-harvesting conditions [76]. Both experimental and theoretical evaluations of these cells have been carried out and the efficiency-limiting factors have been identified [121–123].

9.3.4 Photocatalysis

The photocatalytic properties of anatase TiO_2 particles in degrading undesirable organics from air and water are well documented [15, 17, 18, 124–126]. Organic materials such as hydrocarbons, haloaromatics, phenols, halogenated biphenyls, surfactants, and textile dyes in TiO_2 slurries have been successfully mineralized. For reactor applications it is convenient to immobilize the semiconductor particles on a suitable substrate. Few studies have been reported with TiO_2 particles immobilized on glass substrates [125, 127, 128].

When semiconductor particles are subjected to bandgap excitation, charge separation occurs in each of these particles. One can thus utilize these charge carriers to carry out oxidation and reduction on the same particle. In aqueous solutions the holes at the TiO_2 surface are scavenged by surface hydroxyl groups to generate $^{\bullet}OH$ radicals, which then oxidize the dissolved organics. One of the disadvantages of such a system is the high degree of recombination between photogenerated charge carriers within the individual particles. This is usually overcome by scavenging electrons with a sacrificial electron acceptor such as dissolved oxygen so that the holes can participate in the oxidation of the organics. Therefore scavenging of electrons becomes a limiting factor in controlling the photocatalytic oxidation of organics [129].

Thin semiconductor particulate films coated on a conducting surface provide a convenient way of manipulating the photocatalytic reaction by electrochemical methods [84, 125, 130]. A better charge separation in these films can be achieved by applying an anodic bias to the immobilized semiconductor nanocrystallites [59, 84, 131]. This principle is similar to the one employed by Fujishima and Honda [132] for splitting water at a single TiO_2 crystal using an anodic bias. In an electrochemically assisted photocatalytic process (ECAP) the externally applied anodic bias greatly improves the efficiency of charge separation by driving the photo-generated electrons via the external circuit to the counterelectrode compartment. The degradation of 4-chlorophenol and azo dyes provides a representative example of the use of ECAP in elucidating the mechanism of degradation [59, 84, 85, 131]. By controlling the applied bias potential it is possible to control the degradation rate. The effect of applied bias on the degradation of 4-chlorophenol is shown in Figure 9.12.

The photocatalytic degradation occurs at a faster rate when the applied potential is maintained at $+0.6$ V, while little degradation is seen when the potential is

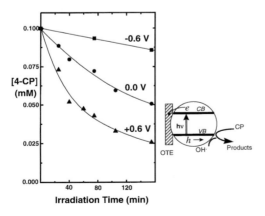

Figure 9.12. Dependence of 4-chlorophenol degradation rate on the externally applied bias. The nanocrystalline TiO_2 film was maintained at constant potentials of (a) -0.6, (b) 0.0, and (c) $+0.6$ V vs. SCE during the photolysis, and the solution was continuously bubbled with a slow stream of nitrogen. (From Reference [84]. Reprinted with permission from the American Chemical Society.)

maintained at -0.6 V vs. SCE. Since the charge separation in the TiO_2 particulate film is maximum when an anodic bias is applied to the OTE/TiO_2 electrode, one observes a higher efficiency for photocatalytic degradation. At potentials close to the flat band potential (-0.6 V vs. SCE) all the electron–hole pairs are lost in the recombination process and hence it is not possible to carry out the oxidation of 4-chlorophenol. In a slurry system, the irradiated particles behave as short-circuited microelectrodes and thus the interfacial charge transfer competes with the charge recombination process. This situation closely resembles the experimental condition in which the OTE/TiO_2 is maintained at 0.0 V (curve b in Figure 9.12). Thus, nanostructured semiconductor films are useful in carrying out electrochemically assisted photocatalysis and overcome the limitation of the electron scavenging process that one encounters in the slurry system.

The advantage of using such an electrochemically assisted photocatalytic (ECAP) technique is not just limited to the faster degradation rates. The electrochemical arrangement also provides a unique opportunity to separate the anodic and cathodic processes and thereby isolate the various reactions occurring in photocatalytic systems. The use of an anodic bias to separate the charge carriers obviates the need for oxygen as an electron scavenger and makes it possible to carry out the photocatalytic reaction in anaerobic conditions.

9.4 Mechanism and Electron Transfer in Semiconductor Thin Films

9.4.1 Charge Injection from Excited Dye into Semiconductor Nanoclusters

The energy difference between the conduction band of the semiconductor and oxidation potential of the excited sensitizer is the major driving force for the excited state charge transfer [133, 134]. Similarly, back electron transfer between the injected electron and oxidized sensitizer is also a controlling factor in maximizing net electron transfer efficiency. Figure 9.13 shows an illustration of energy levels of conduction and valence bands of metal oxide semiconductors and the oxidation potential of ground and excited states of ruthenium(II) polypyridyl complex (sensitizer).

Different approaches have been considered for studying the energy gap dependence of the photosensitization efficiency. Hashimoto *et al* [135, 136] have shown that the excited state lifetime of $Ru(bpy)_3^{2+}$ adsorbed on a metal oxide semiconductor is dependent on the conduction band energy of the semiconductor. Tani [137, 138] has made an effort to establish the energy gap dependence of the electron transfer between silver halides and J aggregates of the dye. Spitler and his coworkers [139, 140] have varied the pH to examine the energetic threshold for dye-sensitized photocurrent generation at $SrTiO_3$ and TiO_2 electrodes. The energy gap

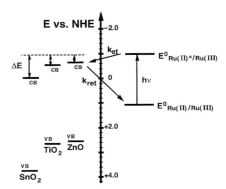

Figure 9.13. Schematic diagram illustrating the energy levels of semiconductor and redox potentials of sensitizer, $(Ru(bpy)_2(dcbpy)^{2+})$.

is dependent on pH since the conduction band of metal oxide semiconductor shifts by 0.059 V pH^{-1}.

9.4.2 Kinetics of the Charge Injection Process

A sensitizer adsorbed on a semiconductor surface can undergo heterogeneous electron transfer in addition to the radiative and radiationless deactivation processes. The excited state processes of a sensitizer (S) on a semiconductor surface (SC) are summarized in Eqs. (9.3–9.5).

$$S + h\nu \rightarrow S^* \rightarrow S + h\nu' \tag{9.3}$$

$$S^* \rightarrow S + heat \tag{9.4}$$

$$S^* + SC \rightarrow S^{+\cdot} + SC(e) \tag{9.5}$$

The emission lifetime of the adsorbed sensitizer serves as a good probe for studying the kinetics of heterogeneous electron transfer between the semiconductor and excited sensitizer (Eq. 9.5). On the basis of the luminescence spectra and lifetimes of $Ru(bpy)_3^{2+*}$ on various metal oxides, Hashimoto et al. [135, 136] have concluded that the interaction between the sensitizer and semiconductor as well as the energetics of the semiconductor and oxidation potentials of the sensitizer control the rate of heterogeneous electron transfer. It has been shown that the carboxylic acid group of $Ru(bpy)_2(dcbpy)^{2+}$ in the present case can provide a strong charge transfer or ester-type linkage between the semiconductor and sensitizer.

When the sensitizer is adsorbed on a nonreactive surface such as alumina, the excited state of the sensitizer is long lived. A slight deviation from exponential behavior, which is seen at short times, is attributed to the excited state annihilation process. The major component of this decay had a lifetime of 259 ns and is similar to the one observed in aqueous solutions. The long-lived excited state on the Al_2O_3 surface rules out any direct participation of the oxide support in deactivating excited sensitizer.

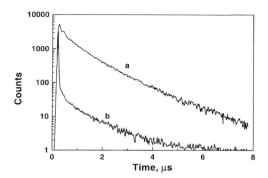

Figure 9.14. Emission decay profiles of Ru(bpy)$_2$(dcbpy)$^{2+*}$ adsorbed on (a) Al$_2$O$_3$ and (b) TiO$_2$ nanoclusters. (From Reference [141]. Reprinted with permission from the American Chemical Society.)

While the influence of surface interaction on the excited state deactivation is minimal for the Al$_2$O$_3$ sample, it is significant for the TiO$_2$ sample. When adsorbed on the TiO$_2$ surface, the emission decay of Ru(bpy)$_2$(dcbpy)$^{2+*}$ significantly deviates from the single exponential behavior. The observed nonexponential behavior is attributed to the existence of multiple injection/adsorption sites on the TiO$_2$ surface. Accordingly one would expect a wide range of lifetimes for the sensitizer adsorbed on TiO$_2$. A simple biexponential kinetic analysis can provide an estimate of the upper and lower limits of emission lifetimes, τ'_s $\alpha\nu\delta$ τ''_s. The fit of Ru(bpy)$_2$(dcbpy)$^{2+*}$ decay to the biexponential decay is shown in Figure 9.14. The lifetimes determined from this analysis are 1.69 ± 0.006 and 9.85 ± 0.044 ns for the fast and slow components respectively. If the decrease in lifetime observed on the TiO$_2$ surface is entirely due to charge injection process (Eq. 9.5), one could express lifetimes, τ'_s or τ''_s such that

$$\tau'_s = 1/(k_r + k_{nr} + k'_{et}) \qquad \tau''_s = 1/(k_r + k_{nr} + k''_{et}) \qquad (9.6)$$

where k_r and k_{nr} are the rate constants for radiative and nonradiative processes and k'_{et} and k''_{et} are the upper and lower limits for the heterogeneous electron transfer process respectively. The lifetime of Ru(bpy)$_2$(dcbpy)$^{2+*}$ in the absence of electron transfer quenching can be estimated from the measurements on the alumina sample ($\tau_0 = 259$ ns). If we assume that $(k_r + k_{nr})$ is the same on both oxide surfaces, one can obtain the rate constant for heterogeneous electron transfer from Ru(bpy)$_2$(dcbpy)$^{2+*}$ into TiO$_2$ particles from Eq. (9.7),

$$k'_{et} = 1/\tau'_s - 1/\tau_0 \qquad k''_{et} = 1/\tau''_s - 1/\tau_0 \qquad (9.7)$$

By substituting the values of τ'_s (1.69 ns) and τ''_s (9.85 ns) we obtain the values for k'_{et} and k''_{et} as 5.5×10^8 and 1.0×10^8 s^{-1} for the fast and slow component of the charge injection process.

The kinetic evaluation of the multiexponential luminescence decay of excited Ru complex adsorbed on SnO$_2$ suggests that multiple injection/adsorption sites exist on the surface of a semiconductor nanocrystallite. Evidence for the heterogeneity of the

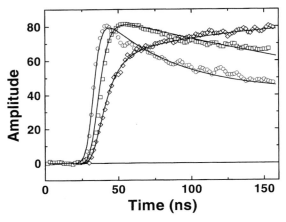

Figure 9.15. Microwave absorption amplitudes as a function of time for nanocrystalline semiconductor film excited at 355 nm (○), the Ru(II) compound on a SnO$_2$ nanocrystalline semiconductor film excited at 532 nm (□), and the Ru(II) complex on a ZnO nanocrystalline film also excited at 532 nm (◇). The solid lines are the calculated kinetic fits with an apparatus time constant of 3.5 ns. (From Reference [144]. Reprinted with permission from the American Chemical Society.)

injection site has been presented by Xie and his coworkers using far-field emission microscopy [142].

The rate constants for heterogeneous electron transfer between excited Ruthenium polypyridyl complex (Ru(II)) and semiconductor crystallites such as SnO$_2$, TiO$_2$, and ZnO lie in the range 10^7–10^9 s^{-1} [55, 141, 143]. Independent microwave absorption and luminescence measurements have been carried out to monitor the charge injection from excited Ru(bpy)$_2$(dcbpy)$^{2+}$ into SnO$_2$, ZnO, and TiO$_2$ nanocrystallites [144]. The growth of microwave absorption (Figure 9.15) was delayed from the laser pulse by a process showing a similar rate constant to the fast decay portion of the luminescence.

The appearance of microwave conductivity at rates corresponding to the luminescence directly confirms the fast component of the heterogeneous electron rate constant to be in the range 1–3×10^8 s^{-1}.

The charge injection from singlet-excited sensitizer into the conduction band of a large-bandgap semiconductor is usually considered to be an ultrafast process occurring in the picosecond time domain. The charge injection process in the case of organic dyes such as anthracene carboxylate [145], squaraines [146], and cresyl violet [147, 148] has been shown to occur in less than 20 ps. Similarly fast electron transfer has also been noted for Ru(H$_2$O)$_2$$^{2-}$ on a TiO$_2$ surface at very low coverage [147]. Recent femtosecond transient spectroscopy of coumarin- [149] and perylene- [150] modified TiO$_2$ systems has confirmed that the sensitized charge injection is completed within a few hundred femtoseconds. For cresyl violet aggregates adsorbed on SnO$_2$ nanocrystallites, a rate constant of $\sim 4 \times 10^{11}$ s^{-1} has been reported for the charge injection process [151].

Relatively smaller charge injection rate constants (10^8–10^9 s^{-1}) have been reported for the charge injection from excited ruthenium complexes into semiconductor nanoparticles [55, 135, 136, 141, 143, 152–155]. The only exception is a recent study which reports a charge injection process for the TiO_2/Ru(II)(dcbpy)$_2$(SCN)$_2$ system to occur within the timescale of < 150 fs to 1.2 ps [156]. It is likely that electron trapping at surface defects becomes a major contributing factor in controlling the heterogeneous electron transfer at the semiconductor surface [157].

The charge injection from the triplet excited dyes into TiO_2 [139] and ZnO [158] colloids has also been shown to occur on a slower timescale. Similarly, the triplet excited states of cresyl violet aggregates inject electrons into SnO_2 nanocrystallites with a rate constant of 5×10^8 s^{-1} [159]. The electron transfer rate constant between the Ru(II) complex and the metal oxide semiconductor is thus comparable to the triplet excited organic dyes. It should be noted that the excited state of Ru(bpy)$_2$(dcbpy)$^{2+}$ involves a metal-to-ligand charge transfer state and implications are that such an electronic configuration of the excited state plays an important role in controlling the electron injection rates. The possibility of a connection between the multiplicity of the excited sensitizer and the rate constant for charge injection deserves more careful study.

9.4.3 Modulation of Electron Transfer at the Semiconductor–Dye Interface

Spectroelectrochemical measurements of metal oxide films have shown that externally applied electrochemical bias causes electron accumulation in semiconductor nanocrystallites [67, 74, 80, 82, 83, 102–104]. The onset potential at which the electron accumulation is seen corresponds to the flat band potential of the semiconductor. In the case of InP semiconductor the applied potential was shown to influence the hot electron injection process. Several researchers have observed an increase in the quenching of the excited state of the sensitizer adsorbed onto an n-type semiconductor electrode or increased production of oxidized sensitizer by biasing the electrode at positive potentials [139, 160, 161]. Recently resonance Raman spectroscopy [162] and transient absorption spectroscopy [163] have been employed to monitor the changes that occur on the nanocrystalline semiconductor surface at positive and negative bias potentials. O Regan et al. [164] have investigated the influence of externally applied bias on the charge injection efficiency and reverse electron transfer process in a TiO_2/Ru(II) system.

It is possible to carry out spectroelectrochemical experiments in a conventional laser flash photolysis set up by modifying the sample holder to accommodate the cell containing OTE/SnO$_2$/sensitizer as the working electrode(WE), Pt wire as the counterelectrode (CE), and Ag/AgCl as the reference electrode (RE) (Figure 9.16). In the example discussed here, a ruthenium-polypyridyl-complex-(Ru(II)) modified SnO$_2$ film cast on a conducting glass surface was used as the working electrode. The excitation of the Ru(II)-modified SnO$_2$ film was carried out in a front face geometry with a 532 nm laser pulse.

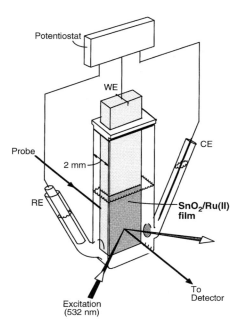

Figure 9.16. Spectroelectrochemical cell employed for probing the photochemical events in a dye-modified semiconductor thin film.

The transparency of the OTE/SnO$_2$/Ru(II) electrode facilitates direct monitoring of the transient absorbance following laser pulse excitation. The electrode potential is maintained constant at a desired value during the laser flash photolysis experiment.

The photoprocesses that follow the laser pulse excitation of Ru(II) complex on SnO$_2$ nanocrystallites are summarized in Eqs. (9.8) and (9.9):

$$Ru(II) + h\nu \rightarrow Ru(II)^* \rightarrow Ru(II) + h\nu' \tag{9.8}$$

$$Ru(II)^* + SnO_2 \rightarrow Ru(III) + SnO_2 \ (e) \tag{9.9}$$

The time-resolved transient absorption spectra recorded following the 532 nm laser pulse excitation of Ru(II)-modified SnO$_2$ electrode are shown in Figure 9.17. The only difference between the two sets of experiments is the applied potential, which was held constant at +0.2 V and −0.7 V respectively. The spectra recorded at early times exhibit spectral features similar to the spectrum of Ru(II)* observed on a SiO$_2$ surface. However, the applied bias influences the spectral characteristics of transients recorded at longer times. The formation of Ru(III) is evident only with the spectrum recorded with +0.2 V bias. The absence of Ru(III) and the appearance of excited Ru(II) (abs. max. at 380 nm) as the only observable transient at −0.7 V indicates the failure of Ru(II)* to participate in the charge injection process.

At potentials more negative than the flat band potential of SnO$_2$, Eq. (9.9) is suppressed, thus extending the lifetime of Ru(II)*. On the other hand, the applied positive bias facilitates electron transfer quenching of Ru(II)* on the SnO$_2$ surface (Eq. 9.8). The applied positive bias shifts the pseudo-Fermi level of the SnO$_2$

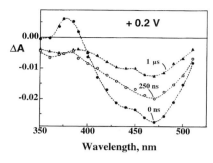 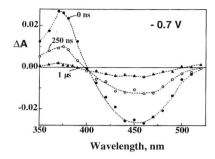

Figure 9.17. Time-resolved absorption spectra recorded following the laser pulse excitation of Ru(II)polypyridyl-complex-modified SnO_2 film. The applied potential was held constant at +0.2 V (left) and −0.7 V vs. Ag/AgCl during the laser flash photolysis experiment. (See Reference [165] for details. Reprinted with permission of the American Chemical Society.)

nanocrystallites in such a way that it provides the necessary driving force for the heterogeneous electron transfer. These results provide an explanation for the earlier photocurrent and luminescence measurements that the applied potential directly controls the heterogeneous electron transfer at the semiconductor interface.

The example discussed above shows that the electron transfer quenching of excited sensitizer on SnO_2 nanocrystallites can be modulated with an externally applied electrochemical bias. Since the rate constant of back electron transfer (k_{ret}) is slower than the charge injection rate constant, k_{et}, we can correlate Ru(III) yield to the fraction of Ru(II)* quenched via the electron transfer route:

$$\{Ru(III)\} = (k_{et}/(k_r + k_{nr} + k_{et}))\{Ru(II)*\} \tag{9.10}$$

where k_r and k_{nr} are rate constants for radiative and nonradiative decay of Ru(II)*.

The dependence of the electron transfer rate constant on the applied bias is shown in Figure 9.18. The apparent rate constant for the charge injection process was constant at potentials greater than −0.2 V with a maximum k_{et} of 4×10^8 s^{-1}. At potentials more negative than −0.2 V, a sharp decrease (more than three orders of magnitude) in k_{et} is observed. At −0.7 V the charge injection process is almost completely suppressed as the deactivation of the excited state is dominated only by the radiative route (i.e., k_{et} becomes smaller than k_r). It should be noted that the applied potential at which we observe this effect is close to the oxidation potential (0.72 V vs. Ag/AgCl) [166, 167] of the excited sensitizer.

Application of an external bias alters the pseudo-Fermi level of the nanocrystalline film. At negative bias the electrons are accumulated within the SnO_2 nanocrystallites, thus shifting the pseudo-Fermi level to more negative potentials. Under these conditions the difference in energy (ΔE) between E'_F *and* $E^0_{Ru(II*/III)}$ that acts as a driving force for the heterogeneous electron transfer decreases. Tani [137, 138], who observed a similar dependence for the quenching of cyanine dyes on AgBr crystals, has employed Marcus theory [168] to explain the initial steep threshold in the $-\Delta E$ region. But the failure to observe the inverted region was attributed to the

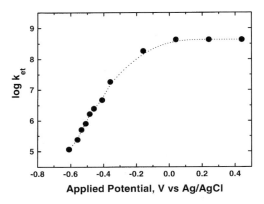

Figure 9.18. The dependence of the heterogeneous charge transfer rate constant on the externally applied electrochemical bias. The Ru(II)-complex-modified SnO_2 film was excited with a 532 nm laser pulse while the electrode was maintained at a desired potential. (See Reference [165] for details. Reprinted with permission of the American Chemical Society.)

semiconducting nature of the acceptor which is composed of continuously distributed electronic energy levels. Similarly, the inverted region was not observed for the heterogeneous electron transfer between Ru(II)* and SnO_2 nanocrystallites. Although the electron transfer theory [168–171] predicts that electron transfer rates should decrease with increasing thermodynamic driving force, this phenomenon has been rarely demonstrated for interfacial or electrochemical processes [172].

9.4.4 Back Electron Transfer

Although the redox couple such as I^{3-}/I^- in solution reacts with the oxidized sensitizer, the competing back electron transfer can significantly decrease the efficiency of net electron accumulation within the particles. Therefore the back electron transfer plays a major role in achieving high incident photon-to-photocurrent generation efficiency (IPCE) of photochemical solar cells. For example, the high IPCE observed in the case of ruthenium-polypyridyl-complex/TiO_2-based photochemical cells has been attributed to the slower back electron transfer process [76, 164, 165, 173]. On the other hand, organic-sensitizer-based photoelectrochemical cells perform very poorly because back electron transfer processes occur at significantly higher rates than the one observed in Ru(II) polypyridyl complex/semiconductor systems. In order to assess the role of back electron transfer, we have monitored the fate of the electrons injected into the semiconductor nanocrystallites and their reaction with oxidized sensitizer molecules.

The photoinjected electrons, if not utilized quickly to generate photocurrent, undergo back electron transfer to regenerate the sensitizer:

$$S^{+\bullet} + SnO_2(e) \rightarrow S + SnO_2 \tag{9.11}$$

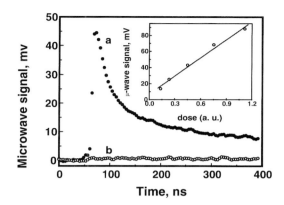

Figure 9.19. Microwave signal (mV) vs. time profiles of (a) $SnO_2/(CV)_2^{2+}$ and (b) $SiO_2/(CV)_2^{2+}$ films under 532 nm laser excitation. The inset shows the dependence of the microwave signal of $SnO_2/(CV)_2^{2+}$ on the laser dose. (From Reference [95]. Reprinted with permission of the American Chemical Society.)

In our previous studies we have demonstrated the usefulness of time-resolved microwave absorption measurements for studying the photosensitization aspects of metal oxide semiconductor particles [57, 144, 174]. Visible excitation of the dye-modified semiconductor sample led to an increase in the microwave conductivity of the sample. Since the observed microwave conductivity directly reflects the mobile charge carriers accumulated in SnO_2 nanocrystallites, we would expect its decay to reflect the back electron transfer process. This technique is complementary to measurement of optical absorption by $S^{+\bullet}$, in that the reaction partner is observed.

We have employed both microwave conductivity and transient absorption measurements to probe the back electron transfer in cresyl-violet-aggregate-capped SnO_2 nanocrystallites [95]. Typical transient decay profiles of cresyl-violet-aggregate-capped nanocrystalline SnO_2 and SiO_2 films are shown in Figure 9.19.

The films were cast on a fused silica plate to facilitate microwave absorption experiments. Excitation of the dye-modified SiO_2 film with a 532 nm laser pulse does not produce any significant microwave signal. On the other hand, excitation of the dye-modified SnO_2 film with a 532 nm laser pulse resulted in the prompt appearance of the microwave signal. This indicated that the charge injection process was completed within the laser pulse duration. The dependence of the initial magnitude of the microwave absorption on the laser intensity was linear (Figure 9.19, inset), as one would expect from a monophotonic charge injection behavior. The similarity of this result with that observed in the transient absorption measurements confirms the complementary nature of these two techniques in probing the charge injection process. The multiexponential behavior of the back electron transfer mainly arises from the presence of an inhomogeneous energy distribution of the trapping/detrapping sites at the semiconductor interface.

Detrapping of electrons from shallow traps is thermodynamically and kinetically favorable and is a rate-determining step in the back electron transfer. Hence it plays an important role in controlling the kinetics of back electron transfer. On the other hand, the activation energy needed to detrap electrons from deeper traps is significantly higher, and hence their role in the back electron transfer is considered to be negligible. The influence of trapping and detrapping on the photosensitization behavior of a ruthenium complex adsorbed on an n-type anatase TiO_2 electrode has been shown by Willig and his coworkers [157]. The multiexponential decay of the cation radical, which extends up to several hundred nanoseconds, suggests the existence of inhomogeneous trap sites.

9.4.5 Charge Transport in Semiconductor Films

Electron transport within the nanocrystalline semiconductor film is an important aspect of the development efficient electrodes for photochemical solar cells. Recently, significant attention has been given to understand the charge transport properties of nanocrystalline semiconductor films [20, 48, 61, 63, 84, 109, 118, 175–178]. Although conclusive evidence that could justify a quantitative model has yet to be established, a few qualitative explanations have already been invoked. One such explanation is the formation of a potential gradient within the semiconductor film that facilitates charge transport towards the collecting surface [20, 48, 61, 65, 84]. The varying degree of electron accumulation alters the quasi-Fermi level in such a way that a potential gradient is created within the thin film (Figure 9.20).

Formation of such a potential gradient provides the necessary driving force for the electron transport to the collecting surface of the OTE. Since this potential gradient is not an ideal type of Schottky barrier, a significant loss of electrons is encountered during the transit because of their recombination with $CV^{3+\bullet}$ at the grain boundaries. Since more grain boundaries are encountered in thicker films, one would expect greater loss of electrons in the back electron transfer at these grain boundaries. This increased loss of electrons during the transport is evident from the lower IPCE observed with films thicker than 0.4 μm in Figure 9.20. The measured IPCE in thicker films with excitation from the solution side was also smaller than that obtained with OTE side excitation. Since most of the absorption of incident light in thicker films occurs near the side of excitation, one would expect fewer encounters of grain boundaries when illuminated from the OTE side.

In the case of SnO_2/Chlorohyll-*a*-based systems it has been shown that more than 70% of photoinjected charge carriers are lost during their transit to the collecting surface of the electrode when the cell is operated under unbiased conditions [118].

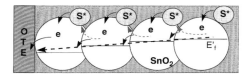

Figure 9.20. Transport of injected charge across semiconductor nanocrystallites. A significant number of electrons are lost as they recombine with the cation radical of the sensitizer at the grain boundaries during their transit to the collecting surface of OTE. E_f: refers to the quasi-Fermi level of the semiconductor nanocluster.

Thus electron transport within the nanocrystalline semiconductor film is a crucial aspect when one seeks to optimize the performance of the photoelectrochemical cells.

9.5 Conclusion

This review has sought to highlight photoinduced electron transfer processes in nanostructured semiconductor systems. These thin films exhibit unusual photocatalytic and photoelectrochemical properties. The porous morphology and high surface area make these thin films important for many practical applications. A variety of spectroscopic and electrochemical techniques are being employed to investigate the mechanism of heterogeneous electron transfer at the semiconductor-electrolyte interface as well as charge transport across the thin film. Kinetic study of these heterogeneous systems will aid in designing novel microheterogeneous assemblies to carry out efficient artificial photosynthesis. The ability to construct semiconductor microelectrode arrays in the form of thin films has demonstrated their potential use in the conversion and storage of solar energy and in the mineralization of chemical pollutants.

Acknowledgments

The work described herein was supported by the Office of Basic Energy Sciences of the US Department of Energy. I would like to thank my colleagues and collaborators (Drs. R. W. Fessenden, K. A. Gray, S. Hotchandani, K. Vinodgopal) and students (Mr. Idriss Bedja, Mr. Di Liu, Mr. C. Nasr), who have made a significant contribution to the research work presented here. This is contribution No. 3982 from the Notre Dame Radiation Laboratory.

References

[1] A. Henglein, *Chem. Rev.* **1989**, *89*, 1861–1873.
[2] A. Henglein, H. Weller, *Photochemical Energy Conversion* (Eds.: J. Norris, D. Meisel), Elsevier, New York **1989**.
[3] P. V. Kamat, *Chem. Rev.* **1993**, *93*, 267–300.
[4] H. Weller, *Angew. Chem., Int. Ed. Engl.* **1993**, *32*, 41–53.
[5] P. V. Kamat, *Progr. React. Kinetics* **1994**, *19*, 277–316.
[6] L. Brus, *Colloid and Interface Science* **1996**, *1*, 197–201.
[7] L. Brus, *J. Phys. Chem.* **1994**, *98*, 3575–3581.
[8] R. W. Siegel, *Scientific American* **1996**, 74–79.

[9] P. Alivisatos, *J. Phys. Chem.* **1996**, *100*, 13226–13239.
[10] P. V. Kamat, in :*Molecular Level Artificial Photosynthetic Materials. Progress in Inorganic Chemistry Series* (Ed.: J. Meyer), Wiley, New York **1997**, Vol. 44, pp. 273–243.
[11] P. V. Kamat, D. Meisel, "Semiconductor nanoclusters – physical, chemical and catalytic aspects," in: *Studies in Surface Science and Catalysis*, Elsevier, Amsterdam **1997**, p. 474.
[12] A. Hagfeldt, M. Grätzel, *Chem. Rev.* **1995**, *95*, 49–68.
[13] P. V. Kamat, *IAPS Newsletter* **1996**, *19*, 14–23.
[14] D. F. Ollis, *Environ. Sci. Technol.* **1985**, *19*, 480–484.
[15] M. A. Fox, *Chemtech* **1992**, *22*, 680–685.
[16] N. Serpone, D. Lawless, R. Terzian, C. Minero, E. Pelizzetti, in: *Photochemical Conversion and Storage of Solar Energy* (Eds.: E. Pelizzetti, M. Schiavello), Kluwer, Dordrecht **1991**.
[17] K. Rajeshwar, *J. Appl. Electrochem.* **1995**, *25*, 1067–1082.
[18] U. Stafford, K. A. Gray, P. V. Kamat, *Heterogeneous Chemistry Reviews* **1996**, *3*, 77–104.
[19] K. Vinodgopal, P. V. Kamat, *Chemtech.* **1996**, *April*, 18–22.
[20] G. Hodes, *Isr. J. Chem.* **1993**, *33*, 95–106.
[21] G. J. Meyer, P. C. Searson, *Interface* **1993**, 23–27.
[22] P. V. Kamat, *Chemtech* **1995**, *June*, 22–28.
[23] M. Krishnan, J. R. White, M. A. Fox, A. J. Bard, *J. Am. Chem. Soc.* **1983**, *105*, 7002–7003.
[24] A. W. H. Mau, C. B. Huang, N. Kakuta, A. J. Bard, A. Campion, M. A. Fox, J. M. White, S. E. Webber, *J. Am. Chem. Soc.* **1984**, *106*, 6537–6542.
[25] J. P. Kuczynski, B. H. Milosavljevic, J. K. Thomas, *J. Phys. Chem.* **1984**, *88*, 980–984.
[26] Y. Nosaka, K. Yamaguchi, H. Yokoyama, H. Miyama, "Formation of ultrafine semiconductor particles and their photoresponsive characteristics in polymer film," *Photoresponsive Materials*, MRS Int. Meet. Adv. Mater. **1989**.
[27] F. R. F. Fan, H. Y. Liu, A. J. Bard, *J. Phys. Chem.* **1985**, *89*, 4418–4420.
[28] K. Honda, A. Kuwano, K. Chiba, A. Ishikawa, H. Miyama, *Chem. Lett.* **1988**, 195–198.
[29] M. Kaneko, T. Okada, S. Teratani, K. Taya, *Electrochim. Acta* **1987**, *32*, 1405–1407.
[30] D. Gningue, G. Horowitz, J. Roncali, F. Garnier, *J. Electroanal. Chem. Interfacial Electrochem* **1989**, *269*, 337–349.
[31] R. Tassoni, R. R. Schrock, *Chem. Mater.* **1994**, *6*, 744–749.
[32] O. V. Salata, P. J. Dobson, P. J. Hull, J. L. Hutchison, *Thin Solid Films* **1994**, *251*, 81–93.
[33] K. R. Gopidas, P. V. Kamat, *Mater. Lett.* **1990**, *9*, 372–378.
[34] M. Gao, Y. Yang, B. Yang, F. Bian, J. Shen, *J. Chem. Soc., Chem. Commun.* **1994**, 2779–2780.
[35] K. C. Yi, J. H. Fendler, *Langmuir* **1990**, *6*, 1519–1521.
[36] N. J. Geddes, R. S. Urquhart, D. N. Furlong, R. Lawrence, K. Tanaka, Y. Okahata, *J. Phys. Chem.* **1993**, *97*, 13767–13772.
[37] D. V. Paranjape, M. Sastry, P. Ganguly, *Appl. Phys. Lett.* **1993**, *63*, 18–20.
[38] R. Rafaeloff, Y. M. Tricot, F. Nome, J. H. Fendler, *J. Phys. Chem.* **1985**, *89*, 533–537.
[39] J. H. Fendler, *Chem. Rev.* **1987**, *87*, 877–899.
[40] L. Moriguchi, H. Maeda, Y. Teraoka, S. Kagawa, *J. Am. Chem. Soc.* **1995**, 1139–1140.
[41] L. Spanhel, M. A. Anderson, *J. Am. Chem. Soc.* **1990**, *112*, 2278–2284.
[42] Q. Xu, M. A. Anderson, *J. Mater. Res.* **1991**, *6*, 1073–1081.
[43] S. Tunesi, M. Anderson, *J. Phys. Chem.* **1991**, *95*, 3399–3405.
[44] J. Sabate, M. A. Anderson, M. A. Aguado, J. Gimenez, M. S. Cervera, C. G. J. Hill, *J. Mol. Catal.* **1992**, *71*, 57–68.
[45] G. Dagan, M. Tomkiewicz, *J. Phys. Chem.* **1993**, *97*, 12651–12655.
[46] M. N. Kamalasanan, N. D. Kumar, S. Chandra, *J. Appl. Phys.* **1993**, *74*, 679–686.
[47] T. Jiang, A. L. Lough, G. A. Ozin, D. Young, *Chem. Mater.* **1995**, *7*, 245–248.
[48] G. Hodes, I. D. J. Howell, L. M. Peter, *J. Electrochem. Soc.* **1992**, *139*, 3136–3140.
[49] S. Hotchandani, P. V. Kamat, *Chem. Phys. Lett.* **1992**, *191*, 320–326.
[50] S. Hotchandani, P. V. Kamat, *J. Electrochem. Soc.* **1992**, *139*, 1630–1634.
[51] S. Hotchandani, P. V. Kamat, *J. Phys. Chem.* **1992**, *96*, 6834–6839.
[52] N. Uekawa, T. Suzuki, S. Ozeki, K. Kaneko, *Langmuir* **1992**, *8*, 1–3.
[53] H. Weller, A. Eychmuller, R. Vogel, L. Katsikas, A. Hasselbarth, M. Giersig, *Isr. J. Chem.* **1993**, *33*, 107–114.

[54] D. Liu, P. V. Kamat, *J. Electroanal. Chem. Interfacial Electrochem* **1993**, *347*, 451–456.
[55] I. Bedja, S. Hotchandani, P. V. Kamat, *J. Phys. Chem.* **1994**, *98*, 4133–4140.
[56] R. Vogel, P. Hoyer, H. Weller, *J. Phys. Chem.* **1994**, *98*, 3183–3188.
[57] I. Bedja, S. Hotchandani, R. Carpentier, R. W. Fessenden, P. V. Kamat, *J. Appl. Phys.* **1994**, *75*, 5444–5456.
[58] I. Bedja, S. Hotchandani, R. Carpentier, K. Vinodgopal, P. V. Kamat, *Thin Solid Films* **1994**, *247*, 195–200.
[59] K. Vinodgopal, U. Stafford, K. A. Gray, P. V. Kamat, *J. Phys. Chem.* **1994**, *98*, 6797–6803.
[60] U. Bjorksten, J. Moser, M. Graetzel, *Chem. Mater.* **1994**, *6*, 858–863.
[61] A. Hagfeldt, S. E. Lindquist, M. Graetzel, *Sol. Energy Mater. Sol. Cells* **1994**, *32*, 245–257.
[62] S. Hotchandani, S. Das, K. G. Thomas, M. V. George, P. V. Kamat, *Res. Chem. Intermed.* **1994**, *20*, 927–938.
[63] D. Liu, P. V. Kamat, *J. Phys. Chem.* **1993**, *97*, 10769–10773.
[64] C. Nasr, S. Hotchandani, P. V. Kamat, *Proc. Ind. Acad. Sci.* **1995**, *107*, 699–708.
[65] C. Nasr, S. Hotchandani, P. V. Kamat, S. Das, K. George Thomas, M. V. George, *Langmuir* **1995**, *11*, 1777–1783.
[66] C. Liu, H. Pan, M. A. Fox, A. J. Bard, *Science Washington, D.C.* **1993**, *261*, 897–899.
[67] P. Hoyer, R. Eichberger, H. Weller, *Ber. Bunsenges. Phys. Chem.* **1993**, *97*, 630–635.
[68] H. Hada, Y. Yonezawa, H. Inaba, *Ber. Bunseges. Phys. Chem.* **1981**, *85*, 425–429.
[69] S. Sakohara, L. D. Tickanen, M. A. Anderson, *J. Phys. Chem.* **1992**, *96*, 11086–11091.
[70] G. Redmond, A. O Keeffe, C. Burgess, C. MacHale, D. Fitzmaurice, *J. Phys. Chem.* **1993**, *97*, 11081–11086.
[71] G. Redmond, D. Fitzmaurice, M. Grätzel, *Chem. Mater.* **1994**, *6*, 686–691.
[72] Y. Harima, Y. D. Wang, K. Matsumoto, K. Yamashita, *J. Chem. Soc., Chem. Commun.* **1994**, 2553–2554.
[73] H. Yoshiki, H. K. A. Fujishima, *J. Electrochem. Soc.* **1995**, *142*, 428–432.
[74] B. O'Regan, M. Graetzel, D. Fitzmaurice, *Chem. Phys. Lett.* **1991**, *183*, 89–93.
[75] L. Kavan, T. Stoto, M. Graetzel, D. J. Fitzmaurice, V. Shklover, *J. Phys. Chem.* **1993**, *97*, 9493–9498.
[76] M. K. Nazeeruddin, A. Kay, I. Rodicio, B. R. Humphry, E. Mueller, P. Liska, N. Vlachopoulos, M. Graetzel, *J. Am. Chem. Soc.* **1993**, *115*, 6382–6390.
[77] X. Marguerettaz, D. Fitzmaurice, *J. Am. Chem. Soc.* **1994**, *116*, 5017–5018.
[78] D. D. Dunuwila, C. D. Gagliardi, K. A. Berglund, *Mater. Lett.* **1994**, *6*, 1556–1562.
[79] N. A. Kotov, F. C. Meldrum, J. H. Fendler, *J. Phys. Chem.* **1994**, *98*, 8827–8830.
[80] A. Hagfeldt, N. Vlachopoulos, M. Graetzel, *J. Electrochem. Soc.* **1994**, *141*, L82–L87.
[81] D. H. Kim, M. A. Anderson, *Environ. Sci. Technol.* **1994**, *28*, 479–483.
[82] I. Bedja, S. Hotchandani, P. V. Kamat, *J. Phys. Chem.* **1993**, *97*, 11064–11070.
[83] S. Hotchandani, I. Bedja, R. W. Fessenden, P. V. Kamat, *Langmuir* **1994**, *10*, 17–22.
[84] K. Vinodgopal, S. Hotchandani, P. V. Kamat, *J. Phys. Chem.* **1993**, *97*, 9040–9044.
[85] K. Vinodgopal, I. Bedja, P. V. Kamat, *Chem. Mater.* **1996**, *8*, 2180.
[86] Y. Wang, in: *Semiconductor Nanoclusters – Physical, Chemical and Catalytic Aspects.* (Eds.: P. V. Kamat, D. Meisel), Elsevier, Amsterdam **1997**, pp. 277–295.
[87] R. Vogel, K. Pohl, H. Weller, *Chem. Phys. Lett.* **1990**, *174*, 241–246.
[88] S. Kohtani, A. Kudo, T. Sakata, *Chem. Phys. Lett.* **1993**, *206*, 166–170.
[89] K. Rajeshwar, R. O. Lezna, N. R. deTacconi, *Anal. Chem.* **1992**, *64*, 429–441.
[90] Y. Golan, L. Margulis, G. Hodes, I. Rubinstein, J. L. Hutchison, *Surface Science* **1994**, *311*, L633–L640.
[91] S. Gorer, G. Hodes, in: *Semiconductor Nanoclusters – Physical, Chemical and Catalytic Aspects* (Eds: P. V. Kamat, D. Meisel), Elsevie, Amsterdam **1997**, pp. 297–320.
[92] J. H. Fendler, *Chem. Mater.* **1996**, *8*, 1616–1624.
[93] J. J. Fendler, in: *Semiconductor Nanoclusters – Physical, Chemical and Catalytic Aspects* (Eds.: P. V. Kamat, P. D. Meisel), Elsevier, Amsterdam **1997**, pp. 261–276.
[94] D. Liu, P. V. Kamat, *J. Electrochem. Soc.* **1995**, *142*, 835–839.
[95] D. Liu, R. W. Fessenden, G. L. Hug, P. V. Kamat, *J. Phys. Chem.* **1997**, *101 B*, 2583–2590.
[96] A. Kay, M. Graetzel, *J. Phys. Chem.* **1993**, *97*, 6272–6277.
[97] C. M. Lampert, *Solar Energy Mater.* **1984**, *11*, 1–27.

[98] J. N. Yao, K. Hashimoto, A. Fujishima, *Nature* **1992**, *355*, 624–626.
[99] T. Torimoto, R. J. Fox III, M. A. Fox, *J. Electrochem. Soc.* **1996**, *143*, 3712–3717.
[100] M. T. Nenadovic, T. Rajh, O. I. Micic, A. J. Nozik, *J. Phys. Chem.* **1984**, *88*, 5827–30.
[101] P. V. Kamat, K. Vinodgopal, *Langmuir* **1996**, *12*, 5739–5741.
[102] B. O'Regan, M. Graetzel, D. Fitzmaurice, *J. Phys. Chem.* **1991**, *95*, 10525–10528.
[103] G. Redmond, D. Fitzmaurice, M. Graetzel, *J. Phys. Chem.* **1993**, *97*, 6951–6954.
[104] G. Redmond, D. Fitzmaurice, *J. Phys. Chem.* **1993**, *97*, 1426–1430.
[105] L. Butler, G. Redmond, D. Fitzmaurice, *J. Phys. Chem.* **1993**, *97*, 10750–10755.
[106] C. Liu, A. J. Bard, *J. Phys. Chem.* **1989**, *93*, 3232–3237.
[107] C. Bechinger, S. Ferrere, A. Zaban, J. Sprague, B. A. Gregg, *Nature* **1996**, *383*, 608–610.
[108] A. Hagfeldt, U. Bjorksten, S. E. Lindquist, *Sol. Energy Mater. Sol. Cells* **1992**, *27*, 293–304.
[109] S. Sodergren, A. Hagfeldt, J. Olsson, S.-E. Lindquist, *J. Phys. Chem.* **1994**, *98*, 5552–5556.
[110] H. Rensmo, H. Lindstrom, S. Sodergren, A.-K. Willstedt, A. Solbrand, A. Hagfeldt, S.-E. Lindquist, *J. Phys. Chem.* **1996**, *143*, 3173–3178.
[111] A. J. Nozik, B. R. Thacker, J. A. Turner, J. M. Olson, *J. Am. Chem. Soc.* **1985**, *107*, 7805–7810.
[112] A. J. Nozik, B. R. Thacker, J. M. Olson, *Nature* **1985**, *316*, 51–3.
[113] A. J. Nozik, B. R. Thacker, J. A. Turner, J. Klem, H. Morkoc, *Appl. Phys. Lett.* **1987**, *50*, 34–36.
[114] A. J. Nozik, B. R. Thacker, J. A. Turner, M. W. Peterson, *J. Am. Chem. Soc.* **1988**, *110*, 7630–7637.
[115] A. J. Nozik, J. A. Turner, M. W. Peterson, *J. Phys. Chem.* **1988**, *92*, 2493–501.
[116] C. A. Parsons, M. W. Peterson, B. R. Thacker, J. A. Turner, A. J. Nozik, *J. Phys. Chem.* **1990**, *94*, 3381–3384.
[117] C. A. Bignozzi, R. Argazzi, T. Indelli, F. Scandola, *Solar Energy Mater.* **1994**, *32*, 229–244.
[118] I. Bedja, P. V. Kamat, S. Hotchandani, *J. Appl. Phys.* **1996**, *80*, 4637–4643.
[119] Y.-S. Kim, K. Liang, K.-Y. Law, D. G. Whitten, *J. Phys. Chem.* **1994**, *98*, 984–988.
[120] C. Nasr, D. Liu, S. Hotchandani, P. V. Kamat, *J. Phys. Chem.* **1996**, *100*, 11054–11061.
[121] G. Smestad, C. Bignozzi, R. A. Argazzi, *Solar Energy Mater.* **1994**, *32*, 259–272.
[122] G. Smestad, *Solar Energy Mater.* **1994**, *32*, 273–288.
[123] S. C. Martin, J. M. Kesselman, D. S. Park, N. S. Lewis, M. R. Hoffmann, *Environ. Sci. Technol.* **1996**, *30*, 2535–2532.
[124] D. F. Ollis, E. Pelizzetti, N. Serpone, *Environ. Sci. Technol.* **1991**, *25*, 1523–1529.
[125] W. A. Zeltner, C. G. Hill Jr., M. A. Anderson, *Chem. Tech.* **1994**, 21–27.
[126] M. R. Hoffmann, S. T. Martin, W. Choi, D. W. Bahnemann, *Chem. Rev.* **1995**, *95*, 69–96.
[127] R. W. Matthews, *Sol. Energy* **1987**, *38*, 405–13.
[128] H. Al-Ekabi, N. Serpone, *J. Phys. Chem.* **1988**, *92*, 5726–5731.
[129] H. Gerischer, A. Heller, *J. Phys. Chem.* **1991**, *95*, 5261–5267.
[130] H. Hidaka, Y. Asai, J. Zhao, K. Nohara, E. Pelizzetti, N. Serpone, *J. Phys. Chem.* **1995**, *99*, 8244–8248.
[131] K. Vinodgopal, P. V. Kamat, *Environ. Sci. Technol.* **1995**, *29*, 841–845.
[132] A. Fujishima, K. Honda, *Nature* **1972**, *238*.
[133] H. Gerischer, F. Willig, *Top. Curr. Chem.* **1976**, *61*, 31–84.
[134] M. A. Ryan, M. T. Spitler, *J. Imaging Sci.* **1989**, *33*, 46–49.
[135] K. Hashimoto, M. Hiramoto, T. Kajiwara, T. Sakata, *J. Phys. Chem.* **1988**, *92*, 4636–4640.
[136] K. Hashimoto, M. Hiramoto, A. B. P. Lever, T. Sakata, *J. Phys. Chem.* **1988**, *92*, 1016–1018.
[137] T. Tani, *J. Imag. Sci.* **1990**, *34*, 143–148.
[138] T. Tani, T. Suzumoto, K. Ohzeki, *J. Phys. Chem.* **1990**, *94*, 1298–1300.
[139] M. A. Ryan, E. C. Fitzgerald, M. T. Spitler, *J. Phys. Chem.* **1989**, *93*, 6150–6156.
[140] L. P. Sonntag, M. T. Spitler, *J. Phys. Chem.* **1985**, *89*, 1453–1457.
[141] K. Vinodgopal, X. Hua, R. L. Dahlgren, A. G. Lappin, L. K. Patterson, P. V. Kamat, *J. Phys. Chem.* **1995**, *99*, 10883–10889.
[142] H. P. Lu, X. S. Xie, *J. Phys. Chem.* **1997**, *101 B*, 2753–2757.
[143] W. E. Ford, M. A. J. Rodgers, *J. Phys. Chem.* **1994**, *98*, 3822–3831.
[144] R. W. Fessenden, P. V. Kamat, *J. Phys. Chem.* **1995**, *99*, 12902–12906.
[145] P. V. Kamat, *Langmuir* **1990**, *6*, 512–513.

[146] P. V. Kamat, S. Das, K. G. Thomas, M. V. George, *Chem. Phys. Lett.* **1991**, *178*, 75–79.

[147] R. Eichberger, F. Willig, *Chem. Phys. Lett.* **1990**, *141*, 159–173.

[148] F. Willig, R. Eichberger, N. S. Sundaresan, B. A. Parkinson, *J. Am. Chem. Soc.* **1990**, *112*, 2702–2707.

[149] J. M. Rehm, G. L. McLendon, Y. Nagasawa, K. Yoshihara, J. Moser, M. Graetzel, *J. Phys. Chem.* **1996**, *100*, 9577–9578.

[150] B. Burfeindt, T. Hannappel, W. Storck, F. Willig, *J. Phys. Chem.* **1996**, *100*, 16463–16465.

[151] I. Martini, G. Hartland, P. V. Kamat, *J. Phys. Chem.* **1997**, *101 B*, 4826–4830.

[152] H. Takemura, T. Saji, M. Fujihira, S. Aoyagui, K. Hashimoto, T. Sakata, *Chem. Phys. Lett.* **1985**, *122*, 496–502.

[153] K. Hashimoto, M. Hiramoto, T. Sakata, H. Muraki, H. Takemura, M. Fujihira, *J. Phys. Chem.* **1987**, *91*, 6198–6203.

[154] R. Argazzi, C. A. Bignozzi, T. A. Heimer, F. N. Castellano, G. J. Meyer, *Inorg. Chem.* **1994**, *33*, 5741–5749.

[155] T. A. Heimer, G. J. Meyer, *J. Lumin.* **1996**, *70*, 468.

[156] Y. Tachibana, J. E. Moser, M. Graetzel, D. R. Klug, J. R. Durrant, *J. Phys. Chem.* **1996**, *100*, 20056–20062.

[157] K. Schwarzburg, F. Willig, *Appl. Phys. Lett.* **1991**, *58*, 2520–2523.

[158] B. Patrick, P. V. Kamat, *J. Phys. Chem.* **1992**, *96*, 1423–1428.

[159] D. Liu, P. V. Kamat, *J. Chem. Phys.* **1996**, *105*, 965–970.

[160] P. A. Breddels, G. Blasse, *Ber. Bunsenges. Phys. Chem* **1982**, *86*, 676–680.

[161] A. Kay, R. Humphry-Baker, M. Grätzel, *J. Phys. Chem.* **1994**, *98*, 952–959.

[162] A. H. Goff, P. Falaras, *J. Electrochem. Soc.* **1995**, *142*, L38–L41.

[163] I. Bedja, S. Hotchandani, P. V. Kamat, *J. Electroanal. Chem.* **1996**, *401*, 237–241.

[164] B. O Regan, J. Moser, M. Anderson, M. Grätzel, *J. Phys. Chem.* **1990**, *94*, 8720–8726.

[165] P. V. Kamat, I. Bedja, S. Hotchandani, L. K. Patterson, *J. Phys. Chem.* **1996**, *100*, 4900–4908.

[166] C. M. Elliott, E. J. Hershenhart, *J. Am. Chem. Soc.* **1982**, *104*, 7519–7526.

[167] A. Launikonis, P. A. Lay, A. W.-H. Mau, A. M. Sargeson, W. H. Sasse, *Aust. J. Chem.* **1986**, *39*, 1053–1062.

[168] R. A. Marcus, *J. Chem. Phys.* **1993**, *43*, 679.

[169] R. A. Marcus, *J. Phys. Chem.* **1990**, *94*, 1050–1055.

[170] J. J. Jortner, *J. Chem. Phys.* **1976**, *64*.

[171] B. S. Brunswig, N. Sutin, *Comments Inorg. Chem.* **1987**, *6*, 209.

[172] H. Lu, J. N. Prieskorn, J. T. Hupp, *J. Am. Chem. Soc.* **1993**, *115*, 4927–4928.

[173] B. O'Regan, M. Grätzel, *Nature (London)* **1991**, *353*, 737–740.

[174] R. W. Fessenden, P. V. Kamat, *Chem. Phys. Lett.* **1986**, *123*, 233–238.

[175] F. Cao, G. Oskam, G. J. Meyer, P. C. Searson, *J. Phys. Chem.* **1996**, *100*, 17021–17027.

[176] F. Cao, G. Oskam, P. C. Searson, J. M. Stipkala, T. A. Heimer, F. Farzad, G. J. Meyer, *J. Phys. Chem.* **1995**, *99*, 11974–11980.

[177] H. Lindstrom, H. Rensmo, S. Sodergren, A. Solbrand, S. Lindquist, *J. Phys. Chem.* **1996**, 3084–3088.

[178] K. Vinodgopal, P. V. Kamat, *Solar Energy Mater. Solar Cells* **1995**, *38*, 401–410.

Chapter 10

Template Synthesis of Nanoparticles in Nanoporous Membranes

John C. Hulteen and Charles R. Martin

10.1 Introduction

Many methods for the fabrication of nanoparticles have been developed, ranging from lithographic techniques to chemical methods [1, 2]. Our research group has been exploring a fabrication method termed template synthesis for the preparation of a variety of micro- and nanomaterials [3–30]. This process involves synthesizing a desired material within the pores of a porous membrane. Because the membranes that are used have cylindrical pores of uniform diameter, a nanocylinder of the desired material is obtained in each pore. Depending on the properties of the material and the chemistry of the pore wall, this nanocylinder may be solid (a nanofibril) or hollow (a nanotubule).

The template method has a number of interesting and useful features. First, it is very general with respect to the types of materials prepared. We have used this method to prepare both nanotubules and nanofibrils composed of conductive polymers [3–13], metals [14–25], semiconductors [26, 27], carbon [28–30], and other materials. Tubular and fibrillar nanostructures with extremely small diameters can be prepared. For example, conductive polymer nanowires with diameters as small as 3 nm have been prepared using this method [31]. It is difficult to make nanowires with diameters this small by lithographic methods. In addition, because the pores in these membranes have monodisperse diameters, analogous monodisperse nanostructures are obtained. Finally, the tubular or fibrillar nanostructures synthesized within the pores can be freed from the template membrane and collected. Alternatively, an ensemble of micro- or nanostructures that protrude from a surface like the bristles of a brush can be obtained.

The intent of this chapter is to provide an overview of the template method. We will start with a brief description of the types of membranes used for template synthesis. Next, the different types of chemistries that have been used to prepare template-synthesized nanostructures will be reviewed. Finally, we will discuss fundamental properties and applications of template-synthesized metal and semiconductor nanostructures. While there has been a significant amount of research in the area of template synthesis of conductive polymer nanostructures, this has been recently reviewed elsewhere [32, 33].

10.2 Membranes Used

Most of the work in template synthesis, to date, has entailed the use of two types of nanoporous membranes, "track-etch" polymeric membranes and porous alumina membranes. However, there are a variety of other templates that could be utilized.

10.2.1 "Track-etch"

A number of companies (such as Nucleopore and Poretics) sell microporous and nanoporous polymeric filtration membranes that have been prepared by the track-etch method [34]. This method entails bombarding a nonporous sheet of the desired material (standard thickness range from 6 to 20 μm) with nuclear fission fragments to create damage tracks in the material, and then chemically etching these tracks into pores. The resulting membranes contain randomly distributed cylindrical pores of uniform diameter (Figure 10.1(a, b)). The commercial membranes are available with pore diameters as small as 10 nm (pore density approximately 10^9 pores per square centimeter). These commercial membranes are prepared from polycarbonate or polyester; however, a number of other materials are also amenable to the track-etch process [34].

Owing to the random nature of the pore-production process, the angle of the pores with respect to the surface normal can be as large as 34° [35]. Therefore, depending on the specific pore diameter and pore density of the track-etched membrane, a number of pores may actually intersect within the membrane. This is a problem when theoretically modeling the optical properties of template-synthesized nanometals, a topic of great interest to our group [18–20]. For example, theory predicts a specific wavelength maximum in the absorption band of isolated metal nanoparticles [18–20]. However, physical contact between the metal nanoparticles synthesized within the pores can shift this absorption maximum by 200 nm or more [36].

10.2.2 Porous Alumina

Porous alumina membranes are prepared via the anodization of Al metal in an acidic solution [37]. These membranes contain cylindrical pores of uniform diameter arranged in a hexagonal array (Figure 10.1(c,d)). However, unlike the track-etch membranes, the pores in these membranes have little or no tilt with respect to the surface normal, resulting in an isolating, nonconnecting pore structure. Although such membranes are sold commercially (Whatman), a very limited number of pore diameters are available. We have, however, prepared membranes of this type with a broad range of pore diameters [18, 20]. We have made membranes with pore diameters as large as 200 nm and as small as 5 nm, and we believe that even smaller pores can be prepared. Pore densities as high as 10^{11} pores per square centimeter

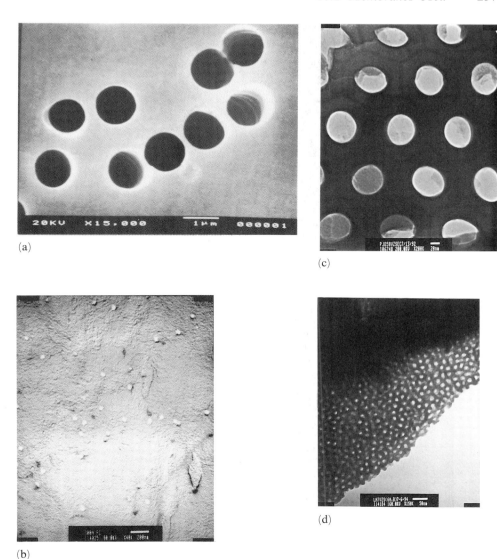

Figure 10.1. Electron micrographs of polycarbonate (a, b) and alumina (c, d) template membranes. For each type of membrane, an image of a larger-pore membrane is presented (a, c) so that the characteristics of the pores can be clearly seen. An image of a membrane with extremely small pores is also presented (b, d). (a) Scanning electron micrograph (SEM) of the surface of a polycarbonate membrane with 1 μm-diameter pores. (b) Transmission electron micrograph (TEM) of a graphite replica of the surface of a polycarbonate membrane with 30 nm-diameter pores. The pores appear "ragged" owing to the artifact of using a graphite replica. (c, d) TEMs of microtomed sections of alumina membranes with 70 nm- (c) and 10 nm- (d) diameter pores.

can be achieved, [38], and typical membrane thickness can range from 10 to 100 μm. The higher pore density is important if one wanted to mass-produce a nanomaterial by the template method. Membranes with high pore density would allow a greater number of nanostructures to be produced per unit area of template membrane.

10.2.3 Other Nanoporous Materials

Tonucci et al. have recently described a nanochannel array glass with pore diameters as small as 33 nm and pore densities as high as 3×10^{10} pores cm^{-2} [39]. Beck et al. have prepared a new class of mesoporous zeolites with large pore diameters [40]. Douglas et al. have shown that the nanoscopic pores in a protein derived from a bacterium can be used to transfer an image of these pores to an underlying substrate [41]. Ghadiri et al. have prepared arrays of polypeptide tubules [42]. Finally, both Ozin [1] and Schollhorn [43] have discussed a wide variety of nanoporous solids that could be used as template materials.

10.3 Template Synthetic Strategies

The limits to which materials can be used in template synthesis are defined by the chemistry required to synthesize the material. Nearly any material can in principle be synthesized within these nanoporous membranes provided a suitable chemical pathway can be developed. Typical concerns that need to be addressed when developing new template synthetic methods include the following: (i) will the precursor solutions used to prepare the material "wet" the pore (i.e., hydrophobic/ hydrophilic considerations)?; (ii) will the deposition reaction proceed too fast, resulting in pore blockage at the membrane surface before tubule/fiber growth can occur within the pores?; (iii) will the host membrane be stable (i.e., thermally and chemically) with respect to the reaction conditions. The following is a general outline of five representative chemical strategies that have been used in our laboratory to conduct template synthesis within the alumina and polymeric template membranes.

10.3.1 Electrochemical Deposition

Electrochemical deposition of a material within the pores is accomplished by coating one face of the membrane with a metal film (usually via either ion sputtering or thermal evaporation) and using this metal film as a cathode for electroplating [17–22, 44, 45]. This method has been used to prepare a variety of metal nanowires, including copper, platinum, gold, silver, and nickel in both track-etch and alumina templates. Typical gold nanowires are shown in Figure 10.2(a). The lengths of these nanowires can be controlled by varying the amount of metal deposited. By deposit-

(a)

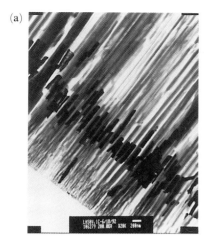

(b)

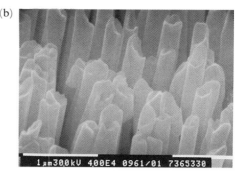

(c)

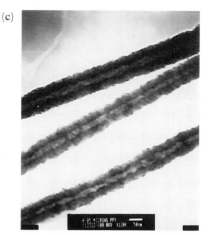

Figure 10.2. Electron micrographs of tubules and fibrils.
(a) TEM of a microtomed section of an alumina template
mebrane showing Au nanofibrils that are 70 nm in
diameter within the pores. (b) SEM of an array of Au
microtubules. (c) TEM of three polypyrrole nanotubules.
The outside diameter is ∼90 nm; the inside diameter is
∼20–30 nm.

ing a small amount of metal, short wires can be obtained; alternatively, by depositing large quantities of metal, long needlelike wires can be prepared [18–20]. This ability to control the length or aspect ratio (length to diameter) of the metal nanowires is especially important in our optical investigations because the optical properties of nanometals are dependent on aspect ratio [18–20, 24].

Hollow metal tubules can also be prepared via this method (Figure 10.2(b)) [17, 22]. To obtain tubules, one must typically chemically derivatize the pore walls so that the electrodeposited metal preferentially deposits on the pore wall; that is, a molecular anchor must be applied. For example, gold tubules have been prepared by attaching a cyanosilane to the walls of the alumina template membrane prior to metal depositions [17, 22, 46]. Owing to the large number of commercially available silanes, this method can provide a general route for tailoring the pore walls in the alumina membranes.

Electrochemical deposition can also be used to synthesize conductive polymers (such as polypyrrole, polyaniline, or poly(3-methylthiophene)) within the pores of these template membranes [10, 13]. When these polymers are synthesized within the pores of track-etched polycarbonate membranes, the polymer preferentially nucleates and grows on the pore walls, resulting in polymeric tubules at short polymerization times (Figure 10.2(c)). By controlling the polymerization time, we can produce thin-walled tubules, thick-walled tubules, or solid fibrils.

The reason that the polymer preferentially nucleates and grows on the pore walls is straightforward [8]. Although the monomers are soluble, the polycationic forms of these polymers are completely insoluble. Hence, there is a solvophobic component to the interaction between the polymer and the pore wall. There is also an electrostatic component because the polymers are cationic and there are anionic sites on the pore walls [8].

10.3.2 Electroless Deposition

Electroless metal deposition involves the use of a chemical reducing agent to plate a metal from solution onto a surface [47]. This method differs from electrochemical deposition in that the surface to be coated need not be electronically conductive. We have developed methods by which Au and other metals can be plated from solution onto the surfaces of both the plastic and alumina membranes [15]. This method involves applying a sensitizer (typically Sn^{2+}) to the membrane surfaces (pore walls and faces). The sensitizer binds to the surfaces via complexation with surface amine, carbonyl, and hydroxyl groups. This sensitized membrane is then activated by exposure to Ag^+, resulting in the formation of discrete nanoscopic Ag particles on the membrane's surfaces. Finally, the Ag-coated membrane is immersed into a Au plating bath containing Au(I) and a reducing agent, which results in Au plating on the membrane faces and pore walls.

The key feature of the electroless deposition process is that metal deposition in the pores starts at the pore wall. Therefore, after short deposition times, a hollow metal tubule (Figure 10.3) is obtained within each pore while long deposition times result in solid metal nanowires. Unlike the electrochemical deposition method, where the length of the metal nanowire can be controlled at will, electroless depo-

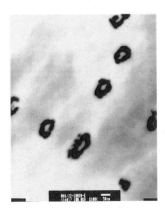

Figure 10.3. TEM showing a microtomed section of an Au-nanotubule-containing membrane. The Au tubules are the black rings. The elliptical appearance is caused by the microtoming process. Pore diameter was 50 nm; plating time was 10 minutes.

sition yields structures that run the complete thickness of the template membrane. However, the inside diameter of the tubules formed via electroless deposition can be controlled at will by varying the metal deposition time [15, 16]. Of course, the outside diameter is determined by the diameter of the pores in the template membrane.

10.3.3 Chemical Polymerization

Chemical template synthesis of a polymer can be accomplished by simply immersing the membrane into a solution containing the desired monomer and a polymerization reagent. This process has been used to synthesize a variety of conductive polymers within the pores of various template membranes [6, 9, 48, 49]. As with electrochemical polymerization, the polymer preferentially nucleates and grows on the pore walls, resulting in tubules at short deposition times and fibers at long times.

Conventional (electronically insulating) plastics can also be chemically synthesized within the pores of these template membranes. For example, polyacrylonitrile tubules can be prepared by immersing an alumina template membrane into a solution containing acrylonitrile and a polymerization initiator [28, 29]. The inside diameter of the resulting polyacrylonitrile (PAN) tubules is varied by controlling the time the membrane remains in the polymerization bath. These PAN tubules have been further processed to create conducting graphitic carbon tubules and fibrils in alumina membranes (Figure 10.4) [28, 29]. This is accomplished by heating

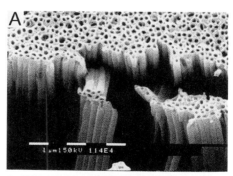

Figure 10.4. SEM images of carbon tubules (a) and fibrils (b) with an outside diameter of 200 nm prepared in an alumina template membrane; the membrane was removed for imaging purposes.

the PAN tubules/alumina membrane composite to 700 °C under argon flow or under vacuum.

10.3.4 Sol–Gel Deposition

Sol–gel chemistry typically involves hydrolysis of a solution of a precursor molecule to obtain first a suspension of colloidal particles (the sol) and then a gel composed of aggregated sol particles. The gel is then thermally treated to yield the desired product. We have recently conducted various sol–gel syntheses within the pores of the alumina membranes to create both tubules and fibers of a variety of inorganic semiconducting materials, including TiO_2, ZnO, and WO_3 [26]. First, an alumina template membrane is immersed into a sol for a given period of time, and the sol deposits on the pore walls. After thermal treatment, either a tubule or fibril of the desired semiconductor is formed within the pores (Figure 10.5). As with other template synthesis techniques, longer immersion times yield fibers, while brief immersion times produce tubules.

The formation of tubules after short immersion times indicates that the sol particles adsorb to the alumina membrane's pore walls. This is expected because the pore walls are negatively charged while the sol particles used to date [26] are positively charged (a similar situation to what was described for conductive polymers). It has also been found that the rate of gelation is faster within the pore than in bulk solution [26]. This is most likely due to the enhancement in the local concentration of the sol particles owing to adsorption on the pore walls.

10.3.5 Chemical Vapor Deposition

A major hurdle in applying chemical vapor deposition (CVD) techniques to template synthesis has been that deposition rates are often too fast. As a result, the surface of the pores becomes blocked before the chemical vapor can traverse the length of the pore. We have, however, developed two template-based CVD syntheses that circumvent this problem. The first entails the CVD of carbon within porous alumina membranes, which has been achieved by our group [30] and others [50]. This involves placing an alumina membrane in a high temperature furnace ($\sim$700 °C) and passing a gas such as ethylene or propylene through the membrane. Thermal decomposition of the gas occurs throughout the pores, resulting in the deposition of carbon films along the length of the pore walls (i.e., carbon tubules are obtained within the pores). The thickness of the walls of the carbon tubes is again dependent on total reaction time and precursor pressure.

The second CVD technique utilizes a template-synthesized structure as a substrate for CVD deposition [51]. For example, we have used a CVD method to coat an ensemble of Au nanotubules with concentric TiS_2 outer nanotubules. The first step of this process requires the electroless plating of Au tubules or fibrils into the pores of a template membrane. The Au surface layer is removed from one face of the plated membrane, and the membrane is dissolved away. The resulting structure

Figure 10.5. SEM images of TiO$_2$ tubules and fibrils prepared in an alumina membrane with 200 nm diameter pores. The sol was maintained at 15 °C, and the immersion time varied from 5 to 60 seconds. (a) Immersion time = 5 s; remnants of the TiO$_2$ surface layer can be seen in this image. (b) Immersion time = 25 s. (c) Immersion time = 60 s.

Figure 10.6. SEM images of an ensemble of Au tubules before (a) and after (a) CVD deposition of the outer TiS$_2$ tubules. The tubules are protruding from the substrate Au surface layer.

is an ensemble of Au tubules or fibrils protruding from the remaining Au surface layer like the bristles of a brush (Figure 10.6(a)). This structure is exposed to the precursor gases used to do CVD synthesis of TiS$_2$. As indicated in Figure 10.6(b), the Au tubules become coated with outer TiS$_2$ tubules.

10.4 Composite Nanostructures

We have shown that a large number of different chemical techniques can be used to prepare tubules or fibrils that are composed of a single material. However, one could imagine a host of applications where composite tubular nanostructures would be necessary. Examples might include concentric nanocapacitor or nanobattery tubules. We have recently developed chemical strategies for preparing such concentric tubular nanostructures [51]. These composites have very high interfacial surface areas between concentric layers of materials. High interfacial areas are obtained because the interfaces are parallel to the long axis of the composite tubular nanostructure.

The fabrication of a semiconductor/conductor tubular nanocomposite will introduce this concept of sequential tubular synthesis [51]. This composite was prepared in a 60 μm-thick alumina template membrane with 200 nm-diameter pores. First, TiO$_2$ tubules are synthesized within the pores of the alumina membrane via the sol–gel process discussed above (Figure 10.7(a)). After thermal treatment of

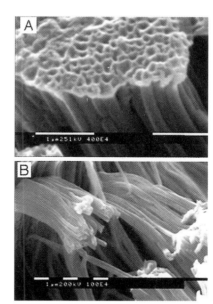

Figure 10.7. SEM images of TiO$_2$ nanotubules prepared by sol–gel methods before (a) and after (b) filling with the polypyrrole nanowires. Outer diameter of tubular composite is 200 nm.

the TiO$_2$ tubules, conductive polypyrrole nanowires were grown using the chemical polymerization method inside the semiconductor tubules (Figure 10.7(b)).

TiO$_2$ is a promising material for photoelectrochemical energy production, and it has been shown that high-surface-area TiO$_2$ has a higher photo efficiency [52]. Therefore, these TiO$_2$/polypyrrole nanocomposites should be excellent photo-catalysts because these template synthesized structures have very high surface area. One problem in using high-surface-area TiO$_2$ as a photocatalyst is the low electrical conductivity of the material. However, this tubular nanocomposite structure should solve this problem because each TiO$_2$ tubule has its own current-collecting electrode inside.

Another method for the construction of a two-component concentric composite has already been described in the CVD synthetic methods section, Sect. 10.5, Figure 10.6 [51]. Au tubules are electrolessly synthesized within the template membrane pores. The membrane is dissolved away, and a thin film of TiS$_2$ is synthesized on the surface of the Au tubules via CVD. TiS$_2$ is a Li$^+$-intercalation material for Li-based rechargeable batteries. We have recently shown that template-synthesized Li$^+$-intercalating materials can provide higher discharge capacities than conventional electrodes made from the same material [53]. As with the photoconductor materials, many Li$^+$-intercalation materials have low electrical conductivities. However, the current-collecting Au electrode inside each TiS$_2$ tubule should again solve this problem. We have shown that the TiS$_2$/Au composite nanostructures reversibly intercalate and deintercalate Li$^+$, and we are currently investigating the charge–discharge kinetics and capacities of these tubular composite battery electrode materials.

An alternative set of chemistries was used to fabricate a conductor/insulator/

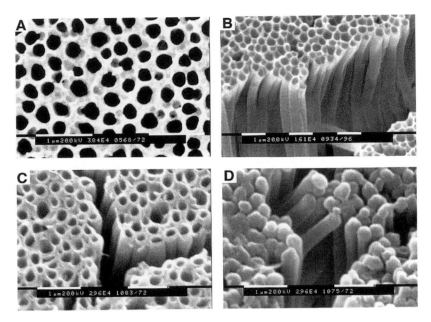

Figure 10.8. SEM image of the surface of the alumina template membrane (a). (b) The carbon tubules obtained after dissolution of the template membrane, and (c) as per (b) but after polymerization of a PAN tubule within each carbon tubule. (d) After electrodeposition of a Au nanowire within each PAN tubule. As noted, the carbon/PAN/Au composites were prepared by doing the appropriate chemistries in sequence, leaving the alumina membrane intact; however, it is easier to image these extremely small structures by dissolving the membrane.

conductor composite consisting of carbon/polyacrylonitrile/gold concentric tubules (Figure 10.8) [51]. Initially, polyacrylonitrile (PAN) tubules were chemically polymerized within the pores of an alumina membrane followed by thermal carbonization resulting in conductive carbon tubules (Figure 10.8(b)). The PAN polymerization step was then repeated creating insulating PAN tubules within the carbon tubules (Figure 10.8(c)). A Au film was then sputtered onto one face of the membrane. Using this film, Au nanowires were electroplated within the inner PAN tubules resulting in the desired concentric tubular C/PAN/Au composite structures (Figure 10.8(d)). We are currently using this synthetic strategy and others to prepare ensembles of nanocapacitors where all of the capacitors are connected in parallel from the surfaces of the template membrane. This will require that all of the electronically conductive outer tubules be electronically insulated from the conductive inner nanowires.

Finally, self-assembly chemistry [54] can also be used as a synthetic step to prepare tubular composites. For example, Au tubules were synthesized within the 1 μm pores of a polycarbonate template membrane via the electroless deposition method. The inside diameter of these tubules was approximately 500 nm, and the length of the tubules was 1.0 μm. The Au-tubule-containing membrane was then immersed in a solution of hexadecyl thiol, causing the thiol to self-assemble onto the inner surfaces of the Au tubules. The template membrane was dissolved away and the freed tubules were collected by filtration.

When these Au/thiol tubules were placed in water, they floated at the air–water interface owing to the presence of the hydrophobic thiol on the inside of the tubule. In contrast, tubules that were not treated with the thiol filled with water and sank [51]. Because self-assembly provides a general way to apply a large number of different chemical functionalities to the inner (and outer) surfaces of such tubules, composite tubules with diverse inner and outer chemistries should be possible.

These have been just a few examples of the types of composite structures that can be fabricated with template synthesis. Composites composed of a variety of different conducting, insulating, semiconducting, photoconducting, and electroactive materials have been prepared. The only limitation on how many different components each composite can contain are the initial diameter of the template pore and the rate of material deposition.

10.5 Optical Properties of Gold Nanoparticles

We [18–20] and others [38, 55] have been investigating the properties of nanometals prepared within the pores of alumina membranes. Through confinement of metals to a nanosized dimension, a variety of changes occur in the optical [18–20, 55], electronic [56], and magnetic [38, 57] properties. The first demonstration of template synthesis for the creation of nanometal fibrils was by Possin in 1970 [58]. Earlier work in which nanometals were used to color alumina is also of interest [59]. Nanometal-containing membranes of this type have also been used as selective solar absorbers [60]. Finally, magnetic metals have been deposited within the pores of such membranes to make vertical magnetic recording media [61].

Our research group [18–20] and others [55] have been primarily interested in the fundamental optical properties of nanocylinders of Au imbedded into alumina membranes. The colors of colloidal suspensions of Au can range from red to purple to blue depending on the diameter of the particle [62], and we have been able to demonstrate analogous colors for Au particles electroplated into the alumina template membrane [18–20]. These colors result from shape-induced changes in the plasmon resonance band of the Au nanoparticle, which corresponds to the wavelength of light that induces the largest electric field on the nanoparticles.

10.5.1 Fabrication

The Au nanoparticles are prepared using the electrodeposition method discussed above (Figure 10.9) [18–20]. First, Ag is deposited onto one face of an alumina

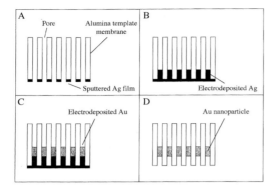

Figure 10.9. Fabrication procedure for Au nanoparticle/alumina composite. (a) Ag is sputtered on one side of the host alumina membrane. (b) The membrane is placed sputtered side down onto a glass plate, and Ag foundation is deposited electrochemically. (c) Au is electrochemically deposited onto the Ag foundations. (d) Ag is removed with nitric acid.

template membrane to provide a conductive film for electrodeposition (Figure 10.9(a)). The membrane is placed Ag film side down on a glass plate and covered with a Ag plating solution. Then, short Ag "plugs" or "posts" are electrochemically grown into the pores (Figure 10.9(b)). These Ag nanoposts are used as foundations onto which the Au nanoparticles are electrochemically grown (Figure 10.9(c)). Finally, the Ag foundations are removed with a nitric acid wash, resulting in an array of Au nanoparticles imbedded within the pores of the alumina membrane (Figure 10.9(d)).

10.5.2 Structural Characterization

The diameter of the electroplated Au nanoparticles is equivalent to the pore diameter of the alumina template membrane. Thus, Au nanoparticles with different diameters can be fabricated in alumina membranes containing different pore diameters. The aspect ratio is controlled by changing the amount of Au electrochemically deposited into the pores. However, we have found that it is not possible to quantitatively predict the aspect ratio of the Au nanoparticles because the plating current efficiency varies from membrane to membrane [24]. Hence, it is not possible to calculate the aspect ratio of the Au nanoparticle obtained from the known quantity of Au deposited and the pore diameter and density. Therefore, transmission electron microscopy (TEM) analysis of the Au nanoparticles synthesized in each membrane is necessary to determine the lengths (and aspect ratios) of the nanoparticles [24]. A TEM image of a transverse section of a Au nanoparticle/alumina composite is shown in Figure 10.2(a). When different amounts of Au are electrodeposited within the pores of the template membrane, we can produce Au nanoparticle shapes that are prolate, spheroid, or oblate [24].

10.5.3 Optical Characterization

The differences in the shapes of the Au nanoparticles result in changes in the optical absorption properties of the composite [18–20, 24]. Such changes are clearly visible as a membrane's color can vary from a bright red to deep blue to turquoise depending on the particle shape [18–20, 24]. The alumina membranes are optically transparent, so the colors are predominantly due to the Au nanoparticles. It should also be noted that the parallel orientation of the pores in the alumina membrane confines the Au particles to a single dimensional alignment. Correspondingly, there is no ambiguity in particle orientation, which is a necessary feature for theoretical modeling of the absorption spectrum.

Figure 10.10 shows the experimental absorption spectra for a variety of Au nanoparticle/alumina composites. The Au particle aspect ratio (length/diameter) varies from 7.7 to 0.38, and the diameter of each Au particle is constant at approximately 52 nm. The reduction in absorption intensity with decreasing aspect ratio is expected owing to the decrease in the metal volume fraction of the composites. The shift in the absorption maximum from 518 nm (aspect ratio = 7.7) to 738 nm (aspect ratio = 0.38) is predicted from simulated spectra obtained using a dynamic Maxwell–Garnett theory [24].

This review shows that the template method can be used to fabricate Au nanoparticles with various diameters and aspect ratios. Shifts in the absorption maximum and changes in the absorption intensity of the Au nanoparticle/alumina composites has been studied as a function of particle diameter and aspect ratio. Current work involves determining the effects of heating the Au nanoparticle/ alumina composite. Changes in the Au nanoparticles aspect ratios, optical properties, and crystal structure have been observed.

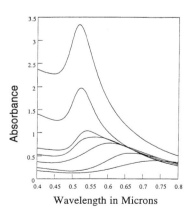

Figure 10.10. Experimental absorption spectra for the Au-nanoparticle-containing membranes. The spectrum with the highest absorbance maximum is for the membrane containing the aspect ratio (length/diameter) of 7.7, then followed by 2.7, 1.3, 0.77, 0.54, 0.46, and 0.38 respectively.

10.6 Nanoelectrode Ensembles

One very exciting application of template synthesis is in the area of electrochemistry. Nanoelectrodes offer opportunities to do electrochemistry in highly resistive media [63, 64] and to investigate the kinetics of redox processes that are too fast to measure at conventional macroscopic electrodes [65–68]. (By macroscopic electrodes we mean disk-shaped electrodes with diameters of the order of 1 mm.) We have used the template method to prepare ensembles of Au nanodisk electrodes where the diameter of the Au disks are as small as 10 nm.

10.6.1 Fabrication

Using the electroless Au deposition procedure, Au nanowires are synthesized within the pores of a polycarbonate track-etch membrane. In addition, both faces of the membrane become coated with thin Au films. If one of these surface Au films is removed, the disk-shaped ends of the Au nanowires traversing the membrane are exposed. These nanodisks can be used as active elements in an ensemble of nanoelectrodes. Figure 10.11 shows a schematic of such a nanoelectrode ensemble (NEE) [14]. Electrical contact is made to the remaining surface layer, which acts as a common current collector for all the nanoelectrode elements.

A consistent problem associated with micro- and nanoelectrodes is achieving an efficient seal between the conductive element and the host material. If a good seal is not achieved, solution can creep into this junction, resulting in significantly higher values of the background or double-layer charging currents. In the case of the NEE, the polycarbonate is stretch-oriented during fabrication to improve mechanical properties. Upon being heated above the glass transition temperature ($\sim 150\ ^\circ$C), the membrane relaxes, shrinks, and seals the junction between the Au nanowires and the polymer membrane [14, 15].

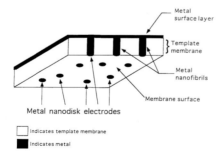

Figure 10.11. Schematic of an edge view of a nanoelectrode ensemble. The nanometal fibrils running through the pores of the template membrane are shown. The lower ends of the fibrils define nanodisks, which are the electrodes. The opposite (upper) ends of the nanofibrils are connected to a common metal film, which is used to make electrical contact to the nanodisks.

10.6.2 Current Response of the NEE

Two different electrochemical response limiting cases can be observed at an NEE, the "total overlap" and "radial" response [14]. Which limiting case is achieved strongly depends upon the distance between the electrode elements and the time-scale (e.g., scan rate) of the electrochemical experiment. When the electrode elements are in close proximity and the scan rate is relatively low, the diffusion layers at each electrode element overlap (Figure 10.12(a)). This overlap results in a single diffusion layer that covers the total geometric area of the NEE. Linear diffusion occurs to the entire NEE surface, and conventional peak-shaped voltammograms are obtained. Also, the total Faradaic current is equivalent to that obtained at an electrode of equivalent geometric area whose entire surface area is Au.

If the electrode elements are located far apart and the timescale of the experiment is relatively fast, the diffusion layers at each electrode act independently, resulting in a radial diffusion field at each individual electrode element (Figure 10.12(b)). The voltammogram in this case has a sigmoidal shape, and the predicted total Faradaic current is equivalent to the sum of the current generated at each individual electrode element within the NEE.

Figure 10.13 shows a series of SEM images of NEEs with varying average distances between the electrode elements [14]. The NEEs were fabricated from polymer template membranes with different pore densities but similar pore diameters. Figure 10.14 presents the Faradaic response of an electroactive species (trimethylaminomethyl ferrocene ($TMAFc^+$)) at each of these NEEs [14]. The NEE with the highest electrode element density (Figures 10.13(a), 10.14(a)) shows a peak-shaped voltammogram, indicative of the total overlap response. In contrast, the NEE with the lowest electrode element density (Figures 10.13(d), 10.14(d)) shows the expected sigmoidal voltammogram. The other two NEEs have an intermediate nanoelectrode density (Figures 10.13(b, c), 10.14(b, c)) and show an intermediate response.

We can quantitatively demonstrate that the NEEs in Figure 10.13(a, d) are operating in the total overlap and radial response modes by comparing experimental and simulated voltammograns. Such a comparison is shown in Figure 10.15. The simulated voltammogram in Figure 10.15(a) is based on the reversible total overlap limiting case, and the experimental voltammogram is the same as in Figure 10.14(a)

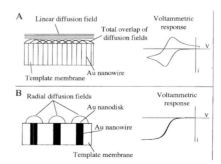

Figure 10.12. Schematic of a side view of NEEs and the corresponding diffusion fields for the total overlap (a) and radial (a) limiting electrochemical response.

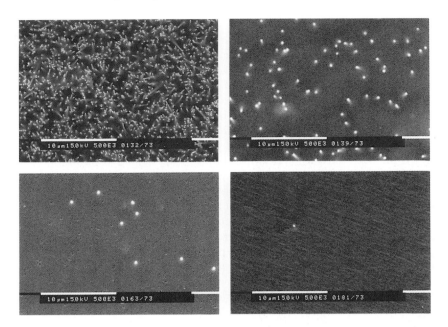

Figure 10.13. SEM images of the surfaces of NEEs showing the disklike electrode elements prepared from membranes with varying pore densities. The average distances between pores are (a, top left) 0.25 μm, (b, top right) 1.1 μm, (c, bottom left) 3.5 μm, and (d, bottom right) 17.5 μm. The diameters of the electrode elements are 100 nm (a, d) and 200 nm (b, c).

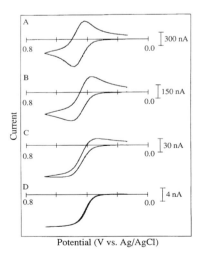

Figure 10.14. Cyclic voltammograms (50 mV s^{-1}) for 50 μm TMAFc$^+$ in 5 mM NaNO$_3$ for NEEs prepared from the membranes shown in Figure 10.13.

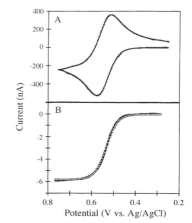

Figure 10.15. Simulated and experimental voltammograms for NEEs prepared from Figure 10.13(a, d). The scan rate and solution are the same as in Figure 10.14.

[14]. The quantitative agreement between the simulated and experimental voltam-mograms confirms that the NEEs at this pore density and scan rate are in the total overlap electrochemical response. It is important to point out that there are no ad-justable parameters in this simulation.

The simulated voltammogram in Figure 10.15(b) assumes a single 100 nm-diameter disk electrode, but the total current is multiplied by the number of elec-trodes within the geometric area of the NEE. The experimental voltammogram is equivalent to Figure 10.14(d). The quantitative agreement between the simulated and experimental voltammograms proves that the radial electrochemical response has been achieved at this NEE. Again, there are no adjustable parameters in this simulation.

10.6.3 Detection Limits

A possible application of these NEEs is the ultra trace detection of electroactive species. We have recently shown that NEEs with 10 nm-diameter disks operating in the total overlap mode show electroanalytical detection limits that are three orders of magnitude lower than detection limits obtained at macroscopic Au disk elec-trodes of comparable geometric area [15]. This occurs because in the total overlap mode, the total Faradaic signal generated at the NEE is equivalent to that obtained at the conventional macroelectrode of equivalent geometric area. However, the background double-layer charging current is significantly less because these currents are proportional only to the active Au area. The ratio of active area to geometric area for a 10 nm NEE is approximately 0.001 [15]. As a result, the background current is reduced by three orders of magnitude, and detection limits can be im-proved by three orders of magnitude.

An example of this enhancement in detection limits at an NEE is shown in Figure 10.16 [15]. Figure 10.16(a) shows voltammograms at a conventional Au macro-electrode at various low concentrations of TMAFc$^+$. As expected, the Faradaic

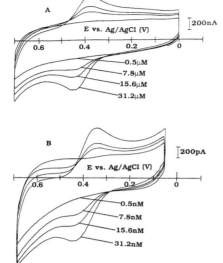

Figure 10.16. Cyclic voltammograms at 100 mV s^{-1} in aqueous TMAFc$^+$ at (a) a gold macrodisk electrode in 50 mM NaNO$_3$, and (b) a 10 nm NEE in 1 mM NaNO$_3$. TMAFc$^+$ concentrations are as indicated.

signal eventually vanishes into the double-layer charging currents as the concentration of TMAFc$^+$ decreases. Figure 10.16(b) shows voltammograms at a NEE with 10 nm-diameter electrode elements and a geometric area equivalent to that of the macroelectrode at various low concentrations of TMAFc$^+$. While the voltammograms essentially look identical to those obtained at the macroelectrode, the concentrations of the electroactive species at the NEE are three orders of magnitude lower than those for the macroelectrode. The detection limit at the macroelectrode was determined to be approximately 2 µm, while the detection limit at the NEE was approximately 2 nm [15].

Template synthesis has been shown to provide a simple means of creating ensembles of nanoelectrode ensembles. These NEEs can achieve electroanalytical detection limits that are three orders of magnitude lower than detection limits obtained at conventional macroelectrodes. We are currently investigating fabrication processes that allow the use of NEEs in nonaqueous solvents.

10.7 Metal Nanotube Membranes

We close our discussion of metal nanostructures with an interesting new type of membrane consisting of Au nanotubes that span the complete thickness of the membrane. We have previously mentioned that by controlling the electroless Au deposition time, the inside diameters of these tubes can be controlled at will. We recently asked the question –can tubes with inside diameters that approach the sizes of molecules be prepared, and if so, what applications might exist for such nanotubule containing membranes.

10.7.1 Fabrication

Typical templates used to prepare the metal nanotubule membranes were 6 μm thick polycarbonate membranes with 50 nm pore diameters and 6×10^8 pores cm^{-2}. Au was electrolessly plated onto the walls of the pores, yielding a Au nanotube within each pore. Variation in the plating time has been shown to produce Au tubules with internal diameters ranging from 34 to 1.4 nm [16]. The diameter of these Au tubules was determined from measurements of gas (He) flux across the membrane [16]. Because the electroless process plates on the membrane surface as well as within the pores, electrical contact with the surface allows electrical control of the potential inside the pores.

10.7.2 Ion-Selective Membranes

The ion transport properties of these Au-nanotubule-containing membranes was studied using a U tube concentration cell where the membrane separates two differing aqueous solutions (Figure 10.17) [16]. In an initial experiment, differing concentrations of KCl were placed on each side of the membrane, and reference electrodes were inserted into each solution to measure the membrane potential (E_m). When the diameters of the Au nanotubules approached 2 nm or less, the membranes displayed near-ideal cation-permselective behavior, i.e., these membranes transport cations but reject anions [16]. This behavior occurs because Cl$^-$ adsorbs strongly to Au, and as a result, the Au tubules have an excess of negative charge (Cl$^-$) on their inner surfaces. This causes anions to be excluded from the pores.

Ion permselectivity can also be controlled by directly changing the potential applied to the Au nanotubules. For this work, it was essential to use an anion that does not adsorb to Au because we wanted to control the charge in the Au tubes and not have it predetermined owing to excess charge from counterion adsorption. Because F$^-$ does not adsorb to Au, KF was chosen as the electrolyte. The U tube as-

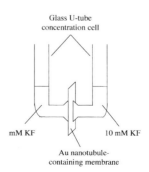

Figure 10.17. Schematic of a U tube concentration cell.

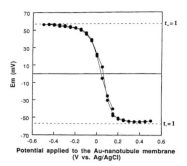

Figure 10.18. Variation of E_m with potential applied to the membrane (1 mM KF on the low-concentration (l) side, and 10 mM KF on the high-concentration (h) side of the membrane; tubule radius ~ 1.1 nm). The potential of the membrane was controlled with a potentiostat vs. a Ag/AgCl reference electrode immersed in the side-h solution. E_m was measured with the membrane under potentiostatic control.

sembly was used again, but this time the membrane was connected to the working electrode lead of a potentiostat. The potential applied to the Au nanotube membrane varied from -0.5 to $+0.5$ V vs. Ag/AgCl. The membrane was placed between solutions of 10 mM and 1 mM KF, and E_m values were measured at each applied potential.

The dashed lines at the top and bottom of Figure 10.18 are the E_m values that would be achieved if the nanotubule membrane showed ideal cation and ideal anion permselectivity respectively. At negative applied potentials, the nanotubule membrane shows ideal cation permselectivity, whereas at positive applied potentials the membrane shows ideal anion permselectivity. This selectivity occurs because at negative applied potentials, an excess negative charge is present on the walls of the Au tubes. This results in exclusion of anions from the tubes. At positive applied potentials, the opposite situation occurs – cations are excluded and anions are transported.

For any combination of metal and electrolyte, there is a potential called the potential of zero charge (pzc) where there is no excess charge on the metal. At this potential the nanotubule membranes should show neither cation nor anion permselectivity, and E_m should approach 0 mV. E_m for the tubule-containing membrane does go from the ideal cation permselective value, through zero, to the ideal anion permselective value. Furthermore, the potential at which E_m approaches zero is close to the reported pzc $(-4$ mV) [69].

We have demonstrated that these Au-nanotubule-containing membranes can be cation permselective, anion permselective, or nonselective, depending on the potential applied to the membrane. These membranes can be as permselective, like the commercially available Nafion polymer, and should have applications in both fundamental and applied electrochemistry. Because the Au tubules have dimensions on the order of molecular sizes and are quite monodisperse, we have been exploring the possibility of separating molecules based upon differences in their physical dimensions.

10.8 Semiconductor Nanotubules and Nanofibers

Electrochemical methods were previously used as a means of depositing semi-conductor materials into the pores of a template membrane [27]. However, this in section we will discuss the properties of semiconductor tubules and fibrils synthesized by a much more versatile deposition method: sol–gel chemistry [26].

10.8.1 Structural Characterization

Upon the confinement of a semiconductor to nanoscopic dimensions, the first two questions that arise are: can we see evidence for quantum confinement and what is the crystal structure of the material? TiO_2 fibrils have been synthesized within the pores of both 200 nm- and 22 nm-pore-diameter alumina membranes [26]. The sol–gel fabrication of TiO_2 fibrils within the pores of alumina membranes has been described earlier in Sect. 10.4. An absorption spectrum of the template alumina membrane containing these fibers showed an abrupt increase in absorbance at a wavelength of approximately 389 nm. This corresponds to the bandgap of bulk TiO_2 [70]. This suggests that the diameter of these fibrils is too large to show evidence for quantum confinement in the absorption spectrum. We are capable of preparing alumina template membranes with pore diameters approaching 5 nm or smaller. Correspondingly, we are currently attempting to prepare fibrils small enough to provide evidence for quantum confinement.

Electron diffraction has been employed to determine the crystal structure of the template-synthesized TiO_2 fibrils [26]. Figure 10.19 shows a TEM image of 22 nm-diameter TiO_2 nanofibers with the membrane dissolved away. The small fibers are

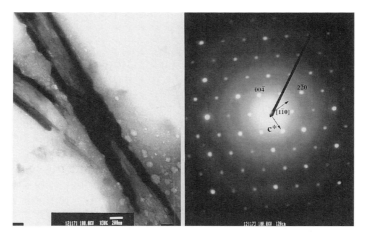

Figure 10.19. (a) TEM image of a bundle of 15 nm-diameter TiO_2 fibrils. (b) The corresponding electron diffraction pattern.

arranged in bundles that can contain anywhere from 2 to 10 or more fibers. Figure 10.19(b) shows the indexed electron diffraction pattern obtained from the center of the fibril bundle on the left side of the main feature in Figure 10.19(a). The orientations of the images are the same, i.e., the c* axis in Figure 10.19(b) is parallel to the fibril bundle axis in Figure 10.19(a). These data show that the fibrils are highly crystalline anatase-phase TiO_2, with the c* axis of the anatase oriented along the long axis of the fibril. Small fibril bundles throughout the sample display the same crystalline orientation; i.e., the reciprocal lattice direction [110] is almost always parallel to the electron beam, and the c* axis is along the fibril axis. We have concluded that these fibrils crystallize as long, prismatic crystals with the rare, and metastable, anatase mineralogical orientation [001] with {110} [71].

10.8.2 Photocatalysis

A standard application of TiO_2 has been as a photocatalyst for the decomposition of organic molecules [72–76]. This is a surface reaction that is thought to involve absorption of a UV photon by TiO_2 to produce an electron–hole pair that reacts with water to yield hydroxyl and superoxide radicals. These radicals can then oxidize the organic molecule. Template-synthesized TiO_2 structures should increase the TiO_2 surface area and correspondingly increase the decomposition reaction rates. For example, TiO_2 fibrils can be synthesized within the pores of a 60 μm-thick alumina membrane with 200 nm-diameter pores [26]. The TiO_2-fibril-containing membrane is attached to an epoxy surface, and the membrane is dissolved away. The calculated surface area of the immobilized fibrils is 315 cm^2 of TiO_2 surface area per cm^2 of planar geometric area. This suggests that, in principle, an enhancement of 315 in the catalytic rate of organic decomposition on template-synthesized TiO_2 fibers is possible vs. a thin-film TiO_2 catalyst. Through the use of tubular structures and/or template membranes with smaller-diameter pores (with correspondingly higher pore densities and surface areas) even larger increases in the rate would be predicted.

We have studied the decomposition of salicylic acid over time on an array of immobilized TiO_2 fibers (Figure 10.5(c)), with exposure to sunlight (Figure 10.20(a)) [26]. The upper curve follows the concentration of salicylic acid for a solution containing no TiO_2 catalyst, and no significant decomposition is observed. The small increase in salicylic acid concentration has been ascribed to the evaporation of water during the exposure to sunlight. The middle curve follows salicylic acid decomposition on a thin film of TiO_2, and the bottom curve shows a marked increase in decomposition of salicylic acid for the template-synthesized TiO_2 fibers.

The decomposition data can be used to quantitatively determine the rate of photodecomposition. If a pseudo-first-order rate law with respect to the salicylic acid concentration is graphed vs. reaction time, rate constants for the decomposition of salicylic acid can be determined (Figure 10.20(b)) [73, 75–77]. The slope of these lines provides the decomposition rate constant. The thin-film catalyst has a rate constant of 0.003 min^{-1} while the fibrillar catalyst shows an increased rate constant of 0.03 min^{-1}. This order of magnitude increase in reaction rate is much smaller than the 315-times enhancement predicted. This is not surprising because the thin-

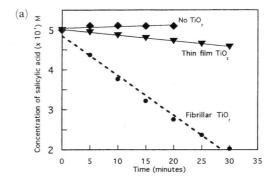

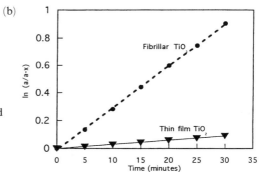

Figure 10.20. (a) Photodecomposition of salicylic acid in sunlight. Data for no photocatalyst, the thin film TiO₂ photocatalyst, and the fibrillar (200 nm) TiO₂ photocatalyst are shown. (b) First-order kinetics of the photodecomposition of salicylic acid with both the thin film and fibrillar TiO₂ photocatalyst.

film TiO_2 undoubtedly has some degree of surface roughness, resulting in higher surface areas and higher decomposition rates than predicted. Also, scanning electron microscopy (SEM) analysis of the fibrilar TiO_2 (Figure 10.5(c)), shows that the fibers "lean" against each other, possibly shading large portions of the surface from the sunlight, resulting in lower decomposition rates than predicted.

This section has shown that single-crystal TiO_2 fibrils can be fabricated via template synthesis and sol–gel chemistry. Also, owing to the increased surface area of the TiO_2 fibril array, the decomposition rate of an organic molecule increases. However, this prototype fibrillar catalyst is not optimal. We are currently working on processes to optimize the fibril arrays by varying the fibril diameter and aspect ratio and the distance between the fibrils. We are also exploring additional applications of these TiO_2 nanofibers, including electrochemistry, battery research, photoelectrochemistry, and enzyme immobilization.

10.9 Conclusion

The template method has become a very simple yet powerful process for the synthesis of nanomaterials. This review has described a host of chemistries that are now available for the template synthesis of a wide variety of nanomaterials, including metals, polymers, carbon, and semiconductors. Applications have ranged from fundamental optical studies to ultra trace molecular detection to high-surface-area catalysis.

What does the future hold for template synthesis? From a fundamental viewpoint, our group is interested in fabricating nanostructures with significantly smaller diameters in order to further explore the effects of size on the properties of materials. We are also developing new chemistries so that tubules and fibrils composed of an even larger variety of materials are available. New applications for template-synthesized nanomaterials are also being developed. We are exploring applications in photocatalysis, chemical analysis, bioencapsulation, biosensors, bioreactors, molecular separations, and electronic and electro-optical devices. Finally, it is clear that if practical applications are to be realized, methods for mass producing template-synthesized nanostructures will be required.

Acknowledgments

This work would not have been possible without the efforts of a number of hard-working and highly motivated graduate students and postdocs. They include Vinod P. Menon, Zhihua Cai, Junting Lei, Wenbin Liang, Ranjani V. Parthasarathy, Charles J. Brumlik, Gabor L. Hornyak, Leon S. Van Dyke, Colby Foss, Matsuhiko Nishizawa, Reginald M. Penner, Charles J. Patrissi, Veronica M. Cepak, Brinda B. Lakshmi, Guangli Che, and Kshama B. Jirage. Financial support from the Office of Naval Research and the Department of Energy is also gratefully acknowledged. We also wish to thank the Colorado State University Electron Microscopy Center.

References

[1] G. A. Ozin, *Adv. Mater.* **1992**, *4*, 612–649.
[2] *Engineering a Small World: From Atomic Manipulation to Microfabrication*, special section of *Science* **1991**, *254*, 1300–1342.
[3] V. P. Menon, J. Lei, C. R. Martin, *Chem. Mater.* **1996**, *8*, 2382–2390.
[4] R. V. Parthasarathy, C. R. Martin, *J. Polm. Sci.* **1996**, *62*, 875–886.
[5] C. R. Martin, R. V. Parthasarathy, *Adv. Mater.* **1995**, *7*, 487–488.
[6] R. Parthasarathy, C. R. Martin, *Nature* **1994**, *369*, 298–301.
[7] C. R. Martin, R. Parthasarathy, V. Menon, *Synth. Met.* **1993**, *55–57*, 1165–1170.
[8] C. R. Martin, *Adv. Mater.* **1991**, *3*, 457–459.
[9] Z. Cai, J. Lei, W.V. Liang, Menon, C. R. Martin, *Chem. Mater.* **1991**, *3*, 960–966.
[10] L. S. Van Dyke, C. R. Martin, *Langmuir* **1990**, *6*, 1123–1132.
[11] W. Liang, C. R. Martin, *J. Am. Chem. Soc.* **1990**, *112*, 9666–9668.
[12] Z. Cai, C. R. Martin, *J. Am. Chem. Soc.* **1989**, *111*, 4138–4139.
[13] R. M. Penner, C. R. Martin, *J. Electrochem. Soc.* **1986**, *133*, 2206–2207.

[14] J. C. Hulteen, V. P. Menon, C. R. Martin, *J. Chem. Soc., Faraday Tran.* **1996**, *92*, 4029–4032.

[15] V. P. Menon, C. R. Martin, *Anal. Chem.* **1995**, *67*, 1920–1928.

[16] M. Nishizawa, V. P. Menon, C. R. Martin, *Science* **1995**, *268*, 700–702.

[17] C. J. Brumlik, V. P. Menon, C. R. Martin, *J. Mater. Res.* **1994**, *9*, 1174–1183.

[18] C. A. Foss Jr., G. L. Hornyak, J. A. Stockert, C. R. Martin, *J. Phys. Chem.* **1994**, *98*, 2963–2971.

[19] C. A. Foss Jr., G. L. Hornyak, J. A. Stockert, C. R. Martin, *Adv. Mater.* **1993**, *5*, 135–136.

[20] C. A. Foss Jr., G. L. Hornyak, J. A. Stockert, C. R. Martin, *J. Phys.Chem.* **1992**, *96*, 7497–7499.

[21] C. J. Brumlik, C. R. Martin, K. Tokuda, *Anal Chem.* **1992**, *64*, 1201–1203.

[22] C. J. Brumlik, C. R. Martin, *J. Am. Chem. Soc.* **1991**, *113*, 3174–3175.

[23] R. M. Penner, C. R. Martin, *Anal. Chem.* **1987**, *59*, 2625–2630.

[24] G. L. Hornyak, C. R. Martin, *J. Phys. Chem.* **1997**, *101*, 1548–1555.

[25] G. L. Hornyak, C. J. Patrissi, C. R. Martin, *Thin Solid Films,* in press.

[26] B. B. Lakshmi, P. K. Dorhout, C. R. Martin, *Chem. Mater.* **1997**, *9*, 857–862.

[27] J. D. Klein, R. D. I. Herrick, D. Palmer, M. J. Sailor, C. J. Brunlik, C. R. Martin, *Chem. Mater.* **1993**, *5*, 902–904.

[28] R. V. Parthasarathy, K. L. N. Phani, C. R. Martin, *Adv. Mater.* **1995**, *7*, 896–897.

[29] J. C. Hulteen, X. C. Chen, C. R. Martin, to be submitted.

[30] G. Che, C. R. Martin, to be submitted.

[31] C.-G. Wu, T. Bein, *Science* **1994**, *264*, 1757–1759.

[32] C. R. Martin, *Handbook of Conductive Polymers,* in press.

[33] C. R. Martin, *Acc. Chem. Res.* **1995**, *28*, 61–68.

[34] R. L. Fleisher, P. B. Price, R. M. Walker, *Nuclear Tracks in Solids,* University of California Press, Berkeley **1975**.

[35] Poretics Corporation, *Product Guide* **1995**.

[36] M. Quinten, U. Kreibig, *Surf. Sci.* **1986**, *172*, 557–577.

[37] A. Despic, V. P. Parkhutik, in: *Modern Aspects of Electrochemistry* (Eds.: J. O. Bockris, R. E. White, B. E. Conway), Plenum Press, New York **1989**, Vol. 20, Chap. 6.

[38] D. Al Mawiawi, N. Coombs, M. Moskovits, *J. Appl. Phys.* **1991**, *70*, 4421–4425.

[39] R. J. Tonucci, B, L. Justus, A. J. Campillo, C. E. Ford, *Science* **1992**, *258*, 783–785.

[40] J. S. Beck, J. C. Varuli, W. J. Roth, M. E. Leonowicz, C. T. Kresge, K. D. Schmitt, C. T.-W. Chu, D. H. Olson, E. W. Sheppard, S. B. McCullen, J. B. Higgins, J. L. Schlenker, *J. Am. Chem. Soc.* **1992**, *114*, 10834–10843.

[41] K. Douglas, G. Devaud, N. A. Clark, *Science* **1992**, *257*, 642–644.

[42] T. D. Clark, M. R. Ghadiri, *J. Am. Chem. Soc.* **1995**, *117*, 12364–12365.

[43] R. Schollhorn, *Chem. Mater.* **1996**, *8*, 1747–1757.

[44] S. K. Chakarvarti, J. Vetter, *J. Micromech. Microeng.* **1993**, *3*, 57–59.

[45] S. K. Chakarvarti, J. Vetter, J. Micromech, *J. Nucl. Instrum. Methods. Phys. Res. B* **1991**, *62*, 109–115.

[46] C. J. Miller, C. A. Widrig, D. H. Charych, M. Majda, *J. Phys. Chem.* **1988**, *92*, 1928.

[47] *Electroless Plating: Fundamentals and Applications* (Eds.: G. O. Mallory, J. B. Hajdu]) American Electroplaters and Surface Finishers Society, Orlando, FL **1990**, Chap. 1, pp. 1–55.

[48] J. Lei, Z. Cai, C. R. Martin, *Synth. Met.* **1992**, *46*, 53–69.

[49] R. V. Parthasarathy, C. R. Martin, *Chem. Mater.* **1994**, *6*, 1627–1632.

[50] T. Kotani, L. Tsai, A. Tomita, *Chem. Mater.* **1996**, *8*, 2109–2113.

[51] V. M. Cpak, J. C. Hulteen, G. Che, K. B. Jirage, B. B. Lakshmi, E. R. Fisher, C. R. Martin, *Chem. Mater.,* in press.

[52] B. O'Regan, M. Gratzel, *Nature* **1991**, *335*, 737–740.

[53] M. Nishizawa et al., *J. Electrochem. Soc.,* in press.

[54] G. M. Whitesides, J. P. Mathius, *Science* **1991**, *254*, 1312–1314.

[55] C. K. Preston, M. Moskovits, *J. Phys. Chem.* **1993**, *97*, 8495–8503.

[56] J. T. Masden, N. Giordino, *Phys. Rev. B* **1987**, *36*, 4197–4202.

[57] T. M. Whitney, J. S. Jiang, P. C. Searson, C. L. Chien, *Science* **1993**, *261*, 1316–1318.

[58] G. E. Possin, *Rev. Sci. Instrum.* **1970**, *41*, 772–774.

[59] T. Asada, Japanese Patent No. 1960, 310, 401.

[60] R. D. Patel, M. G. Takwale, V. K. Nagar, V. G. Bhide, *Thin Solid Films* **1984**, *115*, 169–184.

[61] S. Kawai, in: *Symposium on Electrochemical Technology in Electronics*, Electrochemical Society, Pennington, NJ **1987**, p. 389.

[62] J. C. van de Hulst, *Light Scattering by Small Particles*, Dover, New York **1981**, pp. 397–400.

[63] S. M. Drew, R. M. Wrightman, *J. Electroanal Chem.* **1991**, *317*, 117–124.

[64] M. F. Bento, M. J. Medeiros, M. L. Montenegro, C. Beriot, D. Pletcher, *J. Electroanal. Chem.* **1993**, *345*, 273–286.

[65] A. Russell, K. Repka, T. Dibble, J. Ghoroghchian, J. J. Smith, M. Fleischmann, C. H. Pitt, S. Pons, *Anal. Chem.* **1986**, *58*, 2961–2964.

[66] A. M. Bond, T. L. E. Henderson, D. R. Mann, W. Thormann, C. G. Zoski, *Anal. Chem.* **1988**, *60*, 1878–1882.

[67] N. Oyama, T. Ohsaka, N. Yamamoto, J. Matsui, O. J. Hatosaki, *J. Electroanal. Chem.* **1989**, *265*, 297–304.

[68] Z. J. Karpinski, R. A. Osteryoung, *J. Electroanal. Chem.* **1993**, *349*, 285–297.

[69] J. Clavilier, C. N. V. Huong, *J. Electroanal. Chem.* **1977**, *80*, 101.

[70] B. Enright, D. Fitzmaurice, *J. Phys. Chem.* **1986**, *100*, 1027–1035.

[71] J. D. Dana, rewritten by C. Polache, H. Berman, C. Frondel, *The System of Mineralogy*, Wiley, New York **1955**.

[72] A. Fujishima, K. Honda, *Nature* **1972**, *238*, 37–39.

[73] R. W. Matthew, *J. Phys. Chem.* **1987**, *91*, 3328–3333.

[74] B. Kraeutler, A. J. Bard, *J. Am. Chem. Soc.* **1978**, *100*, 5985–5992.

[75] D. F. Ollis, C. Hsiao, L. Budiman, C. Lee, *J. Catal.* **1984**, *88*, 89–96.

[76] K. Okamoto, Y. Yamamoto, H. Tanaka, M. Tanaka, A. Itaya, *Bull. Chem. Soc. Jpn.* **1985**, *58*, 2015.

[77] R. W. Mathews, *J. Catal.* **1987**, *97*, 565–568.

Chapter 11

Morphology-Dependent Photocatalysis with Nanoparticle Aggregates

M. Tomkiewicz and S. Kelly

11.1 Introduction

The function–morphology correlation of composite catalysts is an issue that needs to be addressed on a number of different scales. There is extensive experimental evidence that supports the notion that the efficiency of photocatalytic reactions is strongly affected by one or more morphological parameters of the catalysts [1–8]. Previously we [7], and others [2–4] , argued that the simultaneous need to optimize mass transport and adsorption of reactants requires morphology of pore structure that is most likely fractal. Avnir et al. [4] have shown that the particle distribution in many effective porous catalysts can be characterized by a fractal dimension D. These authors also argued [2–3] that many heterogeneous photoelectrochemical reactions that are photocatalyzed on granular or particulate solid follow a scaling law that can be represented by:

$$v = kR^{D_R - 3} \tag{11.1}$$

where v is the reaction rate, k is a constant, R is the radius of an assumed spherical particle and D_R is a characteristic exponent labeled the "reaction dimension". The argument was that many physical processes such as distribution of reactive surface sites, depletion of the reservoir of molecules that diffuse to fractal sites, etc., reflect such scaling.

One way of understanding light-induced charge separation processes on semi-conducting substrates is to arrive at a parametrization that correlates material properties with output parameters. The Gärtner equation and its descendants [9] are classic examples in which we try to correlate output parameters of photovoltaic devices, such as short circuit current and open circuit voltage, with material parameters such as minority carriers' diffusion length and absorption coefficient. These four parameters, and many similar ones, are parameters of the collective system. In the general language of complex systems, they are emerging properties of the aggregates – in our case, in its simplest presentation – of the single crystal substrates. Photoelectrochemistry on semiconducting substrates has now reached the point

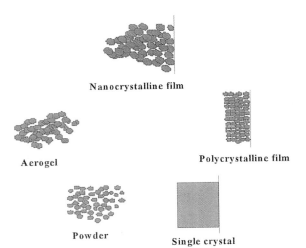

Nanocrystalline film

Aerogel

Polycrystalline film

Powder

Single crystal

Figure 11.1. Different modes of aggregation for particulate matter.

where we need to develop parametrization that spans different aggregation modes. Figure 11.1 demonstrates the central issue; we show there five different modes of aggregation [8]. The specific surface area across these aggregation modes varies between 1000 m^2 g^{-1} for some of the aerogels [10] to 2.5 × 10^{-5} m^2 g^{-1} for the single crystal. There is now a vast and growing literature on photoelectrochemistry with almost every one of these aggregation modes using the same semiconducting material – TiO$_2$. The output parameters are different: photocatalytic products with non-electrode configurations and photovoltaic performance parameters with electrode-style configurations. One can also find considerable overlap – photocatalytic activity is being investigated throughout the series and photovoltaic performance starting with the nanocrystalline electrodes. Subject to their intrinsic limitations (bandgap and band location) almost all the systems are among the most efficient in their respective categories. This brings into focus the need to parametrize these systems in terms that can be applied across the different aggregation modes.

Any attempt at correlating photoelectrochemical properties across the different aggregation modes critically depends on our ability to parametrize, measure, and control the morphologies of these materials and on our ability to measure in a reproducible way the quantum efficiency of output parameters that correlates with the efficiency of the charge separation process. The conversion efficiencies are output parameters that are the end result of a complex chain of events that include absorption of the light, efficiency of carrier separation within a particle, adsorption properties of the electrolyte, charge transfer to the electrolyte, turnover number of the electrolyte, diffusion of reactants and products, etc. In addition to the intrinsic complexity, an important contribution to the balkanized efforts to understand these systems rests on disciplinary segregation imposed by training constraints.

Schematic representation of a photocatalytic activity on a granular substrate is shown in Figure 11.2. In thermal catalysis, the catalyst acts as an extended surface on which the catalytic reaction takes place. The quality and nature of the bulk of

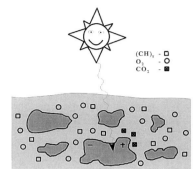

Figure 11.2. Schematic representation of solar photo-
decomposition of water-borne organic pollutants.

the catalyst makes very little difference. Semiconducting photocatalysts are different
in the requirement that the light absorption process takes place in the bulk. The
light absorption induces charge separation and the separate charges need to diffuse
to the solid–liquid interface to participate in the catalytic process. Thus a key additional set of parameters that plays an important role with these materials are the
bulk properties of the individual crystallites.

Many applications that are described in this book involve nanocrystalline semiconductors that are covered with monolayers of dyes. Typically the dye absorbs
light to inject charge carriers into the semiconducting particles. The nature of the
coupling of the dye to the particle is critical for many applications. An effort was
made to try to distinguish between dyes that are physisorbed and dyes that are
chemisorbed [6]. The strength of the interaction is assumed to be specific to the
semiconductor–dye–electrolyte system. Roughness on a scale much larger than the
dye will not make any difference. However, roughness on the scale of the dye can
make a major difference. A technique that at least in principle can provide information on the roughness at this scale is SANS [5].

11.2 TiO₂ Aerogels

We have shown that preliminary comparative photocatalytic activity in photo-
decomposition of salicylic acid with various aerogels, xerogels, and powders of
TiO₂ approximately scales with the total BET surface area [2]. These experiments
were run under conditions not limited by mass transport. Photoelectrochemistry on
aerogels is perhaps the least described among the aggregation modes that were
mentioned in Figure 11.1. Aerogels are highly porous materials that are produced
via sol–gel processing and supercritical drying. The very high surface area and pore
volume of aerogels makes them attractive candidates for catalytic applications.

11.2.1 Morphology

The morphology of the TiO$_2$ aerogels is complex enough for an attempt at a holistic approach at the photocatalytic process that will take into account apparently conflicting requirements. The essence of such an approach is to be able to make meaningful comparisons between different photocatalysts. Such comparisons require sophisticated parametrization that is not easily available. It is not required that such a parametrization be unique, but it is required that it will be consistent. For meaningful comparisons, it is necessary not only to be able to measure the appropriate parameters, but also to develop synthetic techniques for selective control of these parameters.

The morphology of the TiO$_2$ aerogels has been characterized in terms of two length scales. Typical morphology constitutes 5 nm-diameter, rough, crystalline nanoparticles of anatase closely packed into mesoaggregates near 50 nm in size. The mesoaggregates are, in turn, packed to a loosely linked structure with an overall porosity of 80%. The total surface area of these aerogels is attributed to the sum of the surfaces of the nanoparticles [11–13]. Schematic representation of this structure is shown in Figure 11.3. In Figure 11.4 we show the particle and pore size distributions as determined by various experimental techniques [11–12]. Aside from the data obtained from the nitrogen adsorption isotherms at high partial pressure, which measure pore size distribution, the rest of the data describe the particle size

Figure 11.3. Schematic structure of the titanium dioxide aerogel.

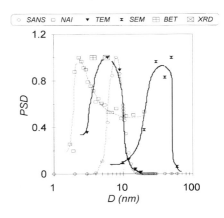

Figure 11.4. Particle size distribution in a typical TiO$_2$ aerogel as determined by various techniques [11].

distribution. The bimodality of the distribution is evident, with SEM the only technique that is sensitive to the mesoparticles. In the original publications [11–13] we emphasized the agreement between the various techniques, taking into account that some of the techniques (SANS) are model sensitive, some are not, and some of the techniques involve averaging from macroscopic samples (XRD, BET) while others are local (TEM, SEM). Within these differences and considering the available dynamic range, a factor of two difference in particle size seems to support the notion that the system is homogeneous and that the particle distribution is approximately uniform.

11.2.2 Control

For the TiO_2 aerogels, the control of the morphological parameters can come at almost every stage of the preparation procedure: At the onset – during the colloidal stages of the sol – appropriate control parameters are water content, temperature, acidity, ionic strength, etc. During the aging process, one can affect the strength of the connectivity between the individual mesoparticles by adjusting the chemistry of the gelation environment. Finally – at the aerogel stage – annealing changes the crystallite size, and at around 700 °C, one can induce a phase change from anatase to rutile. Many of the synthesis–morphology correlations seem to be independent of the oxide, and thus the work on photoactive oxide can benefit from the much larger parallel effort in the SiO_2 aerogels.

11.3 Evolution of Coordination Structure

Successful monitoring of the evolution of the coordination through the synthetic process remains a major challenge. Techniques such as infrared, and in particular Raman, spectroscopies remain the techniques of choice to characterize and parametrize the evolution of the solid network through the condensation reactions that lead to the formation of the solid structures. In many cases the IR work is supported by structural characterization techniques such as EXAFS and NMR spectroscopies that are sensitive to the local environment and can be applied in the liquid and solid phases.

11.3.1 Raman Scattering [14]

Conservation of momentum requires that for a perfect crystal in first-order Raman scattering, only phonons from near the center of the Brillouin zone are involved. In amorphous materials the q vector selection rules do not apply owing to the loss of long-range order and the Raman scattering spectra will resemble the phonon density of states. The microcrystalline case is an intermediate case. Parameters that include position [15], linewidth [16], and intensity [16] were used to follow the evolu-

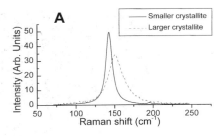

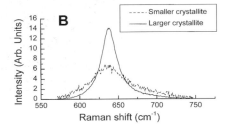

Figure 11.5. Raman spectra for a pair of aerogel samples with different crystallite sizes for the (a) 142 cm^{-1} and (b) 630 cm^{-1} peaks.

tion of the network formation that leads to the crystalline solid. The basic model due to the relaxation of the $q = 0$ selection rule due to the finite size of the crystals yields to the following expression for the Raman Intensity [17]:

$$I(\omega) \propto \int_{BZ} \exp\left(\frac{-q^2 L^2}{8}\right) \frac{d^3 q}{[\omega - \omega(q)]^2 \left(\frac{\Gamma_0}{2}\right)^2} \tag{11.2}$$

where q is expressed in units of $2\pi/a$ (a is the lattice constant), Γ_0 is the intrinsic linewidth, $\omega(q)$ is the phonon dispersion, and L the crystal size. The model predicts asymmetric broadening and redshift with the decreased size of the crystallites. For nonuniform particle distribution and shape, the broadening and the shifts will start to overlap and one often resorts to empirical correlations.

Figure 11.5 shows a comparison of an aerogel with small crystallites compared with one with large crystallites for the (a) 142 and (b) 630 cm^{-1} E_g peaks. It summarizes the main experimental observations that are associated with changes in the particle size: with a decrease in particle size the peak broadens asymmetrically (high frequency), shifts to the blue, and decreases in intensity. These effects are much more pronounced with the 142 cm^{-1} peak. For this study, the main tool that we have used to change the morphology and the particle size distributions is thermal annealing after the supercritical drying. Figure 11.6 shows the FWHM of the 142 cm^{-1} and the 630 cm^{-1} peaks as a function of the corresponding X-ray diffraction crystal size. Figure 11.7 shows the corresponding shifts in peak positions. Figure 11.8 shows the evolution of the 600 cm^{-1} peak during the sol condensation. This peak is the dominant peak in the sol and the gel phases (after subtraction of the solvent lines).

Figures 11.9 and 11.10 summarize the results: they show the theoretical FWHM

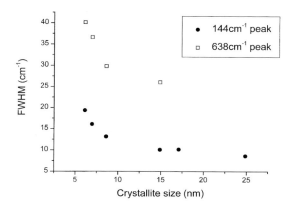

Figure 11.6. Relationship between the FWHM and the crystallite size for the 142 and 630 cm^{-1} peaks.

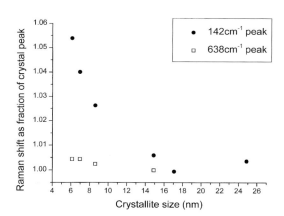

Figure 11.7. The Raman shift vs. crystallite size for the 142 and 630 cm^{-1} peaks.

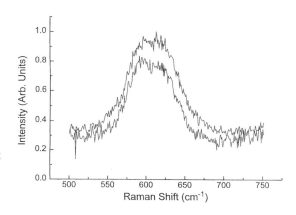

Figure 11.8. Evolution of the 600 cm^{-1} peak during the sol formation. The lower peak was taken 1 min after mixing of the reactants while the upper peak was taken 275 min after mixing (almost gel).

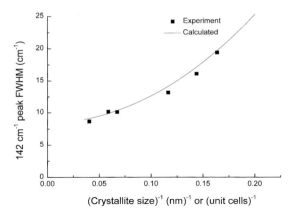

Figure 11.9. Correlation between the crystallite size and the line width of the 142 cm^{-1} peak.

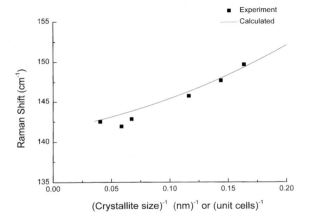

Figure 11.10. Correlation between the crystallite size and the 142 cm^{-1} peak position.

and position of the 142 cm^{-1} peak superimposed on the experimental data, as a function of crystallite size. While the experimental crystalline size was determined from X-ray crystallography, the theoretical fits were presented in terms of multiples of unit cell size. The scaling of the "effective" unit cell size was done so that the slopes the theoretical and experimental data in Figure 11.9 match. The data in Figure 11.10 were drawn without any adjustable parameters. This scaling produced an "effective" unit cell size of 5.55 Å for the 81% positive dispersion. This fits re-markably well with our prediction for an average value for the crystal unit cell somewhere between 3.78 Å and 9.50 Å. Using Figures 11.9 and 11.10, one can fit the data to expressions of the type [18]

$$\Delta v = k_1 \left(\frac{1}{L^\alpha} \right) \tag{11.3}$$

$$\Gamma = k_2 \left(\frac{1}{L^\alpha} \right) + \Gamma_0 \tag{11.4}$$

where Δv is the shift and Γ is the linewidth. α is a scaling parameter that was repeated to the network structure. It was found that for materials with layer structure such as graphite and boehmite $\alpha \approx 1$ while for covalently bonded semiconductors such as Si and GaAs $\alpha \approx 1.5$. From Figure 11.9 α is found to be 1.55.

11.4 Quantum Efficiencies

The difficulty in measurements of the quantum efficiencies of the photo-decomposition of substrates on such aggregates has been discussed in various forums. The spread of quantum efficiencies for oxidation of the same substrate by different commercial sources of TiO_2 [19], combined with other difficulties, was used as an argument to abandon attempts to measure this quantity and substitute instead a concept of "relative photon efficiency" [20], which normalizes the results to the degradation efficiency of a standard compound such as phenol under a given set of experimental conditions. There is no doubt that such a procedure can increase reproducibility and should serve as an important intermediate step in reporting efficiencies. However, it masks the expected variation with the solid state properties of the catalysts, which is the subject of interest here.

Figure 11.11 shows the experimental setup for quantum efficiency measurements that we presently use [8]. This setup is based on the premise that quantum efficiencies can be accurately measured, irrespective of the morphology of the absorber, as long as one is assured of total light absorption of the incident photons. Under these conditions the amount of TiO_2 will be irrelevant. The bottom–up configuration was designed to insure that measurements take place after the samples are settled at the bottom. Although our primary, practical objective in using aerogels was to be able to adjust their density to be below that of water so that they can float, in the absence of a hydrophobic coat, the pores will fill up with the solvent and the aerogels will sink.

We have measured the light intensity at the entrance and the exit (top and bottom) of the reactor. The reactor was constructed from Teflon such that a minimum amount of light would be absorbed by the walls. The light source was a 200 W Hg lamp. Irradiation intensity in the spectral range between 300 and 400 nm was 8 W m^{-2}. All the quantum efficiencies that we quote here are for reactant disappearance and not for product formation. The quantum efficiencies are not optimized. Oxygen was not deliberately added to the reaction vessel. Thus, oxygen used in the

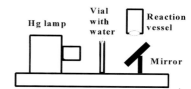

Figure 11.11. Experimental setup for the photo-decomposition of salicylic acid.

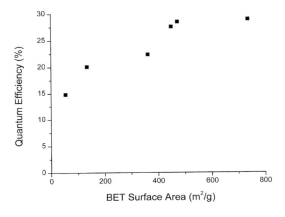

Figure 11.12. Monotonic increase in quantum efficiency with increasing BET surface area.

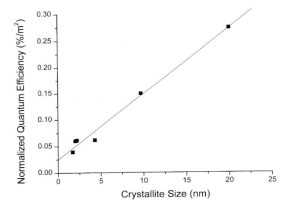

Figure 11.13. Normalized quantum efficiency vs. crystallite size.

reaction is atmospheric oxygen dissolved in the solution. Photodegradation was quenched by filtration of the samples.

Figure 11.12 shows the dependence of the quantum efficiencies on the BET surface area and Figure 11.13 shows the variation of the surface area normalized quantum efficiencies with the crystal size. Figure 11.12 shows a typical monotonic increase of the conversion efficiency with the total surface area that approaches saturation at high surface area. Figure 11.13 shows a linear increase in efficiency with crystal size. It is obvious that the two figures are contradictory in terms of a simple geometrical interpretation. If the only structural features were the nanocrystallites, then a simple geometrical consideration would require that the surface area/unit weight be inversely proportional to the crystallite size. We have argued [8] that the apparent saturation in the efficiency vs. surface area behavior might not only be due to saturation in adsorption but also to the presence of very small crystallites (<1.5 nm) that contribute to the surface area but very little to the charge separation process. An interesting conclusion that emerges out of these results is that the quantum efficiencies seem to be independent of the crystal structure of the crystallites.

We have analyzed in detail the difference sensitivities of BET and X-ray [8] and argued that it is perhaps better, at this stage, to regard the X-ray-measured crystalline radius and the BET surface area as two independent parameters. The crystalline radius can help to quantify the bulk properties such as light absorption and charge separation efficiencies, while the surface area characterizes the surface properties, adsorption, interaction between the charge carriers and substrates, etc. Better correlations and cross correlations will come from better control over the morphology and crystallinity and in particular synthesis of a more uniform particle size and a larger database from which one might draw the correlations.

Acknowledgement

The above work was made possible through the NASA Institutional Award Program at the City University of New York.

References

[1] H. van Damme, in: *Kinetics and Catalysis in Microheterogeneous Systems* (Eds.: M. Grätzel K. Kalyanasundaram), Marcel Dekker, New York **1991**.

[2] D. Avnir, O. Citri, D. Farin, M. Ottolenghi, A. Seri-Levy, in: *Optimal Structures in Heterogeneous Reaction Systems* (Ed.: P. J. Plath), Springer-Verlag, **1990**.

[3] D. Farin, J. Kiwi, D. J. Avnir; *Phys. Chem.* **1989**, *93*, 5851–5854; **1991**, *95*, 6100.

[4] D. Avnir, D. Farin, P. Pfeifer, *Nature* **1984**, *308*, 261–263.

[5] N. Serpone, M. Linder, E. Pelizzetti, in: *Fine Particles Science and Technology. From Micro to Nano Particles* (Ed.: E. Pelizzetti), Reidel **1996**, pp. 657–673.

[6] M. Anpo, T. Kawamura, S. Kodema, K. Maruya, T. Onishi, *J. Phys. Chem.* **1988**, *92*, 438–440.

[7] M. Tomkiewicz, G. Dagan, Z. Zhu, "Morphology and Photocatalytic Activity of TiO_2 Aerogels," in: *Research on Chemical Intermediates* **1994**, *20*, 701–710.

[8] M. Tomkiewicz, S. Kelly, in: *Fine Particles Science and Technology. From Micro to Nano Particles* (Ed.: E. PelizzettI), Reidel **1996**, 403–411.

[9] See M. A. Butler, *J. Appl. Phys.* **1977**, *48*, 1914–1920.

[10] G. Dagan, M. Tomkiewicz, *J. Phys. Chem.* **1993**, *97*, 12651–12655.

[11] Z. Zhu, L. Y. Tsung, M. Tomkiewicz, *J. Phys. Chem.* **1995**, *99*, 15945–15949.

[12] Z. Zhu, M. Lin, G. Dagan, M. Tomkiewicz, *J. Phys. Chem.* **1995**, *99*, 15950–15954.

[13] Z. Zhu, M. Tomkiewicz, *Proc. Mater. Res. Soc.* **1994**, *346*, 353–358.

[14] S. Kelly, F. H. Pollak, M. Tomkiewicz, *J. Phys. Chem.* **1997**.

[15] H. Richter, Z. P. Wang, L. Ley, *Solid State Commun.*, **1981**, *39*, 625–629.

[16] T. W. Zerda, G. J. Hoang, *Non-Crystalline Solids* **1989**, *109*, 9.

[17] See, for example, F. H. Pollak, in: *Analytical Raman Spectroscopy* (Eds.: J. G. Grasselli, B. J. Bulkin), Wiley **1991**, Chap. 6.

[18] C. J. Doss, R. Zallen, *Phys. Rev.* **1993**, *B48*, 15626–15637.
[19] V. Augugliaro, A. Franco, F. Inglese, L. Palmisano, M. Schiavello, in: *Proc. of the 9th. Int. Conf. on Photochem. Conv. and Stor. of Solar Energy* (Eds.: E. Pelizzetti, M. Schiavello) **1992**, 274.
[20] N. Serpone, G. Sauve, E. Pelizzeti, P. Pichat, M. A. Fox, in: *Proc. of the 10th. Int. Conf. on Photochem. Conv. and Stor. of Solar Energy* (Ed.: G. Calzaferri) **1994**, B1.

Chapter 12

Zeta Potential and Colloid Reaction Kinetics

P. Mulvaney

12.1 Introduction

The solid–liquid interface has been the subject of experimental study for some 100 years, beginning with Gibbs, whose work on the thermodynamics of adsorption laid the foundations for interface science. The electrical aspects were investigated by von Helmholtz, Gouy, Chapman, and Stern, as well as other eminent scientists around the turn of the century. Their aim was to explain the structure of the interface and to understand how properties such as the electric potential and surface tension varied across the surface layers. In addition, early theories successfully explained the phenomenon of electrocapillarity, the origins of the Nernst or equilibrium electrode potential, and they could also predict, to within an order of magnitude, the electrical capacitance of an electrode immersed in water. However, it was always recognized that the structure of the electrical double layer (EDL) played an equally important role in electrode kinetics. Butler and Volmer subsequently determined how the kinetics of charge transfer depended on the electrode potential and demonstrated that the equilibrium electrode potential was directly related to the rate of electron transfer.

In the 1940s Derjaguin and Landau, and independently Verwey and Overbeek working in Holland, developed the basic theory of particle coagulation (the DLVO model) in terms of the electrical double layer around each colloid particle in solution [1]. They established that colloid coagulation is about the interaction of electrical double layers. A large body of evidence subsequently accumulated in support of the basic tenets of DLVO theory, and eventually, colloid chemistry adopted the entire electrical double layer structure and its associated thermodynamics as part of its foundations. However redox reactions at particle surfaces could not be readily investigated, and questions about kinetics at colloid surfaces – processes such as redox catalysis, colloid nucleation and dissolution, electron transfer by excited species generated in solution, and charge injection by photosensitizers – all remained largely unanswered. Such redox reactions are central to a plethora of important industrial processes ranging from the photographic process [2], the removal of rust [3], the decontamination of nuclear reactor coolant systems [4], the transport of nutrients such as Mn^{2+} in natural waters [5], solar energy conversion [6], the degradation of paints, the removal of organic pollutants [7], and the electrochemical discharge of the alkaline battery [8]. Over the last 20 years, the use of optically transparent colloids together with the increased availability of stopped-flow spectroscopy, laser flash photolysis, and pulse radiolysis has finally enabled a direct comparison of colloid electrochemical kinetics with standard electrode kinetics to be made. In particular, whilst the equilibrium Nernst potential is fundamental to both areas of surface science, one can now assert that the Butler–Volmer equation (or Tafel equation in its simpler form) will soon be as important to colloid chemists as it is to the electrochemist.

In this chapter, we examine some of the available colloid data on the kinetics of electron transfer and try to highlight the parallels with conventional metal electrochemistry. We will focus on metal oxide particles because they are the most readily understood, and because the majority of the available experimental data have been gleaned from these materials. We begin by presenting a summary of the electrical structure of the metal-oxide–water interface. This enormous subject is covered in many texts in detail, particularly the underlying assumptions inherent in the derivation of the equations describing the double layer [9–13]. The aim here is to explain how the structure of the electrical double layer affects the actual rate of charge transfer at the particle interface. Such understanding will be fundamental to the improved design and exploitation of nanostructured materials [14, 15].

12.2 The EDL around Metal Oxides

12.2.1 The Helmholtz Region

When a conductor is placed into water, the steady state charge that builds up on the solid is usually due to charge transfer between the metal and solution. For example, a platinum electrode usually has an open circuit potential in aerated solution determined by the (largely irreversible) kinetics of the reaction [16]:

$$O_2 + H^+ + e^-(Pt) \rightarrow HO_2 \tag{12.1}$$

For semiconductors or insulators, the amount of charge that can be exchanged is much less, since the mobile charge is due only to impurities. As a consequence, it is usually the preferential loss of lattice cations and anions, or the adsorption of charged species from solution, that determines the amount of surface charge on a particle.

When a metal oxide surface is created in solution, the adsorption of hydroxyl ions or protons leads to the generation of a surface charge, and an electric potential develops between the surface and the bulk solution. Provided the surface activity of these so-called potential-determining ions (H^+, OH^-) remains constant, the surface potential of a metal oxide particle in aqueous solution is given by the familiar

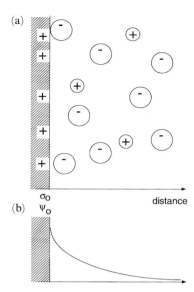

Figure 12.1. The diffuse layer. (a) The charge distribution near a charged surface, and (b) the potential distribution, which follows one of the four equations in Table 12.1.

Nernst equation:

$$\psi_0 = -0.059 \, (\text{pH} - \text{pH}_{\text{pzc}}) \tag{12.2}$$

where pH_{pzc} refers to the pH at which there is equal adsorption of potential-determining cations and anions at the surface. In response to the adsorbed surface charge there will be a local excess of counterions around the particle. These counterions form a diffuse layer around the particle and cause the electric potential to slowly decay to zero as one moves away from the particle surface towards the bulk solution, as shown schematically in Figure 12.1. However, some of these ions may be strongly adsorbed, forming a plane of bound countercharge, which will lower the electric potential immediately adjacent to the particle surface. This region is usually called the Helmholtz or Stern layer and is made up of both strongly polarized water molecules and desolvated ions, as shown in Figure 12.2. The adsorption plane is located at a distance x_1 from the actual surface, and the relative permittivity in the region $0 < x < x_1$ is taken to be ε_1. ε_1 has a value usually taken to be between 2 and 6. Since the distance of these ions from the surface is only of the order of 3 Å, the surface and counter charge may be treated as a parallel plate condenser with a capacitance per unit surface area given by

$$K_I = \sigma_0/(\psi_0 - \psi_1) = \varepsilon_1 \varepsilon_0/x_1 \tag{12.3}$$

where $\varepsilon_1 \varepsilon_0$ is the permittivity of the layer and σ_0 is the surface charge density. Consequently, the potential at x_1 is reduced to

$$\psi_1 = \psi_0 - \sigma_0/K_I = \psi_0 - \sigma_0 x_1/\varepsilon_1 \varepsilon_0 \tag{12.4}$$

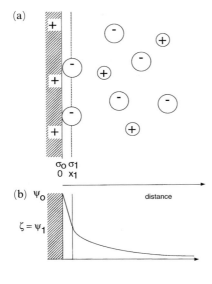

Figure 12.2. The electrical double layer with Stern or Helmholtz layer. (a) Charge distribution is broken up into a layer of specifically adsorbed ions and a diffuse layer. (b) The potential distribution showing the linear decay in the inner region and the diffuse layer potential, which begins at a distance x_1 from the surface and at a lower potential than ψ_0.

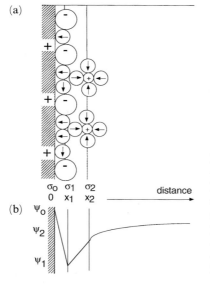

Figure 12.3. The electrical double layer according to the GCSG model. (a) Charge distribution includes an inner layer due to adsorbed ions and a second compact layer due to the finite size of hydrated ions approaching the surface from the diffuse layer. (b) The distribution of potential. For the case shown with superequivalent adsorption, the potential at x_1 or shear plane is the opposite to the intrinsic surface charge. There is also a linear variation in potential between x_1 and x_2.

It is important to realize that although ions adsorbed electrostatically at the Helmholtz plane will not completely neutralize the surface charge, if there is, in addition, a chemical driving energy for adsorption, the adsorbed countercharge may exceed that of the true surface charge, and the overall charge on the particle may be reversed. This is depicted in Figure 12.3 and we see that the potential actually changes sign, before decaying slowly to zero in bulk solution. This situation is often realised with polyelectrolytes such as poly(acrylic acid) or sodium hexametaphosphate, which can superadsorb on colloid surfaces. The ions in the Helmholtz layer are so strongly bound that an electric field applied to the colloid will cause motion of the particle, the adsorbed countercharge, and a monolayer or so of solvent molecules, i.e. the effective charge on the electrokinetic unit is not due solely to the surface charge, but will be reduced (generally) by the plane of countercharge.

In the GCSG (Gouy–Chapman–Stern–Grahame) model [17, 18] a second structural plane is also defined, and this is shown in Figure 12.3. This is at a distance x_2 and is the distance of closest approach of solvated ions to the plane x_1, owing to the finite size of the ion and its hydration shell. This second plane is usually termed the outer Helmholtz plane (OHP), with x_1 then being the inner Helmholtz plane (IHP). In this case, the electric potential decreases further between the inner plane of adsorbed anions and the OHP where counterions reach. It is the potential at x_2 which then orientates other ions in solution and induces the buildup of the space charge layer of counterions around the particle.

12.2.2 The Diffuse Layer

Electroneutrality requires that, overall, the excess charge density around the charged colloid particle in the diffuse layer, σ_d, must equal the charge density on the particle, i.e.

$$\sigma_o + \sigma_1 + \sigma_d = 0 \tag{12.5}$$

The electric potential at any point in the diffuse layer is calculated from Poisson's equation:

$$\mathbf{\nabla}^2 \psi = -\rho/\varepsilon_r \varepsilon_o \tag{12.6}$$

For a symmetrical $z:z$ electrolyte, the charge density at any point in solution is

$$\rho = \zeta e \, (n_+ - n_-) \tag{12.7}$$

where $|z|$ is the absolute electrolyte valency, and n_+ and n_- are the respective ion concentrations, at that point in the solution. Assuming the ions are distributed in the electric field according to the Boltzmann equation, then at any point we can write

$$n_\pm = n^\infty_{\pm} \exp \, (-z_\pm e\psi/kT) \tag{12.8}$$

where n^∞ is the bulk ion concentration, and combining Eqs. (12.8) and (12.9), we obtain the Poisson–Boltzmann (PB) equation:

$$\mathbf{\nabla}^2 \psi = -2n^\infty z e/\varepsilon_r \varepsilon_o \, \sinh \, (ze\psi/2kT) \tag{12.9}$$

The boundary conditions for integration are that in the bulk solution, the potential, and the electric field disappear:

$$\psi \rightarrow 0 \text{ as } x \rightarrow \infty \tag{12.10}$$

and

$$\mathrm{d}\psi/\mathrm{d}x \rightarrow 0 \text{ as } x \rightarrow \infty \tag{12.11}$$

whilst at the start of the diffuse layer, the potential must be ψ_1:

$$\psi_{(x=x_1)} = \psi_1 \tag{12.12}$$

The solution of the PB equation is not straightforward, and the method of solution can be simplified by considering four regimes in turn.

12.2.3   The Diffuse Layer for Micron-Sized Colloid Particles

If the particle radius is large, the double layer can be treated as flat, and Cartesian coordinates used. Further, if the diffuse layer potential at x_1 is small ($\psi \ll kT/ze$), then linearization of the exponential terms leads to

$$\mathbf{V}^2\psi = \mathrm{d}^2\psi/\mathrm{d}x^2 = \kappa^2\psi \tag{12.13}$$

where

$$\kappa^2 = 2n^\infty e^2 z^2 / \varepsilon_r \varepsilon_o kT \tag{12.14}$$

is called the Debye–Hückel parameter or inverse double layer thickness. Eq. (12.13) can be directly integrated using the boundary conditions (Eq. 12.10, 12.11) to yield

$$\psi = \psi_1 \exp{(-\kappa x)} \tag{12.15}$$

This shows that a charged particle has an apparent surface potential ψ_1 which falls off to $1/e$ of its surface value over a distance κ^{-1} in an electrolyte solution. For high potentials, $\psi_1 > 25 \text{ mV}/z$ at 298 K, the linearization is no longer accurate, and Eq. (12.9) must be integrated. The result is

$$\tanh{(ze\psi/4kT)} = \tanh{(ze\psi_1/4kT)} \exp{(-\kappa x)} \tag{12.16}$$

These results for high and low potentials for large particles are summarized in Table 12.1. Eq. (12.16) reveals that even for high potentials the diffuse layer thickness is still κ^{-1}. We can see that the approximation of a flat double layer around a colloid particle will be valid if $\kappa a \ll 1$. To obtain the capacitance of the double layer we note that

Table 12.1. Solution to the PB equation for high and low surface potentials in a symmetric electrolyte for high and low surface curvature.

	Large Particles $\kappa a \gg 1$	Small Particles $\kappa a \ll 1$
low potentials $< 25\,\mathrm{mV}$	$\psi(x) = \psi_1 \exp(-\kappa x)$	$\psi(r) = \dfrac{a\psi_1}{r} \exp(-\kappa(r-a))$
high potentials $> 25\,\mathrm{mV}$	$\tanh(y/4) = \tanh(y_0/4) \exp(-\kappa x)$	$\nabla^2\psi = \dfrac{d^2\psi}{dr^2} + \dfrac{2}{r}\dfrac{d\psi}{dr} = \dfrac{2zn^\infty e}{\varepsilon_r\varepsilon_o} \sinh\left(\dfrac{e\psi}{kT}\right)$
		(numerical solution only)
	where $y = \dfrac{ze\psi}{kT}$ and	$\kappa = \left[\dfrac{2z^2 e^2 n^\infty}{\varepsilon_r\varepsilon_o kT}\right]^{1/2}$

$$(\sigma_1 + \sigma_o) = -\varepsilon_r\varepsilon_o \; \mathrm{d}\psi/\mathrm{d}x_{|x=x1} \qquad (12.17)$$

From (12.12),

$$\mathrm{d}\psi/\mathrm{d}x_{|x=x1} = -\kappa\psi_1, \qquad (12.18)$$

and on substitution we get

$$K_d = (\sigma_o + \sigma_1)/\psi_1 = -\sigma_d/\psi_1 = \varepsilon_r\varepsilon_o\kappa. \qquad (12.19)$$

Thus the diffuse layer around a particle with $\kappa a \ll 1$ behaves like a parallel-plate capacitor, with thickness κ^{-1}, which is why κ^{-1} is called "the diffuse layer thickness."

12.2.4 The Diffuse Layer for Nanosized Particles

Nanosized particles distinguish themselves from their conventional, and larger, micron-sized counterparts by the fact that the double layer must be considered spherical because, for colloid particles with diameters of 100 Å, the assumption of flat double layers is no longer accurate. For example, at 1 mM $NaNO_3$, the Debye length is 100 Å, so $\kappa a = 1$. Integration of Eq. (12.9) must now be carried out in spherical coordinates. For small potentials linearization of Eq. (12.9) yields

$$\psi(r) = \psi_1 a/r \exp\left(-\kappa(r-a)\right) \qquad (12.20)$$

However, an electric potential of the same order as thermal energies is usually insufficient to prevent particle coalescence (see Section 12.2.5). So whilst the simplification renders the solutions more tractable, it does not provide accurate results for stable colloids with higher surface potentials, and the use of the linearized forms is generally inadequate. For highly charged particles, the potential distribution must be solved numerically through, for example, Runge–Kutta methods (see Section 12.2.5 below for an analytic approximation). The presence of the diffuse layer around an electrostatically stabilized colloid particle is essential. It is the repulsion experienced by two colloid particles as their double layers overlap that stabilizes them against coagulation. A high diffuse layer potential and a low electrolyte concentration, which increases the range of repulsion, are necessary for good colloidal stability. (In saying this, we ignore the possibility that the particles may be stabilized by polymers or large surfactants.)

We can now write down the total potential distribution between the particle surface and solution for an insulating or semiconducting metal oxide particle immersed in aqueous solution. For the model shown in Figure 12.2, we have

$$\psi(x_1) = \psi_o - \sigma_o/K_1 \tag{12.21}$$

$$\psi(x > x_1) = f(\kappa, a, \psi_1) \tag{12.22}$$

$$\psi_o = -0.059\,(\text{pH} - \text{pH}_{\text{pzc}}) \tag{12.23}$$

Here, $f(\kappa, a, \psi_1)$ refers to one of the four solutions in Table 12.1. Clearly, even for a model with just a single inner region, there are a number of experimental variables which need to be measured in order to quantify the potential distribution. Given that ε_1 and x_1 are not really directly accessible to experimental verification, simplifications are often advisable. Furthermore, until now, most electrokinetic investigations of colloid systems have been confined to the situation where only indifferent ions such as Na^+ or NO_3^-, are present in solution. In order to carry out electron transfer studies, there must be a redox couple present as well. Further simplification arises if the chemistry in solution can be controlled. By assuming there is no specific adsorption from solution we can set $\sigma_1 = 0$. However, this is clearly a poor approximation if a polyelectrolyte or surfactant has been used to stabilize the particles, or if a strongly chemisorbed ligand such as a thiol or amine derivative has been used to minimize particle growth during preparation. There has been little work done on the specific adsorption of redox couples, or even with simple carboxylic acids such as sodium citrate, which are extensively used to stabilize nanosized metal colloids.

12.2.5 The ZOS Model for Poorly Defined Nanoparticles

Before we discuss the process of electron transfer at colloid surfaces, we will present a simplified analytic version of the standard Stern model shown in Figure 12.2, which will allow us to understand the basic electrochemical kinetics involved, without needing to specify all the parameters of a complete double layer model. The double layer is broken up into a single Helmholtz layer of thickness x_1 and dielectric constant $\varepsilon_I \varepsilon_o$, i.e. with a constant capacitance given by

$$K_I = \varepsilon_I \varepsilon_o/x_1 \tag{12.24}$$

There is no specific adsorption at the plane x_1, which is also taken to be the start of the diffuse layer, i.e.

$$\sigma_1 = 0, \tag{12.25}$$

and

$$\sigma_o = -\sigma_d. \tag{12.26}$$

Finally, to link the electrical double-layer potentials to experimentally accessible data, we assume that the potential at x_1 is identical to the observed, measured "zeta potential."

$$\psi_1 = \zeta. \tag{12.27}$$

For particles in the size regime 30–100 Å, the condition $x_1 \ll a$ still holds and the flat-plate condenser model for the inner layer is justified. However, the diffuse layer thickness, characterised by the Debye–Hückel parameter κ^{-1}, is now much larger than a. The diffuse layer must be considered spherical. In principle, we need to solve the PB equation numerically, but we can save ourselves computational effort by adopting one of the various analytic approximations to Eq. (12.9) that have been developed. Ohshima et al. [19] found that for a sphere of radius a immersed in a 1:1 electrolyte of Debye–Hückel length κ^{-1}

$$\frac{\sigma_d e}{\varepsilon \varepsilon_o \kappa k T} = 2 \sinh\left(\frac{e\zeta}{2kT}\right)\left[1 + \frac{2}{A \cosh^2\left(\frac{e\zeta}{kT}\right)} + \frac{8 \ln\left[\cosh\left(\frac{e\zeta}{4kT}\right)\right]}{A^2 \sinh^2\left(\frac{e\zeta}{2kT}\right)}\right]^{1/2} \tag{12.28}$$

where $A = \kappa a$. The diffuse layer charge, zeta potential, and surface potential are linked by

$$\zeta = \psi_1 = \psi_o - \sigma_o/K_I = \psi_o + \sigma_d/K_I. \tag{12.29}$$

We can now describe the double-layer structure using just one or two parameters, provided we have zeta potential data, which includes the point of zero charge (pH_{pzc}). ψ_o is deduced directly from the pH of the experiment (through Eq. (12.2)). Then from Eq. (12.28), we obtain σ_d using ζ, and a. From Eq. (12.29), this gives us K_I directly. The validity of this approach can be tested using zeta potential data over a wide pH range to determine the average value of the Helmholtz capacitance. This is shown in Figure 12.4, where κa has been fixed at 1 and zeta potential vs. pH data have then been generated for various values of K_I. If no zeta potential data are available for a particular system, then as a last resort, we can use values of K_I determined for micron sized particles of the same material via electrophoresis, and try to create artificial zeta potential vs. pH curves. This model contains the funda-

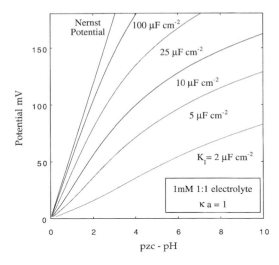

Figure 12.4. The calculated values of ζ vs. pH for nanosized colloid particles with $a = 100$ Å and 1 mM 1:1 electrolyte (i.e. $a = 1$) for different values of the inner layer capacitance, K_I. The diffuse layer charge is calculated from Eq. (12.28), and then ζ is obtained from Eqs. (12.2, 12.29).

mental features required to explain both colloid stability and redox chemistry at a nanoparticle surface. There is a diffuse layer, whose thickness depends upon the electrolyte concentration and surface potential, and an inner layer, the potential across which is controlled via the pH and inner layer capacitance. Because it contains no Stern layer charge and analytical approximations are used for the solution of the PB equation, this model is called the zero order Stern model (ZOS). The way the electric double layer potential is partitioned between the Helmholtz and diffuse layers critically determines both colloid stability and electron transfer kinetics.

12.2.6 The Point of Zero Charge and the Isoelectric Point

All these various double layer models have been designed by colloid chemists to explain the structure of the electrical double layer. In particular, they explain the apparent surface charge density obtained when a suspension is titrated with acid and base, and the observed mobility of the suspension particles when subjected to an electric field at different pH values. Because of the possibility of specific adsorption to colloid particles, there are two possible reference points for the measurement of the electric potential during mobility studies. These are the point of zero charge (pzc) and the isoelectric point (iep) of the solid [13]. The pzc is defined as the concentration of potential determining ions for which the surface charge σ_o is zero. For metal oxides, this corresponds to the pH at which $\Gamma_{H^+} = \Gamma_{OH^-}$, where Γ signifies the adsorption density. The isoelectric point is the concentration of potential determining ions at which the zeta potential is zero. They are often used interchangeably, but this is only the case if, at the pzc, there is no charge at the IHP. Thus, at the i.e.p. $\sigma_o = \sigma_1$ and therefore $\sigma_d = 0$. Conversely, at the p.z.c. $\sigma_o = 0$, and $\sigma_1 = -\sigma_d$.

12.3 Colloid Electron Transfer Kinetics – Theory

We now consider how the kinetics of electron transfer to particles are affected by this electrical double layer structure postulated to explain observed electrokinetic data. The two equations necessary are Fuchs' equation, (also used to describe the kinetics of colloid coagulation,) and the Tafel equation which quantifies the electric field dependence of the electron transfer rate constant.

12.3.1 Mass-Transfer-Limited Reactions

The rate constant for the steady state, diffusion-controlled reaction in solution between two species is given by the familiar Smoluchowski expression,

$$k_{diff} = 4\pi RD, \tag{12.30}$$

where $R = R_{coll} + R_{rad} \sim R_{coll}$ is the combined reaction radii of the electroactive species and the colloid particle, and $D = D_{coll} + D_{rad} \sim D_{rad}$, the combined diffusion coefficient. However charged species will also experience a force due to the electric field around the particle at any pH other than the pzc. The flux of an ionic species with concentration c and charge z_R towards a spherical surface in the presence of a position dependent electric field $\psi(r)$ is [20]

$$J = -D\left[\frac{dc}{dr} + c\frac{z_R e}{kT}\frac{d\psi}{dr}\right] \tag{12.31}$$

The boundary conditions are that $c = 0$ at $r = a$ and $c = c^\infty$ (the bulk radical concentration) at $r = \infty$. Integration yields

$$k_{field} = \frac{4\pi D}{\displaystyle\int_a^\infty \exp\left(\frac{z_R e\psi(r)}{kT}\right)r^{-2}\,dr} \tag{12.32}$$

The integral in the denominator is the reciprocal of the effective reaction radius. When $\psi(r) = 0$, Eq. (12.32) reduces to the Smoluchowski equation; in the presence of a nonzero field, the denominator can be greater than or less than a^{-1}, depending on the signs of z_R and $\psi(r)$. Note that even in the presence of a field, steady state conditions prevail after $\sim 10^{-7}$s, so that the time dependence of the flux can be ignored in almost all colloid systems unless the suspension is very concentrated. e.g. for TiO_2 colloids with a = 50Å, the half-life for reaction with $(CH_3)_2COH$ ($k = 5 \times 10^{10}$ $M^{-1}s^{-1}$, [21]) is 10ns only at 0.5 M TiO_2. It is unusual to work at such concentrations because of particle coalescence or because of the extremely high absorbance of such sols, which renders time-resolved work by spectroscopic means quite difficult.

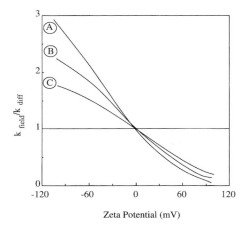

Figure 12.5. The effect of particle zeta potential on the mass transfer limited rate constant for radicals with charge +1 and particle radius 20Å. The ionic strengths are (a) 100 mM, (b) 10 mM and (c) 1 mM. The rate at the pzc is the Smoluchowski limit. Adapted from ref. 23.

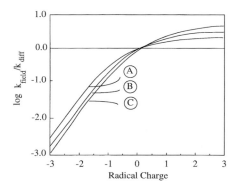

Figure 12.6. The effect of the radical charge on the mass transfer limited rate constant for a zeta potential of -80 mV and for ionic strengths (a) 1 mM, (b) 10 mM and (c) 100 mM assuming a particle radius of 20Å. The rate at $z = 0$ is the Smoluchowski limit. Curve is drawn through nonintegral values of charge for ease of comparison. Adapted from ref. 23.

Solution phase reactants can only approach the surface to the distance x_1 [18, 22]. Hence, the metal oxide can be considered to have a radius a and surface potential ζ, so that the boundary condition required for the integration of Eq. (12.29) is $\psi(a) = \zeta$. Once at the shear plane, transfer or deactivation occurs instantly. Thus for mass transfer limited reactions, we need to know D_{rad} and a, to calculate the Smoluchowski limit. In addition, when there is significant migration we need n^∞ (the bulk electrolyte concentration), z_R (the radical charge) and ζ.

To see how the field affects the mass transfer limited rate constant, calculated values are shown in Figures 12.5 and 12.6 for conditions typical for nanosized colloids in aqueous solution. Figure 12.5 illustrates the dependence on the ζ potential (for $z_R = 1$) for three different electrolyte concentrations and Figure 12.6 the dependence of the mass transfer limited rate constant on the radical charge (at $\zeta = -80$ mV). A particle radius of 20Å was assumed in calculating the potential profile. In each case, the potential distribution was first calculated from the nonlinear PB equation using the given parameters a, κ, ζ. (2000 points out to a distance of $10\kappa^{-1}$). Then for a given value of ζ, and the radical charge, the flux at the surface was calculated by a second integration using Eq. (12.32). As is clear from the results, diffusion controlled radical-colloid interactions are strongly dependent on the double layer properties (salt concentration, ζ, particle radius) and the magnitude of the radical charge. The flux of charged species due to migration rivals that due to the concentration gradient at high potentials and low ionic strength. When the two supplement each other, the rate may triple or quadruple, even at low radical charge. This should be readily discernible using flash photolysis or pulse radiolysis techniques. When the effects of the fluxes are opposing, the effect is far more dramatic, and the net flux to the colloid particle may be retarded by several orders of magnitude. Consequently, the effect of the zeta potential on mass transfer will be most clearly seen when the double layer acts to retard diffusion. In some cases a second order rate constant of just 10^7 M^{-1} s^{-1} may correspond to the mass transfer limit. Given that low ionic strength and high zeta potentials are usually necessary for ionically stabilized sols, the usual criterion that the diffusion limit is reached at a value of $\sim 10^{10}$ M^{-1} s^{-1} will no longer be valid.

Increases in salt concentration will decrease the importance of the migration term for mass transfer limited reactions. The effects of added indifferent electrolyte will be to decrease ζ and to compress the double layer simultaneously. (Since the diffuse layer capacitance is increased, there is a larger potential difference across the Helmholtz region.) However as can be seen from the figure, even in 0.1 M 1:1 electrolyte, pronounced deviations from the Smoluchowski value would be expected.

12.3.2 Activation-Controlled Electron Transfer

For electron transfer into a colloid particle, by a solution species (anodic reactions), the Tafel equation for the anodic electron transfer rate constant k_{et} is given by

$$\log \frac{k_{et}}{k_{et}{}^{pzc}} = \frac{\beta F \Delta \psi}{2.303 RT} \qquad (12.33)$$

where $\Delta \psi$ is the electric potential difference between the particle surface and the plane of electron transfer, and $k_{et}{}^{pzc}$ is the rate constant at $\Delta \psi = 0$. If the ionic strength is high, the diffuse layer capacitance $K_d \to \infty$, and the total double layer field is confined to the Helmholtz layer. The zeta potential then approaches zero. Under these conditions, the entire change in electrode potential can be considered to act on the electrons tunnelling from donor to surface (or surface to donor). However in colloid systems, a high salt concentration will destabilize the sol, since if $\zeta \to 0$, there is no resistance to coagulation. At low ionic strength, the changes in surface potential will not just appear as an overpotential for electron transfer. Some of the electric potential is "lost" in the diffuse layer. The amount "lost" will depend on the relative capacitances of the two layers of the electric double layer. However, the diffuse layer potential governs the local concentration of electroactive ions. Thus, the pH dependence of the rate of electron transfer depends on the zeta potential in two ways. The potential difference $(\psi_o - \zeta)$ alters the rate constant for electron transfer through Eq. (12.33), while the change in ζ potential alters the local concentration of any charged reactants through the Boltzmann equation, Eq. (12.8). When both are included, Eq. (12.33) assumes the form [24, 25]:

$$\log \frac{k_{et}}{k_{et}^{pzc}} = -\beta(\text{pH} - \text{pzc}) - \frac{(\beta + z_R)F\zeta}{2.3RT} \tag{12.34}$$

where is the transfer coefficient for the anodic electron transfer to the colloid and z_R the charge on the reductant. This equation is only valid for simple anodic electron transfer from the OHP. For cathodic electron transfer to an oxidant with charge z_O, the dependence on zeta potential is given by

$$\log(k_{et}/k_{et}^{pzc}) = (1 - \beta)(\text{pH} - \text{pH}_{pzc}) - (z_O + \beta - 1)F\zeta/2.303RT \tag{12.35}$$

The conduction band energy level in the colloid particle is normally the acceptor level for the transferred electron, and at the pzc, this energy level will not be identical to the redox potential of the solution couple. To compare intrinsic rates of electron transfer for the same solution couple with various colloidal semiconductors, it is necessary to decouple this chemical free energy term, $\Delta E_{pzc} = E^{cb}{}_{pzc} - E_{redox}$, which drives the reactions at the pzc. Thus the most useful parameter is $k_{et}{}^{ref}$ given by:

$$k_{et}{}^{ref} = k_{et}{}^{pzc} \exp{(-F\Delta E_{pzc}/RT)} \tag{12.36}$$

where for convenience we assume that the energy levels are potentials on a suitable electrochemical scale.

12.3.3 The Transition between Activation and Mass Transfer Limits

The transition between diffusion and activation control has been discussed by several authors for the case of zero migration [26,27]. The observed rate constant can be readily derived by consideration of the steady state concentration of a reductant at the electron transfer plane to a single colloid particle. Let this be denoted c_{OHP}. The flux due to surface reaction is then $4\pi a^2 k_{et} c_{OHP}$, where k_{et} is the rate constant for electron transfer. This must be balanced by the flux from solution. Integrating Fick's Law with the boundary condition that $c = c_{OHP}$ at $r = a$ rather than $c = 0$ yields

$$k_{obs} c^{\infty} = 4\pi a D(c^{\infty} - c_{OHP}) = 4\pi a^2 k_{et} c_{OHP} \tag{12.37}$$

It can be seen from this equation that the maximum flux to the OHP occurs if $c_{OHP} = 0$, which occurs as k_{et} increases. Conversely, the concentration gradient reduces to zero if $k_{et} = 0$, as expected intuitively. After rearrangement, the observed bimolecular rate per particle is

$$\frac{1}{k_{\text{obs}}} = \frac{1}{4\pi a^2} \left[\frac{1}{k_{\text{et}}} + \frac{a}{D} \right] \tag{12.38}$$

The same manipulations can be used when there is an electric field present around the particle, and in this case the transition from activation control to mass transfer control obeys

$$\frac{1}{k_{\text{obs}}} = \frac{1}{4\pi a^2} \left[\frac{1}{k_{\text{et}}} + \frac{a^2}{D} \int_a^\infty \exp\left(\frac{z_R e\psi(r)}{kT} \right) r^{-2}\, dr \right] \tag{12.39}$$

Clearly, the mass transfer rate constant depends on pH for charged reactants. So, for a reaction which is diffusion controlled over a wide pH range, as might be expected for many colloid reactions with $e^-(\text{aq})$ for example, a pH dependent reaction rate will be found.

12.4 Colloid Kinetics – Experimental Data

12.4.1 The Effect of pH

Various research groups have examined the rate of disappearance of a solution species via electron transfer to colloid particles, or conversely the transfer of electrons to acceptors in solution following photoexcitation of a semiconductor colloid. We cite the studies by Grätzel with TiO_2 [26–28], Darwent and coworkers [29–31], Willner [32] on silica, Bahnemann et al. on ZnO and TiO_2 [33], and Swayambunathan et al. on iron oxide [24, 25]. For many systems the chosen reactants have been radiolytically or photolytically generated, and the rate constants have been found to be close to the mass transfer limit.

Grätzel and Frank initially reported the very dramatic effect of pH on the rate constant for electron transfer using colloidal TiO_2 and methyl viologen as electron acceptor. Their results shown in Figure 12.7, clearly revealed the exponential de-

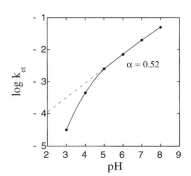

Figure 12.7. The relative rates of electron transfer from photoexcited colloidal TiO_2 to methyl viologen dications as a function of pH. Adapted from ref. 26.

pendence on pH predicted by classical electrochemical kinetics. This work was the first to show that even when the Helmholtz potential difference is governed by ion adsorption from solution and not by the electrical charge provided through an external power supply, the Tafel equation is still applicable. The ionic strength is not mentioned in their paper so it is difficult to reliably assess the importance of diffuse layer corrections [26].

12.4.2 The Effect of Electrolyte Concentration on Electron Transfer

Darwent et al. examined for the first time the role of diffuse layer contributions to the kinetics of electron transfer. They demonstrated that all electron transfer rates depend on ionic strength except at the pzc, as shown in Figure 12.8. They corrected their data for the diffuse layer contribution, using Debye–Hückel theory, modelling the nanosized titania colloids as large charged molecules. By employing weak double layer theory, i.e. low potentials, they showed that the observed transfer coefficient for metal oxide colloids obeys

$$\alpha \sim \alpha_o + (B + C\,I^{0.5})^{-1} \qquad (12.40)$$

where I is the ionic strength and B and C are adjustable parameters. This equation is similar to the one employed in metal electrode kinetics at low overpotentials. The parameters B and C are related to the relative capacitances of the diffuse and Helmholtz layers, and α_o is the transfer coefficient at infinite ionic strength. In Figure 12.9, we have attempted to reanalyse their results using electrophoretic data gleaned from the work of Wiese and Healy [34]. Good agreement is obtained, both for different pH values and for large variations in ionic strength using Eq. (12.35). This clearly illustrates that instead of using B and C as adjustable parameters, experimental zeta potentials can be used to quantify the effects of ionic strength on the rates of electron transfer.

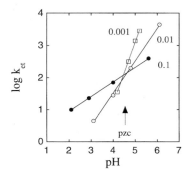

Figure 12.8. The effect of ionic strength and pH on the rate of electron transfer from colloidal TiO_2 to methyl viologen dication in the region around the pzc. Adapted from ref. 29.

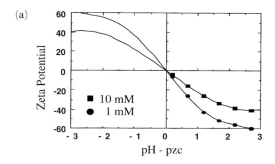

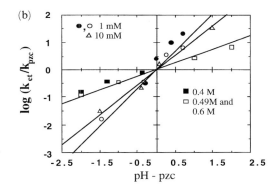

Figure 12.9. (a) Zeta potentials for TiO_2 at 1 mM and 10 mM vs pH from Wiese and Healy [34]. (b) Fits to the data of Darwent et al. using the zeta potential data from Wiese and Healy and Eq. (12.35). Adapted from ref. 25.

12.4.3 The Effect of the Zeta Potential and Radical Charge on the Rate of Electron Transfer

The only electron transfer experiments to date in which zeta potentials have been measured directly on the same nanosized particles is in the work of Swayambunathan et al. [24, 25]. Their results for electron transfer to colloidal iron oxide from both anionic and cationic viologen radicals are shown in Figure 12.10. By using two viologen radicals with opposite charge but virtually identical redox potentials, they confirmed that electrostatic effects dominate the kinetics of e.t. in solution. The rate of electron transfer for both radicals coincides at the pzc, again highlighting the fact that the pzc is the natural reference point for measuring transfer kinetics. However, rather than resorting to Debye–Hückel theory, valid only at low surface potentials, Swayambunathan et al. measured the zeta potential as a function of pH. The reaction becomes mass transfer limited for the anionic viologen radical at low pH as the surface potential of the iron oxide particles becomes very positive. Consequently, the kinetics must include both activation and mass transfer equations.

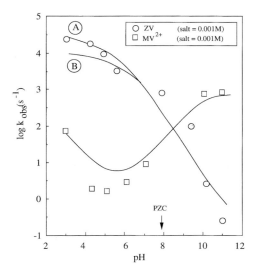

Figure 12.10. The effect of pH on the rate of electron transfer to colloidal ferric oxide particles from cationic and anionic viologen radicals. At low pH, the rate becomes mass transfer limited for the anionic viologen radical. Ionic strength = 1 mM. Particle radius = 20Å. Fits using Eqs. (12.34, 12.38 and 12.39).

$$\frac{1}{k_{obs}} = \frac{1}{4\pi a^2}\left[\frac{1}{k_{et}^{pzc}\exp\left(-\beta(pH - pzc) - \frac{(\beta + z_R)e\zeta}{2.303kT}\right)} + \frac{a^2}{D}\int_a^\infty \exp\left(\frac{z_R e\psi(r)}{kT}\right)r^{-2}\,dr\right]$$

(12.41)

In Figure 12.10, two curves are shown. One of these (curve B) uses the conventional Smoluchowski equation for the calculation of the mass transfer limit, based on Eq. (12.38), together with Eq. (12.34) for k_{et}. From the discussion above, the double layer corrected form, Eq. (12.39), should be better, since the radicals are charged and migration will contribute to the mass transfer of the radical to the colloid surface. Curve A uses Eq. (12.41) which is derived from Eq. (12.39) and Eq. (12.34). However the reaction only becomes mass transfer limited at low pH's, and below pH 3, the increasing solubility of the oxide and increasing electrolyte concentration make comparison with the theoretical values more difficult. The mass transfer limit in this pH range was calculated from Eq. (12.39), using the radius $a = 20$Å, as established by electron microscopy. The inclusion of the migration term does appear to give a better fit to the data than Eq. (12.38) over the limited pH range in which the reaction is diffusion controlled. The observed rate corresponds to a second order rate constant about twice that predicted by the Smoluchowski equation. It is worth noting that the transition to diffusion-migration control takes place over quite a wide pH range, and extends to pH 7, where the reaction is well below the expected mass transfer imposed limit.

It is clear that the entire pH dependence of the rate constants can be unified through the assumption that the zeta potential is close to the potential at the plane of electron transfer. In fact if the zeta potentials were about 10–20 mV higher, the

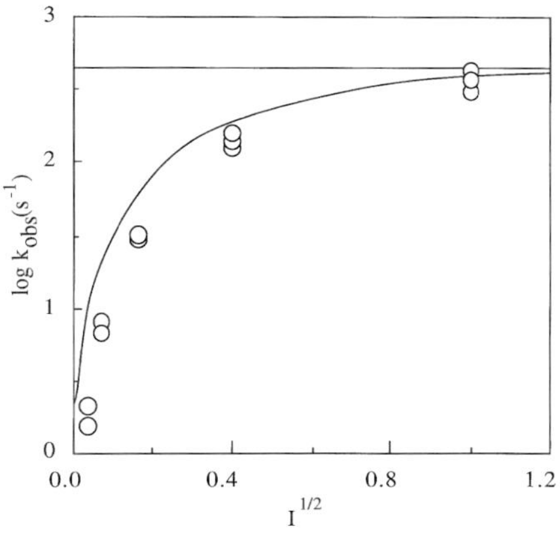

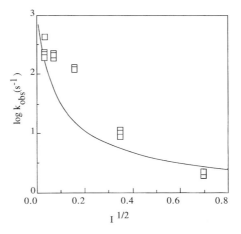

Figure 12.11. The effect of increased ionic strength on the rate of electron transfer to colloidal iron oxide from cationic methyl viologen radicals at pH 5.5. Fit to Eq. (12.34) with $\alpha = 0.5$, using $a = 20\text{Å}$, pzc $= 8.1 \pm 0.3$ using complete solution to the nonlinear PB equation and a Stern layer with $K_I = 400\ \mu\text{Fcm}^{-2}$. Adapted from ref. 24.

Figure 12.12. The effect of increased ionic strength on the rate of electron transfer to colloidal iron oxide from cationic methyl viologen radicals at pH 11. Fit to Eq. (12.34) with $\alpha = 0.5$, using $a = 20\text{Å}$, pzc $= 8.1 \pm 0.3$ using complete solution to PB equation and a Stern layer with $K_I = 400\ \mu\text{Fcm}^{-2}$. Adapted from ref. 24.

agreement would be almost perfect, an indication perhaps that the shear plane lies just beyond the true plane of transfer. Swayambunathan et al. also studied the role of electrolyte concentration [25]. As can be seen in Figure 12.11, the rate at pH 5.5 increases dramatically as salt is added, because of decreased repulsion between the positively charged radical and the positively charged colloid particles. The rate is 400 times faster in 1M electrolyte (NaClO$_4$) at pH 5.5. Conversely, above the pzc, the rate of transfer decreases as the attraction between the now negatively charged sol and the radical is reduced (see Figure 12.12). Unlike the case in purely ionic systems the plot of log k$_{obs}$ versus $I^{1/2}$ is not linear. By extrapolating the rate constant to infinite ionic strength, an estimate can be made of the rate of electron transfer (at pH 5.5 and pH 11.0) when the electric potential is entirely confined to the Helmholtz layer, and all pH changes act as an overpotential for the charge transfer. These data are plotted in Figure 12.13. The rate constants at infinite ionic strength represent the case of electron transfer when $\zeta = 0$, and should fit on the line

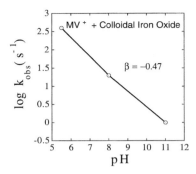

Figure 12.13. Transfer coefficient for electron transfer to iron oxide by methyl viologen radical cations at infinite ionic strength using limiting values at pH 5 and 11 and at the pzc of pH 8.3. Fit to Eq. (12.42), assuming $\zeta = 0$, at infinite ionic strength. Adapted from ref. 23.

represented by the reduced form of Eq. (12.34), namely,

$$\log k_{et} = \log k_{pzc} - \beta(pH - pzc) \tag{12.42}$$

Using the two infinite ionic strength rate constants gives a value of $\beta = 0.47$. This value is the 'true' transfer coefficient for oxidation of methyl viologen radical cations by colloidal iron oxide. Note that at low ionic strength, an experimental analysis of the transfer kinetics over only one or two pH units could easily have led to the conclusion that the radical is either negatively charged or positively charged.

The identification of the zeta potential with the potential at the plane of electron transfer has a further use if, a priori, the transfer coefficient is known. The existence of maxima or minima in the rate of an interfacial charge transfer reaction can then be predicted from Eq. (12.34) following differentiation:

$$\frac{\partial \log k_{et}}{\partial pH} = 0 = -\beta pH - \frac{\partial \zeta}{\partial pH}\frac{(\beta + z_R)F}{2.303RT} \tag{12.43}$$

Hence at the maximum or minimum,

$$0.0592\frac{\beta}{\beta + z_R}pH = \frac{-\partial \zeta}{\partial pH} \tag{12.44}$$

and the slope of the zeta vs. pH curve determines the value of the pH at which a maximum or minimum in the rate of electron transfer occurs. Thus proper characterization of the colloid double layer is essential when attempting to optimize electron transfer.

Another interesting case is the study by Moser et al. of the reduction of a cobaltacenium dicarboxylate anion as a function of pH using photoexcited colloidal TiO_2 [28]. They observed a decrease in the rate with increasing pH, contrary to earlier results with methyl viologen and proposed that the redox active anion was involved in an acid-base equilibrium with a dianion, which was in turn postulated to be electrochemically inert. The effective concentration of the electroactive acceptor then decreased with increasing pH, and this was used to explain the observed

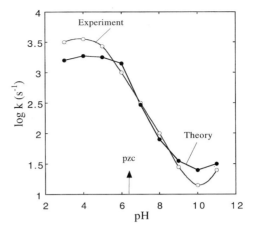

Figure 12.14. Observed dependence of the rate of electron transfer from photoexcited colloidal TiO$_2$ to cobaltacenium dicarboxylate anions on pH and the calculated rates relative to the pzc predicted using the zeta potential data of Wiese and Healy and assuming $z = -1$, and an ionic strength of 1 mM. The entire pH dependence including the slope and absolute value at the minimum near pH 11 is predicted from Eq. (12.35) with $\beta = 0.5$, $z = -1$.

decreased rate of electron transfer. However, a simpler explanation is that the increasingly negative zeta potential at high pH is responsible for the rate decrease. In Figure 12.14, we have fitted their data to Eq. (12.35). This equation accurately predicts both the decrease and the minimum at pH 10.5, and the eventual upturn in the rate, which is not easily explained via the dissociation mechanism.

12.4.4 Non-Nernstian Behavior

Equation 2, the Nernst equation, is clearly fundamental to the interpretation of all the data presented so far, yet we cannot directly measure ψ_o, only a potential at the plane of shear. Charge titration curves obtained from metal oxide suspensions are dramatically different to those obtained on silver halides or mercury, and suggest very large inner layer capacitances [10, 11, 13], implying that the Helmholtz region around metal oxides is a vastly different environment to that around mercury. It now seems clear that for any insulating or semiconducting surface, where the lattice ions themselves are not the potential determining ones, as is the case for AgI where Ag$^+$ and I$^-$ determine the surface potential, an alternative formulation for the surface charging mechanism is required. These are termed "ionizable surface group models". For oxides, the surface is considered to act as an amphoteric acid and base with fundamental surface reactions of the form

$$AH_2^+ \Longrightarrow AH + H^+ \tag{12.45}$$

$$AH \Longrightarrow A^- + H^+ \tag{12.46}$$

determining the surface charge. Here A denotes the surface group on the particle. Each reaction has an associated equilibrium or surface acidity constant, K_{a1} and K_{a2}. Analysis of such surface ionization models suggests that Nernstian behaviour is a limiting form for most surfaces. Healy and White [11] show that deviations from Nernstian behaviour can be characterised by pK, where $pK = pK_{a1} - pK_{a2}$. It defines the difference in acidity of the surface groups. The values can only be determined experimentally, and the resulting equations for the surface potential can only be solved numerically or graphically [10].

The effects of non-Nernstian behaviour on the kinetics of electron transfer have not been examined to date. In principle, if $d\psi_o/dpH < 59$ mV/pH, then the difference must appear as a potential difference within the oxide, but this will only be established by slow proton diffusion through the solid [35]. The two metal oxides for which data are available, TiO_2 and Fe_2O_3, are both reasonably Nernstian, and the fits to the kinetic data are noticeably inferior if less than Nernstian response of ψ_o to pH is assumed in the calculations. Furthermore, flat band measurements on ZnO and TiO_2 prove unequivocally that the bulk energy levels within the metal oxides are shifted by -59 mV/pH change in solution [36, 37]. It is worthwhile noting that ionizable surface group models consistently require large inner layer capacitances (>100 μF cm^{-2}) to reconcile charge titration and electrokinetic data, and the data for electron transfer from viologen radicals to iron oxide can likewise only be reconciled using a large Stern layer capacitance of $450\mu F$ cm^{-2}. So both the e.t. kinetics and charge titration/electrophoresis data indicate that the Helmholtz region of metal oxides is very different to the mercury–water interface.

12.4.5 Extensions to Other Systems

There is a paucity of clear data on e.t. to metal sulphides, or other chalcogenides (MX), as a function of pH or $[H_2X]$. In the case of metal halides, Hoffman and Billings showed that the reduction overpotential of an AgBr electrode varied with pBr [38]. Morrison has reviewed the data for CdS and other sulphide systems, but the conclusions are unclear [37]. Since many workers do not control $[H_2S]$ of the sols after preparation, surface potential control is not possible. Ginley and Butler demonstrated by charge titration that the Fermi level in a CdS electrode is controlled by pH and $[HS^-]$ [39]. van Leeuwen and Lyklema have reported on AgI electrode measurements in which they examine both ion adsorption and electron transfer; their review also discusses processes such as double layer relaxation [40].

12.5 The Effect of Zeta on Radical Scavenging Yields

The viologen radical does not undergo recombination at a perceptible rate, and so it is possible to examine the effect of the double layer upon mass transfer and activation controlled reactions with this radical using quite simple modifications of the Tafel and Smoluchowski equations. In general however, excited species generated by either photolysis or radiolysis undergo various deactivation pathways in addition to reaction with substrates such as colloidal particles. In the case of photolysis, these are usually first-order radiative or nonradiative energy losses, and these are readily incorporated into the equations above. A more common situation in radiolysis is that the radical undergoes self-reaction, i.e. second order loss. Furthermore a number of radicals have pK_a's in the common range of solution acidities. The charged, anionic form will interact with the colloid double layer. Rao and Hayon [41] have made extensive measurements of radical pK_a's by spectrophotometric means, and Henglein and colleagues have measured many radical pK_a values by pulse radiolysis polarography [42]. The radical anion is a better reductant than the neutral, 'acidic' form [41], and often recombination of the charged anionic form is slower than recombination of the neutral radical. Trying to unravel these various effects is an arduous one. In the following, we describe some model calculations on how the double layer parameters control scavenging yields of radicals by colloidal iron oxide particles. We summarize a typical scenario in Figure 12.15, where we show how the radical speciation and particle charge might change with pH.

The scavenging of the radicals under steady state conditions will depend upon both the pK_a of the radical, the pzc of the oxide and whether the reaction is diffusion controlled or activation controlled. For activation controlled processes, it is necessary to know k_{pzc}, the intrinsic rate of transfer at the point of zero charge, for both the acid and base forms of the radical. For activation controlled electron transfer, double layer corrections are also required for neutral radicals. This follows from Eq. (12.34) with $z_R = 0$.

For mass transfer limited reactions of radicals with colloid particles, the position is slightly simpler. Given a well defined acidity constant K_a for the dissociation,

$$RH \Longleftrightarrow R^- + H^+ \qquad (12.47)$$

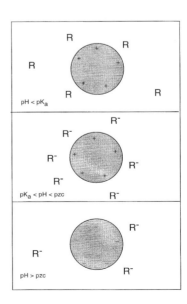

Figure 12.15. Diagram illustrating the changing speciation of radicals and charge around a metal oxide colloid particle. (Top): pH < pK_a and all radicals neutral and oxide positively charged. (Middle): pH raised until pzc > pH > pK_a. Radical anion now predominates and local concentration around oxide particles is enhanced. (Bottom): At higher pH, the oxide particle becomes negatively charged and radical anions are depleted near particle surface.

where RH and R^- are the acid and basic forms of the radical respectively, the fraction of radicals initially present in the protonated form is $\alpha = [RH]/[R]_t = 1/(1 + K_a/[H^+])$, and those in the deprotonated form is $(1 - \alpha)$, where the total radical concentration is $[R]_t$. The rate of disappearance at any pH is then due to recombination of both protonated and deprotonated radicals as well as to colloid encounters by both charged and uncharged radicals.

$$d[R]_t/dt = G(R)D - k_{diff}[R]_t[colloid] - k_{field}(1 - \alpha)[R]_t[colloid]$$
$$- 2\alpha^2 k_1[R]_t^2 - 2\alpha(1 - \alpha)k_2[R]_t^2 - 2(1 - \alpha)^2 k_3[R]_t^2 \qquad (12.48)$$

where k_{field} is the encounter rate constant using Eq. (12.32), k_{diff} is the field-free diffusion controlled rate constant given by Eq. (12.30), while k_1, k_2 and k_3 are the recombination rate constants for radical–radical deactivation, and $G(R)D$ is the production rate of the radical, which for radiolytically generated radicals, is the dose rate, D, times the G value for the species R. Using the steady state approximation, Eq. (12.48) becomes quadratic in $[R]_t$ and the steady state radical concentration is readily found to be

$$[R]_{ss} = \frac{-K_d + (K_d^2 - 4K_r G(R)D)^{1/2}}{2K_r} \qquad (12.49)$$

where

$$K_r = 2\alpha^2 k_1 + 2(1 - \alpha)k_2 + 2(1 - \alpha)^2 k_3 \qquad (12.50)$$

and

$$K_d = \{(1 - \alpha)k_{diff} + k_{field}\alpha\}[colloid]. \qquad (12.51)$$

The value $[R]_{ss}$ can then be inserted into Eq. (12.49) to determine the fraction disappearing by recombination and the fraction scavenged by the colloid. The scavenging efficiency, λ, is then defined as

$$\lambda = \frac{K_d}{K_d + K_r[R]_{ss}} \qquad (12.52)$$

To get a feel for the size of the double layer effects, we have taken the formic acid radical with a pK_a of 3.4 as the reductant, and colloidal iron oxide as the colloid. In Figure 12.16, the speciation and colloid charge as a function of pH are shown. In Figure 12.17, λ is plotted as a function of the colloid concentration for a number of pH's. In the calculations, it was assumed that $a = 20\text{Å}$ and $k_1 = k_2 = k_3 = 1 \times 10^{10}$ $M^{-1}s^{-1}$. As is very clear, the scavenging shows a strong pH dependence, and a maximum occurs at a pH where the radical is deprotonated, but the oxide is still positively charged. As the pH is increased through the pzc, the efficiency decreases dramatically, because both the sol and the radical become negatively charged (see Figure 12.18). Clearly, a primary prerequisite for achieving high effi-

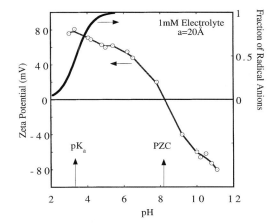

Figure 12.16. The measured zeta potential for nanosized iron oxide particles vs pH at 1 mM electrolyte and the relative population of radicals and radical anions assuming a pK$_a$ of 3.4.

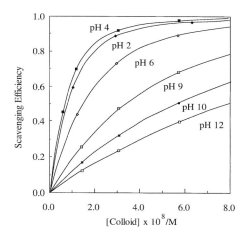

Figure 12.17. The radical scavenging efficiency of colloidal iron oxide as a function of colloid concentration and pH using parameters in Figure 12.16.

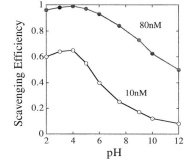

Figure 12.18. The radical scavenging efficiency of colloidal iron oxide as a function of pH at two colloid concentrations. The efficiency peaks at a pH between the pK$_a$ and pzc due to enhanced mass transfer to the colloid particles.

ciency is that pzc > pK_a. However, even if the pzc < pK_a, recombination of the radical anion is often slower than for the neutral form, so higher yields may still be found even with electrostatic repulsion. Thus when a redox reaction is under diffusion control, the double layer may exert a significant effect on the rate of reaction and the efficiency of radical scavenging.

12.6 Colloid Nucleation and Nanoparticle Stability

In this final section, we address briefly the role of stabilizers for nanoparticles. Even colloidal metals which have high Hamaker constants and which should be susceptible to coagulation can be made as sols with quite low zeta potentials that are stable for months at a time. What does the double layer tell us about preparing nanosized particles in water? For the case of low potentials, and small overlap between double layers, the results are quite unexpected. The electrostatic repulsive energy for two spheres of radius a, with low surface potential ψ_o, approaching each other in a medium of Debye length κ^{-1} is given by:

$$V_{rep}(kT) = 4\pi\varepsilon_r\varepsilon_o\psi_o^2 a^2/r \exp(-\kappa a(r/a - 2)) \qquad (12.53)$$

The nonretarded van der Waals attractive energy between particles of radius a is given by

$$V_{att}(kT) = -A/6\{2a^2/(r^2 - 4a^2) + 2a^2/r^2 + \ln(1 - 2a^2/r^2)\} \qquad (12.54)$$

with $r > 2a$, the centre-to-centre distance. According to DLVO theory it is the sum of the two energies that determines particle stability. The usual criterion are that a barrier of $15–20kT$ is sufficient to ensure colloid stability. These two functions are plotted in Figures 12.19 and 12.20 as a function of the particle surface separation and for various particle sizes [43]. It is clear from Eqs. (12.53) and (12.54), that the interaction energy increases with particle radius, a, for both the attractive and repulsive energy. As a consequence, we can see in Figure 12.21 that the barrier height to colloid stability at fixed ψ_o increases as the particle size increases. Consequently,

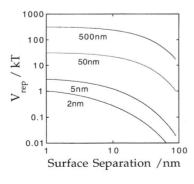

Figure 12.19. The calculated repulsive interaction energy between colloid particles calculated using Eq. (12.53), as a function of the particle separation for a range of diameters from the nanometre to micrometre size regime. Parameters used: $\psi_o = 25$ mV, $\kappa^{-1} = 100$Å.

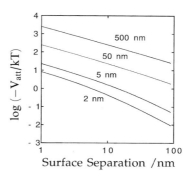

Figure 12.20. The calculated nonretarded van der Waals interaction energy between colloid particles calculated using Eq. (12.54), as a function of the particle separation for a range of diameters from the nanometre to micrometre size regime. Parameters used: $A = 6 \times 10^{-20}$ J.

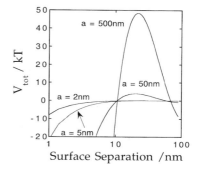

Figure 12.21. The total interaction energy due to both repulsive and attractive forces. Conditions as per Figures 12.19 and 12.20. Critical to nanoparticle nucleation and stabilization in solution is that the repulsive energy is smaller for small particles so a larger zeta potential is required for colloidal stability, but the primary minimum created by attractive dispersion interactions is likewise smaller, so that stabilization by adsorbed polymers, surfactants or chemisorbed complexing agents, such as thiols or small carboxylic acids, is much more efficacious than for larger colloid particles.

there is an automatic tendency for coagulation of particles to slow down as coagulation proceeds. This factor may often determine the final particle size distribution following nucleation. But the primary minimum associated with particle coalescence also becomes deeper as the particle size increases. If two particles > 10 nm in size coalesce in solution, they will not be able to separate again since their thermal energy will be insufficient to allow them out of the primary minimum. Conversely, nanosized particles will peptize relatively easily. It is important to recognize that rapid peptization is essential. An agglomerate of small particles will behave in van der Waals terms like a larger particle and the van der Waals interaction energy with other unpeptized particles will gradually increase if agglomeration is allowed to continue. Provided nanoparticles peptize quickly, the van der Waals potential well around the temporary agglomerate will not have time to deepen further via aggregate–colloid encounters.

It is clear from these figures that only small molecules should be necessary to prevent coalescence and particle coagulation of nanoparticles. Chemisorbed molecules provide a steric barrier, and for particles $<100\text{Å}$ in diameter, this will be sufficient to offset the van der Waals interactions. However, ψ_o should be large to prevent the formation of loose agglomerates. Thus, small stabilizers can be remarkably efficacious in stabilizing nanosized colloid particles.

12.6.1 Some Unresolved Aspects of Colloid Redox Chemistry

The aim of this chapter has been to show how the measured properties of powders and suspensions in liquids are important not just from the thermodynamic or colloid stability viewpoint. The equations describing the electrical double layer around particles also govern the kinetic response to redox disequilibria in solution, and rates of electron transfer can be controlled and optimized once the various factors are understood. Disappointingly, there have been few studies to elucidate how specific adsorption at a colloid surface affects electron transfer, yet most nanosized particles can only be prepared in the presence of strong growth inhibitors such as polyelectrolytes which strongly adsorb to the particle surface. Darwent's work on the effects of sulfate adsorption remains an exception [29], and the PhD work by

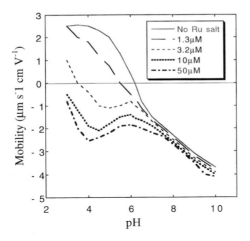

Figure 12.22. Effect of pH on the electrophoretic mobility of colloidal TiO_2 in the presence of tris(2,2'-bipyridine-4,4'-dicarboxylic acid) ruthenium (II). Specific adsorption occurs below the pzc of 6.1. If adsorption occurs at the outer Helmholtz plane (close to the shear plane), then there will be a dramatic increase in the overpotential for electron injection into the titania colloid, which depends exponentially on the potential across the Helmholtz layer. If we assume that ψ_o is constant for a particular pH value then we predict that the rate of electron injection should be enhanced through Eq. (12.34) by an amount $\exp\left(\beta F(\psi_o - \zeta))/RT\right)$. The negative charge on the adsorbed sensitizer not only aids adsorption to the positively charged colloid particles, but simultaneously accelerates injection by creation of an increased Helmholtz potential difference. The degree of enhancement is critically determined by the location of the planes of dye adsorption and electron transfer. Adapted from ref. 45.

Kleijn on the effects of viologen adsorption on RuO_2 electrokinetics is another [44]. These adsorption effects may also play a prominent role in systems such as dye sensitized photoelectron transfer, which has recently been demonstrated as a viable basis for charge separation in solar energy conversion [6]. In the so-called Grätzel cell, high surface area electrodes are synthesized by sintering metal oxide colloid films. The sensitizer is adsorbed primarily electrostatically, but the act of chemisorption modifies the potential distribution at the surface, and this may act to augment or hinder the electron injection rates following illumination. That such effects will be important is immediately apparent from the zeta potential data for titanium dioxide colloids in the presence of the anionic ruthenium dye shown in Figure 12.22 [45]. In this case, since the photoelectron transfer is anodic, the adsorption of the negatively charged dye onto the positively charged metal oxide particles could be beneficial. The zeta potential becomes more negative, but we assume that at any pH, the actual surface potential is fixed by Eq. (12.2), so the adsorption must introduce a large electric field across the Helmholtz layer, driving electron transfer into the particle. This synergistic effect may enhance the rates of e.t. by a factor of 10–100, based on the data in Figure 12.22.

Studies on the effects of complexing agents on rates of electron transfer could well assist in the formulation of additives to improve industrially important redox reactions such as rust removal. The role of extraneous ligands on electron transfer to iron oxide are still speculative. For example, would o-phenanthroline, a potent, neutral complexing agent for Fe(II) slow down e.t. from viologen radicals to colloidal iron oxide by specific adsorption to the surface, thus blocking viologen approach? Or would it conversely aid e.t. by prebinding to selected Fe(III) surface sites, accelerating the actual rate of e.t. to these specific, activated sites? Or would it simply accelerate the rate of Fe(II) desorption following reduction, thereby exposing fresh Fe(III) surface sites more quickly and by this mechanism accelerate particle dissolution? To date only steady state dissolution data are available to help answer these detailed mechanistic questions [46].

What at least should be clear is that simple electron transfer is governed by the overall electrical potential distribution at the colloid surface, with the zeta potential governing the local surface concentration of charged reactants, and the difference $(\psi_0 - \zeta)$ acting as the overpotential for actual transfer from solution to surface. The relative rates at different pH's can be accurately predicted when no specific adsorption occurs if ζ potentials are determined. If the actual data are to be believed, then the plane of electron transfer lies slightly closer to the surface than the plane of shear, which determines the electrokinetic or zeta potential, a conclusion consistent with modern views about the electrical double layer. Frumkin, the discoverer of the diffuse layer effect in electrode kinetics, would have been happy [17, 18], to see the same effects so prominent in colloid redox kinetics too.

In the case of mass transfer limited reactions, there are no data to indicate whether the dramatic effects predicted from the calculations in Figures 12.5 and 12.6 really occur. Such drastic retardation must have important implications for enzyme catayzed reactions as well as colloid redox chemistry. For example, one might expect that the reaction

$$2Fe(CN)_6^{4-} + MnO_2 + 4H^+ \Longrightarrow Mn^{2+} + 2Fe(CN)_6^{3-} + 2H_2O \quad (12.55)$$

would be close to mass transfer limited given the strong $\Delta E^\circ = 1.2V$ for reaction. Yet 250Å MnO_2 sols have zeta potentials of -50 mV at pH 6, so the rate of e.t. is predicted to lie at 10^8 $M^{-1}s^{-1}$, not 10^{11} $M^{-1}s^{-1}$. Likewise reactions such as the disproportionation of superoxide anions by superoxide dismutase, which has a negative mobility at physiological pH, would be hindered by slow migration of the substrate to enzyme active sites at low ionic strength.

In this article, we have not discussed the electrical double layer within colloid particles since this remains a basic unknown in colloid science. Microwave conductivity offers the prospect of determining the concentrations of carriers, at least in nonaqueous systems, but there are again no data except for flat band potentials measured on sintered nanocrystalline electrodes from which to evaluate donor densities, trap energies or internal space charge potentials [6].

Fundamental to the understanding of charge injection into insulating materials is the concept that the potential determining ions regulate the surface potential independently of the charge injected through redox reactions. This can be justified on the basis of the small space charge capacitance compared with the solution phase Helmholtz capacitance. Consequently for a fixed chemical potential of the proton in the bulk solution, the surface potential is fixed, and charge transfer into or out of the particle must be accompanied by proton adsorption or desorption [24, 25]. Because of the facility of these reactions, one can normally assume that an insulating particle retains a constant surface potential during a redox reaction, though obviously after extensive reaction, the chemical potential of the proton within the solid or in the bulk may have changed.

A final, interesting question which appears never to have been systematically investigated is whether the van der Waals forces at the surface have any effect on the rate of electron transfer to and from solution. In principle, the mass transfer rate constant for *all* electroactive species will be enhanced at small separations (<10nm) by dispersion forces, since the molar refractive index of the electroactive species differs from the average refractive index of the medium. The dispersion interaction will not be as important as it is for colloid–colloid interactions because of the small radius of the electroactive species – see Figure 20, but it may still be significant enough to cause perceptible changes in the observed rates of mass transfer to charged surfaces. Whether this effect can be harnessed as a means to further optimize e.t. is still to be determined. Thus, though we have set out to show that the theoretical foundations linking colloid chemistry and electrochemistry have been further bolstered and consolidated through the research on colloid redox chemistry over the last decade or two, many basic questions remain unresolved.

Acknowledgements

This work was supported by an ARC Research Grant. The author is also grateful for the support of the Advanced Mineral Products Research Centre.

References

[1] E. J. Verwey, J. T. Overbeek, *Theory of the Stability of Lyophobic Colloids*, Elsevier, Amsterdam **1948**.

[2] T. H. James, *Theory of the Photographic Process*, 4th edn., MacMillan, London **1977**.

[3] M. A. Blesa, E. B. Borghi, A. J. G. Maroto, A. E. Reggazoni, *J. Coll. Interface Sci.* **1984**, *98*, 295.

[4] D. Bradbury, in: *Water Chemistry of Nuclear Reactor Systems*, British Nuclear Energy Society, London **1978**, p. 373.

[5] J. D. Hem, *Chem. Geol.* **1978**, *21*, 199.

[6] B. O'Regan, J. Moser, M. Anderson, M. Grätzel, *J. Phys. Chem.* **1990**, *94*, 8720.

[7] O. Micic, D. Meisel, in: *Homogeneous and Heterogeneous Photocatalysis* (Eds.: E. Pelizzetti, N. Serpone), NATO ASI Series C174, Reidel, Dordrecht, **1986**.

[8] For example, see F. C. Tye, *Electrochimica Acta* **1985**, *30*, 17; **1974**, *21*, 415.

[9] R. J. Hunter, *Zeta Potential in Colloid Science*, Academic Press, London **1981**.

[10] T. W. Healy, L. R. White, *Adv. Colloid Interface Science* **1978**, *9*, 303.

[11] R. O. James, G. A. Parks, in: *Surface and Colloid Science* (Ed.: E. Matijevic), Vol. 12, Plenum Press, New York **1982**.

[12] G. A. Parks, *Chem. Rev.* **1965**, *65*, 177.

[13] R. J. Hunter, *Foundations of Colloid Science*, Vol. 1, Oxford University Press, Oxford **1989**.

[14] J. H. Fendler, F. C. Meldrum, *Adv. Mater.* **1995**, *7*, 607.

[15] J. H. Fendler, in: *Nanoparticles in Solids and Solutions* (Eds.: J. H. Fendler, I. Dekany), NATO ASI Series, Kluwer, Dordrecht **1996**.

[16] (a) C. E. Zobell, *Bull. Am. Chem. Assoc. Petrol. Geol.*, **1946**, *30*, 477; (b) R. M. Garrels, C. L. Christ, *Solutions, Minerals and Equilibrium*, Jones and Bartlett, Boston **1990**.

[17] R. Parsons, *Modern Aspects of Electrochemistry* **1954**, *1*, 103.

[18] P. Delahay, *Double Layer and Electrode Kinetics*, Wiley, New York **1965**.

[19] H. Ohshima, T. W. Healy, L. R. White, *J. Colloid Interface Sci.* **1982**, *90*, 17.

[20] S. Rice, in: *Comprehensive Chemical Kinetics* (Eds.: C. Bamford, C. Tipper, R. Compton), Vol. 5, Elsevier, Amsterdam **1985**.

[21] A. Henglein, *Ber. Bunsenges. Phys. Chem.* **1982**, *86*, 241.

[22] (a) J. Albery, *Electrode Kinetics*, Clarendon Press, Oxford **1975**; (b) A. J. Bard, L. R Faulkner, *Electrochemical Methods*, Wiley, New York **1980**.

[23] P. Mulvaney, Ph.D. Thesis, University of Melbourne **1988**.

[24] P. Mulvaney, V. Swayambunathan, F. Grieser, D. Meisel, *J. Phys. Chem.* **1988**, *92*, 6732.

[25] (a) P. Mulvaney, V. Swayambunathan, F. Grieser, D. Meisel, *Langmuir* **1990**, *6*, 555; (b) P. Mulvaney, F. Grieser, D. Meisel, in: *Proceedings of the 9th International Congress of Radiation Research*, Toronto, Academic Press, New York **1991**.

[26] M. Grätzel, A. J. Frank, *J. Phys. Chem.* **1982**, *86*, 2964.

[27] D. Duonghong, J. Ramsden, M. Grätzel, *J. Am. Chem. Soc.* **1982**, *104*, 2977.

[28] U. Kölle, J. Moser, M. Grätzel, *Inorg. Chem.* **1985**, *24*, 2253.

[29] J. R. Darwent, A. Lepre, *J. C. S. Faraday Trans.* 2 **1986**, *82*, 2323.

[30] G. T. Brown, J. R. Darwent, *Chem. Comm.* **1985**, 98.

[31] G. T. Brown, J. R Darwent, P. D. I. Fletcher, *J. Am. Chem. Soc.* **1985**, *107*, 6446.

[32] I. Willner, J.-M. Yang, C. Laane, J. W. Otvos, M. Calvin, *J. Phys. Chem.* **1981**, *85*, 3277.

[33] (a) D. W. Bahnemann, C. Kormann, M. R. Hoffmann, *J. Phys. Chem.* **1987**, *91*, 3789; (b) C. Kormann, D. W. Bahnemann, M. R. Hoffmann, *J. Phys. Chem.* **1988**, *92*, 5196.

[34] G. R. Wiese, T. W. Healy, *J. Colloid Interface Sci.* **1975**, *51*, 427.

[35] M. A. Butler, *J. Electrochem Soc.* **1979**, *126*, 338.

[36] M. A. Butler, D. S. Ginley, *J. Electrochem Soc.* **1978**, *125*, 228.

[37] S. R. Morrison, *Electrochemistry at Semiconductor and Oxidized Metal Eectrodes*, Plenum Press, New York **1980**.

[38] A. Hoffman, B. Billings, *J. Electroanal. Chem.* **1977**, *77*, 97.

[39] D. S. Ginley, M. A. Butler, *J. Electrochem Soc.* **1978**, *125*, 1968.

[40] H. P. van Leeuwen, J. Lyklema, in: *Modern Aspects of Electrochemistry*, Vol. 17 (Eds.: J. O'M. Bockris, B. E. Conway, R. E. White), Plenum Press, New York **1986**.
[41] P. Rao, E. Hayon, *J. Am. Chem. Soc.* **1974**, *96*, 1287.
[42] A. Henglein, *Electroanal. Chem.* **1976**, *9*, 163.
[43] G. R. Wiese, T. W. Healy, *Trans. Fara. Soc.*, **1970**, *66*, 490.
[44] M. Kleijn, Ph.D. Thesis, University of Wageningen **1987**.
[45] D. N. Furlong, D. Wells, W. H. F. Sasse, *J. Phys. Chem.* **1986**, *90*, 1106.
[46] V. I. E. Bruyere, M. A. Blesa, *J. Electroanal. Chem.* **1985**, *182*, 141.

Chapter 13

Semiconductor Nanoparticles in Three-Dimensional Matrices

S. G. Romanov and C. M. Sotomayor Torres

13.1 Introduction

The special properties of nanometer-sized semiconductor and dielectric structures promise numerous applications in electronic devices due to the strong nonlinearity of few-electron systems [1] and in optical devices due to the modified light–matter interaction for confined photons [2]. However, since only few electrons are involved in interactions with an external field the output power parameters of a nano-particle ensemble are seriously limited for decoupled nanoparticles. Device applications require measurable current, voltage or light intensity, which makes the design of ordered arrays of synchronously operating semiconductors nanostructures increasingly important in order to enhance their output [3]. To begin with three-dimensional (3D) systems are highly desirable due to a higher density of nanostructures compared to two-dimensional arrays. The next two requirement are the homogeneity of nanostructures in the ensemble and the long-range periodicity of their spatial arrangement. Only by approaching these conditions it becomes possible to expect a resonance interaction between nanostructures in a 3D lattice and the compounded nonlinear response of the whole ensemble. If these requirements are not satisfied, then averaging over the random ensemble will occur and an amorphous-like behaviour will characterise the response of the 3D ensemble. The above considerations emphasize the importance of studying 3D lattices of quantum dots or nanoparticles as electronic and optical materials.

3D arrays of isolated nanoparticles can be realised using a variety of techniques such as coevaporation and metal-organic synthesis [4, 5]. However, one drawback is that not all particles exhibit crystallinity. One successful concept proposed in the early 70s in the A. F. Ioffe Institute (St Petersburg, Russia) was to use a self-organising strategy of nanostructures in 3D by means of structural confinement. It uses a porous crystalline dielectric matrix with an open lattice of structural voids as a host material for impregnation with another substance, a guest material. Using zeolites perfect lattices of nanometer-size clusters were prepared by in-void growth of the guest material [6]. In zeolites voids of ~0.1 nm diameter impose a strong short-range spatial modulation on the guest material dominating over the guest crystalline structure and resulting in the stabilisation of very specific cluster configurations. However, in most cases, with the exception of elemental semiconductors, the close interaction of clusters separated by only a few angstrom led to the formation of a cluster crystal and the loss of cluster individuality. Moreover, the cluster-to-matrix interface was found to modify dramatically electrical and optical properties of the guest material [7, 8, 9].

In order to construct a 3D lattice of nanostructures which preserves the individual properties of the cluster, matrices with large voids are required. Porous matrices with voids in the range of 1 to 5 nm are either largely irregular (like cloverite) or partly perfect (like MCM, chrysotile asbestos). In addition, most of them can be synthesised only in powder form. At present the only matrix which exhibits the desirable properties is the precious stone opal [10], with voids in the range of 50 to 150 nm and the same spacing between adjacent centres. The synthesis of artificial opals was first achieved in mid 1970s in Novosibirsk, Russia [11]. This process is outside the scope of this review, suffice to mention that it consists of the synthesis of identical silica balls, packing the balls by sedimentation under natural or artificial gravity and sintering to increase the solidity of balls package [11, 12, 13].

Opals allowed the design of a large variety of nanostructure arrays: (i) conducting materials, the conductivity of which is controlled by quantum point contacts (or Josephson junctions in the case of superconductors) in between separate nanostructures [14], (ii) dispersed dielectrics [15], (iii) optical materials based upon diffraction properties of opals such as photonic band gap (PBG) structures [16].

In this review we depict the opal structure and some of the methods to infill its voids. We then turn to optical properties of bare and infilled opals and finally we describe and discuss the conductivity regimes of opal-semiconductor composites.

13.2 Material Issues

Opal consists of identical silica balls of diameter D with a size dispersion within 5%. Samples can be prepared with ball diameters ranging from 150 to 350 nm. The size homogeneity of these spheres allows their assembly in a close 3D lattice, usually with *FCC* symmetry. Figure 13.1 shows the *FCC* structure in a scanning electron micrograph (SEM). Empty voids exist between neighboring balls which, in turn, form their own regular lattice [17]. It is instructive to describe the shape of opal voids as polyhedra with sides formed by spherical segments. There are 2 types of interpenetrating voids in the opal lattice: eightfold coordinated large voids each connected with eight fourfold coordinated small voids. A large void has the form of a truncated cube with eight triangular windows that connect it with eight adjacent small voids. The small voids have the form of truncated tetrahedron with four windows to four large voids. The aperture of the window is formed by spherical surfaces of three touching spheres. The large and small voids alternate in position. The size of voids in a *FCC* package of hard balls correlates with D since the diameters of spheres inscribed in the larger and smaller voids are $d_1 = 0.41D$ and

Figure 13.1. Scanning electron micrograph of an accidental cleave of bare opal ((100) planes) showing silica spheres with diameter D = 250 nm.

$d_2 = 0.23D$, respectively. The diameter of a circle inscribed in the triangular window is $d_3 = 0.15D$. During sintering d_3 becomes smaller due to amorphous SiO_2, which partially fills free inner volume. Thus, the tetrahedral voids together with the windows are reduced to a 1D-like channel. The density of voids in opal is typically 10^{14} cm^{-3}. The porosity of the ideal *FCC* package of balls is $\sim 26\%$ of the whole volume. The length of the constriction l may be estimated as $l \leq 0.05D$, where this inequality takes into account the convex shape of channel walls. Opals usually are textured polycrystals with crystallites up to hundred micrometers in size.

To realise 3D arrays of dots and wires of the host material in the opal matrix the in-void synthesis routine was chosen. Two main cases can be distinguished: complete and partial filling. The common requirement for all kinds of chemical treatment of opals is that low-temperature reactions take place in order to preserve the stability of the opal structure and avoid host–guest compounds.

Complete loading of the opal matrix was obtained by forcing molten semiconductors (here InSb and Te) to fill matrix voids under hydrostatic pressure. As an example, the process to prepare an opal–InSb sample is given below [18]. A piece of opal together with an amount of InSb (*n*-type, carrier concentration n $\leq 10^{14}$ cm^{-3}) were placed in a stainless steel container pumped out to remove water from the opal. The process of opal impregnation was carried out in a special high pressure chamber. The opal matrix and the molten semiconductor were held for 5 minutes at a temperature of 600 °C and a pressure of 5 kbar, and then the sample was cooled under pressure. Figure 13.2 shows the SEM image of an accidental cleave of opal–InSb along the (111) plane. The conductive component appears as an ordered network of grains connected with each other via bridges, the silica balls are not seen. The size of grains reflects geometrical considerations with corrections for ball distortions. The recrystallisation of InSb in opal voids results in a slight excess of Sb. In addition, unwanted impurities from the opal and container contaminate the

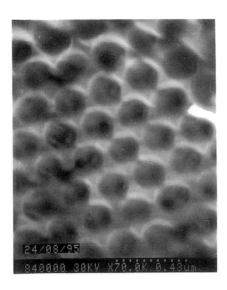

Figure 13.2. Scanning electron micrograph of an InSb infilled opal ((111) plane) showing the InSb grains connected by InSb bridges.

InSb. InSb has a tendency to expand when it solidifies leading to a tight loading of the opal voids and strain of the InSb lattice at the grain-to-wall interface. The density of InSb grains in the opal–InSb nanocomposite is the same as that of voids. This method is suitable for the realisation of composites containing conducting materials, but the selection of the guest material is restricted to those with a 600 °C limit melting temperature. This method allows a replica of the opal void lattice to be made by the InSb grain-bridges lattice, i.e., a regular 3D lattice of quantum dots coupled to each other by bottleneck constrictions.

Another way to infill the opal matrix is the multiple-step chemical synthesis of the "ship-in-the-bottle" type. We have used it to prepare CdS and CdSe guests occupying just a fraction of free internal volume [19]. The synthesis of CdSe starts with the impregnation of the opal matrix with a $Cd(COO)_2$ water solution. After drying, this salt is decomposed at 250 °C to form CdO particles. Exposure of the opal–CdO to a flow of H_2Se results in the formation of CdSe. To vary the content of the guest in opal we changed the dilution rate or, alternatively, repeated the procedure. Typical volume fractions of CdSe are 2 to 3% of the void volume and the average nanoparticle size is ~ 10nm. In Figure 13.3a, the Raman spectrum of opal–CdSe shows the LO phonon at 209 cm^{-1}, which is the bulk phonon of CdSe.

A more sophisticated treatment is surface coatings of the void walls by absorption (for example of S), chemical vapour deposition CVD (for example of TiO_2) or metal organo chemical vapour deposition MOCVD (for example of InP). Here gas phase reactions take place essentially at surface defects of the opal balls. The preparation of opal–InP and opal–TiO_2 are described below.

A standard atmospheric pressure MOCVD reactor was used to grow InP inside the opal matrix. Trimethylindium and phosphine were introduced in the reactor separately in order to extend the diffusion of the reactants into the inner voids of the opal matrix. Trimethylindium was added in a flow of H_2 for up 4 hours. Then phosphine was passed through the reactor for several hours at 350 °C to decompose

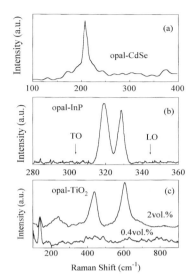

Figure 13.3. Room temperature unanalysed Raman spectra of semiconductor nanocrystals synthesised in voids of opal. (a) opal–CdSe, (b) opal–InP and (c) opal–TiO$_2$.

the hydride. To obtain a higher loading of InP a cyclic growth was carried out. The second cycle increased the InP content by a factor of 5 to 10 depending on the growth conditions. Depending on the coating thickness and the semiconductor compound the resulting opal–semiconductor nanocomposite may be of conducting or insulating type. InP is incorporated in crystalline form as demonstrated by the bulklike phonon Raman spectrum of opal–InP, where Raman lines at 319 and 328 cm^{-1} are related to the TO and LO phonons of bulk InP at 303 and 345 cm^{-1}, respectively, probably shifted by phonon confinement and or layer-to-substrate interaction (see Figure 13.3b). The opal–InP samples were plates with an area between 10 and 100 mm^2 and a thickness between 1 and 0.2 mm. The homogeneity of the guest loading was analysed by electron probe microanalysis (EPMA) scanning a length of 1 to 2 mm with a resolution of ~ 0.5 μm and by SEM for in-void examination. The guest-to-volume fraction obtained from EPMA, wet chemical analysis and specific density measurements were consistent. The guest content was found to be within 0.5 % of the mean value of the sample cross section.

TiO$_2$ was grown by sequential deposition of monolayers [21]. Each step consists of the adsorption of TiCl$_4$ molecules transported in a flow of N$_2$ since TiCl$_4$ substitutes surface OH$^-$ groups in about 30 minutes. This is followed by exposure to water vapour which permits the transformation of TiCl$_4$ into TiO$_2$ in about 30 minutes. With layer thickness up to 20 monolayers TiO$_2$ crystallises as anatase and with further thickness increments the rutile structure dominates. Phonon lines at 143, 447, 612 and 826 cm^{-1} are observed together with a two-phonon band at 200 – 300 cm^{-1} present in the Raman spectrum of opal–TiO$_2$ containing over 3 vol.% of TiO$_2$ (see Figure 13.3c). These phonons compare well with those of the rutile phase at 138, 438, 605 and 819, respectively. These lines are quite distinct from the anatase phase of TiO$_2$, where the 139 cm^{-1} line dominates. The absorption spectra shows the absorption edge near 3.2 eV, which is up-shifted from the 3.05 eV band edge of bulk rutile [22].

The opal–semiconductor composites described above contained crystalline semiconductors in the voids as demonstrated by Raman scattering. Several questions are under investigation concerning the structural properties of this composite. The frequencies measured by Raman scattering here are most likely subjected to the combined effect of phonon confinement and strain of the guest material. This make a quantitative analysis cumbersome. What is the size dispersion of the guest material in a given guest–void volume fraction? Raman scattering cannot rule out the existance of small volumes of amorphous materials and other techniques will have to be used. Moreover, if part of the composite were amorphous how would it modify the optical and electrical properties?. These and other questions are currently being actively pursued.

13.3 Optical Properties

The diffracting properties of opal define its value as a gem stone as well as an optical material. Two points of interest arising from the commensurability of the opal lattice spacing and the wavelengths of the light are: (i) the optical gain for light emitted within a grating (here the opal matrix) due to the distributed positive feedback by Bragg reflection [23] and, (ii) the PBG structure of the grating with a high contrast of the refractive index for media involved [24]. The impact of both effects is the improvement of the emission efficiency and its directionality. Therefore, opal-based PBG materials [13, 16, 25–28] are highly attractive as they offer at present a promising way to realise 3D PBG.

A photonic band structure for photons is analogous to an electronic band structure for electrons since forbidden energy gaps appear as a result of Bragg reflection of the electromagnetic waves describing an electron or a photon. The structure then contains a sequence of forbidden energy bands where no optical propagating mode is allowed. Therefore, the spontaneous emission is suppressed in the gap region, making it possible to channel all the emission in an intentionally selected single mode [24]. Using such a PBG material as laser mirrors could lead, for example, to the realisation of thresholdless lasers [29]. In this context our activity was directed to: (i) improve the PBG in opals and (ii) the study of the emission of semiconductor nanostructures distributed within the opal in the presence of a PBG structure.

The zero approximation of the stop band spectral position is the Bragg law:

$$\lambda = 2\eta_{med}d \sin \alpha \tag{13.1}$$

where λ is the wavelength, d the spacing of grating, α the angle of the incident light and η_{med} the refractive index of the medium surrounding the scattering centres. The Bragg law establishes the strict proportionality between λ and d. However, if the refractive index of the scatterers η_{ball} becomes slightly different from η_{med}, the Bragg law fails. To take η_{ball} into account, one comes to the dynamic diffraction theory to explain the stop-band shift:

$$\lambda_{dyn} = 2d\eta_{ball} \sin \theta \left(\frac{1 + \Psi}{2 \sin^2 \theta}\right) \tag{13.2}$$

where

$$\Psi = 3\phi \frac{(m^2 - 1)}{(m^2 + 2)} \tag{13.3}$$

is the photonic strength characterising the light to matter interaction, $m = \eta_{ball}/\eta_{med}$ and ϕ is the volume fraction of balls to the surrounding medium. This relation fits well the experimental data for η contrast corresponding to e.g. opal impregnated with water [21]. However, the dynamic model does not yield the exact solution of the photonic structure at $\Psi > 0.1$ [30]. It has been shown that the Maxwell–Garnett method to calculate the average η over lattice components (η_{cryst}) is a better approximation for composites with strong light–matter interaction (high Ψ) [31]. This was confirmed experimentally for silica balls in air ($\Psi = 0,6$) [32].

Bare opals exhibit a pronounced stop-band in transmission (T) and reflectance (R) spectra, which shifts in energy for different ball diameters and with varying the angle of incident of the light α for monocrystalline opal immersed in water as shown in Figure 13.4a [21, 25, 26, 33]. The photonic bandgap activity of opal immersed in various liquid media with different refractive indeces shows similar effects [34]. It is possible to span the visible wavelength range by changing D from 180 to 300 nm [21]. Changing α is equivalent to probing different directions in the Brillouin zone, showing stop-bands with well defined dips in the transmission. However, the energy shift of the transmission dip is larger than the dip width. This effect is known as the semimetallic photonic band structure resulting from a weak modulation of η within the grating. Thus, there is no PBG overlap for opal in water (η contrast 1.45:1.33) for different points of the Brillouin zone. In order to achieve such overlap and complete PBG the η contrast should be as high as 2 to 3 [24].

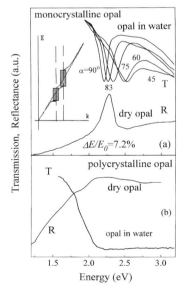

Figure 13.4. (a) Transmission spectra (T) of monocrystalline opal at different angles of light incidence and angle-integrated reflectance spectra (R). (b) Transmission and angle-integrated reflection spectra of polycrystalline opal. Inset: light dispersion in the case of an incomplete PBG.

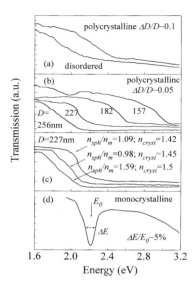

Figure 13.5. Transmission spectra of: (a) disordered and partially ordered opals immersed in water; (b) polycrystalline opal samples with different D in water; (c) the $D = 227$ nm opal samples in different liquid media and thus different n_{cryst}; (d) monocrystalline opal immersed in water.

For polycrystalline bare opal transmission and reflection spectra are of the edge type as shown in Figure 13.4b. This is a result of: (i) the random orientation of the crystallites which mixes all α, (ii) the insufficient η contrast and, (iii) the Rayleigh scattering at grating irregularities with $l < 0.1D$. Nevertheless, this edge has a Bragg diffraction nature since it changes its energy with the refractive index of the liquid in which the opal is immersed (i.e., n_{cryst}) and with the diameter of the silica ball (see Figure 13.5). Furthermore, this edge spreads over a wider spectral range for opals with poorer ordering as has been confirmed by SEM.

There are two ways to increase the η contrast in opals: (i) to fill in the voids and consider the silica-to-semiconductor η contrast or (ii) to coat the surface of the silica spheres and deal with an opal-to-semiconductor-to-air contrast. The first approach was tested in opal–CdS where the CdS ressembled sand grains resting on the silica surface [25, 35, 36]. We explored the second way which yields a higher η contrast and a lower Raleigh scattering. We pursued improvement of the PBG overlap in different directions by modifying the opal balls with a high η coating [26, 37].

Angle-integrated reflection spectra of polycrystalline opal with high η coatings, such as opal–InP, opal– TiO$_2$ and opal–CdS are shown in Figure 13.6 and Figure 13.7. In contrast with the bare opal, spectra of coated opals exhibit two maxima. One of them corresponds to the interband transition in the electronic structure of coating ~ 1.4 eV for InP, ~ 3.2 eV for TiO$_2$ and ~ 2.5 eV for CdS. From these spectra an absorption edge shifted by ~ 0.4 eV for InP and 0.2 eV for TiO$_2$ are measured. These shifts are probably due to two factors: one is the quantum size effect and the other the strain induced by coating the spheres. The precise origin is under investigation. The peak seen in absorption spectra comes from optical interference. These coatings transform the interference edge of bare opal into a peak,

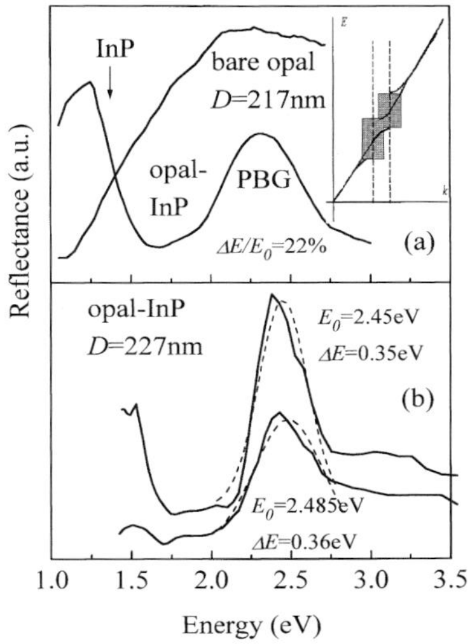

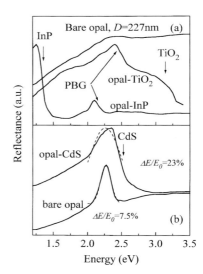

Figure 13.6. (a) Edge-to-peak transformation of the angle-integrated reflection spectrum of polycrystalline opal–InP; the arrow indicates the absorption edge of bulk InP. (b) Shift of the Bragg peak due to the increased coating thickness, upper (lower) spectrum corresponds to thinner (thicker) InP coatings. Dashed lines represent Gaussian fits. Inset: light dispersion in the case of complete PBG.

Figure 13.7. (a) Comparison of angle-integrated reflection spectra of opal with different coatings. (b) Increase of $\Delta E/E$ due to a high-n coating for monocrystalline opal: bare opal (bottom spectrum) and opal–CdS (top spectrum).

independently of whether the electronic gap of the coating lies above or below the photonic gap. This change from an edge to a two-peak absorption spectrum may be interpreted as the overlap of stop bands in all directions due to the increased modulation of the index of refraction.

In general, the position of PBG in the angle-integrated reflection spectrum may be fitted by

$$\lambda = 2\eta_{crys}d \qquad\qquad (13.4)$$

where

$$d = \sqrt{\frac{3}{2}}D \qquad (13.5)$$

is the spacing of (111) planes since these have the highest packing density. Although it seems nontrivial to apply the Maxwell–Garnett procedure to a three-component composite, just a summation of the indices of refraction of the grating components weighted by their volume fraction f (SiO$_2$: $n = 1.45$, $f = 0.74$; air: 1, 0.25; InP: 3.5, 0.01) yields $n_{cryst} \approx 1.33$ which compares fairly well with $n = 1.4$ calculated from the reflectance peak position (see Figure 13.6a). Increasing the content of InP increases n_{cryst} to 1.5. The relative intensity of the Bragg peak decreases with increasing the InP fraction above 2% since InP absorbs in this spectral range. It follows, that the PBG spectral position may be adjusted by controlling n_{cryst} varying the coating thickness (Figure 13.6b) or using a different coating (Figure 13.7a).

The important characteristic of the photonic peak is the ratio $\Delta\omega/\omega$ (ratio of the peak width to its central frequency) or $\Delta E/E$ which should exceed 17% for a complete PBG [24]. Spectra in Figure 13.7b show $\Delta E/E = 7.5\%$ for bare opal and 23% for opal–CdS. Although these spectra were collected using slightly less than a 90% light cone, the do show the overall trend of the PBG variation while exceeding the actual PBG width for any particular direction in the Brillouin zone of the photonic crystal. It is noteworthy that a more realistic value has been achieved using vapour phase synthesis followed by annealling which resulted in opal–CdS with $\Delta\omega/\omega = 10\%$ [25]. This is an encouraging result towards realising a PBG crystal by a relatively simple treatment of opal. Thus in first approximation, the width of the gap depends on the η contrast, and not on the η average (see Figures 13.6b and 13.7b).

The photoluminescence (PL) of opal–semiconductor composites shows that the line shape and intensity depends strongly upon the homogeneity of the semiconductor distribution within the 3D grating. This is expected, since inhomogeneities in the range from 2 to 5 μm result in the superposition of photonic effects from lattices with different photonic band structures. A comparison of EPMA data and PL spectra shows that gratings with infill content deviation higher than ±0.5 at.% have a weak PL intensity, the spectral spread of which covers a wide energy range. In nearly homogeneous samples, the PL line becomes relatively narrow.

The PL line of opal–InP lies 0.3 to 0.4 eV higher than the band edge of bulk InP. This emission energy is consistent with other observations from ~10 nm InP particles [38]. With decreasing temperature the PL intensity increases (see Figure 13.8b), while its half-width remains nearly the same. The remarkable feature is that the PL line in the vicinity of the PBG appears intense only if it is centred at the low-energy side of the PBG, as in the case shown in Figure 13.8a. An efficient photo-excitation of the semiconductor PL is not feasible if the laser line lies within the gap as illustrated in Figure 13.8a.

The PBG of the opal grating strongly affects the emission spectrum of the host semiconductor. Figure 13.9a shows that the PL collected from the excited front surface differs from that collected from the back surface of opal–TiO$_2$ by exactly the PBG transmission. Moreover, the PL intensity reflects the shift of the PBG due to the increase of the coating thickness (see Figure 13.9b).

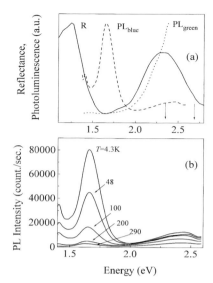

Figure 13.8. (a) Comparison of angle-integrated reflection spectrum and photoluminescence spectra under excitation by 514.5 (short dashed line) and 457.9 (long dashed line) nm radiation. (b) Temperature dependence of the photoluminescence from opal–InP.

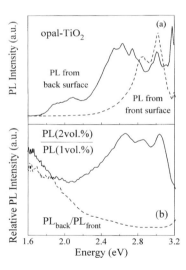

Figure 13.9. (a) Comparison of the opal–TiO$_2$ photoluminescence spectrum from the front and back surfaces. (b) Relative photoluminescence intensity ratio upon changing coating thickness and grating transmission in opal–InP.

The most interesting phenomenon may arise in the case of matching the electronic gap and the PBG, as in opal–CdS shown in Figure 13.7b in that the possibility of optical nonlinearities is enhanced. It is likely that the optical gain observed in the PL intensity upon the incident power in opal–CdSe (see Figure 13.10b) could be explained by this efficient energy gap matching. This is partly supported by the PL line narrowing (Figure 13.10b) as the excitation power is increased, which could be explained as energy relaxation redistribution in favour of PBG-edge emission.

Thus the main results of our activity towards the PBG material design to date are: (i) the preparation of wide-gap photonic materials by deposition of a molecular

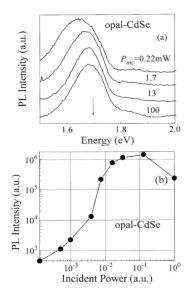

Figure 13.10. (a) Normalised photoluminescence spectra of opal–CdSe upon excitation power chosen for intensity readings. Curves are shifted for clarity (arrow indicates the energy at which the peak intensity was measured. (b) Superlinear dependence of the peak photoluminescence intensity upon excitation power.

thin semiconductor layer on the surface of opal balls, (ii) the separate control of the spectral width and position of the PBG and (iii) the suggestion that to approach the optical nonlinear regime it is necessary to adjust simultaneously the photonic and electronic band structures of the composite grating.

13.4 Transport Properties

Consider the case of a completely infilled opal with a semiconductor material. Then one could argue that the voids from a 3D lattice of grains coupled via bottleneck-like constrictions as shown schematically in Figure 13.11. The SiO_2-semiconductor

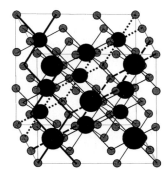

Figure 13.11. Schematic of the semiconductor grain lattice in opal.

interface provides a high potential barrier for electrons of the order of ~ 0.5 eV due to the fundamental energy gap difference which in turn restricts the electron motion within the infilled space of the opal voids. It follows that the carrier distribution in the semiconductor grains can be assumed to be the same as in a usual semiconductor. The most exciting conductivity effects are expected when the grain energy spectrum undergoes size quantization. The role of the electron energy quantization was examined in a comparative study of InSb- and Te-based lattices.

The geometrical modulation of the semiconductor cross section results in a potential relief for charge carriers. Let us consider the representative case of an opal–InSb sample with $D = 227$ nm. First, we estimate the characteristic energies. The confinement energy ΔE is given by

$$\Delta E = \frac{(\pi \hbar)^2}{2m^* d^2} \tag{13.6}$$

where m^* is the effective mass and $\Delta E \approx 30$, 11 and 3 meV for d_3, d_2 and d_1, respectively. It seems reasonably to consider these numbers as the lowermost limit, since the sintering-induced squeezing of void sizes and depletion of the electron population at the grain surface reduces the actual size of grains with corresponding increase of E, which is even higher for the InSb constrictions. Thus d_3 constrictions induce potential barriers along the current path (see Figure 13.12). These barriers localise electrons in grains and leave tunnelling as the only way for them to complete the current path. Since the length of the constriction is just $0.05D$ the barriers are expected to be highly transparent. From Hall measurements, it is known that the typical concentration of electrons is about $n \approx 10^{15}$ cm^{-3} at $T = 150$ K. Hence, around $N \approx 10$ free electrons are found in each in each grain. Correspondingly, the Fermi energy E_F and the Fermi wavelength λ_F in d_2 grain are of order

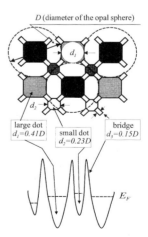

Figure 13.12. Top: (100) plane cross section of the dot lattice in opal. Bottom: corresponding potential profile for charge carriers.

$$E_F = \hbar^2 \frac{\left(\dfrac{3\pi^2 N}{d^3}\right)}{2m^*} \approx 49 \text{ meV} \tag{13.7}$$

and $\lambda_F \approx 47$ nm. Moreover, it is reasonably to assume that electrons are mostly in large grain and that E_F is different for d_1 and d_2 grains. 0D electron quantization conditions are fulfilled for electrons in these grains due to: (i) the size constrain $d_2 \leq \lambda_F$; (ii) the large difference of dielectric constants in InSb ($\varepsilon = 16$) and SiO$_2$ ($\varepsilon = 2.1$) and (iii) the presence of intergrain barriers. The InSb grains in opal may be thought of as QDs with a discrete energy spectrum. The energy level spacing in the 50 nm QD is:

$$\Delta = \rho(E_F)^{-1} = \frac{2\pi^2}{d^3} \left(\frac{2m^*}{\hbar^2}\right)^{-3/2} E_F^{-1/2} \approx 2.8 \text{ meV} \tag{13.8}$$

where $\rho(E_F)$ is the density of states near the Fermi level. In a 3D crystal constructed from QDs the interaction in the lattice splits each energy level into a miniband and the actual level spacing becomes $\Delta \ll kT$ depending on the lattice temperature [39]. Therefore, the QD energy spectrum may be considered as a continuum. In opal–Te the potential relief is less pronounced because the effective mass of holes, the main carriers, is about 10 times larger that in InSb.

The isolation of QDs has an important consequence for the current–voltage characteristics [39], the temperature dependence of resistance ($R(T)$) [40] and the magnetoresistance ($R(B)$) [41]. It is interesting to note, that the d_2 QD in between two d_1 QDs may be considered as a double-barrier tunnel junction [42]. In this case a charging energy

$$E_C = \frac{e^2}{2C_\Sigma} \tag{13.9}$$

is required to place an extra electron on the small dot. For a sphere of 50 nm diameter

$$E_C = \frac{e^2}{d_2 \varepsilon_m} \approx 14 \text{ meV} \tag{13.10}$$

In the case of incomplete miniband filling and $eV, kT > \Delta$ the QD energy spectrum is similar to that of metal island in the Coulomb blockade regime. At low temperatures the QDs are coupled capacitively. Under an external electrical field this lattice may be represented by an array of collinear chains along the field direction. Bearing in mind the structure of the lattice, these 1D arrays of metallic islands are analogous to a multiple-tunnel junction (MTJ) [43, 44] since, as in a MTJ, the conductivity in this QD lattice occurs by sequential single electron tunnelling from one dot to another.

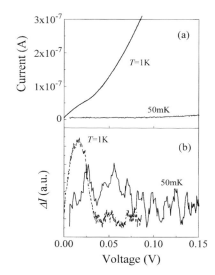

Figure 13.13. (a) I–V characteristics at 1 K and 50 mK of opal–InSb. (b) Excess current obtained from the curves above (see text for explanation).

At low temperatures the opal–InSb exhibites I–V characteristics with no linear dependence over the applied voltage (see Figure 13.13a). They ressemble typical curves for tunnel junction with charging effect [45]: after starting with a very low current below the threshold voltage $V < V_{thr}$ the I–V curves are fitted rather well by the expression $I \sim V^k$, with $k > 2$. Decreasing the temperature causes the current to decrease rapidly and one possible reason is that not all the geometrically available current paths conduct at the lowest temperatures due to lattice imperfections. However, among the large number of possible path configurations, and there are $\sim 10^7$ to 10^8 barriers in parallel on the sample cross section, it is always possible to find a set of low resistance paths. In opal–Te the I–V curve is linear and no Coulomb blockade is observed.

The low-voltage part of the I–V curve deviates from that of the standard Coulomb gap. Figure 13.13b shows the extra current I_{ex} (here represented by ΔI) extracted using a high-voltage fit as the background [46]. This feature is known as high-order tunnelling (cotunnelling) involving several coherent tunnelling events, if the junction resistance R_t is

$$R_t \geq R_Q = \frac{\hbar}{e^2} \approx 26 \text{ k}\Omega \qquad (13.11)$$

The maximum value of the low temperature resistance is around 10^5–10^7 Ohm. For 50 μm spaced potential probes this corresponds to $R_t \geq 10^9$ in the case of a perfect lattice. This ensures that the requirement is fulfilled even when an inhomogeneous current distribution is taken into account. The shift of the maximum of the curves $I_{ex} \propto T$ permits the classification of this effect as inelastic cotunnelling, in contrast to particular resonances in the QD energy spectrum.

Due to the coexistence of many parallel 1D zigzag-like chains of tunnel junctions in the opal-based lattice, their coupling should be taken into account. It was shown, that for electrostatically coupled 1D arrays the "exciton" mechanism of charge transfer may dominate [43]. This "exciton" is made up of an electron in one 1D chain coupled capacitively to a hole, where the hole is the induced charge redistribution in an adjacent 1D chain. Since the interchain capacitance C_0 will be up to K times larger than the capacitance C of each junction (K being the number of junctions in series), the electrostatic energy of this electron–hole pair is much smaller than the energy of unpaired electrons. Consequently, these "excitons" can move along the coupled chains when the voltage bias is much smaller than that necessary for single electron transfer in the Coulomb blockade regime and can transfer current along each chain. The binding energy of this "exciton" is [43]:

$$\varepsilon = \left(\frac{e^2}{4C}\right)\left(1 - \frac{1}{K}\right) \cong \frac{e^2}{4C} \tag{13.12}$$

Therefore, the true Coulomb gap will be of the order of

$$eV < \frac{e^2}{C_0}(\approx 25 \text{ to } 35 \text{ } \mu eV) \tag{13.13}$$

In the range $e^2/C_0 < eV < \varepsilon$ only "exciton" transport is possible, then at $eV > \varepsilon$ "excitons" will be destroyed and at $V > e/2C \approx 14$ mV single electron transport along the chain becomes dominant. In Figure 13.13b it is seen, that at 1 K I_{ex} (I in the figure) has a maximum at $V = 16$ mV. The agreement between the estimated and experimental value of the applied voltage for the destruction of the "exciton" is quite satisfactory.

The nonlinearity of the I–V curves of opal–InSb has a remarkable effect upon the resistance [47] as illustrated in Figure 13.14. The resistance $R(V)$ increases due to the "exciton" destruction reaching a maximum and then gradually decreases as it approaches the single electron transport regime. These changes in resistance exceeds one order of magnitude, and we suggest that $R(V)$ reflects the changes of the transport mechanism. In Figure 13.14 $R(V)$ for two samples with different QD sizes are shown. It can be seen that the overall shape of $R(V)$ curves is not sample specific and, as the temperature increases the magnitude of the resistance decreases and the $R(V)$ maximum moves towards lower voltages. This behaviour is consistent with the model of interacting chains of MTJ, since the total lattice resistance decreases first with increasing temperature due to the increase of the number of interacting tunnel chains and, second, with increasing capacitance which is proportional to d_2.

The dramatic drop of resistance at high voltage is inconsistent with the linear resistance regime predicted by the orthodox theory of single electron trnasport [48]. This discrepancy may be understood if we take into account the multiple coordination of QDs in this 3D lattice: with increasing bias voltage the electric field component projected along the perpendicular direction with respect to the 1D chains becomes sufficiently large to open up a new current path from the same node be-

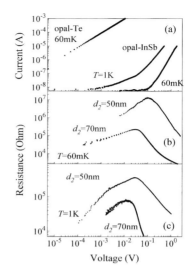

Figure 13.14. (a) I–V characteristics of opal–Te at 60 mK and opal–InSb at 1 K and 60 mK. Resistance against voltage curves for two opal–InSb samples with different dot size d_2 at 60 mK (b) and 1K (c).

cause of its eightfold coordination in the lattice. Therefore the $R(V)$ drops just because the number of parallel circuits increases. We treat this as a 1D to 3D transition in the current distribution since the negative feedback prevents Joule heating.

The excess current $I_{ex}(V)$ at 50 mK was observed to be highly modulated (see Figure 13.13b). A similar trend is observed in the resistance of 50 nm dots in opal–InSb at 60 mK which shows a series of steps of height around 10% of total resistance (Figure 13.15a). Subtracting the smooth background these steps were transformed into peaks spaced by $\Delta V \approx 14$ mV (Figure 13.15b). The power spectrum shown in Figure 13.15c confirms the quasioscillatory behaviour with a periodicity in $1/V$ which corresponds to $\Delta V = 14$ mV. This value is very close to $\Delta V \approx E_C/e$. The charging energy creates a barrier which blocks the entrance of electrons in the chain. Increasing the bias voltage above E_C/e results in another electron entering the chain.

The dependence of the resistance upon temperature $R(V)$ of bulk InSb, opal–InSb having the same impurity content and opal–Te are shown in Figure 13.16. The differences are explained below as a result of the potential landscape. The resistance of all opal–InSb samples can be fitted by

$$R \sim \exp\left(\frac{-T}{T_0}\right) \tag{13.14}$$

in the range 250 to 50 K in contrasts to the usual Arrhenius type dependence (Figure 13.17a). To understand this dependence, tunnelling through thin barriers was considered together with thermal fluctuations which is bound to have an impact in the high-T regime [40]. Assuming adiabatic motion through a barrier with a potential near E_F as

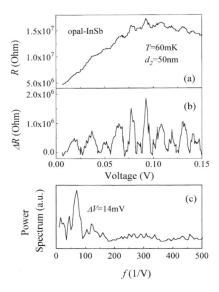

Figure 13.15. (a) Resistance against voltage of opal–InSb at 60 mK (a). The oscillatory nature of $R(V)$ shown here has been obtained subtracting the smooth background of curve (a). The power spectrum of the oscillatory resistance is shown in (c).

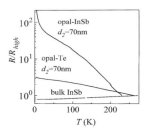

Figure 13.16. Dependence of the resistance upon temperature of bulk InSb, opal–InSb and opal–Te.

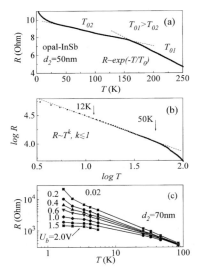

Figure 13.17. The two regimes of conductivity of opal–InSb can be distinguished in the following plots: (a) semilogarithm, (b) a double logarithm coordinates and (c) $R(T)$ data from several I–V curves recorded at different temperatures.

$$\varphi = \frac{U_0}{\cosh^2(\alpha x)} \tag{13.15}$$

where $1/\alpha$ is the barrier width and U_0 is its height, the relationship between resistance and temperature may be derived analytically to be

$$R \sim \exp\left(\frac{2\pi^2 m^*}{\alpha^2 \hbar^2}\right) \tag{13.16}$$

The kink between 150 and 180 K, which is equivalent to ~ 14 meV, corresponds to the decrease of the barrier transparency since $U_0/k = T_0$ increases from 30 to 80 K and could be assigned to the charging effect.

At $T < 50$ K $R(V)$ changes to a power law $R \sim 1/T^k$ with $k \approx 1$ (see Figure 13.17b). The onset of this regime may be due to negligible thermal activation. Following reference [49], $R \sim 1/T$ reflects the squeezing of the width of the overlapping conductance peaks in the Coulomb blockade. At $T < 10$ K a sequential single electron transport becomes the only transport mechanism in the lattice. For $T < 10$ K the slope of $R(V)$ decreases with increasing bias voltage (see Figure 13.17c), since the voltage aligns energy levels arising from different between QDs. The high potential wall formed by the dielectric matrix allows the use of a high bias voltage while keeping electron wavefunctions squeezed within QDs leading to the restoration of a regular current distribution. Thus, hopping in disordered materials, which follow the Mott law, stands in contrasts to macroscopic quantum tunnelling in the ordered lattice displaying a power law dependence.

The investigation of magnetotransport in opal–InSb reveals a number of geometry-related effects. A separation of the classical and quantum mechanical effects is normally accomplished using the high temperature regime. The classical magnetoresistance (MR) can arise as the result of boundary scattering [50]. If the electron motion is restricted in space, the zero-field resistivity is enhanced due to the additional diffuse scattering on the boundary. Under an applied magnetic field the resistance $R(B)$ decreases towards the bulk value unless the cyclotron diameter $2l_{cycl}$ where

$$l_{cycl} = \frac{\hbar k_F}{eB} \tag{13.17}$$

becomes smaller than the constriction [51]. The overall reduction of $R(B)$ is preceded by its initial increase in weak magnetic fields due to the deflection of electrons with velocity directed along the channel axis towards the boundary. This results in a maximum of the MR in the low-field region when $l_{cycl} \approx 0.5d$ [52, 53].

The classical MR size effect is absent if the roughness of the boundaries is on a length scale smaller than λ_F. In this case the resistance becomes field dependent only if a confining potential varies along the path axis, i.e., the chain of QDs with the dot spacing as the characteristic length. The interdot barrier in this case simply reduces the probability of the electron to reach the next dot.

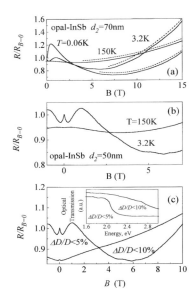

Figure 13.18. Magnetoresistance of opal–InSb. (a) General trend of the high-field magnetoresistance in the case of highly populated quantum dots, (dashed lines are the $R \sim B^2$ fits). (b) Classical negative magnetoresistance (curves are offset for clarity). (c) Correlation between classical negative magnetoresistance and quality of the opal lattice (curves are shifted for clarity). Inset: optical transmission for these bare host matrices.

In Figure 13.18a magnetoresistance curves at different temperatures are shown for an opal–InSb sample with a relatively high electron population. At 150 K (Figure 13.18b) the resistance increases when $B < 0.4$ T followed by a decrease in the range from 0.4 to 3.4 T, while above 3.4 T it shows a quadratic dependence on the magnetic field $R \sim B^2$. This behaviour is considerably different at 3.2K (Figure 13.18b) where the MR trace contains a narrow symmetric peak at zero field decreasing to a minimum at ± 0.24 T. Increasing further the magnetic field, the MR behaviour is qualitatively similar to that at high temperature with the resistance reaching a maximum at 1.04 T. Why does the resistance maximum shift with temperature?. Taking into account the dependence of the cyclotron radius upon the carrier concentration, namely $n \approx 3 \times 10^{13}$ cm^{-3} at 3.2 K and $n \approx 5 \times 10^{14}$ cm^{-3} at 150 K, from Hall effect measurements, $l_{cycl} \approx 16$ nm appears to be temperature independent for a given magnetic field $B(R_{max})$. The estimated channel diameter d is ~ 8 nm and $d_3 = 34$ nm in reasonable agreement with estimates when wavefunction squeezing and carrier depletion are taken into account. Therefore the resistance of the periodically modulated channels is determined by the scattering on constrictions. The negative magnetoresistance NMR around $B = 0$ observed at low temperature can be explained by weak localisation of electrons traversing the loop formed in the network. The magnetic field for threading a single flux quantum $\Phi_0 = h/2e$ through the minimum loop with a surface S is

$$\Delta B = \frac{\hbar}{2eS} = 0.24 \text{ T} \tag{13.18}$$

which agrees well with the minimum in the MR trace at 3.2 K.

The correlation of the NMR and the geometry is obtained from a comparison of ordered and partially disordered lattices. It is seen from Figure 13.18c, that for a partially disordered lattice, i.e., $\Delta D/D < 10\%$, the NMR is reduced down to a weak bend superimposed on the magnetoresistance. Transmission spectra corresponding to the respective bare opals provide a measure of the QD lattice disorder. Thus, the coaddition of classical size NMR over many nonidentical scatters smears out this effect in poorly ordered opal.

The origin of magnetoresistance in the QD lattice and in the disordered bulk material is qualitatively the same, namely, the reduction in the overlap of the wavefunctions from adjacent donors (here quantum dots) under an applied magnetic field [54]. This explains the quadratic dependence of the resistance upon field in samples with higher electron concentration although for this material the critical field $B_c > 4$ to 6 T, is higher than for disordered bulk samples. Generally, geometrical confinement effects can be neglected when considering the density of states, which is then described by Landau levels [55], if $2l_{cycl} \ll d_3$ or the mean free path l is defined by spiral orbits performed by electrons in the magnetic field. This means, that wall scattering dominates over electron–phonon or impurity scattering in QDs while bulk-like scattering appears only in the very high field regime.

In Figure 13.19 magnetoresistance plots of opal–InSb samples with low electron concentration are shown. Comparing the magnetoresistance between 60 and 1020 mK similar features can be found as will be discussed below. It is instructive to differentiate between three scales of resistance variation with magnetic field. The background may be separate into several large scale peaks, i. e., $\Delta B > 1$ T. It is then seen, that the gradual distortion of the magnetoresistance curve is due to the different temperature sensitivity of these peaks. The midscale ($\Delta B \sim 1$ T) and short-scale ($\Delta B < 1$ T) fluctuations were extracted from the background (see Figures 13.10a and 13.10b) and their power spectra (see Figures 13.10c and 13.10d) show that they are quasiperiodical fluctuations (QPF) with a characteristic period 0.96 and 0.28 T, respectively. Midscale patterns demonstrate the correlation through the whole temperature range with a small change at $T < 200$ mK. Near the 1 K limit the oscillations appear more periodic than at 0.05 K. It means that the source of these QPFs is not temperature sensitive. In contrast, short-scale QPFs are less correlated with each other and their correlation energy [56] may be estimated to

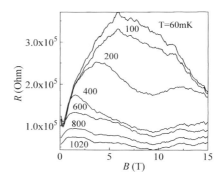

Figure 13.19. Change of magnetoresistance traces of opal–InSb with temperature in the pronounced Coulomb blockade regime.

be $E_c \approx kT = 0.04$ to 0.05 meV. At 0.06 K the short-scale power spectrum peak spreads over a wide range maintaing the periodicity.

To understand this magnetoresistance it is necessary to refer back to the structure of the QD lattice. It has been shown that tunnelling is the only mechanism for charge transfer between QDs. Due to the lattice arrangement QDs within the first coordination sphere may be reached by hops of the same length, then there is a gap up to the next neighbours and so on, i.e. there is only a discrete set of hop lengths available. The probability of interdot tunnelling is affected by the magnetic field since the wavefunction overlap for electrons in different dots decreases and, in turn, the overlap may differ for hops within different coordination spheres. This is assumed to be a source of the large-scale magnetoresistance pattern. It is worth noting the strong contrast between the magnetoresistance of ordered and disordered materials since, for the latter, the magnetoresistance increases exponentially with the field [54].

Considering the similarities of these features to the Aharonov–Bohm oscillations, it seems likely that these QPFs are due to quantum interference, since loop diameters extracted from the oscillation period

$$l = \left(\frac{\Phi_0}{\Delta B} \right)^{1/2} \tag{13.19}$$

where ΔB is the oscillation period, are 46 and 86 nm (see Figure 13.20) which in turn are similar to d_2 and d_1 in this sample. Another contribution to the short-scale

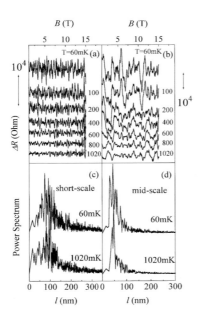

Figure 13.20. Short (a) and midscale (b) quasiperiodic fluctuations at different temperatures and corresponding power spectra (c) and (d), respectively.

QPF may arise from the interference of the electron virtually visiting neighbouring QDs during the hop [57]. For example, with $D = 227$ nm the smallest loop formed in the (100) plane is a square with $D/\sqrt{6} = 92$ nm side. Due to Coulomb barriers the probability of performing Aharonov–Bohm oscillations in large loops in a geometrically ordered lattice is suppressed. Moreover, different orientations of loops with respect to the magnetic field together with lattice imperfections wash out the Aharanov–Bohm oscillations.

The fluctuation patterns superimposed on the smooth magnetoresistance background contain information about the phase-coherence length l_φ due to the quantum interference term arising from the annular structure [58]. An analysis of the power spectrum may be used to estimate the localization length in this network of conducting wires. In the case of noncoplanar loops randomly distributed in 3D the Fourier components F_β of magnetoresistance fluctuations are related to the average phase-breaking length by

$$F_\beta \cong C \exp\left(\frac{-2\langle L_\beta \rangle}{\xi}\right) \tag{13.20}$$

where C corresponds to the number of loops of a given area in the sample and $\langle L_\beta \rangle$ is the length of the shortest path starting and finishing at the same point. Therefore, the power spectrum of magnetoresistance fluctuations should decay exponentially with an exponent inversely proportional to the average length. An order of magnitude of the phase-breaking length can be obtained using the approximation function:

$$2 \log_{10}(F_\beta) = 0.434 \left(c - \frac{8(\pi a)^{0.5}}{\xi}\right) \tag{13.21}$$

where

$$a = \frac{\langle L_\beta \rangle}{4\pi} \tag{13.22}$$

Parameters c and ξ may be estimated from the least squares fit of this function to data shown in Figures 13.20c and 20d. This fit yields ξ to be 320 and 305 nm at 60 mK and 1.02 K, respectively, which is longer than the separation from the first coordination sphere. This length exceeds the circumference of smallest loops $d \approx D/\sqrt{6}$ in the network and the next small loop with $d \approx D$. For the sample with a higher electron population ξ was found to be 10 times larger consistent with lower barriers which result from their greater transparency when the system has a higher Fermi energy.

Finally we consider the Hall resistance behavior (Figure 13.21a) which is also correlated with the Coulomb blockade. The low temperature Hall concentration increases linearly with the current, i.e., $n_H \sim I$. which is a direct consequence of the

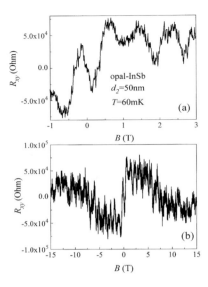

Figure 13.21. Anomalous Hall resistance near zero field (a) and suppression of the Hall effect by the high magnetic field.

increase of the sample volume involved in the charge transfer. The zero-field anomaly of the Hall resistance (see Figure 13.21b) is apparently due to electron focusing in the multiple coordinated QD network. Another result of the blockade is the saturation of the Hall resistance with field because the Hall potential developed across the QD is not enough to overcome the Coulomb barrier. Thus, the electron reservoir is only partly involved in the formation of the Hall voltage across the sample. Increasing further the magnetic field leads to smaller wavefunction overlaps which result in a diminished probability of electrons reaching the Hall probes. This is manifested as a decrease of the Hall resistance.

In conclusion, we have demonstrated that the conductivity of a 3D QD lattice is dominated by interdot tunnelling. This tunnelling is accompanied by thermal activation in the high temperature regime and becomes macroscopic quantum tunnelling under the Coulomb blockade in the low temperature regime. The lattice arrangement of QDs results in a softening of the Coulomb gap in the single electron regime. For highly resistive arrays, the Coulomb blockade appears as a staircase on the I–V curves denoting single electron transport. The classical magneto-size effects due to electron scattering in the periodically modulated confining potential of the 3D lattice, results in a pronounced NMR. At low temperature, the quantization of the magnetic flux penetrating through the lattice induces multiple periodic oscillations which appear superimposed on the smooth magnetoresistance background. Due to the discrete arrangement of QDs in the opal-semiconductor lattice, the high-field magnetoresistance behaviour deviates from the classical law. In addition, saturation of the Hall resistance with respect to the linear behaviour in the large-field scale appears due to the Coulomb blockade, whereas its short-scale deviations near zero field may be attributed to the electron focusing within QDs as multiple-terminal junctions.

13.5 Prospects

We have shown that the nanocomposite based upon artificial opal matrices and semiconductor infills is a functional material with potential applications in both optics and electronics. There are several parameters ameanable to change and these control the composite functionality. The challenges are to be found in the understanding of interface processes and their precise control.

Concerning optical studies opal-based photonic structures making use of the high periodicity and the refractive index contrast of the materials involved is a most promising approach for the control of spontaneous emission in three dimensions.

For their part transport studies reveal a wealth of phenomena concerning electron coherence and quantization which are begging to be studied with scanning probe techniques in order to separate unambiguously processes in each crystallite of the opal–semiconductor composite from the many-crystallite ensemble. These studies could yield key information on the possibilites of quantum networks with view to study experimentally quantum computing.

Acknowledgements

This work has been carried out in a collaboration across several laboratories. The authors are grateful to V. Butko, A. V. Fokin, D. V. Shamshur, H. M. Yates, M. E. Pemble, N. P. Johnson, D. K. Maude and J. C. Portal for their valuable contributions. This work was partly supported by the Royal Society of London, the Russian Foundation for Basic Science grants 94-02-05031 and 96-02-17963, the UK EPSRC grant GR/J90718, the EU ESPRIT project SOLDES 7260, the EU ESPRIT Network of Excellence PHANTOMS 7360 and the Leverhulme Trust grant No F/179/AK.

References

[1] Beaumont, S. P.; Sotomayor Torres, C. M. [Eds.], *Science and Engineering of One- and Zero-Dimensional Semiconductors*, Plenum, New York, USA, **1990**.

[2] Rarity, J.; Weisbuch, C. Eds., *Microcavities and Photonic Bandgaps: Physics and Applications*, Kluwer, Dordrecht, The Netherlands, **1996**.

[3] Bogomolov, V.; Kumzerov, Y.; Romanov, S. G. in *Physics of Nanostructures*, Eds. Davies, J. H.; Long, A. R., **1992**, Institute of Physics, Bristol, p. 317–321.

[4] Salata, O. V.; Dobson, P. J.; Hull, P. J.; Hutchison, J. L. *Appl. Phys. Lett.*, **1994**, *65*, 189–191.

[5] Hamdoun, B.; Ausserre, D.; Joly, S.; Gallot, Y.; Cabuil, V.; Clinard, C. *J. Phys. II France*, **1996**, *6*, 493–501.

[6] Bogomolov, V. N. *Sov. Phys. Uspekhi*, **1978**, *21*, 77–82.

[7] Bogomolov, V. N.; Kholodkevich, S. V.; Romanov, S. G.; Agroskin, L. S. *Solid State Commun.*, **1983**, *47*, 181–183; Bogomolov, V. N.; Efimov, A. N.; Ivanova, M. S.; Poborchii, V. V.; Romanov, S. G.; Smolin Y. I.; Shepelev, Y. F. *Sov. Phys. Solid State* **1992**, *34*, 916–919.

[8] Romanov, S. G. *J. Phys.: Condens. Matter*, **1993**, *5*, 1081–1090.

[9] Romanov, S. G.; Yates, H. M.; Pemble, M. E.;. Agger, J. R;. Anderson, M. W; Sotomayor Torres. C. M.; Butko, V. Y; Kumzerov, Y. A. to appear in *Fizika Tverdogo Tela* **1997**, *39* no. 3, (in Russian) (English Translation *Physics Solid State* **1997**, *39* no. 3.)

[10] Sanders, J. V. *Acta Cryst.* **1968**, *A24*, 427–438.

[11] Deniskina, N. D.; Kalinin, D. V.; Kazantseva, L. K. *Precious Opal, Their Synthesis And Genesis In Nature*, **1980**, Nauka, Novosibirsk, (in Russian)

[12] Salvarezza, R. C.; Vazquez, L.; Miguez, H.; Mayoral, R.; Lopez, C.; Meseguer, F. *Phys. Rev. Letts.* **1996**, *77*, 4572–4575.

[13] Mayoral, R.; Requena, J.; Moya, J. S.; Lopez, C.; Cintas, A.; Miguez, H.; Meseguer, F.; Vazquez, L.; Holgado, M.; Blanco, A. *Adv. Mater.* **1997**, *9*, 257–260.

[14] Bogomolov, V. N.; Kumzerov, Y. A.; Romanov, S. G.; Zhuravlev, V. V. *Physica C: Supercond.*, **1993**, *208*, 371–384; Romanov, S. G.; Fokin, A. V.; Babamuratov, K. *JETP Lett.*, **1993**, *58*, 824; S. G. Romanov, *JETP Lett.*, **1994**, *59* 810–813.

[15] Pankova, S. V.; Poborchii, V. V.; Solov'ev, V. G. *J. Phys.: Cond. Matter*, **1996**, *8*, L203–L206.

[16] Romanov, S. G.; Sotomayor Torres, C. M. in *Microcavities and Photonic Bandgaps*, [Eds. Rarity , J.; Weisbuch, C.], Kluwer, The Netherlands, **1996**, p. 275–282.

[17] Balakirev, V. G.; Bogomolov, V. N.; Zhuravlev, V. V.;. Kumzerov, Y. A; Petranovsky, V. P.; Romanov, S. G.; Samoilovich, L. A. *Crystallogr. Rep.* **1993**, *38*, 348–353.

[18] Romanov, S. G.; Shamshur, D. V.; Sotomayor Torres, C. M. in *Quantum Confinement: Quantum Wires and Dots*, [Eds. Cahay, M.; Bandyopadhjay, S.; Leburton, J. P.; Razeghi, M.], Proc. Electrochemical Society, Pennington, USA **1996**, p. 3 and references therein.

[19] Romanov, S. G; Fokin, A. V.; Tretijakov, V. V.; Butko, V. Y.; Alperovich, V. I.; Johnson, N. P.; Sotomayor Torres, C. M. *J. Cryst. Growth*, **1996**, *159*, 857–860.

[20] Yates, H. M.; Flavell, W. R.; Pemble, M. E.; Johnson, N. P.; Romanov, S. G.; Sotomayor Torres, C. M. *J. Cryst. Growth* , **1997**, *170*, 611–614.

[21] Romanov, S. G.; Fokin, A. V.; Butko, V. Y.; Tretijakov, V. V.; Samoilovich, S. M.; Sotomayor Torres, C. M. to be published in *Fizika Tverdogo Tela* **1997**, *38*, 3347 (in Russian) (English Translation *Phys. Solid State* **1997**, *38*, 3347–3360).

[22] Romanov, S. G.; Fokin, A.; Butko, V.; Johnson, N. P.; Sotomayor Torres, C. M. in *Diagnostic Techniques for Semiconductor Materials Processing II*, [Eds. Pang, S. W.; Glembocki, O. J.; Pollak, F. H.; Celli, F. G.; Sotomayor Torres, C. M.], Mat. Res. Soc. Symp. Proc. *406*, Bellingham, USA **1996**, p. 289–294.

[23] Martorell, J.; Lawandy, N. M. *Optics Commun.*, **1990**, *78* 169–173.

[24] Yablonovitch., E. *J. Modern Optics*, **1994**, *41*, 173–194.

[25] Astratov, V. N.; Bogomolov, V. N.; Kaplyanskii, A. A.; Prokofiev, A. V.; Samoilovich, L. A.; Samoilovich, S. M.; Vlasov, Y. A. *Il Nuovo Cimento*, **1995**, *D17*, 1349–1354; Bogolomov, V. N.; Prokofiev, A. V.; Samoilovich, S. M.; Petrov, E. P.; Kapitonov, A. M.; Gaponenko, S. V. to appear in *J. Luminescence*, **1997**.

[26] Romanov, S. G.; Fokin, A. V.; Butko, V. Y.; Johnson, N. P.; Sotomayor Torres, C. M.; Yates, H. M.; Pemble, M. E. in *Proc. Int. Conf. on The Physics of Semiconductors*, [Eds. Scheffler, M.; Zimmermann, R.], World Scientific, Singapore, **1996**, p. 3219

[27] Astratov, V. N.; Vlasov, Yu. A.; Bogomolov, V. N.; Kaplyanskii, A. A.; Karimov, O. Z.; Kurdjukov, D. A.; Prokofiev, A. V. *in Proc. 23rd Int. Symp. Compound Semiconductors*, St Petersburg, Russia, Institute of Physics Conference Series (in press).

[28] Bogomolov, V. N.; Gaponenko, S. V.; Kapitonov, A. M.; Prokofief, A. V.; Samoilovich, S. M.; Ponyavina, A. N.; Silvanovich, N. I. in *Proc. Int. Conf. on The Physics of Semiconductors*, [Eds. M. Scheffler and R. Zimmermann], World Scientific **1996**, p. 3139–3142.

[29] Hirayama, H.; Hamano, T.; Aoyagi, Y. *Appl. Phys. Lett.*, **1996**, *69*, 791–793.

[30] Yablonovich, E.; Gmitter, T. J.; Leung, K. M. *Phys. Rev. Lett.*, **1993**, *67*, 2295–2298.

[31] Datta, S.; Chan, C. T.; Ho, K. M.;. Soukoulis, C. M *Phys. Rev. B*, **1993**, *48*, 14936–14943.

[32] Vos, W. L.; Sprik, R.; Blaaderen, A.; Imhof, A.; Lagendijk, A; Wegdam, G. H. *Phy. Rev. B*, **1996**, *53*, 16231–16235.

[33] Lopez, C.; Miguez, H.; Vazquez, L.; Meseguer, F.; Mayoral, R.; Ocana, M. to appear in *Superlattices and Microstructure*, **1997**.

[34] Bogomolov, V. N.; Gaponenko, S. V.; Kapitonov, A. M.; Prokofiev, A. V.; Ponyavina, A. N.; Silvanovich, N. I.; Samoilovich, S. M. *Appl. Phys. A*, **1996**, *63*, 613–616.

[35] Astratov, V. N.; Vlasov, Y. A.; Karimov, O. Z.; Kaplyanskii, A. A.; Musikhin, Y. G.; Bert, N. A.; Bogomolov, V. N.; Prokofiev, A. V. to appear in *Superlattices and Microstructure*, **1997**.

[36] Astratov, V. N.; Vlasov, Y. A.; Karimov, O. Z.; Kaplyanskii, A. A.; Musikhin, Y. G.; Bert, N. A.; Bogomolov, V. N.; Prokofiev, A. V. in *Proc. Int. Conf. on The Physics of Semiconductors*, [Eds. Scheffler, M.; Zimmermann, R.], World Scientific, Singapore **1996**, p. 3127–3130.

[37] Romanov, S. G.; Johnson, N. P.; Fokin, A. V.; Butko, V. Y.; Yates, H. M.; Pemble, M. E.; Sotomayor Torres, C. M. to appear in *Appl. Phys. Lett.* (April 1997).

[38] Kurtenbach, A.; Eberl, K.; Shitara, T. *Appl. Phys. Lett.*, **1995**, *66*, 361–363.

[39] Romanov, S. G.; Maude, D. K.; Portal, J. C. to appear in *Appl. Phys. Lett.* in April 1997.

[40] Romanov, S. G.; Shamshur, D. V.; Maude, D. K.; Portal, J. C.; Larkin, A. I. to be published in *Compound Semiconductors 96*, Proc. 23 Int. Symp. on Compound Semicond., [Eds. Shur, M.; Suris, R.], Institute of Physics Conference Series, **1996**.

[41] Romanov, S. G.; Fokin, A. V.; Maude, D. K.; Portal, J. C. *Appl. Phys. Lett.*, **1996**, *69*, 2897–2899.

[42] Romanov, S. G.; Larkin, A. I.; Sotomayor Torres, C. M. unpublished.

[43] Averin, C. M.; Korotkov, A. N.; Nazarov, Yu. A. *Phys. Rev. Lett.*, **1991**, *66*, 2818–2821.

[44] Bahvalov, N. S.; Kazacha, G. S.; Likharev, K. K.; Serdukova, S. I. *Zh. Eksp. Theor Fiz.* **1989**, *95*, 1010 English translation *Sov. Phys. JETP*, **1989**, *68*, 581–592.

[45] Geerligs, L. J. in *Physics of Nanostructures*, [Eds. J. H. Davies and A. R. Long], IOP, Bristol, **1992** p. 171–204.

[46] Averin, D. V.; Nazarov, Yu. V. *Phys. Rev. Lett.*, **1990**, *65*, 2446–2449.

[47] Romanov, S. G.; Fokin, A. V.; Maude, D. K.; Portal, J. C. in *Proc. Int. Conf. on The Physics of Semiconductors*, [Eds. Scheffler, M.; Zimmermann, R.], World Scientific, Singapore **1996**, p. 1585–1588.

[48] Averin, D. V.; Likharev, K. K. *J. Low Temp. Phys.*, **1986**, *62*, 345; K. K. Likharev and A. B. Zorin, *J. Low Temp. Phys.*, **1985**, *59*, 347–382.

[49] Beenaker, C. W. J. *Phys. Rev. B*, **1991**, *44*, 1646–1656.

[50] MacDonald, D. K. C. *Nature*, **1949**, *163*, 637–638.

[51] Blatt, F. J. *Phys. Rev.*, **1954**, *80*, 401–406.

[52] Clocher, D. A.; Skove, M. J. *Phys. Rev. B*, **1977**, *15*, 608–616.

[53] Van Houten, H. et. al., *Appl. Phys. Lett.*, **1986**, *49*, 1781–1783.

[54] Shklovskii, B. I.; Efros, A. L. in *Electronic Properties of Doped Semiconductors*, [Ed. Cardona, M.], Springer Series on Solid State Science, Springer, Berlin **1984**, Vol. 45.

[55] Beenakker, C. W. J.; Van Houten, H. in *Solid State Physics*, Academic Press, **1991**, Vol. 44.

[56] Stone, A. D. *Phys. Rev. Lett.*, **1985**, *54*, 2692–2695.

[57] Raikh, A. D.; Glazman. L. I. *Phys. Rev. Lett.*, **1995**, *75*, 128–131.

[58] Robinson, S. J.; Jeffery, M. *Phys. Rev. B*, **1995**, *51*, 16807–16816.

Chapter 14

Charge Transfer at Nanocrystalline Metal Oxide Semiconductor–Solution Interfaces: Mechanistic and Energetic Links between Electrochromic–Battery Interfaces and Photovoltaic–Photocatalytic Interfaces

B. I. Lemon, L. A. Lyon, and J. T. Hupp

14.1 Introduction

Metal oxide semiconductor films and electrodes displaying only short-range atomic order ("nanocrystallinity") have attracted considerable technological attention because of their typically large effective surface areas, their amenability to dye sensitization, their significant nanoporosity and their often superior charge transport and/or charge storage characteristics [1–3]. Semiconducting oxides of this type typically have been further categorized as either electrochromic/battery materials or as photovoltaic/photocatalytic materials. In particular, those electroactive metal oxides which readily act as intercalation hosts serve as the basis for electrochromic and battery materials [4–5]. On the other hand, those materials for which the conduction band edge lies at a position that is favorable energetically for electron acquisition from light-harvesting species (sensitizing dyes) are candidates for usage in visible-region photovoltaic schemes and UV-region photocatalytic schemes [6–7]. The two types of materials can be prepared by similar methods and in many of the same morphological forms (see description in text). However, as discussed below, widely differing limiting descriptions – based on largely independent historical models – have usually been used to characterize the responses of the two groups of materials to electrochemical and/or photoelectrochemical addition of electronic charge.

For V_2O_5, MoO_3, WO_3 and related materials, electron addition and the attainment of significant dark conductivity is generally accompanied by cation intercalation [8]. As shown specifically in eq. 1 for WO_3, cations are brought into the semiconductor for local charge compensation:

$$W^{VI}O_3 + e^- + M^+ \rightarrow W^VO_3(M^+) \tag{14.1}$$

It is this ability to store cations and thereby stabilize lower metal oxidation states that makes these materials suitable for rechargeable battery applications. When linked to an appropriate counter electrode reaction such as $M^+ + e^- \rightarrow M$, reversal of eq. 1 (i.e. lattice oxidation and cation *de*intercalation) permits electrical energy to be stored.

Often accompanying the electron addition and cation intercalation are changes in the optical absorption spectrum of the material including: a) increases in the apparent optical bandgap (blue shift in the fundamental absorption edge), and b) increases in optical density in the visible and/or near-infrared region [9–10]. Also appearing are dramatic increases in electrical conductivity. While these phenomena clearly are complex, a simplified limiting description emphasizes the role of nominal mixed valency. The availability, at least in a formal sense, of metal centers in multiple oxidation states (V^V/V^{IV}, W^{VI}/W^V, etc.) provides a conceptual basis for electron transport and for intense coloration via long-wavelength light absorption (i.e. optical transitions having intervalence or metal-to-metal charge transfer parentage) [11]. These descriptions can become appreciably altered when account is taken of oxide and/or hydroxide anion mediated metal–metal electronic interactions. The resulting band-type descriptions, however, can still be viewed as descendants of a limiting mixed-valency description.

Wide bandgap materials whose conduction band edges lie at potentials less negative than those of photoexcited dyes are often susceptible to electron injection from surface-bound forms of the dyes (Figure 14.1) and are usually classified as photovoltaic materials. TiO_2, SnO_2, and ZnO are three of the more intensely studied of these semiconductors. Their ability to be prepared in nanocrystalline, high-surface area form allows them to act as particularly efficient light harvesters when employed in dye-sensitized photovoltaic solar cells [1–3]. A typical photovoltaic light-harvesting scheme can be envisaged as shown in Figure 14.1.

Though these semiconductors can be prepared in the same manner as, show similar changes in their optical spectra as, show electrochemistry that can appear identical to, and can exist in similar morphologies as those semiconductors classified as electrochromics, the classical description of how they accommodate excess electronic charge differs dramatically. Indeed, according to the classical model cathodic charging of *n*-type photovoltaic materials results in surface and near-surface electron accumulation (i.e. metal-like behavior) and charge-compensating perturbation of the surrounding electrical double layer (Figure 14.2) [12]. Note that

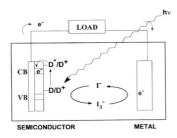

Figure 14.1. A schematic of a dye-sensitized photovoltaic cell. D represents a visible light absorbing dye molecule; D* refers to an electronically excited form of the dye. VB and CB refer to the valence and conduction bands, respectively. I^-/I_3^- is the solution-phase electron "shuttle".

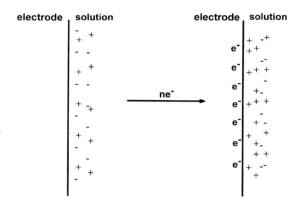

Figure 14.2. Idealized representation of the putative changes in the ionic double layer surrounding a photovoltaic semi-conducting electrode under cathodic bias.

in this description, no ion motion *across* the solution/semiconductor interface is necessary for charge compensation.

Recent studies and a careful survey of both bodies of literature, however, lead to new descriptions in which *both* types of materials undergo charge-compensating cation intercalation when subjected to electron addition. In particular, new spectroelectrochemical [13], electrochemical quartz crystal microbalance (EQCM) [14] and photochemical quartz crystal microbalance (PQCM) [15] studies of nanocrystalline semiconductors link the behaviors of the two types of materials under conditions of cathodic charging. As outlined in section *4*, below, these results also lead to a new mechanistic description of the factors controlling the conduction band energetics at the semiconductor/solution interface in nanocrystalline photovoltaic systems. This alternative description could significantly alter the manner in which electrochemical photovoltaic systems are engineered.

14.2 Electrochromics

14.2.1 V_2O_5

Vanandium oxide (primarily V_2O_5) is a layered metal oxide that can be fashioned via a variety of methods including thermal evaporation [16], sputter deposition [17], electrodeposition [18], chemical vapor deposition [19] and sol–gel condensation [20]. For sol–gel produced films, the structure consists of thin "ribbons" connected side-to-side to form layers. Between the layers is interstitial water; these interstitial layers act as cation conduits giving V_2O_5 its excellent intercalation abilities [20]. Though these sol–gel derived films differ morphologically from films prepared via other methods, all films typically display structured, reversible voltammetric responses where the observed peaks are associated with cathodic intercalation and anodic deintercalation of electrolyte cations [9]. V_2O_5 films offer excellent illustrations of the properties so desired in electrochromic and battery materials. Upon reduction (intercalation) the films show pronounced optical density increases in the visible region [9]. This coloration can be reversible, partially reversible, or irreversible depending on the precise film morphology and the degree of intercalation. Vanadium oxide's ability to store large amounts of cations reversibly (for example, up to 4 Li^+ per formula unit) [21] and to undergo many reversible intercalation/deintercalation cycles [22] (under appropriately controlled conditions) also make it a viable battery material. The details of the intercalation processes have been delineated via various surface analysis, electronic spectroscopy and diffraction investigations, in addition to UV–Vis and cyclic voltammetry studies; the reader is referred to the original literature for further information [8].

14.2.2 MoO_3

MoO_3 is similar to V_2O_5 in many respects. Electroactive films can be prepared by many of the same methods (vacuum and sputter deposition, sol–gel condensation, electrodeposition, anodization, and chemical vapor deposition) [8]. It can exist in either a layered or framework structure; again, the layered structure can intercalate significant quantities of charge- compensating cations [8, 23]. Following V_2O_5, cathodic charging leads to cation intercalation and increases in the visible- region optical absorbance [24] and electrical conductivity [25] of the material. Figure 14.3 shows a voltammetric scan of an electrodeposited MoO_3 film in $LiClO_4$ + propylene carbonate electrolyte solution. A cathodic wave attributed to Li^+ intercalation and an anodic wave representing Li^+ expulsion are noted [24].

MoO_3 is classified as an electrochromic metal oxide due to its reversible and desirable changes in visible-region spectral properties. Figure 14.4 shows the change in the visible absorption spectrum accompanying the above voltammetry. As for V_2O_5, the material becomes highly colored upon reduction and transparent following ion extraction [24], though the changes lie primarily in the visible region. MoO_3's ability to intercalate Li^+ cations also makes it suitable for use in lithium secondary batteries [25].

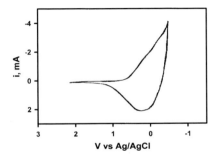

Figure 14.3. Cyclic voltammogram of an electrodeposited MoO_3 film in $LiClO_4$-propylene carbonate. Adapted from ref. 24.

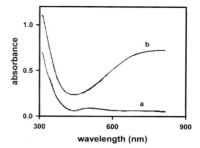

Figure 14.4. UV–Vis spectra of an electrodeposited MoO$_3$ film at under conditions of: (a) anodic bias and (b) cathodic bias. Adapted from ref. 24.

14.2.3 WO$_3$

Tungsten oxide makes up the next class of metal oxides used in electrochromics. As for V$_2$O$_5$ and MoO$_3$, WO$_3$ can be prepared via a wide variety of techniques including chemical vapor deposition [26], rf sputtering [27], sol–gel processing [28], electrochemical deposition [29] and thermal evaporation [30]. WO$_3$ has been extensively examined as an electrochromic material where the same electron-addition/cation-intercalation/change-in-optical-properties scheme is utilized [8].

Cation intercalation (including H$^+$, Li$^+$, and Na$^+$ intercalation) into WO$_3$ has been studied in exhaustive detail via methodologies ranging from cyclic voltammetry [31], chronoamperometry [32], impedance spectrometry [32] and electrochemical quartz crystal microgravimetry [29, 32] to EPR [10], Raman [33] and UV–Vis absorption [28] spectroscopies. Figure 14.5 shows a cyclic voltammogram for a sol–gel derived film of WO$_3$ where the cathodic and anodic waves are attributed to cation intercalation and deintercalation, respectively [10]. Figure 14.6 shows the electro-

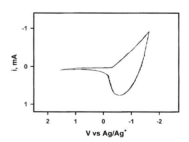

Figure 14.5. Cyclic voltammogram of a sol–gel derived WO$_3$ film in LiClO$_4$-propylene carbonate. Adapted from ref. 10.

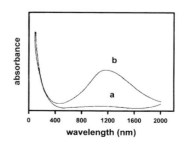

Figure 14.6. UV–Vis spectra of a sol–gel derived WO$_3$ film in LiClO$_4$-propylene carbonate under conditions of: (a) no external bias, and (b) cathodic bias. Adapted from ref. 10.

chromic behavior of the film. Upon reduction/intercalation, the film becomes intensely colored [10].

More recent reports have employed WO_3 in its high surface area, thin-film form for photochemical applications. Surprisingly, in this form tungsten oxide also shows many of the characteristics of more conventional "photovoltaic" metal oxides. It can, for example, be reduced by UV illumination (and subsequent hole scavenging) either to store electrons for solar energy applications [34] or to react with solution-phase molecules in a photocatalytic scheme similar to that in TiO_2 waste-water remediation [6].

14.3 Photovoltaics

14.3.1 General Observations

In contrast to V_2O_5, MoO_3 and WO_3 which can be reductively doped with protons or alkali metals, the so-called photovoltaic metal oxides are usually assumed to respond to electron addition via electron accumulation layer formation and electrical double layer charging (Figure 14.2, above). Recently, however, Meyer and coworkers have pointed that little, if any, *direct* experimental evidence is available in the extant literature to validate the applicability of the cathodic bias/accumulation layer model to metal oxide systems [35]. These authors instead have pointed to the apparently overwhelming importance of trap states or other localized electronic entities. (In contrast, under conditions of anodic bias, compelling evidence does exist (albeit, for single crystals) for the formation of classical depletion layers [36]).

At the same time, various applications-oriented studies have recently shown that nanocrystalline photovoltaic oxides (most notably, titanium dioxide) can functionally mimic their electrochromic/battery oxide counterparts. In addition, as summarized below, nanocrystalline forms of the *n*-type semiconductors TiO_2, SnO_2 and ZnO have all recently been shown to display (to varying degrees) intercalation behavior during cathodic charging. Observation of these phenomena suggests that ion intercalation may be a *general* mode of reactivity for metal oxide semiconductor/solution interfaces.

14.3.2 TiO_2

The electrochemical addition of electrons to titanium dioxide can be accomplished by scanning the electrode potential or sustaining the electrode bias negative of the apparent conduction band edge. (Except where noted, the form of TiO_2 under discussion in this section is nanocrystalline anatase.) Addition is accompanied by striking increases in conductivity and by conversion of the transparent oxide to a deep blue form [37–38]. It is also accompanied by what is typically termed

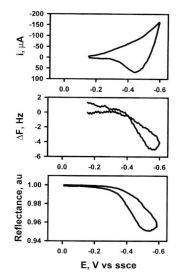

Figure 14.7. EQCM in aqueous media. Panel (a) shows a cyclic voltammogram of a sol–gel derived titanium dioxide film in 1 M HClO$_4$. Panel (b) shows reflected laser intensity at 786 nm. Panel (c) shows the change in frequency of the quartz crystal oscillator. Potentials in all three panels are referenced against SSCE.

"capacitive" current flow (cf. Figure 14.2) [39]. Surprisingly, close inspection of the electron addition process via electrochemical quartz crystal microgravimetry (EQCM) has shown that it is appreciably more complex than expected for a purely capacitive process (cf. Figure 14.2), resembling more closely the batterylike process in eq. 1 [14, 40].

EQCM is a valuable analytical tool for the measurement of small mass changes occurring during electrochemical charging of electroactive materials [41]. Application of this technique to high-area titanium dioxide/water interfaces has led to the observation of proton uptake during electron addition. Figure 14.7 illustrates a typical experiment. In panel (a), the addition of electrons is observable by enhanced current flow at potentials negative of the nominal conduction band edge. Simultaneous monitoring of the electrode mass (panel (b)) indicates that the electrode becomes significantly heavier during the apparent charging process. Conclusive evidence for electron addition (as opposed to faradaic current flow) comes from spectroscopic interrogation of the nanocrystalline film. Panel (c) is a laser reflectance spectroelectrochemical scan acquired simultaneously with the data in the preceding panels. The attenuation in the reflected (786 nm) laser intensity is an indication of an increased absorbance due either to conduction band electrons or, more probably, surface trapped or localized electrons (see below). Quantitative analysis of the EQCM data strongly suggests that for each excess electron added, one proton is intercalated into the material. This is further supported by deuterium isotope experiments where the total change in electrode mass doubles upon substitution of H$_2$O with D$_2$O as solvent [40].

Proton intercalation has also been reported in a photochemical system [15]. Photoelectrochemical quartz crystal microgravimetry (PQCM) allows for simultaneous monitoring of the electrode mass during bandgap illumination of the semi-

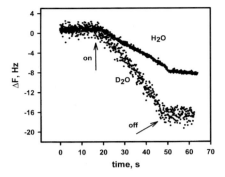

Figure 14.8. PQCM in aqueous media: Change in frequency of the quartz crystal oscillator vs. time for a nanocrystalline titanium dioxide film in contact with H_2O and D_2O. Negative frequency shifts correspond to increases in oscillator mass. Arrows indicate when light is turned on and off.

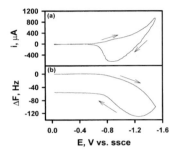

Figure 14.9. EQCM in nonaqueous media: Voltammetry (panel (a)) and microgravimetry (panel (b)) for nanocrystalline titanium dioxide film in scan in acetonitrile containing $LiClO_4$.

conductor. In aqueous media with an appropriate reducing agent (hole scavenger) in solution, proton uptake can be observed as electrons are added to the material. Deuterium isotope experiments showed a mass doubling effect for identical irradiation times (Figure 14.8). These results provide evidence that cation intercalation is necessary for charge compensation, independent of the manner in which electrons are added to TiO_2. Note that evidence (albeit, indirect evidence) for photo intercalation has also been reported for WO_3 [42, 43].

In an early report, Borgarello, et al. described the addition of electrons to colloidal TiO_2 via pulse radiolysis. Subsequent transient conductivity measurements provided evidence for uptake of protons by "reduced" TiO_2 [44].

Perhaps more applicable to the development of charge storage and electrochromic devices are studies of titanium dioxide in nonaqueous media. Of particular interest are voltammetric and EQCM investigations of alkali metal cation intercalation from acetonitrile and propylene carbonate. Grätzel and others have shown via voltammetry and chronoamperometry that nanocrystalline titanium dioxide has some capacity as an intercalation host in both solvents [45–46]. Corroborative of these findings are EQCM measurements performed in nonaqueous media [14, 40]. These again show that cation intercalation accompanies electron addition for the purpose of charge compensation. Figure 14.9 contains representative voltammetric and microgravimetric scans for lithium ion intercalation from acetonitrile. Similar behavior is observed for sodium ion intercalation. The effect is approximately the same in propylene carbonate where Li^+, Na^+ and K^+ have all been found (by EQCM) to intercalate into nanocrystalline TiO_2 (ca. 5 nm diameter anatase particles in sintered film form). In contrast to the first three alkali metals, larger cations such as tetraalkylammonium species are – for obvious steric reasons – unable to intercalate from either acetonitrile or propylene carbonate. (While the steric argument is compelling for tetralkylammonium ions, it is less compelling for ions such as potassium that are only marginally larger in diameter than the channel widths available in bulk anatase. For nanocrystalline electrode/solution interfaces, account must be taken of possible surface and near surface gelation, as well as lattice distortions introduced by electron addition itself.) For tetraalkylammonium species, failure to intercalate is experimentally evidenced, in part, by decreased electrochemical capacities (smaller voltammetric currents). It is also evidenced by nonstoichiometric electrode mass changes where the nonstoichiometry is attributed to ion binding via surface adsorption rather than lattice intercalation [14, 40].

Alkali metal ion intercalation studies have recently been extended to the fashioning of a "rocking chair" battery based on a nanocrystalline (anatase) titanium dioxide cathode and a lithium anode [47]. In this case, the ability of titanium dioxide to store charge as intercalated lithium ions allows for its use as a battery material. Others have prepared titanium dioxide composites for use as cathode materials [48]. (Here, however, the materials may be microcrystalline rather than nanocrystalline.) Surprisingly, under appropriate high-temperature conditions, storage device construction based on rutile has also proven feasible [49]. Following device discharge, x-ray studies [49] have provided further evidence for lithium intercalation into composite electrodes.

The rapid coloration achievable with TiO_2 films has led to at least one investigation of the intercalation phenomenon in the context of fast switching electrochromics [13]. While the coloration effect is sometimes ascribed to excitation of added conduction band electrons [37], the almost universal observation of finite-wavelength absorption maxima (typically between ca. 800 and 1400 nm) is suggestive instead of excitation of partially localized or trapped electrons. Similarities between absorption spectra for Ti^{III} [50] and partially reduced titanium dioxide (films or colloids) have been noted and the nominally forbidden d–d absorption of Ti^{III} has been suggested as an alternative basis for nanoparticle coloration. In view of the typically strong red and near infrared extinction achievable with reduced titanium dioxide, however, more probable chromophoric sources are electric-dipole-allowed transitions having intervalence parentage (i.e., significant Ti^{III}-$Ti^{IV} \rightarrow Ti^{IV}$-$Ti^{III}$ character).

14.3.3 SnO_2

While there are fewer reports of cation intercalation into nanocrystalline tin oxide, the available results strongly suggest a similar mechanism of charge compensation. EQCM in aqueous electrolyte yields nearly identical results to those observed for TiO_2 [51]. Conclusive evidence of proton intercalation is again offered by deuterium isotope experiments where mass doubling is encountered when D_2O replaces H_2O

as the solvent [51]. Intercalation has not been reported from nonaqeous media but it is anticipated that alkali metal cation intercalation would occur in the absence of proton sources.

14.3.4 ZnO

Much as with tin oxide, the body of work on cation intercalation into zinc oxide is not very large. Weller has shown that up to 6 electrons can be reversibly added to a single 5 nm ZnO particle [52, 53]. Related EQCM measurements from our lab have shown that proton uptake, rather than diffuse double layer charging, provides the necessary charge compensation [51]. The contention that H^+ is the intercalant is again supported by isotope experiments.

One difference between ZnO and most other metal oxides in this discussion is ZnO's lack of strong near-IR absorbance following electron addition [54]. Another difference is its comparative instability with respect to reduction to metallic form [51, 53]. Both observations are likely related to the well known instability of Zn^I. The instability provides an obvious driving force for disproportionation into Zn^{II} and Zn^0 following trapping of excess electronic charge. The instability would also preclude the observation of Zn^{II}/Zn^I-based optical intervalence transitions, leaving only lower extinction conduction-band electrons (delocalized or untrapped electrons) as potential sources for electrochromic coloration.

14.4 Energetic Considerations

14.4.1 Potentials

Proton-intercalation-based electrochromics such as iridium oxide/hydroxide [55] or tungsten bronzes (Eq. 14.1) exhibit color transition potentials (E_{tr}) that are strongly dependent upon solution pH [56]. At 298K the dependence can be expressed as:

$$E_{tr} = E_{tr}(pH = 0) - (0.059\ V)(p/n)(pH) \tag{14.2}$$

where p is the proton stoichiometry and n is the electron stoichiometry of the electrochromic reaction. Eq. 14.2, of course, follows in an obvious way from a Nernstian thermodynamic characterization of the redox (or redoxlike) behavior of any system involving formation of a protonated product from a reservoir or solution of variable proton activity. For aprotic systems, such as oxide battery electrodes involving M^+ insertion, Eq. 14.2 can be rewritten with pM and an appropriate open circuit potential in place of pH and E_{tr}, respectively.

Curiously, apparent conduction band edge energies (E_{cb}) for n-type photovoltaic oxides also change systematically with pH [37, 39, 54]. For nanocrystalline titanium dioxide, Nernstian behavior is observable over a truly remarkable range – some

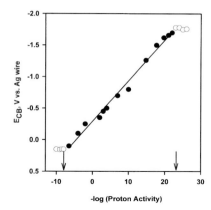

Figure 14.10. Dependence of the conduction band edge energetics (E_{CB}) on proton activity. Arrows at -8 and 24 (open circles) indicate onsets for proton activity independent energy regimes.

32 pH units (Figure 14.10) corresponding to a nearly 2 V variation in E_{cb}! [40] Furthermore, EQCM experiments show that proton intercalation persists at least down to pH $= -5$ and up to pH $= 11$. (At more extreme pH's, solution viscosity effects and porous oxide instability effects prevent EQCM interrogation.) Literature explanations for the ubiquitous Nernstian behavior have tended to focus on redox-independent surface protonation effects, i.e. effects responsible for zeta potentials [39]. While perhaps significant over a limited pH range, these explanations are in-applicable at pH's appreciably above the first pK_a or below the last pK_a of the protonated surface oxide. On the other hand, by analogy to eqs. 1 and 3, electron-addition-driven proton intercalation does provide a physical basis for Nernstian changes in photovoltaic metal oxide E_{cb} values over very wide pH ranges. Evidently at $-\log a_{H^+}$ values below -8 and above $+24$, however, electron addition does ultimately decouple from proton intercalation (see Figure 14.10).

By analogy to battery materials, and in light of recent nonaqueous EQCM findings (Section 2.3), one would also expect (in the absence of proton sources) Nernstian variations in conduction band edge energies with changes in alkali metal ion activity. To the best of our knowledge, experimental studies over an appropriately broad range of activities have not yet been reported. Redmond and Fitzmaurice, however, have described an interesting mixed electrolyte study. They find that lithium perchlorate addition to a dry acetonitrile solution of tetrabutylammonium perchlorate induces large – and apparently non-Nernstian – shifts in E_{cb} for nano-crystalline titanium dioxide [57]. We have observed similar behavior for TiO$_2$ upon addition of a dry proton source to an aprotic acetonitrile electrolyte solution [14]. While conventional cation adsorption effects presumably contribute in both instances, the predominant energy effect is likely associated with changes in the identity of the cation providing charge compensation.

Closely related are studies of the dependence of energies for electron addition on the cation identity in single electrolyte solutions. Redmond and Fitzmaurice have also demonstrated that titanium dioxide conduction band energies at suitably dried nonhydroxylic solvent interfaces are enormously dependent upon cation size, with larger cations yielding more negative apparent conduction band edge potentials

[57]. These important findings have also been corroborated via EQCM measurements. For the first three alkali metals, the observed energy variations apparently are associated with the increasing difficulty of intercalation, although a role for adsorption effects has also been suggested [14].

Finally, Fitzmaurice and coworkers have also described a remarkable correlation between E_{cb} (TiO_2) and the log of the autoprotolysis constant of the solvent [58]. One possible interpretation is that band edge energetics again are governed by proton intercalation and that solvent variations provide yet another method for achieving extensive H^+ activity variations.

14.4.2 Reactivity Implications

Control of photovoltaic/photocatalytic metal oxide semiconductor interface energetics by intercalation-coupled electron addition has a number of interesting reactivity implications. First, kinetic *decoupling* of electron and cation addition in dye-sensitization schemes can lead to unusual and unexpected reactivity patterns. For example, the kinetics of back electron transfer from nanocrystalline titanium dioxide to at least some surface-bound dye species are pH independent, despite substantial pH-induced changes in overall reaction driving force [59]. On the other hand, the kinetic decoupling effect is evidently not universal. Most notably, rates for electron transfer from titanium dioxide to diffusing or adsorbed methyl viologen are strongly pH dependent [60].

A second implication, in this case from Figure 14.10, is that electron addition to titanium dioxide (and presumably other metal oxides) has an enormous positive effect upon its ability in a thermodynamic sense to take up cations. If the break points in Figure 14.10 are interpreted as effective lattice pK_a's (*not* surface pK_a's), then the internal affinity of titanium dioxide for protons is apparently enhanced by more than 30 orders of magnitude by electron addition [40]. The enhancement implies that "reduced" TiO_2 is capable of abstracting protons even from exceedingly weak acids. If so, then the effect may well be usefully exploitable in new or existing photocatalytic/photochemical remediation applications.

14.5 Conclusions

The availability of processable and electrochemically addressable nanocrystalline forms of *n*-type photovoltaic metal oxides such as SnO_2, ZnO and TiO_2 has revealed that these materials can function as intercalation hosts under conditions of electron addition. Apart from cation intercalation capacity differences, the nanocrystalline photovoltaic materials appear to behave in much the same fashion as oxides traditionally used for battery and electrochromic applications (e.g. layered metal oxides). Indeed, both lithium intercalation batteries and electrochromic thin films can be fashioned from nanocrystalline TiO_2.

The coupling of electron addition to cation intercalation also has important fundamental energetic consequences. Most importantly, the phenomenon appears to account for the remarkable sensitivity of photovoltaic metal oxide conduction band edge energies to pH, solvent identity and electrolyte cation identity. While the available intercalation studies have focused on nanocrystalline materials, related defect-based or gel-layer-based chemistry may well account for the energetics of nominally single crystalline or macroscopically polycrystalline photovoltaic semiconductor/solution interfaces, especially under conditions of negative bias and so-called accumulation layer formation.

Acknowledgment

We thank the Office of Naval Research for support of our work on nanocrystalline metal oxides.

References

[1] O'Regan, B.; Grätzel, M. *Nature*, **1991**, *353*, 737–740.
[2] Grätzel, M. Coord. *Chem. Rev.*, **1991**, *111*, 167–174.
[3] Hagfeldt, A.; Grätzel, M. *Chem. Rev.*, **1995**, *95*, 49–68.
[4] Meyer, G. J.; Searson, P. C. *Interface*, **1993**, *2*, 23.
[5] Kamat, P. V. *Prog. Inorg. Chem.*, **1997**, *44*, 273–343.
[6] Hoffmann, M. R.; Martin, S. T.; Choi, W.; Bahnemann, D. W. *Chem. Rev.*, **1995**, *95*, 69–96.
[7] Kamat, P. V. *Chem. Rev.*, **1993**, *93*, 267–300.
[8] Granqvist, C. G. *Handbook of Inorganic Electrochromic Materials*, 1st ed., Elsevier, Amsterdam, 1995.
[9] Cogan, S. F.; Nguyen, N. M.; Perrotti, S. J.; Rauh, R. D. *J. Appl. Phys*, **1989**, *66*, 1333–1337.
[10] Chemseddine, A.; Morineau, R.; Livage, J. *Solid State Ionics*, **1983**, *9–10*, 357–361.
[11] Livage, J.; Jolivet, J. P.; Tronc, E. *J. Non-Cryst. Sol.*, **1990**, *121*, 35–39.
[12] Gerischer, H. *Electrochim. Acta.*, **1990**, *35*, 1677–1699.
[13] Hagfeldt, A.; Vlachopoulos, N.; Grätzel, M. *J. Electrochem. Soc.*, **1994**, *141*, L82–L84.
[14] Lyon, L. A.; Hupp, J. T. *J. Phys. Chem.*, **1995**, *99*, 15718–15720.
[15] Lemon, B. I.; Hupp, J. T. *J. Phys. Chem.*, **1996**, *100*, 14578–14580.
[16] Audiere, J. P.; Madi, A.; Grenet, J. C. *J. Mat. Sci.*, **1982**, *17*, 2973–2978.
[17] Rauh, R. D.; Cogan, S. F. *Solid State Ionics*, **1988**, *28–30*, 1707–1714.
[18] Burke, L. D.; O'Sullivan, E. J. M. *J. Electroanal. Chem.*, **1980**, *111*, 383–384.
[19] Szörényi, T.; Bali, K.; Hevesi, I. *J. Non-Cryst. Sol.*, **1980**, *35–36*, 1245–1248.
[20] Livage, J. *Chem. Mater.*, **1991**, *3*, 578–593.
[21] Le, D. B.; Passerini, S.; Tipton, A. L.; Owens, B. B.; Smyrl, W. H. *J. Electrochem Soc.*, **1995**, *142*, L102–L103.
[22] Pereira-Ramos, J. P.; Baddour, R.; Bach, S.; Baffier, N. *Solid State Ionics*, **1992**, *53–56*, 701–709.
[23] Julien, C.; Nazri, G. A.; Guesdon, J. P.; Gorenstein, A.; Khelfa, A.; Hussain, O. M. *Solid State Ionics*, **1994**, *73*, 319–326.

[24] Guerfi, A.; Paynter, R. W.; Dao, L. H. *J. Electrochem. Soc.*, **1995**, *142*, 3457–3464.

[25] Julien, C.; Nazri, G. A. *Solid State Ionics*, **1994**, *68*, 111–116.

[26] Davazoglou, D.; Donnadieu, A. *Thin Solid Films*, **1988**, *164*, 369–374.

[27] Akram, H.; Kitao, M.; Yamada, S. *J. Appl. Phys.*, **1989**, *66*, 4364–4367.

[28] Bedja, I.; Hotchandani, S.; Carpentier, R.; Vinodgopal, K.; Kamat, P. V. *Thin Solid Films*, **1994**, *247*, 195–200.

[29] Córdoba de Torresi, S. I.; Gorenstein, A.; Torresi, R. M.; Vázquez, M. V. *J. Electroanal. Chem.*, **1991**, *318*, 131–144.

[30] Ashrit, P. V.; Bader, G.; Girouard, F. E.; Truong, V. *J. Appl. Phys.*, **1989**, *65*, 1356–1357.

[31] Hotchandani, S.; Bedja, I.; Fessenden, R. W.; Kamat, P. V. *Langmuir* **1994**, *10*, 17–22.

[32] Bohnke, O.; Vuillemin, B.; Gabrielli, C.; Keddam, M.; Perrot, H. *Electrochim. Acta.*, **1995**, *40*, 2765–2773.

[33] Delichere, P.; Falaras, P.; Froment, M.; Hugot-Le Goff, A.; Agius, B. *Thin Solid Films*, **1988**, *161*, 35–46.

[34] Bejda, I.; Hotchandani, S.; Kamat, P. V. *J. Phys. Chem.*, **1993**, *97*, 11064–11070.

[35] Cao, F.; Oskam, G.; Searson, P. C.; Stipkala, J. M.; Heimer, T. A. ; Farzad, F.; Meyer, G. J. *J. Phys. Chem.*, **1995**, *99*, 11974–11980.

[36] Lantz, J. M.; Baba, R.; Corn, R. M. *J. Phys. Chem.*, **1993**, *97*, 7392–7395.

[37] Rothenberger, G.; Fitzmaurice, D.; Gratzel, M. *J. Phys. Chem.*, **1992**, *96*, 5983–5986.

[38] Marguerettaz, X.; Fitzmaurice, D. *J. Am. Chem. Soc.*, **1994**, *116*, 5017–5018.

[39] Finklea, H. O. in *Semiconducting Electrodes* (Ed.: H. O. Finklea), Elsevier, New York, 1988, Chapter 2.

[40] Lyon, L. A. *Electrochemical and Photochemical Studies of Polymer Modified and Semiconductor Electrodes*, Ph.D. Thesis, Northwestern University, 1996.

[41] Buttry, D. A. in *Electroanalytical Chemistry* (Ed.: A. J. Bard), Marcel Dekker Inc., New York, USA, 1991, Chapter 1.

[42] Nagasu, M.; Koshida, N. *J. Appl. Phys.*, **1992**, *71*, 398–402.

[43] Bechinger, C.; Herminghaus, S.; Leiderer, P. *Thin Solid Films*, **1994**, *239*, 156–160.

[44] Borgarello, E.; Pelizzetti, E.; Mulac, W. A.; Meisel, D. *Chem. Soc., Faraday Trans. 1*, **1985**, *81*, 143–159.

[45] Kavan, L.; Grätzel, M.; Rathousky, J.; Zukal, A. *J. Electrochem. Soc.*, **1996**, *143*, 394–400.

[46] Kavan, L.; Kratochvilova, K.; Grätzel, M. *J. Electroanal. Chem.*, **1995**, *394*, 93–102.

[47] Huang, S. Y.; Kavan, L.; Exnar, I.; Grätzel, M. *J. Electrochem Soc.*, **1995**, *142*, L142–L144.

[48] Ohzuku, T.; Takehara, Z.; Yoshizawa, S. *Electrochim. Acta.*, **1979**, *24*, 219–222.

[49] Macklin, W. J.; Neat, R. J. *Solid State Ionics*, **1992**, *53–56*, 694–700.

[50] Howe, R. F.; Grätzel, M. *J. Phys. Chem.*, **1985**, *89*, 4495–4499.

[51] Lemon, B. I.; Hupp, J. T. *J. Phys. Chem. B*, **1997**, *101*, 2246–2429.

[52] Weller, H.; Eychmüller, A. *Advances in Photochemistry*, **1995**, *20*, 165–217.

[53] Hoyer, P.; Weller, H. *J. Phys. Chem.*, **1995**, *99*, 14096–14100.

[54] Redmond, G.; O'Keefe, A.; Burgess, C.; MacHale, C.; Fitzmaurice, D. *J. Phys. Chem.*, **1993**, *97*, 11081–11086.

[55] Gottesfeld, S.; McIntyre, J. D. E. *J. Electrochem. Soc.*, **1979**, *126*, 742–750.

[56] Pickup, P. G.; Birss, V. I. *J. Electroanal. Chem.*, **1988**, *240*, 185–199.

[57] Redmond, G.; Fitzmaurice, D. *J. Phys. Chem.*, **1993**, *97*, 1426–1430.

[58] Enright, B.; Redmond, G.; Fitzmaurice, D. *J. Phys. Chem.*, **1994**, *98*, 6195–6200.

[59] Yan, S. G.; Hupp, J. T. *J. Phys. Chem.*, **1996**, *100*, 6867–6870.

[60] Moser, J.; Grätzel, M. *J. Am. Chem. Soc.*, **1983**, *105*, 6547–6555.

Chapter 15

Nanoparticle-Mediated Monoelectron Conductivity

S. Carrara

15.1 Introduction

From the origin of networks [1] up to modern digital electronics [2], two centuries have been spent to improve code systems and technological devices useful for data managing. Similar things happened for telecommunications. Starting from the early experiments on electricity and magnetism up to the discovery of radio-communications [3], a lot of effort has been spent, in each epoch to overcome the existing technological limitations. In particular, in the second half of this century, understanding the physics of semiconductors [4, 5] has dramatically increased the system's capabilities both for digital electronics and for applications in communication technology [6]. In this period, the industry of semiconductors had an incredible growth [7], which was paralled by theoretical studies. Let us mention briefly some of the most important contributions:

- The diode research program, launched in 1942 by Henry C. Torrey and Marvin Fox in the Radiation Laboratory of the Massachusetts Institute of Technology, was the first important effort toward a systematic research on semiconductor physics and technology [8].
- John Bardeen and Walter Brattain discovered transistor action in point contact electrodes with germanium [8]. and Bardeen, Brattain and Shockley invented the semiconductor based transistor [6]. The effort towards a satisfactory characterisation and understanding of the underlying physics went on for a long time, and led to the discovery of photoconductive effects, presented at a conference held in 1956 [9]. Immediately afterwards, in 1959 Jack Kilby, at Texas Instruments, and Robert Noyce, at Fairchild Semiconductor Corporation, developed the first semiconductor based integrated circuit [10].
- The sixties witnessed a continuous growth of this field. In 1961, integrated circuits went into commercial production. In 1963, Gunn perfected solid state microwave oscillators. Between 1962 and 1966, high-speed digital communications were introduced.
- In 1964, Moore formulated his law: the number of transistors in a chip would double each year. As a matter of fact, at present, Moore's law is no longer true:

already in 1975, the rate of growth per year reduced to 1.5. A similar rate of size reduction has been made for VLSI (very large scale of integration).

Other significant and entirely new problems are emerging in semiconductor physics. New technologies – based on UV, X-ray or electron-beam methodologies – are needed in to overcome the present limits of lithography. Small transistors are plagued by insuficcient heat dissipation and fluctuation of the electric properties. All these factors seems to limit the size reduction achievable with current semiconductor technology to 0.2 micron [10].

Applications of quantum phenomena to communication processes [11] and information theory [12, 13] is becoming a recognized practice. Quantum devices will play a significant role in the future developments of electronics. Any application related to the quantum phenomena requires that the device's sizes be comparable with the electron wavelength, which, in ordinary voltage conditions, may be estimated to be in the order of a few Ångstroms. This restricts the allowable size of each single device to a few nanometer which cannot be acieved with current technology. Furthermore, even if such size reduction were achievable, it would take, according to Moore's law, several centuries. In other words, the present semiconductor technology seems unsuited to the construction of truly "quantum devices", which are very likely to become the next step in the development of electronics. An important progress may come from the fact that conducting or semiconducting nanoparticles do indeed have the correct sizes for quantum effects, and are already available now!

Among other advantages, the use of quantum mechanical devices would also allow a direct control – electron by electron – of the current flow [14, 15]. This could help overcoming most of the problems arising from thermal dissipation in very small volumes, or from the smallness of the depletion zones. Adoption of a self-consistent mathematical model will allow a simple characterisation of the basic phenomena of monoelectron conductivity.

15.2 Historical Review

Monoelectron conductivity means – by definition – controlling each single charge, i.e. measuring the charge quantization of each individual electron, as in the famous Millikan experiment. During this last decade, much work has been done on single charge phenomena, in order to obtain such control and to understand the related physics.

15.2.1 Single Charge Phenomena

To introduce single charge phenomena from a general point of view, we refer to a box with sizes small enough to allow observation of quantum phenomena involving elementary charged particles: the so called "quantum dot". To get an intuitive picture of a dot, one may consider e.g. a nanometric sized metallic grain separated from the neighbourhood by a tunnelling barrier. In a system of this kind, it is possible to control the charge inside the grain by means of an applied voltage between the dot and the neighbourhood. Essentially, this means that one can inject the electrons inside the dot, one at a time, by controlling the bias voltage of the tunnelling junction. Of course, the process is influenced by the thermal conditions of the dot: in particular, if the thermal energy is much higher than the electrostatic one, the electrons will be expelled. Under such circumstances, no single charge controlling is possible.

A necessary requirement for the trapping of individual charges inside the dot to be successful is, therefore, that the thermal energy be smaller than the electrostatic one. This may be achieved indifferently either by decreasing the temperature or by increasing the capacitance of the dot. Once the single electron has been trapped, some different phenomena are observable in the system. The first one, known as the *Coulomb Gap*, appears as a current suppression near the zero bias voltage. The current characteristics are suppressed at the beginning of the axis, because the junction charge is reduced by a quantity equal to electron charge when an electron is tunnelling. At the same time, the junction voltage is reduced of a quantity (the *Coulomb Gap*) equal to the ratio between the electron charge and the junction capacitance. Therefore, the current is suppressed because the junction needs of an extra voltage to carry the same current.

The second observable effect is due to the electrons already present in the dot: the increase in electrostatic energy results in a repulsion of the next incoming electron. This gives rise to a potential barrier for the incoming electron, as a result of which the tunnelling current is partially blocked. This effect is known as the *Coulomb Blockade* of the tunnelling current. A further phenomenon arises from the fact that each new incoming electron is excluded from the dot by the presence of the previous one. In fact, the new electron must overcome the potential barrier due to the previous one before entering the dot. It therefore needs more energy, which must be supplied by the bias voltage, in order for the electron to tunnel into the dot. But the subsequent one is excluded too, and so on. This exclusion of the next incoming electron – now called *Coulomb exclusion* – results in "staircase behaviour" of the relationship between the electron number and the bias voltage, due to the potential barrier mentioned above. Of course, the current flow from the injecting system to the dot is modulated by these facts and it results in a *monoelectron conductivity*. In fact, suppressing the current after the new incoming electron results in a set of stairs appearing also in the current–voltage characteristics. This phenomenon is known as the *Coulomb Staircase*. Oscillations could also appear in current–voltage characteristics near the voltage steps due to the entering electrons. In such cases, *Negative Differential Resistance* (NDR) regions in the current curves are observable. Basically, these oscillations too are monoelectron phenomena, whose explanation is related to some resonant tunnelling effects mediated by the energy levels of the dot due by the single charge confinement.

As a further single charge effect we shall finally mention the linear relationship between the current and the frequency of the voltage signal. If a sinusoidal signal is superimposed to the bias voltage, the amplitude of the measured current through the dot is proportional to the signal frequency. The ratio between current and frequency is exactly equal to the electron charge. In the last ten years, a lot of efforts have been devoted to organising experimental setups in order to measure such effects, and to understand the theory of the physical phenomena related to the *monoelectron conductivity*.

15.2.2 The Theory

After the early experiments on very small superconductive particles and the introduction of a simple model concerning their observed behaviour [16, 17], a fundamental progress in the theory of single charge phenomena was achieved by D. V. Averim and K. K. Likharev, in their work published in 1986 [18]. Extending some earlier results on quantum oscillations in Josephson junctions [19–21], the Authors developed the theory of *Coulomb Blockade* in a very small tunnelling junction, showing that it was essentially due to the electron–electron interactions. Using the Hamiltonian representation of the system within the framework of second quantization, with the tunnelling effects expressed as a perturbation described by suitable creation and annihilation operators, and performing the analysis of the Von Neuman equation for the density matrix, they succeeded in establishing the fundamental features of single charge phenomena. In particular they proved that, if the thermal excitation of the electrons was smaller than the electrostatic energy associated with charge separation in a junction, the *Coulomb Blockade* effect could occur even in the absence of Josephson tunnelling. Furthermore, they showed that the next incoming electron could pass the junction only after a further increase of the applied voltage, equal to the ratio between the electron charge and the junction capacitance. Finally, they concluded their work with an investigation of the relationship between the single electron oscillations through the junction and the light emission due to inelastic tunnelling. According to this analysis, the condition of vanishing Josephson tunnelling allows the possibility of having normal electrodes forming the junction. In other words, under the stated circumstance, no superconducting electrodes are necessary; therefore, in principle, the observation of *Coulomb Blockade* is possible at room temperature too. This fundamental work switched on the interest in the field of single charge phenomena, as witnessed by the large amount of contributions published since then. In particular, it has been suggested to consider a two junctions system with an electrical island in between, in order to observe single charge conductivity. Such a system allows trapping the electrons inside the island, because of the tunnelling blockade occurring in the two junctions. A steplike current–voltage behaviour was found (i.e. the *Coulomb Staircase*), and the steps were related to the *Coulomb Exclusion* for the next electron entering the island [22]. Furthermore, the phenomenon of *Coulomb Gap* was forecasted, and the suggestion of using the modern *Scanning Tunnelling Microscope* for observing monoelectron conductivity in very small particles was proposed. This was the beginning of studies on monoelectron phenomena in terms of quantum dot concept [23]. Using ballistic theory of charge transport, it was possible to identify the current behaviour of a dot as a peak series with distances equal to those occurring in the *Coulomb Staircase*. Moreover, the application of single charge theory to dot-based systems was useful in the study of semiconductor particles as well as metallic ones [24].

The theoretical studies of these particles or grains indicated that their current behaviour could also include domains with *Negative Differential Resistance* along the current–voltage curves [25]. This fact was related to the presence of resonant phenomena and, in particular, of *resonant tunnelling* through the quantized energy levels due to the change confinement in quantum dots [26–28]. Following these ideas, some studies were devoted to the characterisation of the single electron trap on the resonant energy levels [29], while others were oriented towards the construction of a unified theoretical frame embodying both phenomena [30–32]. This gave rise to interesting models, exhibiting *Staircase* and differential negative resistance regions on the same current–voltage curves [33, 34]. In the meantime, within the last decade, several researchers have been working at the verification of these theories, and at the creation of real devices for the experimental observation of monoelectron conductivity.

15.2.3 Experimental Results

The fundamental work of Averim and Likharev [18] prompted the experimental researches on single charge phenomena. Only one year later, the first measurements on *Coulomb Gap* and *Coulomb Staircase* appeared in the literature.

The technique adopted by the labs was to lower the temperature down to few Kelvin, in order to reach a situation in which the electrostatic energy due to charge separation was effectively higher than the thermal electronic excitation. Different samples were prepared, in order to measure monoelectron conductivity in the different experimental configurations proposed to realise a two junctions system.

The first result came from the group of Likharev: it presented periodic oscillation of the right frequency in the current behaviour of an indium granule monolayer [35]. The first clear observation of the *Coulomb Gap* was published two months later [36]. Oone month later, the first *Coulomb Staircase* measurement appeared in the literature [37]. In the former case the effect was obtained by placing very small junctions between two electrodes [36]. The *Staircase* was found in a particle system [37]. Further experiments were done in the direction suggested by the previous ones. Once again, the experimental configurations were obtained either by means of electrodes in a two or three junctions schemes, or by operating with particle based samples. Within the first group of experiments, the story went on with the observation of the *Coulomb Staircase* [38], the *Coulomb Gap* and the charge oscillations in normal and superconductive electrodes [39], up to the usage of a single charge phenomena for proving the existence of *Cooper pair* in superconductivity [40]. Curves showing *Negative Differential Resistance* werealso observed on particle based systems. This fact was immediately related to the quantized energy levels due to the zero-dimensional sizes of the particles, even if the observations did not rely on single charge phenomena [41]. In these experiments, the use of the *Scanning Tunnelling Microscopes* (STM) for the characterisation of single small particle was first introduced [42]. Subsequent to the observation of the *Coulomb Gap* in a single junction [43], *Coulomb Gaps* and *Coulomb Staircases* were measured on a single metal granule [44]. The presence of nonlinear effects in the current was related to the particle sizes [45]. Finally, some peculiar phenomena due to the quantum levels in the dots were identified [46, 47], even if no resonance effect was directly observed.

A crucial point for a commercial device is to operate at room temperature. In this respect, as already mentioned above, the fundamental work of Averim and Likharev indicated the possibility that single charge phenomena be indeed observable at room temperature. In the last few years, some efforts were spent in this direction. The first evidence of monoelectron charging phenomenon at room temperature was observed in 1991 [48]. It consisted of a very weak *Coulomb Staircase* on single liquid crystal molecules, revealed by the *Scanning Tunnelling Microscope*. In particular, the derivative of the current–voltage characteristics was shown to present periodic oscillations; the analysis indicated how such oscillations could be related to the single electron ionisation of the molecules. One year later, the occurrence of monoelectron phenomena in a nanoparticle of Au, exhibiting both *Coulomb Gap* and *Coulomb Staircase* was reported [49]. The relationship between these effects and a) the symmetry of the two system junctions, and b) the STM tip positioning were also established [50].

The subsequent experimental studies focussed on the choice of the system, trying several different materials to build the junctions. Among others, ZrO_2, polyvinylpyrrolidone [50], dithiol films [51] were tested as insulators; interesting results were also obtained by replacing metallic nanoparticles with semiconductor ones. Indeed, both *Coulomb Staircase* and *Negative Differential Resistance* (NDR) were recently observed in semiconductor nanoparticles [52]. In particular, the occurrence of NDR behaviour in the current–voltage curves was related to the tip role in STM experiment. It was shown that, by varying the allocated current – i.e. the tip–sample distance – NDR becomes manifest in the step like behaviour of the curves [52]. In this work, the presence of NDR was related to the *Coulomb Blockade* phenomena and not to the resonant tunnelling, because of two main considerations: first, the frequency of the observed NDR oscillations was found to depend on the particle sizes and on the tip–particle distance. This indicated that the NDR oscillations were related to the capacitance geometry of the two-junction system. Second, on the basis of a model based on ballistic theory on nanocontacts [33], as well as on simulations on the two-barrier systems [34], the Authors succeeded in interpreting these oscillations as a *Coulomb Blockade* effect. Subsequent experiments indicated one further reason why the NDR should indeed be related to the quantum size of the particles, rather than to their semiconductor intrinsic nature. In fact, essentially identical phenomena were observed both in CdS and in PbS nanoparticles [53], with oscillations and stairs in current bearing no resemblance of the totally different semiconductor behaviour of the corresponding bulk materials. Finally, quite recently, the research on monoelectron phenomena has recorded another important result, namely the production of a room-temperature single-particle junction system not requiring the use of STM microscope. Room temperature *Coulomb Staircase* was observed by synthesising a nanoparticle of CdS directly into a very sharp tungsten tip, and employing a piezo-mover in order to switch on a tunnelling current between the tip and a flat electrode [54]. In this way, the importance of single-dot junctions working at room temperature as possible tools for technical applications was finally established.

15.2.4 Technological Applications

In spite of the efforts spent in this direction, some specifictechnological problems related to the construction of electronic devices involving monoelectron junctions are still open. For example, the behaviour of nanowires is not well understood, while the studies on nanocontacts are beginning just now. Nevertheless, some possible applications have already appeared in the literature in the last few years. Already in the original work of Averim and Likharev, the linear relationship between current and frequency in the Coulomb oscillations was recognised as the basic phenomenon for the construction of a standard current device and the single charge trapping as a conceptually easy digital memory [18]. In the subsequent years, other applications were proposed, and analysed in detail. For example, a single electron tunnelling based device was proposed as a sensitive high-frequency electromagnetic wave detector [14]. In particular, a current from the infrared photon absorption could be obtained by applying to a two-junction system a bias voltage equal to the Coulomb Blockade potential. The sensitivity of the resulting device would turn out to be comparable with the usual far-infrared decoders, thus allowing infrared images to be recorded with a videocamera [15]. In a similar way, following the original idea that monoelectron junctions could be used as logic memories, various digital applications have been examined. Among these, we mention inverting logical circuits, EX-OR devices [14], analogue to digital converters, and voltage–frequency decoders. Highly sensitive electrometers may be obtained, due the fact that single electrons can be trapped into the dot [55]. Quantum dot transistors [56], as well as their operability at room temperature were also considered [57].

All these contributions helped clarifying the possible use of quantum dots in nanoelectronics and quantum computing. This, in turn, rose the question whether the modelling of such devices was consistent with the laws of quantum mechanics [58]. The use of the devices in the construction of a *simulator* for the quantum laws was also considered [59]. Indeed, several industries are presently carrying on research plans on quantum dots or, more generally, on quantum devices [60]. So far, the economic benefits of these projects depend mainly on their usage in electronics. Thus, for example, the reduced size of the junctions themselves ensures a very low heat dissipation, which, in turn, overcomes most problems usually met in ordinary semiconductor technology. In addition to this, various already existing electronic devices could be easily converted to the new technology: the hope is that, in this way, the present schematic configurations may be radically simplified by the use of monoelectron systems.

15.3 Monoelectron Conductivity

After an overview of the history of monoelectron conductivity, let us now examine in greater detail what a quantum dot really is, and how monoelectron conductivity may flow through it. This section will be devoted to the definition of the quantum dot, to a study of its electrostatic behaviour and, finally, to a characterisation of its electrical properties, using the current–voltage curves as an "image" of the system itself.

15.3.1 Semiclassical Models

Starting with the intuitive picture outlined in Section 15.2, we define a quantum dot as *a very small space region in which it is possible to trap an elementary particle, making its quantum nature observable.* A necessary condition for this to happen is that the space region be isolated from the rest of the world, and enabled only to receive particles from the outside. Thus, in pictorial terms, a quantum dot may be thought of as a kind of "electrical island": a gate is needed in order to allow the particles into the island, but it is also necessary to prevent their rejection through the same gate. A tunnelling junction between the island and the rest of the system does the job. A charged particle may be injected inside, by applying an electrical field to the junction. Moreover, the thermal excitation of the particle within the island must be smaller than its electrostatic energy, to prevent the particle rejection. Of course, an "electrical heart" must be connected with the island, in order to apply the electrical field to the gate junction. Unfortunately, this means that, once a charged particle has passed through the tunnelling junction, it will flow into the electrical heart, and disappear: no trapping of charged particles inside the dot would therefore occur. To avoid this effect (i.e., in electrical terms, to avoid a direct ohmic contact with the "heart" electrode), a second junction could be added to the island. This should be large enough to prevent tunnelling current through; more precisely, it should have a different tunnelling probability as compared with the first one. Indeed, different tunnelling probabilities ensure trapping of charged particle in between the junctions, even in the presence of a small tunnelling current through the second one. A configuration of this kind is essentially a two electrode system connecting a quantum dot with a battery through tunnelling junctions (see top part of Figure 15.1).

The situation may be modelled by an electrical scheme in which a series of two tunnelling capacitors are connected to a battery: if the capacitances are different, electrons may be trapped in between (see bottom part of Figure 15.1). From an electric viewpoint, the situation is precisely the same as before: both pictures are therefore suited to describe a system able to trap electrical charged particles inside a small space region. In particular, in order to make the *quantum* behaviour of the trapped particles observable, we must relate it to a suitable set of macroscopically measurable quantities. Since we are dealing with charged particles, a natural idea is

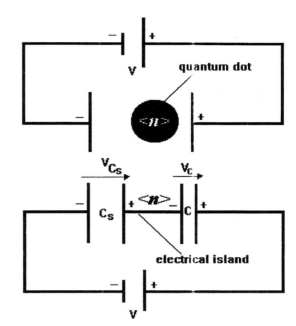

Figure 15.1. Top: schematic representation of a dot in electrical contact with two electrodes, and containing an average number ⟨*n*⟩ of electrons trapped inside. Bottom: equivalent model, with the dot represented schematically as an electrical island between two capacitors.

to analyse the electric behaviour of the system, observing e.g. the total current flowing through it. To achieve this situation, tunnelling events must occur also through the second capacitor, so that the two junctions system has to be modelled as a two barriers system, as shown in Figure 15.2. A two barriers system is characterised by corresponding barrier heights and widths. The current flowing through it is due to tunnelling events in the first barrier, coupled with tunnelling events in the second one. In the meantime, the electrons trapped inside the dot are in stationary states depending on the dot's geometry: their energy is therefore quantized according to the latter.

As a consequence of Pauli's principle, each new electron entering the dot fills a new energy level, just above the higher occupied one. In particular, the mean number of electrons, (or, more formally, the average occupation number ⟨*n*⟩ is related to the bias voltage of the system. Referring to the Fermi–Dirac statistics [61], the average occupation number is given by

$$\langle n \rangle = \frac{\sum_{n=-\infty}^{+\infty} n \dfrac{1}{1 + e^{(E_n(V))/kT}}}{\sum_{n=-\infty}^{+\infty} \dfrac{1}{1 + e^{(E_n(V))/kT}}} \tag{15.1}$$

If we assume that the quantum dot is operating at room temperature, Eq. (15.1) may be approximated as [39]

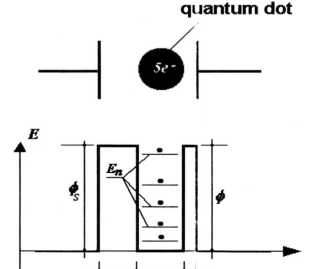

Figure 15.2. Energy diagram of a two-barrier system, with a quantum dot in between. The electrons trapped inside are in stationary states, with the energy levels depending on the size of the dot.

$$\langle n \rangle = \frac{\sum_{n=-\infty}^{+\infty} n e^{-(E_n(V))/kT}}{\sum_{n=-\infty}^{+\infty} e^{-(E_n(V))/kT}} \tag{15.2}$$

consistently with Boltzman's statistics applied to the semiclassical model. As explicitly pointed out by Eqs. (15.1 and 15.2), the energy of n electrons inside the dot is a function of the applied voltage. We shall now examine the nature of this relationship.

15.3.2 Electrostatic Considerations

In the previous section, it was indicated how a double junctions system could trap a number $\langle n \rangle$ of electrons in between. According to Figure 15.1, if the electrons are trapped in between two capacitors, the bias voltage is redistributed between the capacitors. In particular, the capacitor voltages become [34]

$$V_{C_s} = \frac{V}{C_s}\frac{CC_s}{C + C_s} - \langle n \rangle \frac{e}{2C_s} \cong \frac{V}{2} - \langle n \rangle \frac{e}{2C_s}$$

$$V_C = \frac{V}{C}\frac{CC_s}{C + C_s} + \langle n \rangle \frac{e}{2C_s} \cong \frac{V}{2} + \langle n \rangle \frac{e}{2C_s} \tag{15.3}$$

provided that the two capacitances are close enough to make their difference negligible. The two voltages are therefore *different* whenever there are electrons trapped in between. However, this may happen only if the electrostatic energy of the incoming electron is large enough to prevent its rejection through the injecting gate. This energy needs to be higher than the one due to thermal excitation of the electron. In terms of the capacitance C_s, this condition is expressed by the relation (see e.g. [18]).

$$\frac{e^2}{2C_s} > kT \tag{15.4}$$

Equation (15.4) is an important constraint for trapping of electrons inside the electrical island to occur. Decreasing the right-hand side by decreasing the temperature T is a way to satisfy the condition. Alternatively, one may equally well increase the left-hand side by decreasing the capacitance C_s. Both possibilities have been considered in Section 15.2.3, in the description of possible experimental setups for the observation of single-charge phenomena: either devices working at low temperature, or devices working at room temperature, but with a geometry resulting in small enough capacitances. In any case, if the condition in Eq. (15.4) is satisfied, the total energy of the system is related to the bias voltage and to the electron number inside the island. Equating this energy to the one supplied by the power source, we get the relation [39]:

$$E_n(V) = \frac{(C_s V - ne)^2}{2C_s} - \frac{C_s V^2}{2} \tag{15.5}$$

By inserting Eq. (15.5) into Eq. (15.2), we get explicit formula relating the average occupation $\langle n \rangle$ to the bias voltage V, namely

$$\langle n \rangle = \frac{\sum_{\forall n} n e^{-(V - V_n)^2 / \sigma^2}}{\sum_{\forall n} e^{-(V - V_n)^2 / \sigma^2}} \tag{15.6}$$

with

$$V_n = n \frac{e}{C_s}, \qquad \sigma^2 = \frac{2kT}{C_S} \tag{15.7}$$

On the basis of Eq. (15.6), we see that the behaviour of $\langle n \rangle$ is completely determined by the quantities (15.7). The first of these defines the *Voltage Step* required in order for $\langle n \rangle$ to increase by a unity, while the second one characterises the stair's shape. Both quantities depend on the capacitance C_s. The latter plays therefore a crucial role in the construction of devices suited to give rise to well-defined stairs, and thus to make monoelectron conductivity manifest. At the same time, the value of σ^2 depends also on T, thus confirming that the phenomena will be more evident at lower temperatures. To sum up, both the representation (Eq. 15.7) for σ^2 and the (15.4) point out that a good observation of monoelectron conductivity relies on a proper comparison between the temperature and the geometry of the junctions. Figure 15.3 shows how the average number of electrons inside the island looks like for two different values of T. In one case, the phenomenon is more evident, because

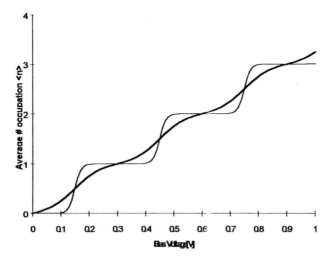

Figure 15.3. Average occupation number $\langle n \rangle$ inside the dot upon the applied bias voltage V. The number $\langle n \rangle$ is showed at different values of the quantity σ^2 but at the same value of the voltage step V_n.

the well defined stairs. Of course, the average number of occupation itself modifies the stair shape and distance by changing the capacitance value. The figure also shows how a controlled storage of electrons inside the dot by means of a proper choice of the bias voltage is possible. For example, referring to Figure 15.3, it is possible to conclude that at 0.9 Volts there are precisely three electrons stored inside.

The trapped electrons need to be in stationary states, because of their wave nature, and of the fact that they are placed inside a closed space region. As mentioned in section 15.2, it was theoretically predicted that monoelectron systems could present both staircase and NDR features in current voltage characteristics. In particular, the second property is directly related to the stationary state concept. According to the latter, in fact, the electrons trapped inside the dot are described by stationary wave functions, whose energy levels – shown in Figure 15.2 – obey the equation [26]

$$E_n = E_0 + \frac{n^2 e^2}{2R} \tag{15.8}$$

NDR in monoelectron conductivity may therefore result from possible resonances between these levels.

15.3.3 Current in a Monoelectron System

In order to account for the conductivity of a monoelectronic system, we assume that the current is entirely due tunnelling of electrons through the junctions barriers. The probability to cross a potential barrier is related to the barrier's height by the equation

$$P \propto e^{-\sqrt{(2m/\hbar^2)\phi d}} \tag{15.9}$$

On the other hand, the height d depends on the voltage V of the junction. In particular, it is decreased by the potential energy due to the bias voltage. Therefore, the probabilities to cross both barriers of a monoelectron system are related to the bias voltages described in Eq. (15.3), and may be written as

$$P_j(V) \propto e^{-\sqrt{(2m/\hbar^2)(\phi_j - eV_j)d}} \tag{15.10}$$

the index $j = 1, 2$ labelling the junctions. Regarding now the current flowing through the whole system as being proportional to the product of the two probabilities, it easy to write [34]

$$I \propto V e^{-\sqrt{(2m/\hbar^2)[\phi_s - e((V/2) - \langle n \rangle(e/2C_s))]d_s}} e^{-\sqrt{(2m/\hbar^2)[\phi - e((V/2) + \langle n \rangle(e/2C_s))]d}} \tag{15.11}$$

The presence of the average number $\langle n \rangle$ at the right hand side of Eq. (15.11) shows that the electrons present inside the dot do indeed influence the conductivity of the system. On the other hand, $\langle n \rangle$ is related to the applied voltage V by Eq. (15.6). In this respect, the functional relationship between current and voltage resulting from Eq. (15.11) is a bit complicated. However, the shape of the current–voltage curves is very easy to understand. For example, when the average occupation number $\langle n \rangle$ vanishes, Eq. (15.11) describes a simple exponential characteristic. Instances of the current–voltage curves for nonvanishing $\langle n \rangle$ are shown in Figure 15.4. Once again, a critical role is played by the capacitance C_s. In fact, from Eq. (15.4) we know that a necessary condition for the average number $\langle n \rangle$ to be different from zero is the validity of the relation

$$C_s < \frac{e^2}{2kT} \tag{15.12}$$

On the other hand, on the basis of equation (11), for nonvanishing $\langle n \rangle$ a current suppression upon an increase in voltage occurs when d_S is much greater than d. This is the phenomenon of *Coulomb Blockade,* as described in Section 15.1, resulting in current–voltage curves shaped in a series of stairs. For example, for barrier heights of a few electron volts, barrier widths of some Ångstroms, and a capacitance $C_S = 2.5 \ 10^{-18}$ F, the curve exhibits typical stairs, as in the thin line of Figure 15.4. If the capacitance C_S is small enough, regions with NDR may also occur in the current–voltage characteristic. This happens whenever a positive variation of the voltage V results in a negative variation of the current in equation (11). With some approximations, under the assumption of almost equal barriers, it may be seen that a choice of the capacitance satisfying the condition [34]

$$C_s < \sqrt{\frac{2md^2}{\phi\hbar^2}e^2\langle n \rangle} \tag{15.13}$$

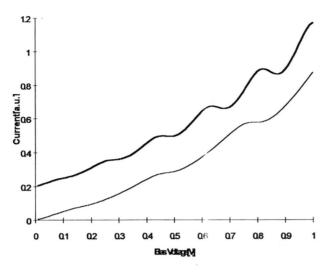

Figure 15.4. Current–voltage characteristics obtained from Eq. (15.11) for different values of the capacitance C_s, but for the same values of the other parameters. The curves are shifted 0.2 Volts for a better understanding. The thin curve exhibits *Coulomb Staircase*, while the thick one presents regions in which NDR occurs.

allows observing NDR in the electrical characteristics. For example, curves similar to the one showed in the thick line of Figure 15.4 are possible for values of $C_s <$ 1.9 10^{-19} F. It is worth noticing that the NDR regions shown in Figure 15.4 are obtained without taking into account any kind of resonance phenomena in the quantum dot. This shows that some nonlinear effects in monoelectron conductivity may be simulated without accounting explicitly for the resonant tunnelling through the levels described in Eq. (15.8). The right hand side of the inequality (Eq. 15.13) depends explicitly on the average occupation number $\langle n \rangle$: this means that the requirement may be met by increasing the value of the latter. For example, the thick line shown in Figure 15.4 presents clear NDR in the fourth step but not in the previous ones. This indicates that the constraint (Eq. 15.13) is satisfied when four electrons are inside the dot, but is violated when their number is reduced to one two or three. Behaviours of this kind had already been pointed out in the past [34] and had been foreseen by other theoretical frames [33]. Furthermore, they were experimentally observed in some specific systems [41, 52].

15.4 Nanoparticle-Mediated Monoelectron Conductivity

We shall now discuss how one can possibly use a nanoparticle in order to obtain a quantum dot in the sense described at the beginning of Section 15.3. As pointed out in the Introduction, elementary particles need to be confined in a space region with sizes not larger than a few Ångstroms, if one is willing to make their quantum behaviour manifest. In this respect, nanoparticles technology is perfectly suited to the purpose. In particular, metallic or semiconductor nanoparticles could be utilised in the observation of the quantum nature of electrons. Of course, the behaviour of the materials at such small sizes is not exactly equal to that in bulk. On the other hand, so far, monoelectron conductivity has been observed only in these kinds of nanoparticles.

15.4.1 Nanoparticles as Traps

In the recent past, a lot of effort has been devoted to the creation and characterisation of nanometer sized structures, resorting to various kinds of methodologies and techniques. For example, nanostructures were grown in different media as solutions [62, 63], zeolites [64] or Langmuir–Blodget films [65]. Or they were built by diffusion-controlled aggregation in surfaces [66], or by evaporation of semiconductors in a gaseous environment [67]. The characterisations of the resulting nanoparticles was performed by using small angle X-ray diffraction [68], electron microscopy [67], optical absorption [69], atomic force [66, 70], scanning tunnelling microscopy [71], ion bombardment [72]. In the course of different experiments, these structures appeared to be shaped like tubules [66, 69], spheres [66, 70] or disks [52, 68]. The last two shapes are especially suited to form quantum dots. The problem, of course, is how to put contacts on the nanoparticles, in order to give rise to a two-junction system. The requirement that the dot be isolated from the rest of the word by means of two junctions is in fact implicit in the definition of the dot itself, as pointed out in Section 15.3. One junction is needed in order to inject the electrons, and another one in order to apply a bias voltage without discharging the dot.

A monoelectron system would therefore require a nanoparticle – or better, a cluster of nanoparticles – with an insulating medium trapping them in between two macroscopical electrodes. The resulting device is indeed a two barriers system: therefore, if the nanoparticles are small enough, monoelectron phenomena could be observable. A layer of nanoparticles may be regarded as a possible accomplishment of such a system, essentially equivalent to a layer of quantum dots. This idea was effectively ensued, and experimentally tested. *Coulomb Gap* and *Coulomb Staircase* were actually observed [37]; investigations on the nanoparticle sizes were also performed [73]. In all cases, however, the measured effects depended on the cooperative conductivity of *all* nanoparticles in between the two electrodes: in order to obtain a clear observation of monoelectron conductivity, it was therefore necessary to operate with systems working at very low temperatures. These kinds of systems are not particularly suited to analysing the conductivity of the single quantum dot. A better choice is to employ single nanoparticles, contacted with a very sharp electrical tip. This possibility was implemented making use of the tip of an STM microscope [44, 45]. More recently, an experimental setup has been proposed, involving a very sharp tip, but avoiding the STM microscope [54]. The resulting single nanoparticle systems have been shown to give rise to monoelectron conductivity at room temperature [48–50], thus offering a possible starting point for the creation of devices suited to quantum electronics applications. Furthermore, typical resonance phenomena, such as the occurrence of NDR regions in current–voltage curves, were observed in such single dot systems [52]. Thus, also the possibility of getting a better understanding of these phenomena is now available [47, 53].

15.4.2 Electrical Capacitance of a Nanoparticle

Employing a nanoparticle in order to set up a two capacitors system as in Figure 15.1 requires evaluating the electrical capacitance C_s involved the representation (Eq. 15.11) of the total current. The value of C_s depends on the particle shape: differently shaped particles will therefore present different capacitance values. Basically, a quantum dot is a zero-dimensional structure, which may be equally well simulated either by a spherical or by a disklike particle. So far, however, it is not completely clear which shapes come from which preparation procedures: some Authors indicate the spherical model [50, 51], while others the disk-shaped one [45, 52, 74]. In particular, in the spherical case, the particle's capacitance has the value [73]

$$C_s = 2\pi\varepsilon_r\varepsilon_0 R \tag{15.14}$$

R denoting the particle's radius and ε_r accounting for the insulating medium. Conversely, in the case of disk particles, electrically separated from a planar plate by means of an insulating layer and contacted by an STM tip (see Figure 15.5), the electrical capacitance of the nanoparticle-tip system is [34]

$$C_s = 2\pi\varepsilon_r\varepsilon_0 \frac{R^2}{\sqrt{R^2 - d^2} - d} \tag{15.15}$$

d and R denoting now the nanoparticle–tip distance and the disk radius. If $d \ll R$, both Eqs. (15.14) and (15.15) provide the same linear relationship between C_s an. Moreover, the capacitance values coming from both equations are very small in the case of nanoparticles. Therefore, it will be quite hard to guess the correct expression for C_s by means electrical measurements based on Eq. (15.7), even if some efforts were spent to explain the $C_s - R$ relationship [73]. On the other hand, monoelectron phenomena were observed in very small capacitance systems. In particular, *Coulomb Gap* was observed in a junction of 10^{-18} Farad [43], *Coulomb Staircase* in systems of 10^{-18}–10^{-19} Farad [45, 50, 54] and current curves presenting NDR with periodicity corresponding to a 10^{-19} Farad [52]. Both Eqs. (15.14) and (15.15) indicate that these are precisely the orders of magnitude of the capacitances arising from nanometer sized particles. Therefore, nanoparticles with capacitance values

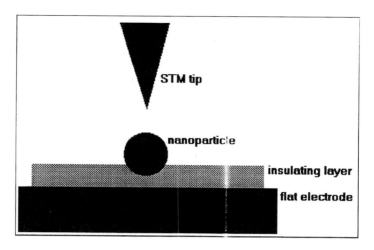

Figure 15.5. Experimental setup in which a nanoparticle is used as a quantum dot. The two-junction system consists of a flat electrode in the bottom side and a very sharp one in the top.

impossible to reach with micron sized usual capacitors became necessary in order to obtain monoelectron conductivity.

15.4.3 The Role of Nanoparticle Size

Equations (15.14, 15.15) indicate which particle sizes are required in order to observe *Coulomb Staircase* or NRD in current curves. More precisely, they relate the sizes to the required capacitance of the system. Furthermore, Eq. (15.7) enables to calculate the steps in current–voltage characteristics. Basically, these equations provide the basic tools for a correct use of nanoparticle technology in order to set up quantum devices with the right properties. Simulations are possible e.g. by applying (15.11) and (15.15) in the presence of barrier heights of some electron volts, with tip–sample distance of one nanometer [43, 45]. Figure 15.6 shows that no monoelectron phenomenon is observable at room temperature for particles greater than 100 nm in diameter. Moreover, *Coulomb Staircase* is realistically observable only for particle diameters smaller than 10 nm, as the simulation presented in Figure 15.7 clearly indicates. Finally, the appearance of NDR in current curves is possible only if the particle's diameters are under 4.5 nm, as shown in Figure 15.8. Of course, the previous values have only an approximate meaning, for two main reasons: first, the model presented here is an approximated semiclassical one; second, the exact values of the other parameters involved in Eqs. (15.11) and (15.15) are not known so far. However, it is known from the literature that monoelectron conductivity was not observed in particles whose sizes exceed 10 nm [35, 45, 73, 74], while measurements of *Coulomb Staircase* at room temperature were done on particles with sizes ranging from 2 up to 5 nm [50, 52–54]. Only in these small structures NDR phenomena were detected [52, 53].

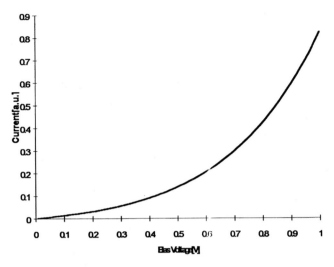

Figure 15.6. Current–voltage characteristic of a system with a particle of 200 nm in diameter showing that no monoelectron conductivity is observable.

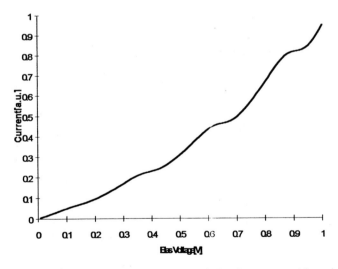

Figure 15.7. Current–voltage characteristic of a system with a particle of 20 nm in diameter, exhibiting a *Coulomb Staircase* effect, with a voltage step approximately equal to 0.22 volts.

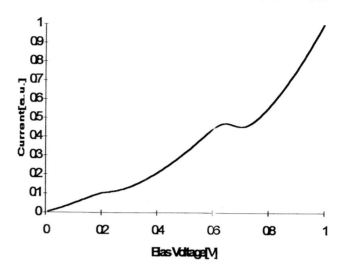

Figure 15.8. Current–voltage characteristics of a system with a particle of 4 nm in diameter, exhibiting NDR regions with a voltage distance approximately equal to 0.6 volts.

15.5 Conclusions

As a concluding comment, we wish to emphasise once again the importance of nanoparticles technology in the future applications in all fields of electronics. In particular, as shown in Section 4.2 of this Chapter, the use of nanoparticles allows setting up devices with capacitance values ranging from 10^{-18} down to 10^{-20} Farad: values unattainable with normal micron sized capacitors. Moreover, nanometer sized particles allow to obtain real configurations for quantum dots. In fact, mono-electron conductivity was observed in systems with nanoparticles. Nanoparticles will surely play a crucial role in solving the miniaturisation problems of micro-electronics and in the creation of new devices based on quantum mechanical effects. Not all problems in this direction are completely solved yet. For example, inter-connecting different quantum devices is a nontrivial job. Furthermore, the speed of signals decreases with the sizes of the wires. In addition, the electromigration could destroy the wire when its size is too much small [10] (even if the possibility of working with very tiny wires is still considered, and some studies are focussed on this point [75]). The consequences of Heisenberg's uncertainty principle must necessarily be taken into account in the application of quantum dot technology to the construction of digital devices [58]. Software tools are also necessary in order to simulate the behaviour of these devices in quantum electronics [76].

In any case, the future work on nanoparticles and related quantum dots is rich of exciting promises, not only in the construction of single-electron transistors [56], but also as a starting point for the creation of new devices [15] which may open a completely new way for computing [59].

References

[1] G. J. Holzmann, B. Pehrson, *The Early History of Data Networks*, IEEE Computer Society Press, Washington, **1995**.

[2] R. G. Middleton, *Understanding Digital Logic Circuits*, Howard W. Sams & Co., Indianapolis, **1982**.

[3] G. R. M. Garratt, *The Early History of radio*, The Institution of Electrical Engineers, IEE Hist. Techn. Series, vol. 20, London, **1994**.

[4] B. I. Bleany, B. Bleany, *Electricity and Magnetism*, Oxford University Press, Hong Kong, **1985**.

[5] N. W. Ashcroft, N. D. Mermin, *Solid State Physics*, W. B. Saunders Company, Orlando, **1976**.

[6] A. B. Carlson, *Communication Systems*, McGraw-Hill, Singapore, **1986**, 12.

[7] P. R. Morris, *A history of the world semiconductor industry*, The Institution of Electrical Engineers, IEE Hist. Techn. Series, vol. 12, London, **1990**.

[8] F. Seitz, *Research on Silicon and Germanium in World War II*, Physics Today, Jan. **1995**, 22.

[9] Burstein, Picus, Sclar in *Photoconductivity Conference*, John Wiley & Sons, New York, **1956**, 275.

[10] R. Turton, *The Quantum Dot – A Journey into the Future of Microelectronics*, W. H. Freeman Spektrum, Oxford, **1995**.

[11] D. M. Greenberg, A. Zeilinger (Eds.), *Fundamental Problems in Quantum Theory*, Annals of the New York Academy of Science, New York, **1995**, vol. 755, part XXII.

[12] S. Lloyd, *Science*, **1993**, *261*, 1569.

[13] I. Cirac, P. Zoller, *Phys. Rev. Lett.* **1995**, *74*, 4091.

[14] H. Grabert, M. H. Devoret (Eds), *Single Charge Tunnelling – Coulomb Blockade Phenomena in Nanostructures*, NATO ASI B, New York, **1992**, vol. 294.

[15] K. K. Likharev, *FED Journal* **1995**, *6*, 5.

[16] I. Giaever, H. R. Zeller, *Phys. Rev. Lett.* **1968**, *20*, 1504.

[17] H. R. Zeller, I. Giaever, *Phys. Rev.* 181 **1969**, *20*, 789.

[18] D. V. Averim & K. K. Likharev, *J. Low Temp. Phys.*, **1986**, *62*, 345–373.

[19] K. K. Likharev, A. B. Zorin, *J. Low Temp.* **1985**, *59*, 347–382.

[20] E. Ben-Jacob, *Phys. Lett.* **1985**, *108A*, 289–292.

[21] D. A. Averim, K. K. Likharev, *Sov. Phys. JEPT* **1986**, *63*, 427–432.

[22] K. Mullen, E. Ben-Jacob, R. C. Jaklevic, Z. Shuss, *Phys. Rev.*, **1988**, *B37*, 98–105.

[23] L. I. Glazmann, R. I. Shechter, *J. Phys. Condens. Matter* **1989**, *1*, 5811–5815.

[24] D. A. Averim, A. N. Korotkov, K. K. Likharev, *Phys. Rev. B*, **1991**, *44*, 6199–6211.

[25] O. V. Gritsenko, P. I. Lazarev: *On the volt–ampere characteristic of molecular monoelectronic elements*, in Molecular Electronics (F. T. Hong, Ed.), Plenum Press, New York, **1989**, 277–288.

[26] F. Guinea, N. García, *Phys. Rev. Lett.* **1990**, *65*, 281–284.

[27] C. W. Beenakker, H. vam Houten, A. A. M. Staring, *Phys. Rev. B*, **1991**, *44*, 1657–1662.

[28] M. Sumetskiî, *Phys. Rev. B*, **1993**, *48*, 4586–4591.

[29] A. Groshev, T. Ivanov, V. Valtchinov, *Phys. Rev. Lett.* **1991**, *66*, 1082–1085.

[30] C. W. Beenakker, *Phys. Rev. B*, **1991**, *44*, 1646–1656.

[31] A. D. Stone, R. A. Jalabert, Y. Alhassid: *Statistical theory of Coulomb Blockade and resonant tunnelling oscillations in quantum dots*, in *Proceedings of the 14th Taniguchi Symposium*, Springer-Verlag, Berlin, **1992**, 39–52.

[32] V. N. Prigodin, K. B. Efetov, S. Iida, *Phys. Rev. Lett.* **1993**, *71*, 1230–1233.

[33] Song He, S. Das Sarma, *Phys. Rev. B*, **1993**, *48*, 4629–4635.

[34] S. Carrara, V. Erokhin, P. Facci, C. Nicolini, *On the role of nanoparticle sizes in monoelectron conductivity*, in J. Fendler, I. Décáni (Eds.): *Nanoparticles in solids and solutions*, Dordrecht NL, NATO ASI Series 3. High Technology – vol. 18 **1996**, pag. 497–503.

[35] L. S. Kuz'min, K. K. Likharev, *JEPT Lett.* **1987**, *45*, 496.

[36] T. A. Fulton, G. J. Dolan, *Phys. Rev. Lett.* **1987**, *59 109*.

[37] J. B. Barner, S. T. Ruggiero, *Phys. Rev. Lett.* **1987**, *59*, 807.

[38] L. P. Kouwenhoven, N. C. van der Vaart, A. T. Johnson, W. Kool, et al.: *Z. Phys. B, Condensed Matter*, **1991**, *85*, 367.

[39] P. Lafarge, H. Pothier, E. R. Williams, D. Esteve et al: *Z. Phys. B, Condensed Matter*, **1991**, *85*, 327.

[40] P. Lafarge, P. Joyez, D. Esteve, C. Urbina, M. H. Devoret, *Nature*, **1993**, *365*, 422.

[41] M. A. Reed, J. N. Randall, R. J. Aggaewal, R. J. Matyi, et al. *Phys. Rev. Lett.* **1988**, *60*, 535.

[42] J. S. Weiner, H. F. Hess, R. B. Robinson, T. R. Hayes, et al. *Appl. Phys. Lett.* **1991**, *58*, 2402.

[43] P. J. M. van Bentum, H. van Kempen. L. E. C. van de Leemput, P. A. A. Teunissen, *Phys. Rev. Lett.* **1988**, *60*, 369.

[44] P. J. M. van Bentum, R. T. M. Smokers, H. van Kempen, *Phys. Rev. Lett.* **1988**, *60*, 2543.

[45] R. Wilkins, E. Ben-Jacob, R. C. Jaklevic, *Phys. Rev. Lett*, **1989**, *63*, 801

[46] M. F. Crommie, C. P. Lutz, D. M. Eigler, *Science* **1993**, *262*, 218.

[47] J. G. A. Dubois, J. W. Gerritsen, S. E. Shafranjuk, E. J. G. Boon, G. Schmid, H. van Kempen, *Europhys. Lett.* **1996**, *33*, 279.

[48] H. Nejoh, *Nature*, **1991**, *353*, 640.

[49] C. Shönenberger, H. van Houten, H. C. Donkersloot, *Europhys. Lett.* **1992**, *20*, 249.

[50] C. Shönenberger, H. van Houten, H. C. Donkersloot, A. M. T. van der Putten, L. G. J. Fokkink, *Physica Scripta*, **1992**, *T45*, 289.

[51] M. Dorogi, J. Gomez, R. Osifchin, R. P. Andres, R. Reifenberger, *Phys. Rev. B* **1995**, *52*, 9071.

[52] V. Erokhin, P. Facci, S. Carrara, C. Nicolini, *J. Phys. D: Appl. Phys.* **1995**, *28*, 2534.

[53] V. Erokhin, P. Facci, S. Carrara, C. Nicolini, *Monoelectron phenomena in nanometer scale particles formed in LB films*, Thin Solid Films **1996** in press.

[54] P. Facci, V. Erokhin, S. Carrara, C. Nicolini, *Proc. Natl. Acad. Sci.* **1996**, *93*, 10556.

[55] R. J. Haug, K. von Klitzing, *FED Journal* **1995**, *6*, 4.

[56] E. Leobandung, L. Guo, S. Y. Chou, *Appl. Phys. Lett.* **1995**, *67*, 2338.

[57] K. Matsumoto, M. Ishii, K. Segawa, Y. Oka, B. J. Vartanian, J. S. Harris, *Appl. Phys. Lett.* **1996**, *68*, 34.

[58] K. K. Likharev, A. N. Korotkov, *Science* **1996**, *273*, 763.

[59] S. Lloyd, *Science* **1996**, *273*, 1073.

[60] S. Kimura, A. Asai, S. Okayama, *FED Journal* **1995**, *6*, 20.

[61] T. L. Hill, *An introduction to Statistical Thermodynamics*, Addison-Wesley, **1960**, par. 22–1, 438–439 (Italian version).

[62] L. E. Brus, *J. Phys. Chem.* **1986**, *90*, 2555.

[63] Y. M. Tricot, J. H. Fendler, *J. Phys. Chem.* **1986**, *90*, 3369.

[64] Y. Wang, N. Herron, *J. Phys. Chem.* **1987**, *91*, 257.

[65] E. S. Smotkin, C. Lee, A. J. Bard, A. Champion, M. A. Fox, T. E. Mallouk, S. I. Webber, J. M. White, *Chem. Phys. Lett.* **1988**, *152*, 265.

[66] H. Röder, E. Hahn, H. Brune, J. P. Bucher, K. Kern, *Nature* **1993**, *366*, 141.

[67] R. Kamalakaran, A. K. Singh, O. N. Srivastava, *J. Phys. Condens. Matt.* **1995**, *7*, L529.

[68] V. Erokhin, L. Feigin, G. Ivakin, V. Klechkovskaya, Yu. Lvov, N. Stiopina, *Makromol. Chem. Macromol. Symp.* **1991**, *46*, 359.

[69] Y. Tian, C. Wu, N. Kotov, J. H. Fendler, *Advanced Materials* **1994**, *12*, 959.

[70] J. Yang, F. C. Meldrum, J. H. Fendler, *J. Phys. Chem.* **1995**, *99*, 5500.

[71] P. Facci, V. Erokhin, A. Tronin, C. Nicolini, *J. Phys. Chem.* **1994**, *98*, 13323.

[72] K. Asai, T. Yamaki, K. Ishigure, H. Shibata, *Thin Solid Films* **1996**, *277*, 169.

[73] S. T. Ruggiero, J. B. Barner, *Z. Phys. Cond. Matt.*, **1991**, *85*, 333.

[74] C. Shönenberger, H. van Houten, J. M. Kerkhof, H. C. Donkersloot, *Appl. Surf. Sci.* **1993**, *67*, 222.

[75] G. Timp, A. M. Chang, P. Mankiewich, R. Behringer, J. E. Cunningham, T. Y. Chang, R. E. Howard, *Phys. Rev. Lett.* **1987**, *59*, 732.

[76] D. V. Averin, K. K. Likharev, in H. Grabert, M. H. Devoret (Eds), *Single Charge Tunnelling – Coulomb Blockade Phenomena in Nanostructures*, NATO ASI B, New York, **1992**, vol. 294, 311.

two Ti^{4+} sites and are oriented parallel to the substrate surface. It noted also, that development of a charge transfer absorption in the visible region of the spectrum accompanies adsorption of salicylate and catechol. The corresponding absorption for benzoate or the phthalates is blue shifted and is not observed in the visible region of the spectrum due to their greater ionisation potential.

Salicylate was therefore identified as the molecule most suitable for assembling a TiO_2 nanocrystallite and a molecular electron acceptor. The salicylate modified viologen was thus prepared and adsorbed at the surface of a TiO_2 nanocrystallite. Adsorption was monitored *via* the resulting visible charge transfer absorption. As stated, salicylate is adsorbed normal to the crystallite surface at a single Ti^{4+} site [6], thus it may be concluded the viologen component is also normal to the crystallite surface as represented in Figure 16.1.

As may be seen from transient in Figure 16.3, assigned principally to the radical cation of the viologen, bandgap excitation of the heterosupermolecule TiO_2–SV results in electron transfer from the TiO_2 nanocrystallite to the viologen. That is, the associated heterosupramolecular function is light-induced vectorial electron transfer [3, 7], see Figure 16.4.

As the formal potential of the viologen in TiO_2–SV may be varied systematically, a rate constant for electron transfer has been determined as a function of the free energy change associated with the electron transfer step. A principal advantage of the above approach is that the distance from, and orientation of the molecular

Figure 16.3. Transient absorption at 600 nm for TiO_2–SV in a deaerated water–ethanol mixture (4:1 by vol.) at 25 °C following bandgap excitation at 355 nm (2 mJ per pulse).

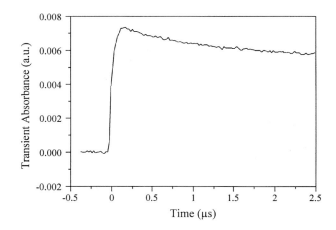

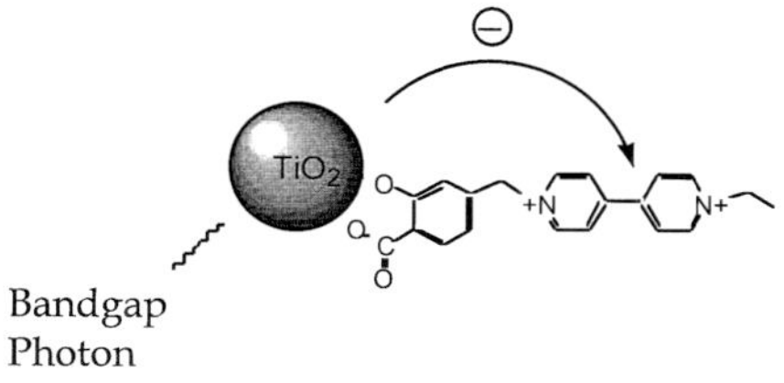

Figure 16.4. The heterosupermolecule formed by covalently assembly of a TiO_2 nanocrystallite and an appropriately modified viologen has as its associated heterosupramolecular function light-induced vectorial electron flow.

Bandgap
Photon

electron acceptor with respect to, the semiconductor nanocrystallite donor is known. The results of these and related studies will be reported in the near future [8].

Finally, recent advances in the preparation of monodispersions of metal and semiconductor nanocrystallites have been accompanied by the development of a wide range of capping agents [9]. It is likely that these molecules, which are strongly adsorbed at the surface of a nanocrystallite, may be used, following modification, to covalently assemble a wide range of molecular and condensed phase components.

16.2.2 Noncovalent Self-assembly of a Heterosupermolecule

A general strategy for the noncovalent self-assembly of condensed phase and molecular components is as follows: Identify a stabiliser or capping agent that may be modified to incorporate a receptor site but still used to prepare a dispersion of the required condensed phase component. Modify the molecular component to incorporate the complementary substrate site. Upon mixing, it is expected that the condensed phase component will recognise and selectively bind the molecular component and the required heterosupermolecule will self-assemble in solution.

To illustrate this strategy, the noncovalent self-assembly of a TiO$_2$ nanocrystallite and a viologen molecule to form the heterosupermolecule shown in Figure 16.5 is described. The condensed phase components, a dispersion of 22Å diameter TiO$_2$ nanocrystallites (anatase) in a chloroform–acetone mixture, were prepared following the method reported by Kotov *et al.* but using the modified diaminopyridine I as a stabiliser [10]. Synthesis of I has been described in detail by Brienne *et al.* [11]. The molecular component, a modified viologen II, was prepared as described in detail elsewhere [12].

It had previously been demonstrated by Lehn and coworkers that diaminopyridine and uracil, appropriately derivatised by addition of alkane chains, self-assemble in chloroform by formation of an array of three complementary hydrogen bonds [11–13]. Therefore, TiO$_2$ nanocrystallites prepared using a stabiliser incorporating a diaminopyridine moiety, TiO$_2$-I, were expected to recognise and selectively binding a viologen substrate incorporating a uracil moiety, II.

That the above expectation is justified, is evident from the ^{1}H NMR spectra shown in Figure 16.6 for TiO$_2$-I, II and TiO$_2$-(I+II) in chloroform-d/acetone-d$_6$. The amidic proton resonances of TiO$_2$-I are observed at δ 8.67. The amidic and

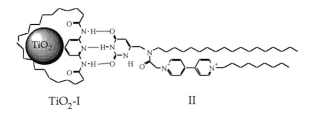

TiO$_2$-I   II

Figure 16.5. The heterosupermolecule TiO$_2$-(I+II) formed by self-assembly of TiO$_2$-I, a TiO$_2$ nanocrystallite stabilised by I, and an appropriately modified viologen II.

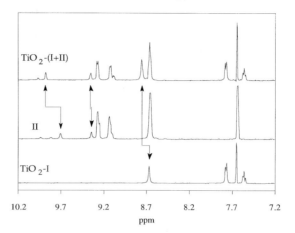

Figure 16.6. ^{1}H NMR spectra of TiO$_2$-I, II and TiO$_2$-(I + II) in a chloroform-d/acetone-d$_6$ mixture (1:1 by volume) at 20°C.

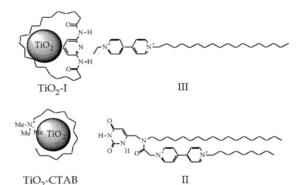

Figure 16.7. Mixtures of the condensed phase and molecular components TiO$_2$-(I and III) and TiO$_2$-(CTAB and II) do not self-assemble to form heterosupermolecules.

imidic proton resonances of II are observed at δ 9.35 and δ 9.70 respectively. The basis for these assignments have been discussed in detail elsewhere [12]. For TiO$_2$-(I + II), the above resonances are observed at δ 8.76, δ 9.36 and δ 9.88 respectively. The measured down field shifts in the resonances assigned to the amidic protons of I and the imidic proton of II are consistent with self-assembly, by complementary hydrogen bonding, of the heterosupramolecular complex TiO$_2$-(I + II). This finding was confirmed by detailed FT–IR studies [12].

^{1}H NMR spectra were also recorded for a 1:1 mixture of TiO$_2$-I and III, denoted TiO$_2$-(I and III), and a 1:1 mixture of TiO$_2$–CTAB and II, denoted TiO$_2$–(CTAB and II), in chloroform-d/acetone-d$_6$, see Figure 16.7. For TiO$_2$-(I and II), the amidic proton resonances of TiO$_2$-I are observed at δ 8.67 both prior to and following addition of III. For TiO$_2$-(CTAB and II), the imidic proton resonance of II is observed at δ 9.79 both prior to and following its addition to TiO$_2$–CTAB. These observations support the assertion that neither TiO$_2$–I and III or TiO$_2$-CTAB and II self-assemble to form a heterosupermolecule by complementary hydrogen bonding. These finding were confirmed by detailed FT–IR studies [12].

In short, both a TiO$_2$ nanocrystallite prepared in the presence of a stabiliser incorporating a diaminopyridine moiety and a viologen incorporating a uracil moiety are necessary to self-assemble the required heterosupermolecule in solution. On this basis, it was expected that bandgap excitation of the TiO$_2$ nanocrystallite in TiO$_2$-(I + II) would result in immediate electron transfer to a viologen, but that bandgap excitation of the TiO$_2$ nanocrystallite in either TiO$_2$-(I and III) or TiO$_2$-(CTAB and II) would not.

That this is the case is clearly seen from the µs transients measured following bandgap excitation of TiO$_2$-I and TiO$_2$-(I + II) in deaerated chloroform/acetone and shown in Figure 16.7. For reasons discussed in detail elsewhere, the transient measured for TiO$_2$-I is assigned to long-lived electrons trapped in the TiO$_2$ nanocrystallite [12, 14]. Also for reasons discussed in detail elsewhere, the transient measured for TiO$_2$-(I + II) is assigned to radical cation of hydrogen bonded II and to long-lived electrons trapped in the TiO$_2$ nanocrystallite of TiO$_2$-(I + II) (7, 12). The slow component of the µs transient for deaerated TiO$_2$-(I + II) is assigned to the radical cation of II formed by diffusion to TiO$_2$-I [8, 15]. The µs absorption transients measured following bandgap excitation of TiO$_2$-(I and III) and TiO$_2$-(CTAB and II) in deaerated chloroform/acetone is shown in Figure 16.9. For reasons discussed in detail elsewhere, these transients are assigned to electrons trapped in a TiO$_2$ nanocrystallite and to the radical cation (slow component) of III formed by diffusion to the surface of TiO$_2$-I [7, 12, 14].

As the optical absorption at 355 nm of the nanocrystallite in TiO$_2$-I nm is 0.1 a.u, the pulse energy is 2 mJ, the cross sectional area for irradiation is 0.4 cm^2 and assumed reflection losses are 20%, it is estimated that fourteen electron–hole pairs are generated in each TiO$_2$ nanocrystallite. From the initial amplitude of the µs transient in Figure 16.8 for deaerated TiO$_2$-(I + II) and the known extinction coefficient for the reduced form of viologen [7], it is estimated that one radical cation of II is formed per particle. That is, the charge separation efficiency is about 6% with the majority of the photogenerated electron–hole pairs being lost by recombination or trapping [14, 16]. From the initial amplitudes of the µs transients for degassed TiO$_2$-(I and III) and TiO$_2$-(CTAB and II), equal to that for degassed TiO$_2$-I, it is clear

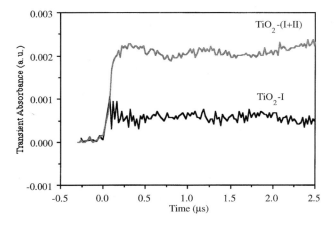

Figure 16.8. Transient absorption at 600 nm for TiO$_2$-I and TiO$_2$-(I + II) in a deaerated chloroform/acetone mixture (1:1 by vol.) at 25 °C following bandgap excitation at 355 nm (2 mJ per pulse).

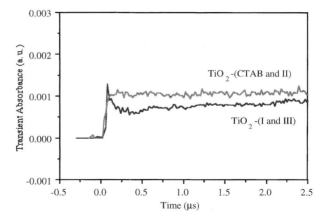

Figure 16.9. Transient absorption at 600 nm for TiO$_2$-(I and III) and TiO$_2$-(CTAB and II) in a deaerated chloroform/acetone mixture (1:1: by vol.) at 25 °C following bandgap excitation at 355 nm (2 mJ per pulse).

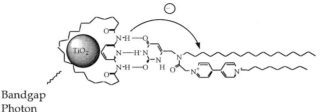

Figure 16.10. Light-induced vectorial electron flow is observed following bandgap excitation of the constituent nanocrystallite of the hetero-supermolecule TiO$_2$-(I + II).

Bandgap Photon

that no radical cations of III and II, respectively, are formed within the laser pulse although, as for TiO$_2$-(I + II), they are subsequently formed on the µs timescale by diffusion.

In short, TiO$_2$-I containing a diaminopyridine moiety recognises and selectively binds the modified viologen II containing a uracil moiety. Light-induced vectorial electron flow is observed for the resulting donor–acceptor complex as shown in Figure 16.10. It is noted that, there is ample precedent for electron transfer over long distances in supermolecules and their organised assemblies [17]. In the absence of a uracil moiety the modified nanocrystallite TiO$_2$-I does not recognise or selectively bind the viologen III. Similarly, in the absence of a diaminopyridine moiety the modified nanocrystallite TiO$_2$-CTAB does not recognise or selectively bind the modified viologen II. In neither case is direct light-induced electron transfer to the viologen moiety observed.

16.2.3 Heterosupermolecules – Are They Necessary?

An example of a covalently assembled and noncovalently self-assembled hetero-supermolecule have been presented. Further, the associated heterosupramolecular function, light-induced vectorial electron flow, has been demonstrated for each.

There are however, a many examples of chemistry in the recent literature which could be described in the same terms as those used above to describe the covalent assembly of the heterosupermolecule in Figure 16.1 [18]. (It appears, no previous examples of noncovalently self-assembled condensed phase and molecule components have been reported.) For example, the constituent nanocrystallites of a colloidal semiconductor dispersion at which are adsorbed sensitiser molecules may well be thought of as a heterosupermolecules [19].

One might therefore ask what purpose is served by the terminology introduced above. The justification offered is that it represents an attempt to describe in a systematic fashion the rapidly expanding activity at the interface between conventional covalent and noncovalent molecular chemistry and colloidal chemistry. Further, adopting and modifying supramolecular terminology not only facilitates a systematic description of this emerging activity, but also provides insights into the challenges and opportunities that have, and are expected to, arise at this interface.

16.3 Heterosupramolecular Assemblies

Having demonstrated general strategies for the covalent assembly and noncovalent self-assembly of heterosupermolecules, general strategies for the organisation of heterosupermolecules are considered. Also considered, are the opportunities presented by the organisation of heterosupermolecules for the preparation of practical nanometer scale devices and materials.

General strategies for the organisation of heterosupermolecules are as follows: Preorganise one of the condensed or molecular phase components and assemble the remaining components. Alternatively, assemble all of the condensed phase and molecular components and organise the resulting heterosupermolecules.

16.3.1 Covalent Heterosupramolecular Assemblies

An example of the first of the strategies outlined above is the following: TiO_2 nanocrystallites are deposited on a conducting glass substrate and fired to ensure an ohmic contact between constituent nanocrystallites of the resulting nanoporous-nanocrystalline film and the conducting glass substrate [4]. The molecular components , a viologen and an anthraquinone, are covalently linked and adsorbed, *via* a salicylic acid, at the surface of a constituent nanocrystallite of the nanoporous-nanocrystalline film [20]. A constituent heterosupermolecule of the resulting heterosupramolecular assembly, denoted TiO_2–SVQ, is shown in Figure 16.11.

It was expected, based on the findings outlined in Section 16.2.1 and related studies in the literature (12, 21), that bandgap irradiation of TiO_2–SVQ would result in viologen mediated electron transfer to anthraquinone and that subsequent protonation would result in long-lived charge separation, see Figure 16.12.

Both expectations were seen to be justified following detailed electrochemical and spectroscopic studies although, electron transfer from the viologen of one hetero-

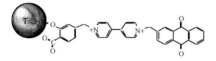

Figure 16.11. The heterosupermolecule TiO$_2$–SVQ formed by covalently assembly of a TiO$_2$ nanocrystallite, a salicylate modified viologen and an anthraquionone SVQ.

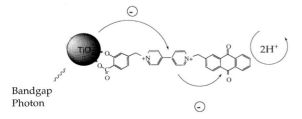

Bandgap
Photon

Figure 16.12. Bandgap excitation of the TiO$_2$ nanocrystallite (donor) results in electron transfer to the viologen (first acceptor) and formation of the corresponding radical cation. Subsequent electron transfer to the anthraquinone (second acceptor) is accompanied by protonation of the corresponding radical and long-lived charge separation.

supermolecule to the anthraquinone of another was found to be an important process, *i.e.* crosstalk is observed [20].

Therefore, the heterosupermolecule in Figure 16.11 has as its associated heterosupramolecular functions light-induced vectorial electron transfer and long-lived charge separation, see Figure 16.12.

It was also expected, following incorporation of the heterosupramolecular assembly TiO$_2$–SVQ as the working electrode in an electrochemical cell, that potentiostatic control of the Fermi level within the constituent nanocrystallites would permit the associated heterosupramolecular functions, light-induced vectorial electron flow and long-lived charge separation, to be modulated [22].

This expectation is seen to be justified following detailed electrochemical and spectroscopic studies. Specifically, the absorbance measured at 600 nm following bandgap irradiation at 355 nm of TiO$_2$–SVQ at the indicated applied potential, and assigned principally to reduced viologen [7, 12, 22], is plotted against irradiation time in Figure 16.13. At the open circuit potential, sigmoidal growth of the absorbance at 600 nm is observed. This behaviour is consistent with viologen mediated reduction of anthraquinone. Following application of a cathodic potential step between 0.00 V and −0.70 V (60 s) and returning to 0.00 V (15 s) formation of reduced viologen at the open circuit potential is significantly faster and growth of the associated absorbance no longer sigmoidal. The increased rate of formation of reduced viologen is likely a consequence of irreversible occupation of electron trap states during the negative potential step [22]. The absence of the sigmoidal feature is consistent with prior reduction of the anthraquinone. At an applied potential

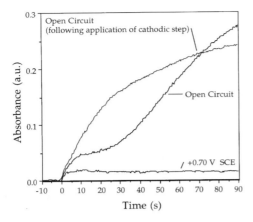

Figure 16.13. Absorbance at 600 nm TiO$_2$–SVQ following bandgap excitation at 355 nm (5.5 mJ pulse^{-1}). The applied potentials were +0.70 V, open circuit and open circuit following application of a cathodic step from 0.00 V to −0.70 V (60 s), returning to 0.00 V (15 s) prior to irradiation.

of 0.70 V there is a small initial increase in absorbance by reduced viologen but the steady state value is significantly less than that measured at the open circuit potential.

It is therefore possible to modulate the function of the heterosupermolecule in Figure 16.11 between the three states as shown in Figure 16.14. Firstly, a state (State One) in which no light-induced vectorial electron flow is observed. Secondly, a state (State Two) in which light-induced electron transfer to the viologen is observed. Thirdly, a state (State Three) in which viologen mediated light-induced electron transfer to the anthraquinone and long-lived charge separation is observed. Further, it is clear that the modulation state of the constituent heterosupermolecules of the heterosupramolecular assembly TiO$_2$–SVQ may be inferred from the potential applied to the TiO$_2$ nanocrystallites *via* the conducting glass substrate.

This and related observations [23], demonstrate one of the principal advantages of heterosupramolecular assemblies when compared to conventional supramolecular assemblies. Namely, in an appropriately organised heterosupramolecular assembly, the constituent condensed phase components yield an intrinsic substrate possessing properties characteristic of the bulk material. Modulation of the bulk properties of the intrinsic substrate, for example potentiostatic modulation of the Fermi level of the TiO$_2$ nanocrystallites constituting the heterosupramolecular assembly in Figure 16.11, will modulate the properties of the constituent hetero-supermolecules of the heterosupramolecular assembly. Further, as stated above, if the bulk property which is being modulated is monitored, then the modulation state may be inferred.

Generally, progress toward realisation of practical devices based on organised assemblies of supermolecules has been slow. Among the reasons for this have been difficulties associated with identifying substrates capable of modulating supramolecular function and providing information concerning modulation state [1, 2].

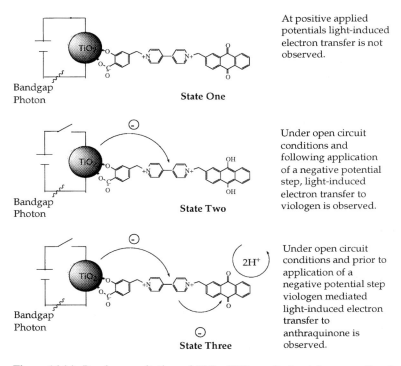

At positive applied
potentials light-induced
electron transfer is not
observed.

State One

Under open circuit
conditions and
following application
of a negative potential
step, light-induced
electron transfer to
viologen is observed.

State Two

Under open circuit
conditions and prior to
application of a
negative potential step
viologen mediated
light-induced electron
transfer to
anthraquinone is
observed.

State Three

Figure 16.14. Bandgap excitation of TiO_2–SVQ results in viologen mediated electron transfer to the anthraquinone component.

Heterosupramolecular assemblies offer, therefore, potentially general advantages in this and related respects.

16.3.2 Noncovalent Heterosupramolecular Assemblies

An example of the second of the strategies outlined above is the following: The condensed phase and molecular components of the heterosupermolecule are non-covalently self-assembled and the resulting heterosupermolecule noncovalently self-organised to yield an ordered array, or superlattice, of semiconductor nano-crystallites.

The required condensed phase components, a dispersion of 22Å diameter TiO_2 nanocrystallites (anatase) in chloroform, were prepared following the method reported by Kotov *et al.* but using the modified diaminopyridine I and uracil IV as stabilisers. Synthesis of I and IV has been described in detail by Brienne *et al.* [11].

It was proposed that TiO_2 nanocrystallites prepared in the presence of the stabiliser incorporating a diaminopyridine moiety I, TiO_2-I, would recognise and selectively bind a TiO_2 nanocrystallite prepared in the presence of the stabiliser

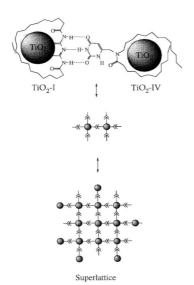

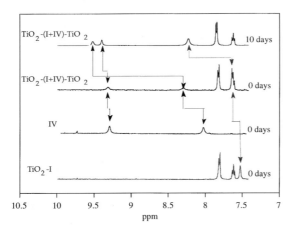

Figure 16.15. Self-assembly of the heterosupermolecule TiO_2-$(I+IV)$-TiO_2 and its subsequent self-organisation to form a superlattice.

Figure 16.16. ^{1}H NMR spectra in chloroform-d at 20°C of TiO_2-I, TiO_2-IV after 0 days and of TiO_2-$(I+IV)$-TiO_2 after 0 and 10 days.

incorporating a uracil moiety IV, TiO_2-IV. It was also proposed that following their self-assembly, the resulting heterosupermolecules, denoted TiO_2-$(I+IV)$-TiO_2, would self-organise to form a superlattice, see Figure 1.15. That this is the case has been demonstrated by ^{1}H NMR and dynamic light scattering studies [24].

The ^{1}H NMR spectra of TiO_2-I, TiO_2-IV and TiO_2-$(I+IV)$-TiO_2 are shown in Figure 16.16. Upon mixing of TiO_2-I and TiO_2-IV to form TiO_2-$(I+IV)$-TiO_2, the resonance at δ 7.53 assigned to the amidic protons in TiO_2-I and the resonance at δ 8.02 assigned to the imidic proton in TiO_2-IV are shifted downfield by less than 0.3 ppm to δ 7.64 and δ 8.29 respectively. Following 10 days ageing however, the resonances assigned to the amidic protons in TiO_2-I and the imidic proton in TiO_2-

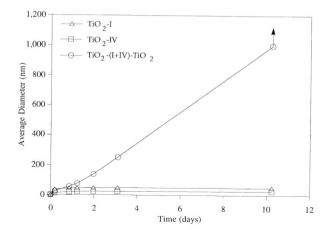

Figure 16.17. Average aggregate diameter in TiO$_2$-I, TiO$_2$-IV and TiO$_2$-(I + IV)-TiO$_2$ (1:1 mixture by vol.) during 10 days ageing in the dark at 25 °C.

IV are shifted downfield to δ 8.23 and δ 9.52 respectively. It is clear therefore, that TiO$_2$-I and TiO$_2$-IV self-assemble by complementary hydrogen bonding to form the heterosupermolecule TiO$_2$-(I + IV)-TiO$_2$ shown in Figure 16.16 [11–13], but that the above hydrogen bonds are formed on a timescale of days and are weaker than those formed immediately upon mixing I and II [12]. It should be noted that this latter observation strongly supports the assertion that I and IV are mostly at the surface of a TiO$_2$ nanocrystallite [12].

The average diameter of the aggregates present in TiO$_2$-I, TiO$_2$-IV and TiO$_2$-(I + IV)-TiO$_2$, as determined by dynamic light scattering, is plotted in Figure 16.17. As expected, some aggregation is observed in both TiO$_2$-I and TiO$_2$-IV during 10 days. Specifically, the average aggregate diameter in TiO$_2$-I increases from an initial value of less than 10 nm to 50 nm, while in TiO$_2$-IV it increases from an initial value of less than 10 nm to 27 nm. In the case of TiO$_2$-(I + IV)-TiO$_2$ however, the extent of aggregation during 10 days is significantly greater. Specifically, the average aggregate diameter increases from an initial value of less than 10 nm, to a value of 260 nm after 3 days and a final value of greater than 1000 nm after 10 days. Low-resolution electron micrographs of samples prepared from the dispersions aged for 10 days confirm that, in the case of TiO$_2$-I and TiO$_2$-IV there is limited aggregation. However, low resolution micrographs confirm that, in the case of TiO$_2$-(I + IV)-TiO$_2$ there is extensive aggregation leading to the formation of mesoaggregates possessing diameters in the range 500 nm–1500 nm.

The light scattering and electron microscopy studies described above support the view that TiO$_2$-I and TiO$_2$-IV self-assemble, by complementary hydrogen bonding as shown in Figure 16.15, to form the heterosupermolecule TiO$_2$-(I + IV)-TiO$_2$. However, they also clearly demonstrate that these heterosupermolecules subsequently self-organise to form mesoaggregates.

Stating that the heterosupermolecule TiO$_2$-(I + IV)-TiO$_2$ self-organises to form a mesoaggregate implies an ordering of the constituent nanocrystallites. That this is the case, or that a superlattice is indeed formed, is clearly seen from the medium

 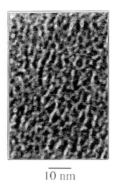

$\overline{20\ nm}$ $\overline{10\ nm}$

Figure 16.18. Medium and high resolution electron micrographs of TiO_2-(I + II)-TiO_2 mesoaggregate in Figure 16.17 after 10 days.

and high resolution electron micrographs of a mesoaggregate shown in Figure 16.18. From the corresponding high-resolution electron micrograph, it is also clear that the superlattice consists of nanocrystallites that are 22Å2 $ in diameter and organised in planes that are separated by 6Å. This, in turn, is consistent with an observed onset for absorption of 350Å10 nm and the presence of the modified stabilisers I and IV at the surface of the constituent nanocrystallites of the superlattice [25].

It is noted that there are numerous recent examples of the noncovalent self-assembly of metal and semiconductor nanocrystallite superlattices [26]. In general, self-assembly in these cases is driven by long-range nonspecific interactions between the alkane chains of the capping agents used to prepare the parent dispersion. In many cases, the resulting superlattices have shown extended order, a consequence of the constituent nanocrystallites possessing a very narrow size distribution. It is noted however, that such an approach has limited potential with respect to the self-assembly of complex structures. This will likely not be a limitation of an approach based on specific, or recognition directed, interactions between nanocrystallites.

Finally, the above and related activities demonstrate another principle advantage of heterosupramolecular assemblies when compared to conventional supramolecular assemblies. Namely, it is possibly to endow an organised heterosupramolecular assembly with architectural properties characteristic of the condensed phase components. As stated, progress toward realisation of practical devices based on organised assemblies of supermolecules has been slow. Among the reasons for this have been difficulties associated with endowing supramolecular assemblies with the required structural or architectural properties [1, 2]. Heterosupramolecular assemblies offer, therefore, potentially general advantages in this and related respects.

16.3.3 Heterosupramolecular Assemblies – What Do We Gain?

It has been demonstrated that the methodologies whose development have accompanied the assembly of heterosupermolecules in solution may be extended to permit their organisation, also in solution.

An advantage of heterosupramolecular assemblies compared to the more conventional supramolecular assemblies, is that the function of the heterosupermolecule is linked to the bulk properties of the condensed phase component or components which may be modulated. That is, by modulating a bulk property of a condensed phase component the heterosupramolecular function of a heterosupermolecule may be modulated. Further, if the bulk property which is being modulated can be monitored, then the modulation state of the heterosupermolecule within the heterosupramolecular assembly may be inferred. Another principle advantage of heterosupramolecular assemblies when compared to conventional supramolecular assemblies is the consequent possibly of endowing an appropriately organised heterosupramolecular assembly with architectural properties characteristic of the condensed phase components.

As stated, progress toward realisation of practical devices based on organised assemblies of supermolecules has been slow. A reason for this, has been the difficulties encountered in identifying substrates capable of modulating supramolecular function and providing information concerning modulation state of the assembly. Another reason, has been the difficulties encountered in attempting to endow supramolecular assemblies with the required structural or architectural properties. Heterosupramolecular assemblies clearly offer the prospect of significant progress in relation to both of these difficulties and therefore offer the prospect of practical molecular scale devices.

Finally, while adressability is, in principle, a consequence of organisation, the heterosupramolecular assemblies described here are not been sufficiently ordered to permit addressing of an individual heterosupermolecule. This limitation is, itself, being addressed [27].

16.4 Heterosupramolecular Chemistry and Molecular-Scale Devices

The self-assembly and self-organisation of complex nanostructures in solution is an important objective of materials chemistry and physics. The importance of this objective is a result of a desire to be able to programme the bottom–up assembly of molecular-scale devices in solution. It is in this context, that we have sought to develop a systematic chemistry, termed heterosupramolecular chemistry, of covalently assembled and noncovalently self-assembled condensed phase and molecular components whose intrinsic properties largely persist but which possess well defined heterosupramolecular functions or properties.

The difficulties encountered to date in developing a systematic heterosupramolecular chemistry, and those likely to be encountered in the future, are justified as it is expected that the development of such a chemistry will facilitate the programmed bottom–up assembly of practical molecular-scale devices in solution.

The use of both condensed phase and molecular components will permit heterosupermolecules and heterosupramolecular assemblies possessing novel functions and properties to be self-assembled and self-organised in solution. In this context, recent advances in the preparation of nanocrystallites of a wide range of materials possessing well defined sizes, surface properties and crystal structures have been important [9]. Also of importance have been the development of strategies for linking molecules, typically capping groups or sensitiser molecules, to the surface of these nanocrystallites [9, 18]. Further, conventional supramolecular chemistry continues to provide an ever increasing number of receptor-substrate pairs that can be used to self-assemble the condensed phase and molecular components of a hetero-supermolecule or heterosupramolecular assembly [28].

The incorporation of condensed phase components in a heterosupermolecule or heterosupramolecular assembly facilitates practical and efficient modulation of function or property. Specifically, as the function or property of the heterosuper-molecule or heterosupramolecular assembly is dependent on the bulk properties of the constituent condensed phase components, modulation of a bulk property will therefore modulate the function or property of the heterosupermolecule or hetero-supramolecular assembly. That this is so, has been demonstrated for a number of cases to date and is expected to be generally true [20, 23].

The use of both condensed phase and molecular components will permit hetero-supermolecules and heterosupramolecular assemblies possessing novel architectures to be self-assembled and self-organised in solution. That this will be necessary is increasingly apparent. For example, Grätzel has described a stable and efficient regenerative photoelectrochemical cell for the conversion of solar energy to electrical energy [4]. Innovative aspects of the Grätzel cell include the use of ruthenium complexes as sensitisers whose absorption spectra overlap well the solar emission spectrum and the use of 10 m thick nanoporous-nanocrystalline semiconductor films with a surface roughness of greater than 1000 as photoanodes. Central to the efficient operation of the Grätzel cell is the fact that the sensitiser molecules, as a consequence of their being adsorbed directly at the photoanode, are effectively stacked and the probability of an incident visible photon being absorbed is close to unity. It is important to note that while the above light-harvesting strategy is clearly based on that of green plants, its practical implementation utilises a heterosupra-molecular assembly and that implementation of the same strategy using only condensed phase or molecular components has not proved possible.

Another desirable feature is that a heterosupermolecule or heterosupramolecular assembly be self-assembled and self-organised in solution using a parallel algorithm. That is, that subcomponents of the heterosupermolecule or heterosupramolecular assembly be self-assembled at the same time in the same solution and that they be subsequently assembled to yield the required device or material. While it is clear that the use of both molecular and condensed phase components does not reduce the difficulties presented by the need to meet this requirement, neither does it add to them. In this context however, we note recent work of Mirkin *et al.* and Alivisatos *et al.* in which the self-assembly of Au nanocrystallites have been directed by use of non-self-complementary oligomers of DNA [29]. These studies can be usefully discussed as heterosupramolecular chemistry.

Finally, for heterosupramolecular chemistry to fulfil its promise it will be necessary to be able to scale-up the process of self-assembly and self-organisation in solution. In this respect the use of DNA based receptor-substrate pairs offers the prospect of utilising related commercial technologies [29].

In the future it is possible that solid state and synthetic chemistry will evolve into a seamless continuum of activities which may be usefully discussed in the framework of a systematic heterosupramolecular chemistry.

References

[1] (a) J.-M. Lehn, *Angew. Chem. Int. Ed. Engl.* **1988**, *27*, 89–112; (b) D. J. Cram, *Angew. Chem. Int. Ed. Engl.* **1988**, *27*, 1009–1112.

[2] (a) J.-M. Lehn, *Supramolecular Chemistry*; VCH, New York **1995**, Chap. 8, (b) V. Balzani, F. Scandola, *Supramolecular Photochemistry*, Ellis Horwood, New York **1991**, Chap. 3.

[3] X. Marguerettaz, R. O Neill, D. Fitzmaurice, *J. Am. Chem. Soc.* **1994**, *116*, 2629–2630.

[4] (a) B. O Regan, M. Grätzel, *Nature* **1991**, *353*, 737–740; (b) M. K. Nazeeruddin, A. Kay, I. Rodicio, R. Humphry-Baker, E. Müller, P. Liska, N. Vlachopoulos, M. Grätzel, *J. Am. Chem. Soc.* **1993**, *115*, 6382–6390.

[5] (a) U. Kölle, J. Moser, M. Grätzel, *Inorg. Chem.* **1985**, *24*, 2253–2258; (b) H. Frei, D. Fitzmaurice, M. Grätzel, *Langmuir* **1990**, *6*, 198–206; (c) J. Moser, S. Punchihewa, P. Infelta, M. Grätzel, *Langmuir* **1991**, *7*, 3012–3017; (d) G. Redmond, D. Fitzmaurice, M. Grätzel, *J. Phys. Chem.* **1993**, *97*, 6951–6954.

[6] J. Moser, M. Grätzel, *Chem. Phys.* **1993**, *176*, 493–500.

[7] (a) B. Kok, H. Rurainski, O. Owens, *Biochem. Biophys. Acta* **1965**, *109*, 347–356. (b) P. Trudinger, *Anal. Biochem.* **1970**, *36*, 222–224; (c) T. Watanabe, K. Honda, *J. Phys. Chem.* **1982**, *86*, 2617–2619.

[8] X. Marguerettaz, D. Fitzmaurice, paper in preparation.

[9] P. Alivisatos, *J. Phys. Chem.* **1996**, *100*, 13226–13239.

[10] N. Kotov, F. Meldrum, J. Fendler, *J. Phys. Chem.* **1994**, *98*, 8827–8830.

[11] M.-J. Brienne, J. Gabard, J.-M. Lehn, *J. Chem. Soc. Chem. Comm.* **1989**, 1868–1870.

[12] (a) L. Cusack, S. N. Rao, J. Wenger, D. Fitzmaurice, *Chem. Mater.* **1997**, *9*, 624–631; (b) L. Cusack, S. N. Rao, D. Fitzmaurice, *Chem. Eur. J.* **1997**, *3*, 202–207; (c) L. Cusack, X. Marguerettaz, S. N. Rao, J. Wenger, D. Fitzmaurice, *Chem. Mater.* **1997**, *9*, 1765–1772.

[13] (a) B. Feibush, A. Fiueroa, R. Charles, K. Onan, P. Feibush, B. Karger, *J. Am. Chem. Soc.* **1986**, *108*, 3310–3318; (b) B. Feibush, M. Saha, K. Onan, B. Karger, R. Giese, *J. Am. Chem. Soc.* **1987**, *109*, 7531–7533; (c) A. Hamilton, D. Van Engen, *J. Am. Chem. Soc.* **1987**, *109*, 5035–5036; (d) A. Bisson, F. Carver, C. Hunter, J. Waltho, *J. Am. Chem. Soc.* **1994**, *116*, 10292–10293.

[14] (a) A. Henglein, *Ber. Bunsenges. Phys. Chem.* **1982**, *86*, 241–246; (b) D. Duonghong, J. Ramsden, M. Grätzel, *J. Am. Chem. Soc.* **1982**, *104*, 2977–2985; (c) U. Kolle, J. Moser, M. Grätzel, *Inorg. Chem.* **1985**, *24*, 2253–2258; (d) G. Rothenberger, J. Moser, M. Grätzel, M. Serpone, D. Sharma, *J. Am. Chem. Soc.* **1985**, *107*, 8054–8059.

[15] H. Frei, D. Fitzmaurice, M. Grätzel, *Langmuir* **1990**, *6*, 198–206.

[16] (a) D. Bahnemann, A. Henglein, J. Lilie, L. Spanhel. *J. Phys. Chem.* **1984**, *88*, 709–711; (b) O. Micic, Y. Zhang, R. Cromack, A. Trifunac, M. Thurnauer, *J. Phys. Chem.* **1993**, *97*, 7277–7283.

[17] (a) C. Chidsey, *Science* **1991**, *251*, 919–922. (b) H. Finklea, D. Hanshew, *J. Am. Chem. Soc.* **1992**, *114*, 3173–3181; (c) G. Cleland, B. Horrocks, A. Houlton, *J. Chem. Soc. Faraday Trans.* **1995**, *91*, 4001–4003.

[18] A. Hagfeldt, M. Grätzel, *Chem. Rev.* **1995**, *95*, 49–68.

[19] (a) A. McEvoy, M. Grätzel, *Sol. Energy. Mater. Sol. Cells* **1994**, *32*, 221–227; (b) T. Gerfin, M. Grätzel, L. Walder, in: *Molecular Level Artificial Photosynthetic Materials* (Ed.: G. Meyer), Wiley, New York, **1997**, Chap. 7.

[20] (a) X. Marguerettaz, D. Fitzmaurice, *J. Am. Chem. Soc.* **1994**, *116*, 5017–5018; (b) X. Marguerettaz, G. Redmond, S. Nagaraja Rao, D. Fitzmaurice, *Chem. Eur. J.* **1996**, *2*, 420–428.

[21] (a) C. Hable, R. Crooks, M. Wrighton, *J. Phys. Chem.* **1989**, *93*, 1190–1192; (b) C. Hable, R. Crooks, J. Valentine, R. Giasson, M. Wrighton, *J. Phys. Chem.* **1993**, *97*, 6060–6065; (c) G. Callabrese, R. Buchanan, M. Wrighton, *J. Am. Chem. Soc.* **1983**, *105*, 5594–5600; (d) A.-M. Brun, S. Hubig, M. Rodgers, W. Wade, *J. Phys. Chem.* **1990**, *94*, 3869–3871; (e) A.-M. Brun, S. Hubig, M. Rodgers, W. Wade, *J. Phys. Chem.* **1992**, *96*, 710–715.

[22] (a) B. O Regan, M. Grätzel, D. Fitzmaurice, *Chem. Phys. Lett.* **1991**, *183*, 89–93; (b) B. O Regan, M. Grätzel, D. Fitzmaurice, *J. Phys. Chem.* **1991**, *95*, 10525–10528; (c) G. Rothenberger, D. Fitzmaurice, M. Grätzel, *J. Phys. Chem.* **1992**, *96*, 5983–5986; (d) G. Redmond, A. O Keeffe, C. Burgess, C. MacHale, D. Fitzmaurice, *J. Phys. Chem.* **1993**, *97*, 11081–11086; (e) G. Redmond, D. Fitzmaurice, *J. Phys Chem.* **1993**, *97*, 1426–1430; (f) B. Enright, G. Redmond, D. Fitzmaurice, *J. Phys. Chem.* **1994**, *98*, 6195–6200.

[23] (a) R. Hoyle, J. Sotomeyer, G. Will, D. Fitzmaurice, *J. Phys. Chem.*, submitted; (b) G. Will, R. Hoyle, G. Boschloo, D. Fitzmaurice *J. Am. Chem. Soc.*, in preparation.

[24] L. Cusack, R. Rizza, A. Gorelov, D. Fitzmaurice, *Angew. Chem. Int. Ed. Eng.* **1997**, *36*, 848–851.

[25] Serpone, N; Lawless, D.; Khairutdinov, R. J. Phys. Chem. 1995, 99, 16646–16654.

[26] (a) J. Heath, C. Knobler, D. Leff, *J. Phys. Chem.* **1997**, *101*, 189–197; (b) C. Murray, C. Kagan, M. Bawendi, *Science* **1995**, *270*, 1335–1337.

[27] A. Merrins, X. Marguerettaz, S. N. Rao, D. Fitzmaurice, paper in preparation.

[28] D. Philp, J. F. Stoddart, *Angew. Chem. Int, Ed, Engl.* **1996**, *35*, 1154–1196.

[29] (a) B. Mirkin, R. Letsinger, R. Mucic, J. Storhoff, *Nature* **1996**, *382*, 607–609; (b) P. Alivisatos, X. Peng, T. Wilson, K. Johnson, C. Loweth, M. Bruchez, P. Schultz, *Nature* **1996**, *382*, 609–611.

Chapter 17

Nanoclusters in Zeolites

J. B. Nagy, I. Hannus, and I. Kiricsi

17.1 Introduction

The preparation of monodisperse metal particles or ionic clusters in zeolites is a very interesting subject. Indeed, the size of the particles is controlled by the size of the zeolitic cages or channels and in addition, a beneficial interaction can occur between the particles and the zeolite walls.

Zeolites appear different from other supports such as amorphous materials, glasses, etc. because the size and shape of the cages together with their interconnectivities also influence the properties of the dispersed particles. The types of particles and ionic clusters formed can be correlated with the structure, the concentration of the cation exchanged and the proton content of the zeolite.

As the zeolites are used as catalysts or catalyst supports, a large part of the literature is devoted to the preparation and catalytic activities of metal clusters in zeolites. A rather comprehensive review was made by P. A. Jacobs up to 1984 [1]. This reference also contains previously published excellent review papers.

By confining insulator, semiconductor and metallic materials inside a zeolite host matrix, one can create three-dimensional periodic arrays of single-size and shape clusters of the guest components stabilized by the zeolitic matrix. Depending on the degree of interaction between clusters, it may be possible to create extended cluster lattices of various sizes, the physical properties of which may be different from both the individual clusters and the bulk material of the guest ions [2].

In the present review emphasis will be put on the preparation and characterization of nanoclusters in zeolites. Some applications will be evoked as molecular wires, nanoporous molecular electronic materials and nonlinear optical materials [3,4]. The examples will be taken either from our own work or they will be selected on a rather personal basis.

17.2 Synthesis of Nanoparticles in Zeolite Hosts

In order to better understand the different systems, we shall first analyze the structure of the host zeolites and then we review the different methods of synthesis used to prepare the composite materials.

17.2.1 Description of Some Common Zeolite Structures

In most of the cases sodalite, zeolite A, X, Y, mordenite, zeolites L, Ω and ZSM-5 were used as host materials. In general, not only the structure of the zeolitic cages and channels will be important, but also their connectivities and the position of counterions neutralizing the framework negative charges.

The general formula of a zeolite, which is an aluminosilicate, is:

$$M_x D_y (Al_{x+2y} Si_z O_{2(x+2y+z)})$$

where M is a monovalent and D a divalent cation neutralizing the negative charges introduced in the structure by replacing a tetravalent silicon atom by a trivalent aluminium atom.

Some typical zeolite structures are given in Figure 17.1.

Zeolites possessing three-dimensional framework structure are constructed by joining together $[SiO_4]^{4-}$ and $[AlO_4]^{5-}$ coordination polyhedra. By definition these tetrahedra are assembled together such that the oxygen at each tetrahedral corner is shared with that in an identical tetrahedron sitting either Si or Al in its center. This corner sharing forms infinite lattice comprised of identical building elements, the unit cells, in a manner common to all crystalline materials.

One possibility to classify zeolite structures would be to relate them to the symmetry of their unit cells. This would be inconvenient and is much simplified by the observation that zeolite structures often have identical repeating subunits which are less complex than their unit cells. These represent only the aluminosilicate skeleton and exclude consideration of water molecules and cations sitting within the cavities and channels of the framework.

The number of cations present in the pore system of zeolites is determined by the number of $[AlO_4]^{5-}$ units. This arises from the isomorphous substitution of Al^{3+} for Si^{4+} in the tetrahedra, resulting in residual negative charge on the oxygen framework. These negative charges are compensated by the cations present in the synthesis medium and held in the interstices of the structure.

The extent and location of water molecules depend on the overall architecture of the zeolite, such as the size and shape of the cavities, channels, and the number and nature of cations present in the structure.

In Table 17.1. are summarized the characteristics of the most common zeolite structures: the IUPAC nomenclature, common name and the typical unit cell content.

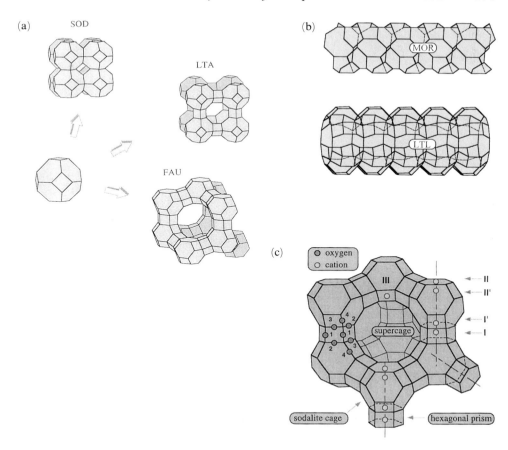

Figure 17.1. (a) Zeolite structures formed by sodalite units assemblies: sodalite (SOD), zeolite Linde type A (LTA) and faujasite – zeolite X or Y (FAU). (b) Zeolite structures containing channels: mordenite (MOR) and zeolite Linde type L (LTL). (c) Structure of faujasite (zeolite X or Y): siting of cations in the hexagonal prisms (I), in the sodalite cages (I' and II') and in the supercages (II and III).

Table 17.1. Classification of some zeolites.

Structure Type	Name	Unit cell composition
LTL	Zeolite L	$K_6Na_3Al_9Si_{27}O_{72} \times 21\,H_2O$
SOD	Sodalite	$Na_6Al_6Si_6O_{24} \times 8\,H_2O$
LTA	Zeolite A	$Na_{12}Al_{12}Si_{12}O_{48} \times 27\,H_2O$
FAU	Zeolite X	$Na_{88}Al_{88}Si_{104}O_{384} \times 235\,H_2O$
	Zeolite Y	$Na_{58}Al_{58}Si_{134}O_{384} \times 210\,H_2O$
MOR	Mordenite	$Na_8Al_8Si_{40}O_{96} \times 24\,H_2O$

Zeolites A, X, Y. Linde type A (LTA) zeolite is one of the most utilized zeolites. Its pore structure consists of truncated octahedra linked to other cavities through six-membered rings and of truncated cuboctahedra linked together through eight-membered rings. The largest utilizations of zeolite A include ion-exchange (water softener) and adsorbents (Figure 17.1(a)).

Synthetic faujasite materials are mainly zeolite X (high Al content, close to 1 Si/Al ratio) and zeolite Y (higher Si/Al ratio). These can be reconstructed from soda-lite (truncated octahedra) cages in a diamondlike array, the sodalite cages being interconnected through a hexagonal prism (Figure 17.1(a)).

In addition to the sodalite cages, the main cavities of faujasite material are about 11 Å diameter, and are interconnected through twelve-membered rings apertures of about 7.4 Å diameter, allowing adsorption of large molecules such as trime-thylbenzene.

The main use of X-type (Al-rich) materials include adsorption and catalysis. Y-type zeolite is used as catalyst or catalyst support, mainly in fluid catalytic cracking and hydrocracking processes.

Mordenite. Its structure is composed of four- and five-membered rings, generating a porous system delimited by eight- and twelve-membered rings. The porous system consists of linear channels, oriented along the crystallographic c axis, and with apertures consisting of eight- and twelve-membered rings (6.7 × 7.0 Å). These channels are linked together through a secondary pore system, along basis, and consisting of eight-membered rings (Figure 17.1(b)).

Compared to zeolites A, X or Y, mordenite is characterized by a higher silica content resulting in higher thermal stability. In addition to a higher Si/Al ratio, the mordenite structure contains five-membered rings thought to enhance the acid site strength of the material.

Zeolite L is characterized by a monodimensional pore system composed of parallel twelve-membered ring linear channels and by a quite low Si/Al atomic ratio.

Some dimensional parameters are collected in Table 17.2. These data are used for estimation of the main molecular sieving parameters of various zeolites.

Table 17.2. Dimensional parameters of some zeolites.

Zeolite	Number of oxygen atoms in the ring	Effective window size [nm]	Void volume*
Zeolite A	6	0.23	0.47
	8	0.45	
Zeolite X	6	0.23	0.53
	12	0.78	
Zeolite L	12	0.71	0.28
Mordenite	12	0.67 × 0.7	0.26
	8	0.29 × 0.57	

* Void volume is expressed as cm^3 liquid H_2O/cm^3 crystal.

To sum up we hope that the structural description given in this short chapter convey the main feature of zeolites as porous media composed of a series of different regular channels and cavities. Access to these interstitial voids is via well defined windows composed of various numbers of tetrahedra, thereby well-constructed three-dimensional space is created. The dimensions of these channels and cavities are critical to the unique properties exhibited by zeolites. The cages possess space for formation of different metal or non metal clusters.

17.2.2 Synthesis of Metal Particles and Ionic Custers in Zeolites

As zeolites are three-dimensional cation exchangers, a straightforward manner to prepare metal containing precursors is by ion exchange with reducible transition ions. The cation exchange capacity (CEC) together with the number and nature of the cationic sites have to be considered as important parameters [7].

The ion exchange can be carried out either in solution [8] or in the solid phase [9]. The ion exchange can be stoichiometric or non stoichiometric. It was also observed that a competitive ion exchange could favour the more uniform distribution of ions in the zeolite crystallites [1].

Another interesting method to prepare the precursors to metal loaded zeolites is the adsorption of compounds which are easily decomposed in a subsequent step. Metal carbonyls were used in many cases, where they were introduced in the pore structure of the zeolite either by distillation or by sublimation [1]. The alkali metal azides were used for the preparation of the alkali metal precursors [10]. The sorption of labile organic complexes, such as trialkylrhodium, nickel dithiophosphate or bis-(toluene) metal (0) were used to transport metal atoms into zeolites [1]. Polycyano inclusion compounds were also used as transition metal precursors in zeolites.

Where the ion exchange is not possible, as for example in the case of high silica zeolites, the incipient wetness technique is used [11].

Bimetallic or trimetallic zeolites can also be prepared using several methods of preparation either simultaneously or successively [1].

Table 17.3 illustrates the various methods of transformation of the metal-containing precursors into metal particle or ionic cluster loaded zeolites.

In most of the cases, the precursors have to be dehydrated before final treatment. As the hydrated ions are located in accessible sites, a redistribution of bare ions occur among the different sites during a slow dehydration. The site population are governed by the energetic and coordinative differences as well as the competition for a site with other cations. For example in the hexagonal prism of faujasite, octahedral coordination to lattice oxygen atoms is possible, whereas in a six-membered ring, only a one-sided trigonal coordination may exist. The stability of this site is improved when extra-lattice species formed from water hydrolysis remain in the zeolite, changing thereby the site symmetry from trigonal to tetrahedral. Hydrolysis of metal ions will affect the nature and the location of the species which will be reduced afterwards [1].

Since most ion exchanges are not complete, at least one other cation is competing

Table 17.3. Synthesis of metal particles or ionic clusters loaded zeolites.

Precursor	Zeolite	Method of preparation	Final system	References
Ag^+	A	Autoreduction	Ag_3^{2+} cluster/A	1, 12
Ag^+, Na^+	Y	γ-irradiation	Ag_3^{2+} and Ag_3^0/Y	13
$AgX(X = Cl, Br, I)$	SOD	Ion Exchange	$Ag_{8-2n}Na_{2n}X_2$/SOD	2, 5, 14, 57, 59, 60
Ag, C_2O_4	SOD	Direct synthesis	Ag_4OX^{2+}, Ag_4^{2+}, Ag_4^{3+}	2, 14
M^+, Na^+	X	NaN_3 decomposition	Na_x^0, Na_4^{3+}; Cs_x^0	10, 16, 19–23
Ag^+	Y	γ-irradiation	Ag_2^0	15
Ag^+	Y	H_2 reduction	Ag^0	1, 100, 101
$M^+(Li^+, Na^+, K^+, Rb^+, Cs^+)$	X	MVD		
Na^+	SOD, Y	NaN_3 decomposition	Na_6^{5+}, K_3^{2+}	17
	Gallosilicate	MVD	Na_4^{3+}	18
M	A	MVD	Ferromagnetic M_x	24–27
Cd^{2+}	A, X, Y	H_2S	CdS, Cd_4S_4	4, 28, 29
Cd^{2+}	A, X, Y, MOR, ALPO$_4$	H_2Se	Se_8, CdSe, Se_x	4, 30, 31, 58
Se, Te melt	MOR	VPD	Se_x^0, Te_x^0	32, 33
Zn^{2+}	Boralite	H_2S	Zn_4S	4
	SOD		Zn_4S_4	4
Organometallic Compound	Y	MOCVD $(CH_3)_3Ga + PH_3$	$Ga_{28}P_{13}$	34
Ru^{3+}, Pd^{2+}, Pt^{2+} Rh^{3+}, Ir^{2+} ammine complexes	Y	Autoreduction	Ru_x^0, Pd_x^0, Pt_x^0 Rh_x^0, Ir_x^0	1, 102
Pt^{2+}	KL	H_2, $NaBH_4$	Pt^0	96
$Ni(CO)_4$	NaY, HY	Thermal decomposition	Ni_x^0	35, 36
$Fe(CO)_5$	HY	Thermal and photochemical decomposition	Fe_x^0	36, 37
$Fe(CO)_5$	NaY, HY	Thermal decomposition	Fe_x^0	1
Ag^+	NaY, NaMOR NaCHA	H_2 reduction	Ag°	1

Precursor	Zeolite	Method	Product	Ref.
Cu^{2+}	NaY	H_2 reduction	Cu°	1
Ni^{2+}	NaY	H_2 reduction	Ni°, NiO	1
		Pt/H_2 reduction	Ni°	1
Cu^{2+}	Y	CO reduction	Cu°	1
$Rh(NH_3)^{3+}$	NaY	CO reduction	$Rh^+(CO)_2$	1
		H_2 reduction	Rh°	44
Ni^{2+}, Co^{2+}	NaY, NaA	Metal reduction	Ni°, Co°	1
$Pt(NH_3)_4^{2+}$	NaY, NaX	H_2 reduction	Pt°	38, 49
$Mn(NO_3)_2$, $Rh(NH_3)_5^{3+}$	NaY	H_2 reduction	MnO/Rh	39, 45
$Pd(NH_3)_4^{2+}$, $FeSO_4$	NaY	H_2 reduction	Fe^{2+}/Pd^0	40, 54, 89, 90
$Pt(NH_3)_4^{2+}$, $FeSO_4$	NaY	H_2 reduction	Fe^{2+}/Pt^0	41, 45
$Co(NO_3)_2$, $Pd(NH_3)_4^{2+}$	NaY	H_2 reduction	Co^{2+}/Pd^0	42, 51, 53
$Co(NO_3)_2$, $Pt(NH_3)_4^{2+}$	NaY	H_2 reduction	Co^0/Pt^0	97–99
$Pt(NH_3)_4^{2+}$, $Cu(NO_3)_2$	NaY	H_2 reduction	$Pt_x^0 Cu_y$	43, 48–50
$Rh(NH_3)_5^{3+}$, $Cr(NO_3)_3$	NaY	H_2 reduction	Cr^{3+}/Rh^0	44
$Rh(NH_3)_5^{3+}$, $FeSO_4$	NaY	H_2 reduction	Fe^{2+}/Rh	46, 55
$Pt(NH_3)_4^{2+}$, $Re_2(CO)_{10}$	NaY	H_2 reduction, decomposition	$Pt^0\text{-}Re^0$	47, 106
$Pd(NH_3)_4^{2+}$, $Ni(NO_3)_2$	NaY	H_2 reduction	$Pd^0 Ni_x^0$	52
$(CH_3)_2Zn$, $(CH_3)_2Cd$	NaY	Surface reaction	CH_3-Zn-Zeo CH_3-Cd-Zeo	56
Si_2H_6	NaY	CVD followed by surface reaction	Si_{60}	61, 62
SnS_2	–	Direct hydrothermal synthesis	R-SnX-1 (X = S, Se)	63–65
Ge, S, CsOH, Fe^{2+}	–	Direct hydrothermal synthesis	Open framework $Cs_2FeGe_4S_{10}$	66
Ge, S, TMAOH, Fe^{2+}	–		$TMA_2FeGe_4S_{10}$	
Ge, S, TMAOH, Mn^{2+}			$TMA_2MnGe_4S_{10}$	
K_2S, Sn°	–	Direct hydrothermal synthesis	$K_2Sn_4Se_8$	67

MVD: metal vapour deposition; VPD: vapour phase deposition; CVD: chemical vapour deposition; MOCVD: molecular chemical vapour deposition
CHA: chabazite; TMA: tetramethylammonium; TMA: tetramethylammonium

with the transition metal cation. We will see subsequently how the presence of two or three metal ions influence each other in the zeolitic pores.

In the autoreduction processes the zeolite itself is the reducing agent, as for example the formation of Ag in AgA, X, Y, CHA (chabazite) or MOR [1,100]:

$$2(Ag^+ZO^-) \rightarrow \tfrac{1}{2}O_2 + 2Ag° + ZO^- + Z^+$$

where ZO^- is the negatively charged zeolite lattice and Z^+ is a Lewis acid site.

The autoreduction of transition metal ions occurs with the ammonia of the ammine complexes. In the case of $Pt(NH_3)_6$ Y zeolite [1]:

$$Pt(NH_3)_6^{2+} 2\, ZO^- \rightarrow Pt° + 2HY + \tfrac{1}{3}N_2 + 5\tfrac{1}{3}NH_3$$

The metal carbonyls are decomposed either thermally or photochemically, leading to quasi atomically dispersed species [1]. The stepwise decomposition of $Fe(CO)_5$ on NaY is:

$$Fe(CO)_5 \xrightarrow{\text{slow}} Fe(CO)_2 \xrightarrow{\text{fast}} Fe(CO)_{0.25} \rightarrow Fe$$

while on HY zeolite:

$$Fe(CO)_5 \xrightarrow{\text{slow}} Fe_3(CO)_{12} \xrightarrow{\text{fast}} Fe(CO) \rightarrow Fe$$

An interesting observation was made during the thermal and photochemical decomposition of $Fe(CO)_5$ adsorbed on HY. The reaction was followed by ^{13}C–NMR of the adsorbed species. The initial chemical shift (δ) was equal to 209.9 ppm vs TMS. During the thermal decomposition a low field shift (higher δ values) was observed, while a high field shift occurred during the photochemical decomposition. This was interpreted by a stronger interaction between the intermediate $Fe(CO)_x$ species and the surface during the photochemically activated process [36, 37]. The size of the particles was larger than the size of the supercage (ca 11 Å), showing that sintering occurred and the larger particles were deposited at the external surface of the crystallites.

The reduction of the transition metal ions by molecular hydrogen yields the metal in the zero oxidation state and the zeolite in the hydrogen form:

$$M^{n+}nZO^- + \frac{n}{2}H_2 \rightarrow M° + nZOH$$

The reducibility of the various transition metal ions M^{2+} was explained by Klier et al. [68]. The potential for reduction into neutral atoms is (ΔE):

$$\Delta E = I_1 + I_2 + E_c$$

where I_1 and I_2 are the first and second ionization potentials and E_c is the stabilization energy of the divalent metal ion in a site with D_{3h} symmetry. A good correlation was obtained between ΔE and the electrochemical potentials for Cr^{2+}, Fe^{2+}, Co^{2+}, Ni^{2+} and Cu^{2+}.

The reduction by atomic hydrogen is easier. Indeed, an additional decrease of free energy occurs by 360–440 kJ mol^{-1} within the range of 300–700 K. The reduction with CO includes metal carbonyls as intermediate species. The reduction with ammonia is more effective than with molecular hydrogen, but it leads to higher metal sintering. For metal ions which are difficult to reduce, alkali metal vapour was also used, as for example for Ni^{2+} and Co^{2+} in zeolites A and Y [1].

A particularly interesting method of preparation of alkali metal clusters in zeolites is the thermal decomposition of metal azides first proposed by Fejes et al [69]:

$$2MN_3 \rightarrow 2M^\circ + 3N_2$$

The zero valent metal atoms can form higher clusters and it can also form ionic clusters with the cations present in the zeolite. A more detailed characterization using both EPR and NMR will be given below.

Generally speaking, the reducibility of transition metal ions in determined [1] by:

1. the structure and chemical composition of the zeolite matrix,
2. the nature and amount of cocation,
3. the site location in the structure,
4. the presence of oxidizing sites such as surface hydroxyl groups, and
5. the presence of residual water of hydration.

It is important to emphasize that the formation of zero valent atoms is accompanied by the rearrangement of cations on the different sites. In addition, the formation of metal or ionic clusters also leads to a perturbation of the cations provoking a novel rearrangement between the various sites.

The presence of surface hydroxyl groups is responsible for the reoxidation of the highly reactive metal atoms.

$$M^{n+}nZO^- + \frac{n}{2}H_2 \rightleftharpoons nZOH + M^\circ$$

For Ni^{2+}, a critical –OH concentration exists (40 to 50% of the Cation Exchange Capacity), above which the Ni^{2+} ions are almost irreducible [1].

The dynamic rearrangement of the cations was studied during the reduction of Ni^{2+} ions in partially dehydrated Y zeolite. The supercage species are reduced first, followed by those in the sodalite cages or the hexagonal prisms [1].

The influence of irreducible polyvalent cations (such as Ca^{2+}, La^{3+} and Ce^{3+}) may be rationalized in terms of competition for the same sites between transition metal ions and the polyvalent ions. In mixed $Ni^{2+} - La^{3+}$ cationic forms of NaX, the Ni^{2+} ions preferentially occupy the sites in the hexagonal prisms under completely anhydrous conditions. During reduction, the hexagonal prisms are progressively emptied and the La^{3+} ions occupy the SI sites in the sodalite cages, since for electrostatic reasons both types of sites cannot be fully occupied at the same time (Figure 17.1(c)). It is assumed that the formation of mixed cation forms of the

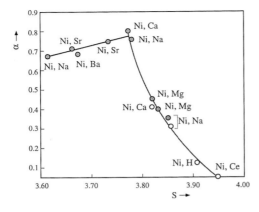

Figure 17.2. Variation of the degree of reduction α, of Ni^{2+} in Y zeolite with the nature and number of the irreducible cations as quantified by the Sanderson electronegativity, S [1].

type La–O–Ni during the sample dehydration is responsible for the decreased reducibility [1].

The $Ni^{2+} - Ce^{3+}$ cationic form of NaX behaves quite differently. Upon dehydration, the Ce^{3+} ions are preferentially fixed in the hexagonal prisms, thus forcing the Ni^{2+} ions to be located to a greater extent in the supercages and hence increasing their reducibility [1].

The overall reducibility as expressed by the degree of reduction α varies in a complex manner with the Sanderson electronegativity S (Figure 17.2). It is possible, that for low values of S, the compositional parameters (sites of location) play a greater role. For higher S values, a monotonous decrease of α is observed with increasing S values.

Bimetallic alkali metal particles are formed in the decomposition of sodium azide in MY zeolite, where M = Li, K, Rb, Cs (Table 17.2). Pt, Pd and Rh metal clusters are prepared in presence of Mn^{2+}, Fe^{2+}, Co^{2+}, Cu^{2+}, Cr^{3+} and Ni^{2+} using the ammine complex of the noble metals and the nitrate or sulfate of the transition metals. In most of the cases, the transition metals could not be reduced with molecular hydrogen. Alloys of nanoparticles could only be obtained for Pt–Cu and Pd–Ni (Table 17.3).

Finally, semiconducting clusters such as CdS, CdSe, Zn_4S, etc.were obtained from Cd^{2+} or Zn^{2+} containing zeolites using H_2S or H_2Se as precipitating agent.

It is worthwile to mention the recent preparation of silicon nanoclusters in NaY zeolite, using a chemical vapour deposition of Si_2H_6 followed by a surface reaction [61, 62].

17.3 Characterization of Nanoparticles in Zeolite Hosts

A great number of physicochemical techniques were used to charcterize the metal clusters and ionic aggregates formed in the zeolites. Table 17.4 summerizes the most

Table 17.4. Physicochemical methods used for characterization of the nanoparticles.

Technique	Information	Examples	References
XRD	Location of cations, isolated atoms, clusters	AgX, SOD	1, 2, 66, 67
X-ray line broadening	Aggregates or large particles (>2 nm)	PtY	1
SAXS[a]	Particle size distribution down to atomic size	PtY	1
Radial electron distribution (RED)	Structures of aggregates or clusters	PtY	1
Transmission electron microscopy (TEM)	Particle sizes >0.5 nm	PtY, PtPdY	1, 88
EXAFS[b]	Structures of aggregates or clusters	PtY, $Zn_6S_4^{4+}$, PtPdY	1, 58, 88, 103
XPS[c]	Metal location inside or outside the zeolite	Pt, Pd-LTL	1, 79
Multi NMR	Position of cations and nature of the clusters	PtX, AgXSOD, NaRbY	2, 21, 34, 38, 57, 77
NMR of adsorbed Xenon	Average number of atoms per aggregate or particle	PtNaY, Na_x^0Y	1, 22, 43, 50, 83–85, 104
EPR	Nature of paramagnetic clusters	MX, MY	2, 10, 15, 16, 23, 74, 105
Ferromagnetic resonance	Interaction with neighbouring particles	K-LTA, Rb-LTA	24, 26
UV-visible absorption spectroscopy	Surface plasmon or individual particle transitions	AgXSOD, NaX KX, RbX, NiY	2, 4, 15, 23–25, 34, 58

[a] Small Angle X-ray scattering.
[b] Extended X-ray absorption Fine Structure Spectroscopy
[c] X-ray Photoelectron Spectroscopy

frequently used methods, together with the information one may obtain. Instead of describing all the different characterization techniques separately, we shall analyze a few systems with reference giving to special techniques.

17.3.1 Silver and Silver Halide Zeolites

Sodium silver sodalites (SOD) were extensively studied by Ozin and coworkers [2]. The host material SOD led to confined clusters of insulators, semiconductors and metals. The silver halide zeolites were obtained by ion exchange of their sodium form (Table 17.3).

The regular, all-space filling framework of SOD provides a temperature stable homogeneous nanoporous matrix of sodalite cavities. This is suitable for stabilizing small isolated or interacting molecules, atoms or clusters. Only small cations and molecules up to the size of water can pass through the six rings (Figure 17.1(a)) [2].

Several types of silver and silver halide inclusion compounds can be synthesized. First, Na^+, Ag^+ and Br^- may act as independent ions occupying sites in the unit cell. Second, these ions may interact to form $(Na_x Ag_{4-x}Br)^{3+}$ clusters with molecular behaviour. Each cluster behave like an isolated molecule within the cage, limiting the cluster spatial confinement to ca 6.6 Å. If the clusters interact with each other for example, through the framework, they can form larger units. In this case, the confinement volume could go up to the size of the sodalite crystallite. The Na_8X_2SOD were obtained by direct synthesis. Combined results from powder XRD, Rietveld refinement, ^{23}Na DOR/MAS-NMR, far IR and mid-IR indicate that the anions are randomly distributed throughout the sodalite lattice, forming a homogeneous solid solution [2].

Sodium ions were exchanged by silver ions using either an aqueous solution or the silver salt melt procedure. The unit cell size of sodalite varies with the type of halide and the types and relative concentrations of the cations present (Figure 17.3). Table 17.5 contains the interatomic distances in the various MX sodalites. The halide effect is mainly a space-filling effect, the large anion causing an expansion of the sodalite cage (Figure 17.4). Replacement of Na^+ by Ag^+ causes a contraction of the cage. As Na^+ ($r = 1.13$ Å) and Ag^+ ($r = 1.14$ Å) radii are quite similar, the contraction of the sodalite cage is explained by a more covalent bond between Ag^+ and the central halide X-ions (Table 17.5) [2]. This strong interaction results in

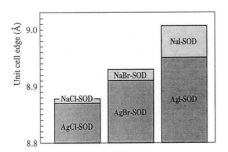

Figure 17.3. Block diagram showing the unit cell sizes of sodium and silver halosodalites [2].

Table 17.5. Interatomic distances, Å Units [2].

Sample	M-X	M-O	Ag-Ag	Ag-Ag (next)	M-X (bulk)	Ag-X[a] (mol.)	Ag-X (next)
NaCl-SOD	2.734	2.372	–	–	2.820	–	–
AgCl-SOD	2.537	2.475	4.142	4.920	2.775	2.28	5.146
NaBr-SOD	2.888	2.356	–	–	2.989	–	–
AgBr-SOD	2.671	2.444	4.361	4.859	2.887	2.393	5.047
NaI-SOD	3.089	2.383	–	–	3.237	–	–
AgI-SOD	2.779	2.576	4.539	4.821	3.040	2.545	4.918

[a] Gas phase molecule

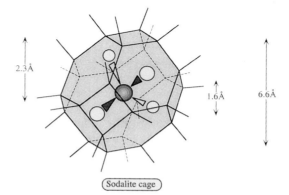

Figure 17.4. Br⁻ encapsulated within a sodalite cage in Na_8Br_2-sodalite: sodalite cage with a central anion and four tetrahedrally disposed cations in the six-ring C_3 sites [57].

shorter Ag–X distances compared to Na–X and an increase in the Ag-framework oxygen separations (Table 17.5).

In Na, AgBr sodalites at loading levels near 2,5-3 Ag^+/u.c. an abrupt break occurred in the magnitude not only for the unit cell edges of the cubic sodalite cage, but also for the peak positions of far-IR absorptions associated with a translational mode of Na^+ near the sodalite six-ring site. Both effects may be related to a percolation threshold for connectivity between AgBr units [2].

Table 17.5 lists the cation–anion, cation–oxygen and silver–silver distances in halosodalites, in the bulk halides and gas phase silver halides. The silver halide distances in sodalite fall between those of the vapour phase molecules and the bulk semiconductor solids. The distance between silver and a halide anion in the same cage increases going from Cl, Br to I, while the separation of the silver to anion in an adjacent cage decreases on going through the halide group. The Ag–Ag distances within each cage increases from Cl, Br to I. Extended Hückel molecular orbital calculations have shown that the orbital overlap, and thus the atomic interaction, is significant for silver at these distances, even though the Ag–Ag separations are much longer than in the metal. At high loading AgX level, one can thus consider the organized array of Ag_4X units as expanded silver halide semiconductors [2].

The UV–visible reflectance spectra were reported for silver halide sodalites. For completely silver-exchanged samples, the optical absorptions were assigned to a) broad 245–250 nm band, Si 3s, 3p $\Leftarrow$ Ag 4d; b) sharp 245–250 nm band, Si 3s, 3p, Ag 5s $\Leftarrow$ Xnp, Ag5s, 5p, 4d [2].

In many host matrices containing quantum size particles, an increase in the loading of the semidonctuctor material results in a red shift as the particle size increases [2, 70]. Inside the sodalites the I–VII cluster nuclearity is limited to five and non significant absorption band shifts occur at higher loadings for the Cl^- and Br^- series. Significant overlap of atomic orbitals within a cluster, as well as overlap of Ag 5p orbitals between adjacent clusters and through framework atoms (see below) allows electronic communication between cages and the contents of the cages [2].

The absorption edges of silver halides and other semiconductors have been fitted to equations for the energy dependence of the absorption index for both direct and indirect, allowed and forbidden electronic transitions, to determine the allowedness of the interband transitions. Despite the fact that the characteristics of Ag_4X^{3+} clusters are close to a molecular nature, the absorption band broadening due to the connectivity between the clusters suggests that the formation of narrow bands may also be possible in silver halosodalites. Narrow bands may occur in insulators and metals when valence electrons have both localized and band like characteristics. The band edges are equal to 3.78 eV for the three AgX sodalites $(X = Cl, Br, I)$ values which are significantly lower than the optical bandgaps in sodium sodalites, influenced by the nature of the halide ion (5.2–6.1 eV) [2].

The interband transitions in the silver halosodalites appear to be indirect, i.e., the bottom of the conduction band is at a different k value than the top of the valence band.

Density of state diagrams show that already for one cluster in a cage the cluster band is clearly visible. At high cluster loadings, the band broadening becomes more pronounced and the orbital mixing is more extensive. The HOMO has Ag4d, Cl3p and Ag5s characteristics, while the LUMO is formed by framework Si3s and Si3p orbitals with some Ag5s mixed in [2] (see above).

Halogen (^{81}Br and ^{35}Cl) NMR provides invaluable clues about percolation thresholds and electronic coupling between clusters. Figure 17.5 shows the MAS–^{81}Br NMR spectra of sodalites $Na_{8-n}Ag_nBr_2$–SOD with n = 0, 0.8, 2,4, 5.2 and 8. The sodium sodalite Na_8Br_2–SOD shows a single NMR line at −219 ppm vs 0.1 M aqueous NaBr solution, ascribed to Na_4Br clusters. The line appears upfield with respect to bulk NaBr (−7 ppm). The shielding indicates that the charge density around Br^- anion is higher for the Na_4Br tetrahedra in Na_8Br_2–SOD than for the Na_6Br octahedra in bulk [2]. This leads to a decrease in the paramagnetic terms of the chemical shift [71]. In presence of Ag^+ ions a new NMR line appears at 214 ppm, which is identical to the line of bulk AgBr. Indeed, this line disappears upon washing the solid sample with diluted $Na_2S_2O_3$ and hence it is attributed to nanocrystalline AgBr at the external surface of the crystallites. A new NMR line appears at −550 ppm at high AgBr loadings (n = 4, 5.2 and 8). (Figure 17.5). This highly shifted NMR line is attributed to Ag_4Br clusters [2, 57]. (In mixed $Na_{4-x}Ag_xBr$ (x = 0–3) clusters the quadrupolar broadening of the ^{81}Br–NMR line is too large and the lines are unobservable. Indeed, the total intensity of the NMR line de-

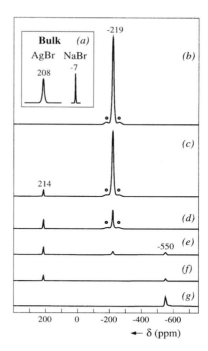

Figure 17.5. MAS ^{81}Br-NMR spectra of (a) bulk NaBr and AgBr; and of $Na_{8-n}Ag_nBr_{2-}$ SOD with (b) n = 0; (c) n = 0.8; (d) n = 2; (e) n = 4; (f) n = 5.2 and (g) n = 8.

creases upon additions of Ag^+ ions to the sample). The increased shielding of the Br^- nuclei in the encapsulated $Na_{4-x}Ag_xBr$ clusters can only be explained on the basis of intercavity electronic coupling – directly through space and/or indirectly through the sodalite framework – which introduces orbital overlap between adjacent $Na_{4-x}Ag_xBr$ clusters [2, 57]. Indeed, the increased covalency in Ag_4Br clusters in comparison with Na_4Br clusters would reduce the density of electronic charge around the ^{81}Br nuclei, leading to an increase in the paramagnetic contribution and hence a more positive chemical shift. This is observed in bulk AgBr (d = 208 ppm) compared to bulk NaBr (d = −7 ppm). An additional proof is the absence of the −550 ppm line in sodalites containing empty sodalite cavities, where the percolation could not take place [2, 57]. The ^{35}Cl–NMR spectra confirm the interpretation of the ^{81}Br–NMR results.

The sodalites containing also empty cavities show interesting results in the UV–visible spectra. At low silver and bromide loadings a very sharp UV spectrum resembles that of an isolated gas phase silver bromide monomer (Br^-(4p), Ag^+(4d) → Ag^+(5s)) (Figure 17.6). The line broadening and red shift of the band edge observed when either the silver or the bromide concentration is increased indicates electronic coupling between Ag_4Br clusters. Broadening to phonon coupling is an additional factor. Indeed, upon cooling Ag, Br–SOD to 27 K three major components could be resolved completely and further components partially [2].

These data provide evidence for the existence of collective electronic and vibrational coupling interactions between $Na_{4-x}Ag_xBr$ clusters over the full Ag^+ and

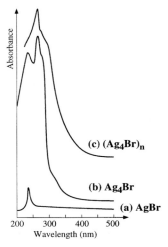

(c) $(Ag_4Br)_n$

(b) Ag_4Br

(a) AgBr

Figure 17.6. UV-visible spectra of the sodalite encapsulated isolated AgBr molecule, the isolated Ag_4Br cluster and the extended $(Ag_4Br)_n$ quantum supralattice [2].

Br^- loading ranges. They also reveal the genesis of the Br^- (4p), Ag^+ (4d) minivalence band and the Ag^+ (5s) miniconduction band on passing from embryonic $Na_{4-x}Ag_xBr$ clusters to a quantum lattice built of the same clusters, and ultimately to the parent bulk mixed sodium–silver halides [2].

In certain sodalites, containing oxalate or formate anions and trapped water, intrasodalite chemical reactions can be carried out in these 6.6 Å diameter sodalite cages, playing the role of nanoreactors [2]. For silver sodalites, such redox reactions lead to interesting chromic responses, such as pressure sensitivity, light and X-ray sensitivites, water and heat sensitivity, all resulting in the colour or the fluorescence. These properties can be exploited to create materials for high density reversible optical data storage [2, 72].

Silver oxalatosodalites contain Ag_4Ox^{2+}, Ag_4^{4+} and Ag_3^{3+} clusters and the latters predominate. Oxalate acts as an internal reducing agent for both photoreduction and thermal reduction of silver:

$$2Ag^+(Z) + C_2O_4^{2-}(Z) \rightarrow 2Ag^\circ(Z) + 2CO_2$$

During photoreduction Ag_4^{q+} clusters with short Ag–Ag bond lengths are formed, with q = 2 or 3. These clusters are entrapped within the sodalite cages. Upon continued UV irradiation, the optical spectra suggest the formation of an expanded metal or semiconductor, as the Ag_4^{q+} cluster density increases (Figure 17.7). Indeed the UV–visible spectra show the development of a broad background resembling a silver plasmon absorption [2]. During sample irradiation, more and more cages are filled with partially reduced silver clusters. Although they are surrounded by a sea of Ag_3^{3+} triangles which are not reduced, they may communicate electronically forming various bands [2]. (Note, that in Y zeolite, Ag_3^{2+} and Ag_3^0 clusters could be formed upon γ irradiation [13]).

As a conclusion, the unique structural properties of the sodalite host and the

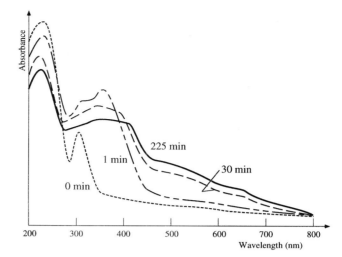

Figure 17.7. Optical absorption bands of silver oxalatosodalites after various UV irradiation times [2].

variety of guests that can be encapsulated make this material an ideal model system for probing the physicochemical properties of clusters built up of the components of bulk insulators, semiconductors and metals, as well as a promising candidate for advanced materials research [2].

17.3.2 Alkali Metal and Ionic Clusters in Zeolites

The physicochemical properties of small metal particles and ionic clusters can be more easily explored when they are encapsulated and stabilized in the zeolite cages or channels. The conduction electron energy spectrum of a metal is usually considered to be a continuum. However, for small metal particles having a small number of conduction electrons, the discreteness of the spectrum has an important consequence on the thermodynamic properties such as magnetic susceptibility or thermal relaxation. These properties should strongly depend upon whether the number of electrons is even or odd when the thermal energy is smaller than the average spacing of the electron energy levels [21].

Since the first description of the generation of charged and neutral sodium clusters in faujasite zeolites by Rabo et al. [23], their preparation and physicochemical characterization were in focus using XRD, EPR, NMR, IR, UV–visible and quantum chemistry [10].

During the decomposition of NaN_3 on CsY faujasite the following reactions could be distinguished and the relative concentrations were followed as a function of time:

$$2NaN_3 \rightarrow 2Na + 3N_2$$

The outercrystalline Na clusters (Na_{out}) were characterized by a broad EPR line

Table 17.6. EPR parameters for alkali-metal-loaded alkali-metal cation-exchanged X-zeolite prepared by alkali-metal azide decomposition [17].

Sample	g value	$\Delta H/G$	$A\mathrm{iso}/G$
Li/LiX	2.0029 ± 0.0005	2.8 ± 0.2	–
Na/NaX	2.0013 ± 0.0005	4.2 ± 0.2	25.5 ± 0.2
K/KX	1.9997 ± 0.0005	6.1 ± 0.2	12.8 ± 0.2
Rb/RbX	1.9924 ± 0.0005	8.3 ± 0.2	–
Cs/CsX	1.9685 ± 0.0005	22.2 ± 0.2	–

at $g = 2.076$. The migration of outercrystalline Na clusters into the faujasite cages (Na_{int}) was followed:

$$Na_{out} \rightarrow Na_{int}$$

The latter showed a narrow ESR line at $g = 2.003$.

Finally, the sodium metal could reduce the Cs^+ ions leading to Cs_x clusters, with $g = 1.998$ [10]. In NaY zeolite, Na_4^{3+} clusters were also identified, showing the hyperfine structure [16, 19].

The most systematic work was carried out by the group of Kevan et al [17]. They have synthesized the alkali metal clusters by both azide decomposition and vapour deposition in X zeolites. Both methods gave similar EPR spectra. The EPR parameters for the various M/MX zeolites, i.e. the g factors, linewidths and hyperfine coupling constants are reported in Table 17.6.

The Li/LiX sample is red brown, the Na/NaX and the K/KX samples are dark blue and the Rb/RbX and Cs/CsX samlples are also blue. The g factor decreases, while the linewidth increases from Li to Cs. Indeed, the g values of conduction electrons in alkali-metal particles deviate more below the free-electron value ($g = 2.0023$) as the spin–orbit coupling increases with increasing atomic weight. The parallel relationship between the increasing linewidth and the increasing g shift with increasing atomic number reflects an effect of the spin–orbit coupling constant on the relaxation time. The small spin–orbit coupling constant for lithium gives a narrow linewidth for its metal particle and a larger spin–orbit coupling constant for caesium gives a broader linewidth for its particle. The fact that the linewidths are approximately temperature independent suggests that surface scattering dominates the relaxation rate. Therefore, the size of the alkali-metal particles is sufficiently small (<10 nm) to show small metal particle properties [17].

The Na/NaX and K/KX samples show hyperfine structure. The alkali metal clusters can be formed either in the α cage (12.5 Å diameter) or in the β cage (6.5 Å diameter) (Figure 17.1(c)).

Note that the alkali-metal atoms can enter the α cage, while they are too large to enter the β cage with an aperture of ca 2.5 Å. However, they can transfer an electron to produce a paramagnetic species in the β cage.

The 19-line signal of the Na/NaX sample was previously attributed to an ionic cluster species, Na_6^{5+} [23], consisting of an upaired electron interacting with six equi-

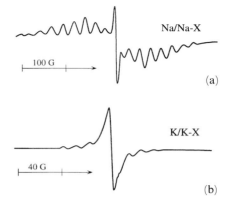

Figure 17.8. EPR spectra at 300 K of (a) Na/NaX and (b) K/KX zeolite samples [74].

Table 17.7. Cation site distribution per unit cell of Na- and K-exchanged zeolites X and Y and ionic clusters formed in these zeolites [17].

Number of sites	Zeolite Y		Zeolite X	
	Na^+	K^+	Na^+	K^+
16 SI	7.5	5.4	4	9.2
32 SI'	19.5	18.1	32.3	13.6
32 SII	30	26.8	30.8	25.6
32 SII' supercage		4.4		38.2
cluster formed	Na_4^{3+}	K_4^{3+}	Na_6^{5+}	K_3^{2+}

valent sodium nuclei ($I_{Na} = 3/2$) (Figure 17.8). The 10-line multiplet of the K/KX sample is attributed to K_3^{2+} clusters, consisting of an unpaired electron interacting with three equivalent potassium nuclei ($I_K = 3/2$) [17]. On the other hand Harrison et al determined Na_4^{3+} and K_4^{3+} ionic clusters in NaY and KY zolites, respectively [73]. The formation of these different species can be explained on the basis of available cations in both X and Y zeolites (Table 17.7). The number of SII sites is about the same in X and Y zeolites (Figure 17.1(c)). In NaX zeolites, the number of cations on SI' sites is larger than in Y zeolite. In KX zeolites, however, the number of cations in site SI' is smaller than in Y zeolite. It seems that the number of cations in site SI', i.e. the number of cations inside the β cage is an important factor that affects the number of nuclei in the ionic clusters formed. Indeed, Na_4^{3+} is formed in NaY zeolite and Na_6^{5+} in NaX zeolite, while K_4^{3+} is formed in KY and K_3^{2+} KX zeolite (Table 17.7). Moreover, these results suggest that the nuclei in the ionic clusters come from the originally exchanged cations which exist in the zeolite and that the clusters are formed inside the β cages [17]. Finally, by comparing the diameters of the sodium atom (3.7 Å), sodium cation (2.3 Å), potassium atom (4.6 Å) and potassium cation (3.0 Å) with the diameter of the β cage (6.5 Å), one concludes that these ionic clusters must be highly positively charged so that the formed cluster can fit into the β cage [17].

In another study a complete matrix of experiments involving the reaction of all five alkali metal vapours with all five alkali cation exchanged X zeolites was reported [74].

Except for the Li/MX samples which have a purple colour, all the other samples are blue or dark blue depending on the concentration of the alkali metal. All the samples give a single EPR line at ca $g = 2.0$ which is assigned to an alkali metal particle formed within the zeolite. Hyperfine patterns are observed in Na/LiX, M/NaX and M/KX samples which are assigned to alkali metal ionic clusters formed in the β cage (see above).

Alkali metal particles formed in M/MX systems were characterized by conduction electron EPR lines and have distinctly different g values (Table 17.8). The diagonal M/MX values allowed one to assign the nature of the metal particles. For example, the g value in K/RbX sample is very close to the value of rubidium metal particle and is so assigned. All the metal particles are formed in the α cages of the zeolite.

Table 17.9 shows the gas phase enthalpies for electron transfer from alkali metal atoms to alkali metal cations [74]. Positive values indicate that electron transfer

Table 17.8. Room temperature EPR g values of alkali metal particles in alkali metal cation exchanged X zeolites reacted with alkali metal vapour and assignment of metal particle formed [74].

Metal vapor	Li-X	Na-X	K-X	Rb-X	Cs-X
Li	2.0034	2.0053	2.0004	1.9939	1.9712
	Li	Li	Li or K	Rb	Cs
Na	2.0032	2.0016	1.9998	1.9936	1.9776
	Li	Na	Na or K	Rb	Cs
K	2.0008	1.9999	1.9997	1.9939	1.9722
	K	K	K	Rb	Cs
Rb	1.9932	1.9929	1.9930	1.9929	1.9736
	Rb	Rb	Rb	Rb	Cs
Cs	1.9640	1.9686	1.9998	1.9931	1.9686
	Cs	Cs	K	Rb	Cs

Table 17.9. ΔH values for gas-phase electron transfer from alkali metal atoms to alkali metal cations (kJ/mol) [74].

ΔH	Li^+	Na^+	K^+	Rb^+	Cs^+
Li^0	0	244	1014	1172	1445
Na^0	-244	0	770	928	1201
K^0	-1014	-770	0	158	431
Rb^0	-1172	-928	-158	0	273
Cs^0	-1445	-1201	-431	-273	0

from the alkali atoms with smaller atomic number to the alkali cations with larger atomic number is endothermic in the gas phase. The gas phase thermochemistry seems to apply to Na/LiX, Cs/KX and Cs/RbX samples in which electron transfer from the metal vapour atom occurs, and to Li/NaX, Li/KX and Na/KX samples where no electron transfer occurs (Table 17.8).

It may also be expected that the gas phase thermochemistry is significantly modified by the highly ionic nature of the zeolite lattice. This can result either in no electron transfer when expected or reverse electron transfer from the smaller atomic number atom to the larger atomic number cation. The following systems show no electron transfer although such would be exothermic in the gas phase: K/LiX, Rb/LiX, Cs/LiX, K/NaX, Rb/NaX, Cs/NaX and Rb/KX.

The third category of systems is where electron transfer occurs in the reverse direction from that predicted by the gas phase thermochemistry: Li/RbX, Na/RbX, K/RbX, Li/CsX, Na/CsX, K/CsX and Rb/CsX. All these samples include electron transfer from a smaller atomic number metal vapour atom to the large Rb^+ or Cs^+ cations. Reverse electron transfer only occurs for RbX and CsX zeolites. These largest cations may partially block the 12-ring apertures to the α cages and inhibit the diffusion of metal vapour into the α cages (Figure 17.1(c)). However, if electron transfer occurs, the Rb and Cs atoms formed have a much weaker interaction with the 12-ring window and migrate more in the α cages. This open the way for the Li^+, Na^+ ... cations formed by reverse electron transfer to migrate into the α cages to find a coordination site. These size factors probably influence the overall energetics for this reverse electron transfer [74].

Similar reverse electron transfer was observed on Y zeolite, where MN_3 (M = Li, Na and Cs) decomposition on MY (M = Li, Na, K, Rb and Cs) zeolites was systematically explored [16]. Moreover, not only pure M_x^0 particles, but also mixed alkali metal particles such as Na_xK_y, Na_xRb_y or Na_xCs_y were observed [10, 16].

Nuclear Magnetic Resonance was also used to characterize the alkali metal particles encapsulated into the zeolite cages [10, 21, 75–77]. The ^{23}Na NMR spectrum of the Na loaded NaY zeolite shows a broad and unshifted line (vs NaCl) at room temperature that is assigned to sodium metal particles with even number of atoms with an uncomplete spin pairing due to spin orbit coupling. In the Rb loaded NaY zeolite the ^{23}Na NMR spectrum shows a broad unshifted line and a largely shifted narrow line at 1660 ppm at room temperature (Figure 17.9) [77]. The insert shows the temperature dependence of the shifted line between 280 and 260 K. By decreasing the temperature, the resonance frequency of this line decreases slightly and its intensity decreases and vanishes at 260 K, while a new narrow line appears at 1122 ppm. At high Rb loading Na–Rb alloy could be formed, characterized by the NMR line at 1660 ppm (6350 ppm for ^{87}Rb NMR line vs Rb_2CO_3). The Knight shift of the bulk sodium metal is 1120 ppm. The low field shift of the Na–Rb alloy can be explained by the more electronegative character of sodium with respect to rubidium, causing an increase in the spin density of Na nuclei and a decrease on Rb nuclei. This leads to an increase in the Na Knight shift (6530 ppm for bulk rubidium metal). The temperature dependence of the observed shifted lines described above can be interpreted as a consequence of a phase transition of the alloy at 260 K from a liquid alloy to two separated Na and Rb phases. Indeed, the resonance

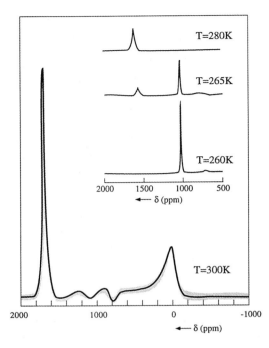

Figure 17.9. Room temperature ^{23}Na NMR spectra of NaY zeolite loaded with Rb. Insert: temperature dependence of the highly shifted NMR line [77].

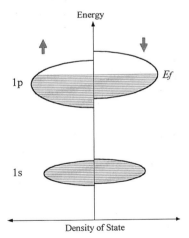

Figure 17.10. Schematic representation of the model of the itinerant electron ferromagnetism in K-LTA zeolite [26].

positions of the new lines correspond to the Knight shifts of bulk Na and Rb, indicating the formation of separated metal aggregates.

Note that ^{129}Xe NMR was also used to probe the presence of sodium metal particles in zeolite Y. This method will be described in more detail below [22].

Novel electronic properties can be expected in mutually interacting arrayed clusters. The zeolite provides a periodic nanoscale space for guest materials and an array of clusters can be stabilized there [24]. Ferromagnetism is the result of electron correlation. If electrons are localized at each cluster, they exhibit paramagnetism and diamagnetism in open and closed shells of localized electronic states, respectively [26]. Free electron metals such as potassium exhibit Pauli paramagnetism and Landau diamagnetism but no ferromagnetism at any electron density. However, when alkali metal atoms are loaded into the dehydrated zeolite cages the s electrons are delocalized over many cations resident in the space of framework and cationic clusters are generated. Ferromagnetism was observed in K–LTA zeolite, although no magnetic element is contained there. The most remarkable ferromagnetism was observed at ca 5K atoms per α cage. From the point of view of optical properties, the 4s electrons of guest K atoms occupy 1s- and 1p-like orbitals. The doped electrons are essentially confined in the α cage, because the resultant energy of the localized state in the α cage is lower than that in the β cage. The α cage have six windows with a diameter of 5Å and these windows are shared with an adjacent α cage. The framework potential for electrons may be lowered at windows because of the weakness of the framework-repulsive potential. If cations are distributed near windows, the potential for electrons is lowered further. Therefore an overlap between molecular orbitals in adjacent α cages is expected at the windows and the electrons become partly itinerant [26]. The intercluster electron transfer leads to the energy band originating from respective quantum electronic state of the cluster. Two and six electrons per α cage from guest alkali atoms fill the 1s and 1p energy bands, respectively. Figure 17.10 represents the model of the itinerant electron ferromagnetism in K–LTA zeolite. The density of state is shown for up- and down-spin electrons in energy bands originating form 1s and 1p molecular orbitals of the K cluster. If the Fermi energy E_f is located at the energy of a sufficiently high state density for the 1p band, a finite population difference between up and down spin minimizes the total energy, resulting in ferromagnetism [26].

According to theoretical expectations, the ferromagnetism as well as the optical properties are very sensitive to the intercluster transfer energy for electron. This can be varied by changing the nature of the alkali atom, i.e. changing its ionization potential.

The optical and magnetic properties for Na–, K– and Rb–LTA zeolites will be compared for 4.9, 5.4 and 5 alkali atoms per cage, respectively. The samples were prepared by vapour phase deposition [24]. Figure 17.11 shows the reflection spectra for the three samples. Dotted curves indicate the region where the transmission through each powder particle cannot be neglected. The surface plasmon excitation at ca 2 eV dominates both the Na and K cluster optical spectra, because the confinement potential depths for electrons is sufficiently large. However, in Rb clusters, the cluster potential is shallower, because of the smaller ionization energy of the Rb atom. The surface plasmon band becomes indistinct and the individual excitation dominates the spectrum. They are assigned to 1p–1d transitions at 1.6 and 2 eV in the Rb–LTA sample. Note also that the oscillator strengths for the 2 eV bands are equal to ca 4 and 3.5 for the Na–LTA and the K–LTA samples, respectively [24]. These large values are due to surface plasmon enhancement effects, because the internal electric field at the surface plasmon energy is resonantly excited by the external electromagnetic wave.

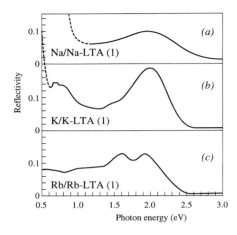

Figure 17.11. Reflection spectra of Na-LTA (4.9/α cage) (a), K-LTA (5.4/α cage) (b) and Rb-LTA (5/α cage) (c) [24].

The magnetic susceptibilities of the K– and Rb–clusters follow the Curie–Weiss law, showing ferromagnetic behaviour. The Weiss temperatures are estimated to be 6 and 2.7 K, respectively. On the other hand the magnetic susceptibility of Na clusters is $-1 \pm 1 \times 10^{-6}$ emu/cm^3 and shows that the Na clusters have no magnetic moment and are diamagnetic [24].

The ferromagnetism of the K and Rb clusters can be explained by the theory of itinerant electron magnetism. The Curie temperature increases with the increase of the Stoner factor, which is given by the product of the on-site Coulomb energy and the density of state at the Fermi surface. As the electron densities are ca 5 per α cage, the Fermi surface is located at the center of the 1p energy band (Figure 17.10). The value of the intercluster transfer energy of electrons is higher, while the density of state at the Fermi surface is lower for the Rb cluster with respect to the K clusters. This may result in a smaller Stoner factor and hence a smaller ferromagnetism in the Rb clusters, as it is observed experimentally [24].

In Na clusters, the intercluster transfer energy is smaller and the density of state at the Fermi surface is higher than in the K cluster. Hence, a greater ferromagnetism could be expected for the Na cluster. However the Na clusters show neither ferromagnetism nor paramagnetism but diamagnetism. This can be explained by a strong electron–phonon interaction in the Na clusters leading to the pairing of up- and down-spin electrons. The present electron–phonon interaction is mainly caused by the electronic interaction with cation displacement, because the framework is rigid. When an electron moves in α cage, cations will change position in the cage in order to decrease the total energy. If the intercluster transfer energy of electron is sufficiently smaller than the energy gain due to electron–phonon interaction, an electron can be localized within the α cage. If the second electron comes to the same cage, the potential depth will further increase. Hence, two electrons in the same cage decrease the energy, compared with those in different cages. Indeed, the electron–phonon interaction has to be greater than the repulsive Coulomb interaction between two electrons. As a result, as the homogeneous distribution of electrons is less stable, the electrons gather together to generate stable clusters containing even number of electrons. In the case of Na clusters only the diamagnetic case is observed. In the present case the average electron number is 5 per α cage. Hence, half of the clusters should have four electrons and half six electrons.

17.3.3 Transition Metal Clusters in Zeolites

Platinum particles incorporated in X or Y zeolites were extensively studied by Gallezot et al [in ref. 1] and Sachtler et al [49], using various physicochemical methods for their characterization (Table 17.4). As a result, the location, particle size distribution and the structure of the particles are reasonably understood. The size and structure of the particles in Y zeolite were determined by crystal structure analysis.

When the precursor Pt^{2+} ions are in the supercages, small Pt particles ($d = 0.6–1.3$ nm) with an average coordination number of 7.7, as determined from EXAFS measurements, are formed in the supercages. If Pt^{2+} ions are distributed in both supercages and sodalite cages, Pt particles are formed by adding Pt atoms escaping from the sodalite cages onto the surface of Pt particles already formed in the supercages.

The location and size of the Pt particles is determined by the temperature of oxidation before reduction is carried out. In addition, Pt agglomeration in the zeolite crystals can take place to give aggregates with sizes that exceed that of the supercages (ca 2.0 nm diameter). The structure of the small Pt particles occupying the supercages (Figure 17.1(c)) and having a diameter of ca 1.0 nm was determined by radial electron distribution method [1]. The interatomic distances are contracted with respect to the fcc (face-centered-cubic) structure of bulk Pt, although upon adsorption of hydrogen, relaxation towards the normal fcc structure of the bulk metal occurs. From EXAFS analysis of the first-neighbour distances, it was concluded for the 0.7–0.8 nm particles, that a mixture of icosahedra and cuboctahedra with contracted lattice parameter existed, whereas for the larger extrasupercage particles of 2.0 nm, the bulk cubic model better described the results.

The electron deficient nature of Pt particles in zeolites is now firmly established. The early evidence for this behaviour was shown by the IR frequency variation of adsorbed probe molecules such as CO or NO [1]. Indeed, v_{CO} shifts to higher frequencies when fewer electrons are available for back-donations of the d-metal orbitals into the empty π^*-antibonding orbitals of CO. An increase of v_{CO} was observed with decreasing particle size.

Several explanations have been proposed in order to account for the metal electron deficiency. One of the explanations suggests a partial electron transfer from the metal to the support [78]. This was recently confirmed on Pd–LTL samples by both XPS and IR measurements [79]. In XPS, the binding energy was shown to be particle size dependent, increasing by ca 0.5 eV as the particle size decreased form 5.0 to 1.0 nm. However, (see ref. 79), the Pd particle size remained quasi constant (ca 1.3 nm) and the 1.5 eV binding energy variations from acidic to alkaline samples could only be attributed to the interaction between the metal particle and the support. The Pd particles can either be electron deficient (on acidic supports) or electron rich (on alkaline supports) compared to bulk Pd. The ν_{CO} frequency variations confirmed the metal-support electron transfer.

Sachtler et al suggested that Pd in acidic zeolite Y can form postively charged Pd_n-H^+ adducts, where the proton charge is delocalized over the Pd cluster [80]. Finally, Van Santen et al have proposed that electrons are attracted by metal atoms of the particle near the cations, leading to electron-deficient metal atoms that are at the opposite side of the cations [81] (i.e. polarization of the metal particle by a nearby cation).

The first ^{195}Pt NMR spectra of Pt particles in zeolites were reported by van der Klink et al [38, 82]. The NMR spectra were taken point-by-point by scanning the frequency with a spin-echo sequence. Between 240 000 and 320 000 scans were accumulated for each point, at repetition rate between 33 and 50 s^{-1} [82]. The diameter of the Pt particles varied from 1.0 to 1.6 nm in Y zeolite (the 1.0 nm particle contains ca 30 Pt atoms).

Most of the Pt–Y samples contained particles too big to fit into the supercages without steric problem (d = 1.25 nm). As the TEM observations clearly showed that they were inside the matrix, the local framework structure of the zeolite was damaged during the Pt particle growth.

Figure 17.12 shows the spin echo point-by-point ^{195}Pt NMR spectra of the clean Pt metal particles (a) encapsulated in NaX zeoite, under oxygen (b) and hydrogen

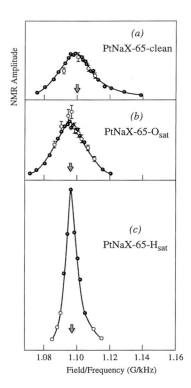

Figure 17.12. ^{195}Pt NMR spectra of PtNaX-65 (65% Pt dispersion), obtained at 80 K by the point by point spin echo method (a) Clean surface, (b) Saturated with oxygen, (c) Saturated with hydrogen. The arrows indicate the centers of gravity (first moments) of the spectra [38].

(c) adsorption [38]. The NMR line shape does not vary markedly with temperature as the Knight shift is not very sensitive to the temperature-dependent changes in the local density of states at the Fermi level. The effect of oxygen chemisorption is less marked than the hydrogen chemisorption. The line maximum shifts from 1.100 G/kHz to 1.096 G/kHz under oxygen or hydrogen chemisorption.

The ^{195}Pt nuclear spin-lattice relaxation time (T_1) was studied as a function of temperature. At temperature above 80K all clean-surface samples show the Korringa relaxation mechanism with $T_1 T = C'$, characteristic of nuclei in the metallic state.

In Pt NMR, both the Knight shift K and the Korringa product $T_1 T$ depend on contribution from the s-like and d-like local density of states (LDOS) at the Fermi energy. Many combinations of s- and d-like LDOS may give the same Knight shift, but different $T_1 T$ products and therefore the spin-lattice relaxation curve, measured at a given point (fixed K) in a small particle NMR spectrum is a more sensitive probe for the metallic nature of the Pt atoms. The recovered signal amplitudes A_{jk} were measured at temperature Tj for different values τ_k of the relaxation interval after saturation of the nuclear magnetization. The A_{jk} vs τT curves were then scaled in order to obtain a single multiexponential curve (Figure 17.13) [38].

The PtNaY data show that the size independence observed at 80 K (Figure 17.13(a)) is lost at lower temperatures and that smaller and smaller fractions of the signal obey time–temperature scaling when the temperature is lowered. The scaling fraction at fixed temperature is seen to diminish with decreasing particle size [38]. On the average, a fraction of ca 0.3 of the signal obeys time–temperature scaling down to 22 K. After chemisorption of oxygen or hydrogen, deviations from scaling behaviour are seen at relatively high 225–250 K temperatures.

As a conclusion, the metallic character of the Pt atoms decreases with decreasing partical size as well as with oxygen or hydrogen chemisorption. At 22K only particles with diameters longer than 1.6–1.9 nm are still metallic [38].

^{129}Xe NMR is a powerful technique to characterize the metal particles incorporated into the zeolite cavities [43, 50, 83–85]. The NMR chemical shift of ^{129}Xe is very sensitive to the void volume, the nature of cations, the nature and size of metal particles.... [83].

Xenon atoms are exchanging very rapidly on the NMR time scale between supported metal clusters in large cavities, the cage walls and adjacent zeolite crystals. As a result of this fast exchange only a single NMR line is obtained. In general, a hyperbolic shaped relationship is observed for the ^{129}Xe NMR chemical shift and the Xenon pressure (or the concentration of physisorbed Xe in moles of Xe per gram of zeolite) for PtNaY zeolites. If one assumes that four Xe atoms can approach the Pt particle in the supercage through the 4 windows and if z is the maximum number of Xe atoms that can fit an empty supercage at high pressure, one arrives at the expression [43] of the chemical shift at a certain Xe pressure $\delta(p)$:

$$\delta(p) = \delta_{Pt} N_{Pt} \frac{4}{[N_{Pt}4 + (1 - N_{Pt})zap]} + \delta_{NaY}(p)$$

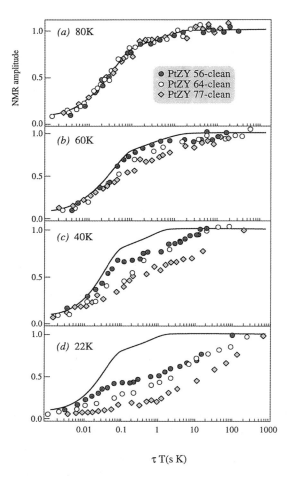

Figure 17.13. Time-temperature scaled spin-lattice relaxation data for three clean-surface Pt NaY samples taken at spectral position 1.100 G/kHz. The full curve in all panels is a double-exponential fit to the 80 K data [38].

where δ_{Pt} is the Xe chemical shift of xenon in contact with the metal particle, $\delta_{NaY}(p)$ is the chemical shift of xenon in contact with the zeolite wall at a certain pressure, N_{Pt} is the fraction of supercages containing a metal cluster, $N_{NaY} = 1 - N_{Pt}$ is the fraction of empty supercages and a is the adsorption constant of xenon in empty NaY supercages. (Note that N_{NaY} is considered close to unity.) If $\Delta(p) = \delta(p) - \delta_{NaY}(p)$, a linearized form of the above equation can be obtained:

$$\frac{1}{\Delta(p)} = \frac{1}{\delta_{Pt}} + \frac{za}{4\delta_{Pt}} \frac{1 - N_{Pt}}{N_{Pt}} p$$

The product za was obtained from adsorption isotherms and is equal to 0.0033 Torr^{-1} at room temperature [86].

Table 17.10 shows the fraction of cavities occupied by Pt or PtCu alloy particles, together with the number of atom per particle. The formation of alloy particles

Table 17.10. Number of mono- or bi-metallic clusters per supercages ($N = N_{Pt}$ or $N = N_{PtCu}$) and number of atoms per clusters ($m = m_{Pt} = n_{Pt}/N_{Pt}$ or $m = (m_{Pt} + m_{Cu}) = (n_{Pt} + n_{Cu})/N_{PtCu}$) with n_{Pt} and n_{Cu} being the number of atoms per supercages for samples Pt_xCu_y/NaY.

Sample	N	m
$Pt_{1.75}$	0.078	22
$Pt_{0.45}$	0.025	18
$Pt_{0.26}$	0.020	13
$Pt_{0.087}$	0.0029	30
$Pt_{0.89}Cu_{0.89}$	0.038	47
$Pt_{0.26}Cu_{0.10}$	0.034	11
$Pt_{0.26}Cu_{0.20}$	0.026	18
$Pt_{0.26}Cu_{0.26}$	0.027	19

during the reduction is strongly favoured if the precursor of the less reducible metal is mobile over the surface of the support [50]. In the case of PtCuY, platinum ions will be reduced first and the subsequent reduction of Cu^{2+} ions will be catalyzed by the Pt particle. As a consequence, a copper outer shell is formed covering the inner core Pt particle. This is confirmed by ^{129}Xe NMR and EXAFS data. The latter show clearly that the Pt–Pt distance did not change from pure Pt particles to the PtCu alloys encapsulated in the zeolite supercages.

Extended X-ray absorption fine structrue (EXAFS) is the most commun physical method for determining the coordination number of the supported metal and thus for obtaining information about the particle size [103].

For example, EXAFS confirmed the existence of small PtPd bimetallic particles in NaY zeolite [88]. It also showed that Pt and Pd atoms are randomly mixed. From the average total coordination numbers diameters of ca 1 nm were calculated for the bimetallic particles. The particle size of the pure Pd particles was ca 2.5 nm. Transmission electron microscopy (TEM) showed average diameters of less than 2 nm for bimetallic particles and of 5.8 nm for the monometallic Pd particle. The discrepancy between the EXAFS and the TEM values can be explained as follows. EXAFS underestimates the coordination number for very small particles. On the other hand, it is difficult to distinguish very small (<1 nm) particles by TEM, from the zeolite background because of lack of contrast and therefore the number averaged diameter is overestimated [88].

Supported bimetal particles consisting of iron and a platinum group metal (Ru, Rh, Pd, Ir or Pt) have been studied extensively (Table 17.1). These studies led to the general conclusion that the reducibility of the iron group metals is enhanced by the platinum group metals. Whereas the reduction of, for example, Fe^{3+} or Fe^{2+} ions to $Fe°$ inside a zeolite such as Y is known to be more difficult than reduction of Fe_2O_3 particles, it is probable that some of these ions can be reduced to $Fe°$ in the presence of platinum group metals, but the extent of this reduction will depend on the relative position of the ions, their mobility, the concentration of protons and the temperature [54]. It was shown that Pd or Pt particles strongly interact with iron ions and the chemical anchoring of these particles by iron ions led to an enhanced dispersion of the particles. Reduction of ion-exchanged PdFe/NaY with H_2 results in the formation of Fe^{2+} ions and PdFe alloy clusters. The extent of Pd-enhanced reduction of Fe^{2+} to $Fe°$ decreases with increasing proton concentration in the zeolite. Data obtained form Mössbauer spectroscopy, ferromagnetic resonance and FTIR of adsorbed CO indicate prevalent formation of PdFe alloy clusters [54]. An average composition of $Pd_{16}Fe$ and $Pd_{10}Fe_3$ was found for the reduced clusters in PdFe/NaHY and neutralized PdFe/NaY, respectively. A strong ferromagnetic resonance signal from $PdFe_x$ particles isolated in different zeolite cages and containing only one or two Fe atoms suggests that Pd atoms surrounding an Fe atom become polarized. Long-range magnetic coupling results from strong spin polarization of the Pd matrix, which aligns itself in the direction of the magnetic moment of the Fe atoms [54].

17.3.4 Miscellaneous

17.3.4.1 New Forms of Luminescent Silicon

There is current interest in devising physical and chemical methods for converting silicon into an efficient room temperature light emitter [61]. The well known diamond lattice of bulk silicon makes it an indirect band gap semiconductor with electric dipole forbidden luminescence. Finding a method to trick silicon into circumventing this strict selection rule and evoke useful luminescence from silicon, while at the same time maintaining the good electrical transport, mechanical and thermal stability of the crystalline form of silicon, would provide opportunities for introducing light emitting silicon into integrated circuitry. This could lead to the development of silicon-based optoelectronic devices [61].

Approaches that are currently used for engineering photoluminescence or electroluminescence into crystalline silicon, involve defect engineering and band structure engineering [61,62]. The former method includes substitution of isoelectronic or isovalent impurities for silicon and rare earth doping. The latter method includes alloying, such as $Si_{1-x}Ge_x$ superlattices and the use of quantum size effects, exemplified by quantum well, wire and cluster forms of silicon [61, 62].

The chemical vapor deposition of disilane Si_2H_6 within the diamond lattice of 13 Å supercages in the acid form of zeolite Y (HY) has been employed to assemble an array of silicon nanoclusters. In this process, Si_2H_6 molecules are reactively anchored to Brönsted acid sites in dehydrated HY. Controlled thermal treatment of these samples in a closed system induces a series of H_2 elimination reactions to yield encapsulated capped silicon nanoclusters. The resulting materials are air and water stable and display orange–red photoluminescence at room temperature. The intensity and lifetime of this photoluminescence is temperature dependent and the intensity is proportional to the disilane loading [61, 62] (Figure 17.14). The optical absorption edges and luminescence energies of these materials display monotonic red shifts with increasing disilane loading. They display no temperature dependence of their absorption edges which are blueshifted with respect to bulk silicon (Figure 17.14).

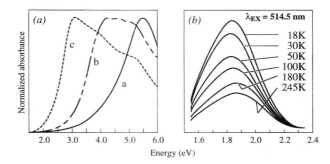

Figure 17.14. Left: changes in the optical absorption spectra of the silicon nanoclusters in zeolite Y with respect to Si_2H_6 loading, (a) 2 Si_2H_6/supercage, (b) 5 Si_2H_6/supercage, (c) 8 Si_2H_6/supercage. Right: emission spectra of silicon nanocluster in Y zeolite, obtained from 4 Si_2H_6/supercage [61, 62].

An interesting series of correlations was reported relating such parameters, as average Si coordination numbers for clusters of different sizes and shapes, inverse characteristic length scales, total number of silicon atoms contained in a cubic Si cluster of a particular size and the corresponding measured luminescence lifetimes and peak luminescence energies [91] (Figure 17.15). This study enables one to conveniently ascribe a measured luminescence to a specific size silicon nanostructure regardless of how the sample was prepared. For example, it provides a foundation for ascribing the measured 1.8 eV luminescence energy to ca 13 Å diameter nanoclusters containing around 60 atoms. It can be shown by molecular models that this cluster can fit the Y zeolite supercage [61].

The chemical vapour deposition of Si_2H_6 was also used to form silicon clusters in the 35 Å hexagonal symmetry mesoporous channel of MCM-41. The TEM images of the material point to a dominant intrachannel silicon cluster growth process rather than one in which a hollow silicon cylinder forms and inwardly grows from the channel walls of the host [61]. This is consistent with the behaviour of the optical absorption edge which monotonically red shifts and asymptotically converges to a limiting energy that corresponds to a silicon cluster of about 35 Å in size [61, 91]. It appears that silicon cluster growth is indeed spatially constrained by the 35 Å channel dimensions of the mesoporous silica host.

Studies have also been initiated that are aimed at synthesizing $Si_{1-x}Ge_x$ binary nanoclusters by topotactic chemical vapour deposition of disilane–digermane mixtures within the diamond lattice of 13 Å diameter supercages of zeolite Y [61]. Noteworthy is the composition edges that has been observed for the encapsulated $Si_{1-x}Ge_x$ binary nanoclusters which are consistently blue shifted with respect to their bulk $Si_{1-x}Ge_x$ alloys. Trends in the composition dependence of the optical absorption spectra of the binary $Si_{1-x}Ge_x$ nanoclusters are consistent with an EXAFS derived structural model based on a diminishing diamondlike four coordinate Ge cluster core and a thickening Si outer coating, the latter being oxide and hydride capped [61].

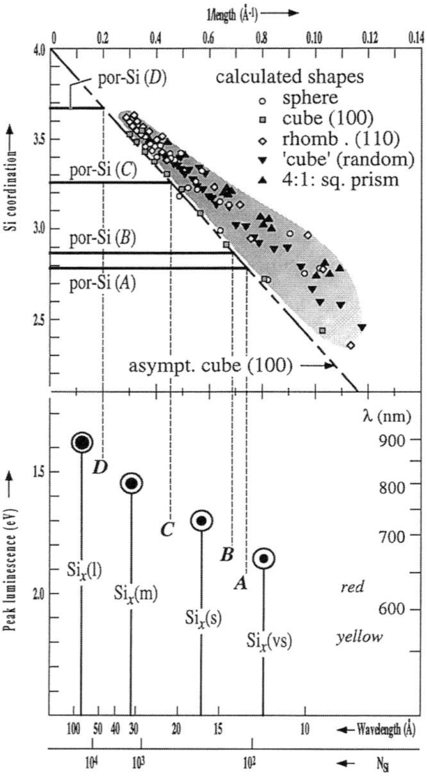

Figure 17.15. Top: Correlation between average Si coordination number for particles of different shapes versus their characteristic lengths. Bottom: experimental EXAFS determined coordination numbers for differently prepared porous silicon samples labeled A, B, C and D, average silicon particle size, total number of silicon atoms Ni_{Si} contained in a cubic silicon particle and measured peak luminescence energy [91].

17.3.4.2 Semiconductor Nanoclusters in Zeolites

A promising approach involves synthetic methods that creatively blend conventional semiconductor technology with three dimensional structure controlled topotactic growth of semiconductor nanoclusters [56, 58]. A method is utilized to assemble and organize II–VI semiconductor nanoclusters from molecular chemical vapour deposition (MOCVD) type reagents within the diamond lattice of 13 Å spherical cavities found in zeolite Y (Figure 17.1(c)). Specifically, these experiments involve the room temperature reaction of volatile $(CH_3)_2M$ reagents (M = Zn, Cd) with Brönsted acid sites in $H_nNa_{56-n}Y$. This yields materials containing $ZOMCH_3$ (ZO represents the zeolite oxide framework) moieties anchored at α-cage (or supercage) sites of zeolite Y. Exposure of $(CH_3M)_{48}Na_8Y$, containing six $ZOMCH_3$ precursors per α cage, to H_2X (X = S, Se) induces a subsequent transformation to materials the composition of which is close to $(M_6X_4)_8H_{16}Na_8Y$ [58]. From IR, EXAFS and Rietveld PXRD structure analyses, an ideal structural model based on M_4X_4 cubane-type clusters, anchored through two of their chalcogenide vertices to the oxide framework of the zeolite by site II and III α-cage M^{2+} cations (see Figure 17.1(c)), namely $M_2(M_4X_4)^{4+}$ is favoured over an alternative $M_6X_4^{4+}$ adamantane geometry (Figure 17.16) [58].

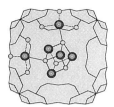

Figure 17.16. Model of the proposed structure of Zn_2 $(Zn_4S_4)^{4+}$ anchored cubane cluster in the α-cage of zeolite Y based on EXAFS, Rietveld refinement of PXRD and minimum energy calculations [58].

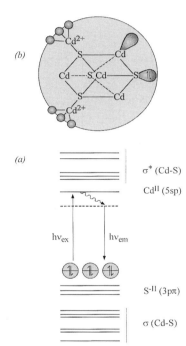

Figure 17.17. (a) Schematic energy level diagram depicting the HOMO (S^{-II}, 3p), LUMO (Cd^{II}, 5sp) and midgap states arising from nonterminated dangling bonds (surface states) in (b) the anchored Cd_2 $(Cd_4S_4)^{4+}$ cluster located in the α-cage of zeolite Y [58].

Six $(CH_3)_2Cd$ species per α cage can also be loaded into partially exchanged $H_nNa_{56-n}Y$ having $n = 0, 8, 16, 24, 32$ and 40. These react with, for example, H_2Se to yield nanoclusters of the type Cd_6Se_6 for $n = 0$ and 8, $Cd_6Se_5^{2+}$ for $n = 16, 24$ and 32 and $Cd_6Se_4^{4+}$ for $n = 40$ [58].

The optical reflectance spectra of the $M_6X_4^{4+}$ nanoclusters display a blue shift of their absorption edges with respect to the bulk II–VI semiconductors, characteristic of quantum confinement. The order of the absorption edges of these nanoclusters, namely $Zn_6S_4^{4+} > Zn_6Se_4^{4+} > Cd_6S_4^{4+} > Cd_6Se_4^{4+}$ follows that of the bandgap energies of the parent bulk semiconductors. The luminescence spectrum of $Cd_6Se_4^{4+}$ nanocluster at 625 nm originates from localized surface states associated with improperly terminated dangling bonds (Figure 17.17). Following excitation at 400 nm (cluster) or 240 nm (zeolite framework) the electron hole pair can relax non-radiatively into the midgaps traps states. It is probable that this process is at the origin of the Stokes-shifted red luminescence observed around 625 nm. The temperature dependence of the observed luminescence quenching of the 625 nm emission is consistent with a multiphonon radiationless relaxation process. Possible vibrational modes that could contribute to the average phonon frequency of 378 cm^{-1} value, determined from the simulation of the relative intensity of the luminescence as a function of temperature, are a combination of both $Cd_6Se_4^{4+}$ cluster skeletal modes (for bulk CdS, ca 270 cm^{-1}) and δTO_4 deformation and pore-opening modes of the zeolite, which occur in the regions of 500 and 300 cm^{-1}, respectively. These results are consistent with a structural model involving the $Cd_6S_4^{4+}$ cluster anchored in the α cage of the zeolite host [58].

17.3.4.3 Quantum Chains

The production of selenium chains with controlled lengths and conformations in a zeolite host provides an interesting opportunity to probe quantum size effects in 1D semiconductor [3, 31–33]. Isolated selenium chains could bridge the behaviour between atomic–molecular forms of selenium and the bulk semiconductor trigonal form. Selenium is of interest, because it has an intermediate electrical conductivity with a negative coefficient of resistivity in the dark. It is markedly photoconductive and is used in photoelectric devices and xerography [3].

Single selenium and tellurium chains were stabilized in the unidimensional channels of mordenite [32]. Isolated Se chains in mordenite channels are strongly photosensitive [33, 92]. The chains form the same helical structure as in the crystalline trigonal Se [3, 31] (Figure 17.18). The combination of EXAFS and solid state NMR spectroscopy with optical absorption techniques revealed the effect of spatial confinement of Se in mordenite [31]. Encapsulation of Se causes the Se–Se bond length (2.34 Å) to shorten relative to the distance observed in helical chains of trigonal Se$_n$ (2.373 Å). No interchain correlation remains if Se is adsorbed in the channels, indicating the presence of isolated, single molecular Se$_n$ units [31].

Se has also been successfully loaded in A, X, Y and AlPO-5 molecular sieves [31]. Se is predominantly in the trigonal from in X, Y and AlPO-5, while only Se$_8$ crown

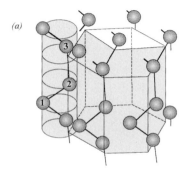

Se

Figure 17.18. (a) Structure of crystalline Se; (b) A projection of the Se chain along the c-axis [3].

ring was found in zeolite A. A mixture of allotropes and helical chains occupy the large 3D-pore systems of X and Y zeolites. A single helical chain occupies the AlPO-5 channels as it was shown for mordenite [3, 31].

17.3.4.4 Microporous Semiconductors

The synthesis of crystalline nanoporous metal sulfide or selenide materials was recently carried out using template based hydrothermal methods [63, 67, 93, 94]. These materials can be viewed as having a lattice of quantum antidots. In this case, the electrons are constrained to the regions between quantum dots [67, 94]. Note, that this new field was initiated by the group of E.M. Flanigen [95].

A noteworthy structural feature discovered in the R-SnX-1 ($R_2Sn_3S_7$, $R_2Sn_3Se_7$) and R-SnS-3 ($R_2Sn_4S_9$) nanoporous Sn (IV) chalcogenides is the unique ability of their open frameworks to undergo elastic deformations in response to variation of the organic template or adsorbed molecular guests [63]. For example, R-SnS-1 isostructures were synthesized with occluded NH_4^+, Me_4N^+, Et_4N^+, $^nPr_3NH^+$, QH^+, $^tBuNH_3^+$ and $DABCOH^+$ guestes (QH^+ = quinuclidinium and $DABCOH^+$ = diaza – [2,2,2]-bicyclooctanium). These structures are formed by Sn_3S_4 broken-cube cluster building blocks and basic 24-atom ring porous layers exist in their unit cells but with slightly different pore sizes, shapes and interlamellar spacings [63] (Figure 17.19). For example, when viewed orthogonally to the layers, the regular hexagonal-shaped 24 atom ring pores in TMA-SnS-1 are distorted away from hexagonal in TEA-SnS-1 through a kind of shearing deformation (Figure 17.19(a, b)) ($TMA = Me_4N^+$ and $TEA = Et_4N^+$). Similarly, the regular elliptical shaped 32-atom ring pores in TBA-SnS-3 become distorted elliptical shapes in TPA-SnS-3 through an analogous type of shearing motion (Figure 17.19(c, d)) ($TBA = Bn_4N^+$ and $TPA = {}^nPr_4N^+$) [63].

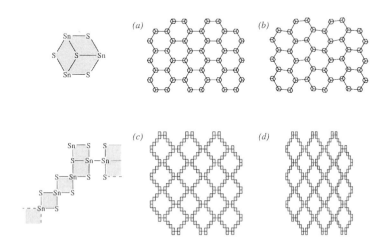

Figure 17.19. Pore size and shape distortions in (a) TMA-SnS-1, (b) TEA-SnS-1, (c) TBA-SnS-3, and (d) TPA-SnS-3 [63].

The TMA-SnSe-1 system can well illustrate the effect that different loadings of an adsorbed molecular guest can have on the framework structure of a nanoporous Sn(IV) chalcogenide. The small change from just three to two H_2O molecular guests in a pore causes the whole nanoporous layer structure to flex with a concomitant redistribution of the template cations and remaining water molecules within the pores, that is a kind of cooperative reconstruction of the entire host–guest system (similar to an 'inorganic enzyme mimic'). Significantly, this structure-bonding change is accompanied by a ca 75 nm redshift on the optical absorption edge of these materials [63].

The synthesis, structures and optical properties of the essentially phase-pure, isostructural and compositionally tunable TMA-SnS$_x$Se$_{1-x}$-1 series of nanoporous Sn(IV) thioselenides were also reported [94]. It was demonstrated that the possibility existed to chemically fine tune the electronic and optical properties of a crystalline nanoporous material, built from the components of a bulk semiconductor, with void spaces of molecular dimensionality. The discovery of this new class of nano-materials is a significant step towards the development of tunable semiconductor quantum antidot lattices, molecular discriminating nanoporous semiconductor and electronically tunable membranes [94].

17.4 Prospects

The way exists to design advanced materials for quantum electronics, nonlinear optics, photonics, chemoselective sensing, size- and shape selective electrocatalysis and redox processes [67].

The unique advantages of zeolites are obvious. The route is now open to synthesize zeolite based electronic, optical, magnetic and dielectric materials. The marked proton conductivity of certain acid zeolites makes possible the fabrication of electro-chromic zeolites. The molecular size pore and channel structure of zeolites provides a novel medium to create ultrahigh resolution images and package and process information at high storage densities. The unidimensional channel architecture of certain zeolites permits the assembly of low dimensional zeolite conductors and the control of energy transfer and energy migration of restricted dimensions [3].

For example, sodalite nanostructures could be used as nonlinear optical materials for optical transistors in optical computing, frequency doubling and other optical applications. Sodalites fulfill one prerequisite for second-harmonic generation, namely, they are noncentrosymmetric. However, the noncentrosymmetry applies only to the guest portion of the material. It should be necessary to increase the dipole moment of the guest, by combining cations with large differences in electronegativity (for example – sodium–silver cations) and aligning them by carrying out the ion exchange in a strong electric field. The possibility exists to use silver soda-lites for applications in high density imaging and reversible optical data storage. In this field, compositional fine tuning is required to increase image stability and contrast, as well as optimize reading, writing and erasing response times, threshold powers and wave lengths [2].

Nonoxide frameworks may also be found having their characteristic challenges, pitfalls and technological relevance, but they are also part of the investigation of self assembly employed to form open framework structures for advanced materials applications [67].

Acknowledgments

The authors are indebted to Mr Francis Valette for the nice drawings and Mrs Dorina Popa and Mr Mohamed Hammida for technical assistance.

References

[1] P. A. Jacobs, Stud. Surf. Sci. Catal. **1984**, *29*, 357.
[2] A. Stein, G. A. Ozin, in: *Proc. Ninth Int. Zeolite Conference*, (Eds.: R. vonBallmoos et al.), Butterworth-Heinemann, Boston **1993**, Vol. 1 p. 93.
[3] G. A. Ozin, A. Kuperman, A. Stein, *Angew. Chem.* **1989**, *101*, 373.
[4] G. D. Stucky, J. E. MacDougall, *Science* **1990**, *247*, 669.
[5] D. W. Breck, *Zeolite Molecular Sieves. Structure, Chemistry and Use*, Wiley-Interscience, New York **1974**.
[6] *Atlas of Zeolite Structure Types*, (Eds.: W. M. Meier, D. H. Olson, C. Baerlocher), Elsevier, London, **1996**.
[7] W. J. Mortier, *Compilation of Extra Framework Sites in Zeolites*, Butterworth, London **1982**.
[8] R. M. Barrer, *Hydrothermal Chemistry of Zeolites*, Academic Press, London **1982**.
[9] H. G. Karge, H. K. Beyer, *Stud. Surf. Sci. Catal.* **1991**, *69*, 43.
[10] I. Hannus, J. B. Nagy, I. Kiricsi, *Hyperfine Interaction* **1996**, *99*, 409.
[11] K. H. Rhee, F. R. Brown, D. H. Finseth, J. M. Stencel, *Zeolites* **1983**, *3*, 394.
[12] L. R. Gellens, W. J. Mortier, R. A. Schoonheydt, J. Uytterhoeven, *J. Phys. Chem.* **1981**, *85*, 2783.
[13] E. Gachard, J. Belloni, M. A. Subramanian, *J. Mater. Chem.* **1996**, *6*, 867.
[14] G. A. Ozin, A. Kuperman, A. Stein, *Angew. Chem. Int. Ed. Engl.* **1989**, *28*, 359.
[15] D. R. Brown, L. Kevan, *J. Phys. Chem.* **1986**, *90*, 1129.
[16] I. Hannus, A. Béres, J. B. Nagy, J. Halasz, I. Kiricsi, *J. Mol. Structure*, **1997**, *43*, 410–411.
[17] B. Xu, L. Kevan, *J. Chem. Soc., Faraday Trans.* **1991**, *87*, 2843.
[18] R. E. H. Breuer, E. de Boer, G. Geisman, *Zeolites* **1989**, *9*, 336.
[19] L. R. M. Martens, P. J. Grobet, P. A. Jacobs, *Nature* **1985**, *315*, 568.
[20] L. R. M. Martens, P. J. Grobet, W. J. M. Vermeiren, P. A. Jacobs, *Stud. Surf. Sci. Catal.* **1986**, *28*, 935.
[21] E. Trescos, F. Rachdi, L. C. de Ménorval, F. Fajula, *Int. J. Mod. Phys. B* **1992**, *6*, 3779.
[22] E. Trescos, L. C. de Ménorval, F. Rachdi, *J. Phys. Chem.* **1993**, *97*, 6943.
[23] J. A. Rabo, C. L. Angel, P. H. Kasai, V. Schomaker, *Discuss. Faraday Soc.* **1966**, *41*, 328.
[24] Y. Nozue, T. Kodeira, S. Ohwashi, N. Togashi, T. Monji, O. Terasaki, *Stud. Surf. Sci. Catal.* **1994**, *84*, 837.
[25] T. Kodaira, Y. Nozue, S. Ohwashi, T. Goto, O. Terasaki, *Phys. Rev. B* **1993**, *48*, 12245.
[26] Y. Nozue, T. Kodaira, S. Ohwashi, T. Goto, O. Terasaki, *Phys. Rev. B* **1993**, *48*, 12253.
[27] T. Kodaira, Y. Nozue, O. Terasaki, H. Takeo, *Stud. Surf. Sci. Catal.*, **1997**, *105*, 2139.
[28] Y. Wang, N. Herron, *J. Phys. Chem.* **1988**, *92*, 4988.

[29] N. Herron, Y. Wang, M. M. Eddy, G. D. Stucky, D. E. Cox, K. Moller, T. Bein, J. Am. Chem. Soc. **1989**, *111*, 530.
[30] K. Moller, T. Bein, M. Eddy, G. D. Stucky, N. Herron, Report 1988, TR-11.
[31] J. B. Parise, J. E. MacDougall, N. Herron, R. Farlee, A. W. Sleight, Y. Wang, T. Bein, K. Moller, L. M. Moroney, *Inorg. Chem.* **1988**, *27*, 221.
[32] V. N. Bogomolov, S. V. Kholodkevich, S. G. Romanov, L. S. Agroskin, *Solid State Commun.* **1983**, *47*, 181.
[33] K. Tamura, S. Hosokawa, H. Endo, S. Yamasaki, H. Oyanagi, *J. Phys. Soc. Jpn.* **1986**, *55*, 528.
[34] J. E. Mac Dougall, H. Eckert, G. D. Stucky, N. Herron, Y. Wang, K. Moller, T. Bein, D. Cox, J. Am. Chem. Soc. **1989**, *111*, 800.
[35] E. G. Derouane, J. B. Nagy and J. C. Védrine, *J. Catal.* **1977**, *46*, 434.
[36] J. B. Nagy, M. Van Eenoo, E. G. Derouane, J. C. Védrine, in: *Magnetic Resonance in Colloid and Interface Science* (Eds.: J. P. Fraissard, H. A. Resing), Reidel, Dordrecht **1980**, p. 591.
[37] J. B. Nagy, M. Van Eenoo, E. G. Derouane, J. Catal. **1979**, *58*, 230.
[38] Y. Y. Tong, D. Laub, G. Schulz-Ekloff, A. J. Renouprez, J. J. van der Klink, *Phys. Rev. B* **1995**, *52*, 8407.
[39] H. Trevino, W. M. H. Sachtler, *Catal. Letters* **1994**, *27*, 251.
[40] S. T. Homeyer, L. L. Sheu, Z. Zhang, W. M. H. Sachtler, R. R. Balse, J. A. Dumesic, *Appl. Catal.* **1990**, *64*, 225.
[41] V. R. Balse, W. M. H. Sachtler, J. A. Dumesic, *Catal. Letters* **1988**, *1*, 275.
[42] Z. Karpinski, Z. Zhang, W. M. H. Sachtler, *Catal. Letters* **1992**, *13*, 123.
[43] G. Moretti, W. M. H. Sachtler, *Catal. Letters* **1993**, *17*, 285.
[44] M. S. Tzou, B. K. Teo, W. M. H. Sachtler, *Langmuir* **1986**, *2*, 773.
[45] H. Trevino, G.-D. Lei and W. M. H. Sachtler, *J. Catal.* **1995**, *154*, 245.
[46] V. Schünemann, H. Trevino, G. D. Lei, D. C. Tomczak, W. M. Sachtler, K. Fogash and J. A. Dumesic, *J. Catal.* **1995**, *153*, 144.
[47] C. M. Tsang, S. M. Augustine, J. Butt, W. M. H. Sachtler, *Appl. Catal.* **1989**, *46*, 45.
[48] G. Moretti, W. M. H. Sachtler, *J. Catal.* **1989**, *15*, 205.
[49] M.-S. Tzou, M. Kusunoki, K. Asakura, H. Kuroda, G. Moretti, W. M. H. Sachtler, *J. Phys. Chem.* **1991**, *95*, 5210.
[50] D. H. Ahn, J. S. Lee, M. Normura, W. M. H. Sachtler, G. Moretti, S. I. Wo, R. Ryoo, *J. Catal.* **1992**, *133*, 191.
[51] Z. Zhang, L. Xu, W. M. H. Sachtler, *J. Catal.* **1991**, *131*, 502.
[52] J. S. Feeley, A. Y. Stakheev, F. A. P. Cavalcanti, W. M. H. Sachtler, *J. Catal.* **1992**, *136*, 182.
[53] Y. -G. Yin, Z. Zhang, W. M. H. Sachtler, *J. Catal.* **1992**, *138*, 721; **1993**, *139*, 444.
[54] L. Xu, G.-D. Lei, W. M. H. Sachtler, R. D. Cortright, J. A. Dumesic, *J. Phys. Chem.* **1993**, *97*, 11517.
[55] V. Schünemann, H. Trevino, W. M. H. Sachtler, K. Fogash, J. A. Dumesic, *J. Phys. Chem.* **1995**, *99*, 1317.
[56] M. R. Steele, P. M. Macdonald, G. A. Ozin, *J. Amer. Chem. Soc.* **1993**, *115*, 7285.
[57] R. Jelinek, A. Stein and G. A. Ozin, *J. Amer. Chem. Soc.* 1993, *115*, 2390.
[58] G. A. Ozin, M. R. Steele, A. J. Holmes, *Chem. Materials* **1994**, *6*, 999.
[59] S. Oliver, G. A. Ozin, L. A. Ozin, *Adv. Materials* **1995**, *7*, 948.
[60] G. A. Ozin, *Adv. Chem. Series* **1995**, *245*, 335.
[61] O. Dag, A. Kuperman, G. A. Ozin, *Adv. Materials* **1995**, *7*, 72.
[62] O. Dag, A. Kuperman, P. M. Macdonald, G. A. Ozin, *Stud. Surf. Sci. Catal.* **1994**, *84*, 1107.
[63] H. Ahari, C. L. Bowes, T. Jiang, A. Lough, G. A. Ozin, R. L. Bedard, S. Petrov, D. Young, *Adv. Materials* **1995**, *7*, 375.
[64] P. Enzel, G. S. Henderson, G. A. Ozin, R. L. Bedard, *Adv. Materials* **1995**, *7*, 64.
[65] T. Jiang, A. J. Lough, G. A. Ozin, D. Young, R. L. Bedard, *Chem. Materials* **1995**, *7*, 245.
[66] C. L. Bowes, A. J. Lough, A. Malek, G. A. Ozin, S. Petrov, D. Young, *Chem. Ber.* **1996**, *129*, 283.
[67] C. L. Bowes, G. A. Ozin, *Adv. Materials,* **1996**, *8*, 13.
[68] K. Klier, P. J. Hutta, R. Kellerman, *ACS Symp. Ser.* **1977**, *40*, 108.
[69] P. Fejes, I. Hannus, I. Kiricsi, K. Varga, *Acta Phys. Chem.* **1978**, *24*, 119.

[70] A. Henglein, *Top. Curr. Chem.* **1988**, *143*, 113.

[71] B. Lindman, S. Forsen, *Chlorine, Bromine and Iodine NMR, NMR Basic Principles and Progress* (Eds.: P. Piehl, E. Fluck, R. Kosfeld), Springer, Berlin **1976**, Vol. 12.

[72] G. A. Ozin, A. Stein, G. Stucky, J. Godber, in: *Inclusion Phenomena and Molecular Recognition* (Ed.: J.-L. Atwood), Plenum, New York **1990**, p. 379.

[73] M. R. Harrison, P. P. Edwards, J. Klinowski, J. M. Thomas, D. C. Johnson, C. J. Page, *J. Solid State Chem.* **1984**, *54*, 330; *J. Chem. Soc., Chem. Commun.* **1984**, 982.

[74] B. Xu, L. Kevan, *J. Phys. Chem.* **1992**, *96*, 2642.

[75] P. J. Grobet, L. R. M. Martens, W. J. M. Vermeiren, D. R. Huybrechts, P. A. Jacobs, *Z. Phys. D* **1989**, *12*, 37.

[76] P. A. Anderson, P. P. Edwards, *J. Am. Chem. Soc.* **1992**, *114*, 10608.

[77] E. Trescos, F. Rachdi, L. C. de Ménorval, F. Fajula, T. Nunes, G. Feio, *J. Phys. Chem.* **1993**, *97*, 11855.

[78] R. A. Dalla Betta, M. Boudart, *Proc. 5th Int. Cong. on Catal.* (Ed.: J. W. Hightower), North Holland **1973**, *2*, 1329.

[79] B. L. Mojet, M. J. Kappers, J. C. Muijsers, J. W. Niermantsverdriet, J. T. Miller, F. S. Modica, D. C. Koningsberger, *Stud. Surf. Sci. Catal.* **1994**, *84B*, 909.

[80] S. T. Homeyer, Z. Karpinski, W. M. H. Sachtler, *J. Catal.* **1990**, *123*, 60.

[81] A. P. J. Jansen, R. A. Van Santen, *J. Phys. Chem.* **1990**, *94*, 6764.

[82] Y. Y. Tong, J. J. van der Klink, G. Clugnet, A. J. Renouprez, D. Laub, P. A. Buffat, *Surf. Sci.* **1993**, *292*, 276.

[83] L. C. de Ménorval, T. Ito, J. Fraissard, *J. Chem. Soc., Faraday Trans. 1* **1982**, *78*, 403.

[84] R. Schoemaker, T. Apple, *J. Phys. Chem.* **1987**, *91*, 4024.

[85] S. J. Cho, S. M. Jung, Y. G. Shul, R. Ryoo, *J. Phys. Chem.* **1992**, *96*, 9922.

[86] C. Tway, T. Apple, *J. Catal.* **1990**, *123*, 375.

[87] M. Boudart, M. G. Samant, R. Ryoo, *Ultramicroscopy* **1986**, *20*, 125.

[88] T. Rades, C. Pak, M. Polisset-Thfoin, R. Ryoo, J. Fraissard, *Catal. Lett.* **1994**, *29*, 91.

[89] B. M. Choudary, K. Lazar, I. Bogyay, L. Guczi, *J. Chem. Soc., Faraday Trans.* **1990**, *86*, 419.

[90] L. Guczi, *Catal. Lett.* **1990**, *7*, 205.

[91] S. Schupper, S. L. Friedman, M. A. Marcus, D. L. Adler, Y. H. Xie, F. H. Ross, T. D. Harris, W. L. Brown, Y. J. Chabal, L. E. Brus, P. H. Citrin, *Phys. Rev. Lett.* **1994**, *72*, 2648.

[92] Y. Katayama, M. Yao, Y. Ajiro, M. Inui, H. Endo, *J. Phys. Soc. Jpn* **1989**, *58*, 1811.

[93] T. Jiang, G. A. Ozin, R. L. Bedard, *Adv. Materials* **1995**, *7*, 166.

[94] H. Ahari, G. A. Ozin, R. L. Bedard, S. Petrov, D. Young, *Adv. Materials* **1995**, *7*, 370.

[95] R. L. Bedard, L. D. Vail, S. T. Wilson, E. M. Flanigen, US Patent 4,800,761 (**1989**) and 4,933,068 (**1990**).

[96] I. Manninger, Z. Paal, B. Tesche, U. Klengler, J. Halasz, I. Kiricsi, *J. Molec. Catal.* **1991**, *64*, 361.

[97] L. Guczi, G. Lu, Z. Zsoldos, Z. Koppany, *Stud. Surf. Sci. Catal.* **1994**, *84*, 949.

[98] G. Lu, T. Hoffer, L. Guczi, *Appl. Catal. A: General* **1992**, *93*, 61.

[99] L. Guczi, K. V. Sarma, I. Borko, *Catal. Lett.* **1996**, *39*, 43.

[100] H. K. Beyer, P. A. Jacobs, *Metal Microstructures in Zeolites* (Eds.: P. A. Jacobs et al.) Elsevier, Amsterdam, **1982**, p. 95.

[101] P. A. Jacobs, W. De Wilde, R. A. Schoonheydt, J. B. Uytterhoeven, H. K. Beyer, *J. Chem. Soc., Faraday Trans. 1* **1976**, *72*, 1221.

[102] D. C. Tomczak, G. D. Lei, V. Schuenemann, H. Trevin, W. M. H. Sachtler, *Microporous Mater.* **1996**, *5*, 263.

[103] Z. C. Zhang, G. D. Lei, W. M. H. Sachtler, *Ser. Synchrotron Radiat. Techn. Appl.* **1996**, *2*, 173.

[104] E. Trescos, L. C. de Ménorval, F. Rachdi, *Appl. Magn. Réson.* **1995**, *8*, 489.

[105] V. I. Srdanov, N. P. Blake, D. Markgraber, H. Metiu, G. D. Stucky, in: *Advanced Zeolite Science and Applications* (Eds.: J. C. Jansen et al.) Elsevier, Amsterdam **1994**, p. 115.

Chapter 18

Nanoparticles and Nanostructured Films: Current Accomplishments and Future Prospects

J. H. Fendler and Y. Tian

18.1 Introduction

Properties of solid materials undergo drastic changes when their dimensions are reduced to the nanometer size regime. For semiconductors this transition occurs when the particle sizes are comparable to the de Broglie wavelength of the electron, to the mean free path of the exciton, or to the wavelength of the phonons. Particles in the nanometer range, i.e., nanoparticles, are said to be size quantized. Size-quantized particles are prepared by both physical and chemical methods. The more traditional approach involves physical processing (grinding or ball-milling of larger particles, crystal growing, ion implantation, and molecular epitaxy, for example) rather than chemical syntheses. Advantage, on the other hand, is taken of versatile inorganic, organic, and electrochemical synthetic methodologies in the chemical approach to construct nanoparticles with distinct electronic structures from their constituent atoms and molecules, and crystal and surface structures from their bulk materials. Additionally, preparative, polymer, and colloid chemical techniques are employed to control the sizes and monodispersities of the incipient nanoparticles and to stabilize them in the solid state and in aqueous and nonaqueous dispersions [1, 2]. Significantly, biomineralization, mother nature's construction of nanoparticles in cells [3, 4], has inspired many nanoparticle preparations [5, 6].

The unique properties of metallic, semiconducting, magnetic, and ferroelectric nanoparticles have prompted the burgeoning interest in these systems. The size-dependent behavior of nanoparticles manifests itself in markedly altered physical and chemical behavior. In particular, changes in mechanical, optical, electrical, electro-optical, magnetic, and magneto-optical properties have been investigated [7–10].

It is important to keep in mind that the smaller the particles are, the larger the portion of their constituent atoms are located at the surface. For example, in a 2 nm-diameter gold (or CdS) particle, approximately 60% of the atoms (or molecules) are located at the surface. In semiconductors, this arrangement facilitates electron and/or hole transfers to and from acceptors and/or donors localized at the nanoparticle surface. In metallic nanoparticles, a large surface-area-to-volume ratio permits effective charge transfer and elicits charge-transfer-dependent changes in the optical absorption spectra [11]. It is becoming increasingly recognized that the properties of nanoparticles are strongly influenced by the physics and chemistry of their surface states. The surface states, in turn, are affected by the crystal morphology of the nanoparticles, by the chemicals attached to them (capping reagents, for example [12]), and by the media which surround them [13].

Nanoparticle research has reached a well-deserved maturity. Ample summaries, reviews, and books document the accumulated information [14, 15]. Nanoparticle research continues to be the subject of national and international symposia and workshops [16]. The current state of the art activities focus upon treating nanoparticles as large macromolecules and linking them, by electrostatic interactions or covalent bonds, into heterostructured supramolecules, two-dimensional arrays, or three dimensional networks. Alternatively, and additionally, nanoparticles are being self-assembled to hierarchically more complex structures [17]. The integration of nanoparticles into nanostructured films has, in fact, been reflected in many of the chapters assembled in the present volume.

The goals of the present closing chapter are (i) to summarize the current activities in nanoparticle preparations, (ii) to enable the neophyte to launch desired experiments by providing data on the properties of the most frequently used bulk semiconductors and on the preparation and characterization of the corresponding nanoparticles, and (iii) to speculate on future research directions related to nanoparticles and nanostructured films.

18.2 Preparations of Nanoparticles and Nanostructured Films: Current State of the Art

18.2.1 Definitions

Strictly speaking, nanoparticles are uniformly constituted from identical atoms or molecules, nonagglomerated and monodisperse in some liquid. In reality, they often agglomerate into larger irregular entities and are rarely monodispersed. Chemically prepared nanoparticles seldom have uniform purity. Furthermore, nanoparticles are stabilized by surfactants or large polyions or are embedded in some matrix. This, inevitably, alters their surface states.

Deposition or transfer of the nanoparticles to solid support results in the formation of nanoparticulate (nanostructured) films in which the thickness, the packing density, and the orientation of the nanoparticles are potentially controllable. Nanoparticulate films are interesting since they are size quantized even though the individual nanoparticles are in physical contact with each other. Thus, lateral conductivity becomes measurable in nanostructured films prepared from conducting or semiconducting materials.

18.2.2 Chemical Preparations of Nanoparticles

Metallic, semiconducting, and magnetic nanoparticles have been prepared in a wide variety of different media by wet chemical techniques. The available literature, up to 1992, has been summarized in a recent review [5]. More recent developments in semiconductor nanoparticle preparations are surveyed (albeit not exhaustively) in Tables 18.1–18.6. The salient features that appear from these tables are as follows.

- The preparation of highly monodisperse colloidal semiconductor nanoparticle dispersions has become a routine matter [18–22]

In a benchmark method 13, 14, 16, 19, and 23 Å-diameter CdS nanoparticles were prepared as fully redispersible powders [19]. The method involved the addition of H_2S to a vigorously stirred aqueous solution of $Cd(ClO_4)_2 \cdot 6H_2O$ (1.97 g, 4.70 mmol) and 1-thioglycerol (1 mL, 11.53 mmol), adjusted to pH = 11.2 (1 M NaOH). The desired size of CdS nanoparticles was obtained by the judicious control of the quantity of H_2S introduced and by changes of the temperature. Low-molecular-weight contaminants were removed by dialysis. Replacing the perchlorate anion by acetate or nitrate ions had a profound influence on the mean diameters of the nanoparticles produced [19]. In an alternative and versatile method, injection of appropriate organometallic reagents into a hot coordinating solvent (tri-n-octylphosphine, TOP) led to the production of nearly monodisperse CdS, CdSe, and CdTe [20]. Production of monodispersed particles is believed to be the consequence of a temporary discrete nucleation, attained as the consequence of an abrupt supersaturation (upon reagent injection) and a subsequent controlled particle growth by Ostwald-ripening. Typically, nanoparticles are stabilized by bulky anions (hexametaphosphate, for example), by polyions (polyvinyl alcohol, for example), or by capping by nucleophilic reagents (thiophenol, for example). The function of the capping agent is to inhibit the nanoparticle growth at the desired size and to ensure monodispersity [20]. Importantly, given populations of nanoparticle dispersions can be dried and stored as solids. The solid nanoparticles can then subsequently be redispersed without any alteration of their sizes and size distributions.

- Size-selected precipitation provides a convenient means for producing monodisperse nanoparticles [18–20].

Advantage is taken in size-selected precipitation of preferential flocculation of the larger nanoparticles by increasing the polarity of the dispersing liquid. Enhancing the polarity of the media reduces the attractive forces between the nanoparticles and the larger particles experience the greatest change in interparticle interactions and thus they precipitate first. This results, in turn, in a narrowing of the size distribution of the nanoparticles, which remain in the supernatant. Size-selected precipitation has been successfully employed, for example, for the isolation of highly monodisperse 6.4, 7.2, 8.8, 11.6, 19.4, 25, and 48 Å-radius CdS nanoparticles [20]. Addition of judicious amounts of ethanol, 2-propanol, or acetone to aqueous dis-

Table 18.1. Preparations of CdS nanoparticles.

Method	Media/Stabilizer	Comments	References
Cluster synthesis	Aqueous/thiol derivatives	Structurally defined clusters with sizes ~1.0–1.6 nm in diameter were synthesized. NMR, X-ray single crystallographic characterization were carried out. Optical adsorption showed a strong exciton transition at about 290 nm.	[23–27]
Clay platelet compartmentalization	Aqueous/silicates nanoreactor	Size-quantized particles were prepared in nanophase reactors provided by binary liquids adsorbed at layered silicates. Optical absorption, XRD were used to characterize the particles.	[13, 28]
Hydrolysis in zeolites	Zeolite-X	CdS clusters were synthesized in zeolite-X through hydrolysis to form capped semiconductor nanoclusters. Optical properties were examined.	[29, 30]
Solid solution	Aqueous/silicates, silica	Growth of nanocrystals was studied in silicate glasses and in thin SiO_2 films in the initial stages of the phase separation of a solid solution, colloidal silica–CdS nanocomposites. Particles and clusters of semiconductor–metal sulfides were grown *in situ* in porous silica pillared layered phosphates.	[31–33]
Ultrasound	Aqueous/hexametaphosphate	Q-state CdS particles were generated by ultrasound at 20 kHz upon aqueous cadmium thiols.	[34]
Colloid synthesis, arrest precipitation, and size-selective partition	Aqueous or nonaqueous/ small molecule stabilizers: aminocalixarene, TOPO, water-soluble thiols	Organically capped, highly monodisperse quantum-sized particles were made by colloidal route followed by size-selective precipitation. TEM, XRD, fluorescence emission, optical absorption characterization were carried out. Size-dependent oscillator strength, temperature shift of the excitonic transition energy, and reversible absorbency shift were observed.	[18–22, 35, 36]
Thermolysis	4-ethylpyridine	CdS and CdSe nanoparticles were synthesized by thermolysis of diethyldithio- or diethyldiseleno-carbamates of cadmium. Absorption spectra and TEM were used for characterization.	[37]

Electrodeposition	Aqueous electrolytes/Au single crystal on electrodes	Quantum dots were electrically deposited on Au(111) single-crystal domains coated on electrode. Epitaxial CdS nd CdSe nanocrystals were characterized by TEM. Scanning tunneling microscopy of electrochemically grown CdS monolayers on Au(111) was done. Quantum size effects were investigated in the chemical solution deposition mechanisms of semiconductor films.	[38–42]
Polymer templates	Biopolymers, block polymers and porous polymers as compartments	II–VI semiconductor nanoparticles were synthesized by the reaction of a metal alkyl polymer adduct with hydrogen sulfide. Q-CdS clusters were stabilized by polynucleotides. Nanoparticles were also made within an ordered polypeptide matrix. Size control of nanoparticles in styrene-based random ionomers was studied. Spherical ionic microdomains were observed in styrene-based diblock ionomers. A Q-dot condensation polymer in chelate polymer microparticles was formed and polymer size dependence of the optical properties was studied. Amorphous polysilene quioxanes were employed as a confinement matrix for quantum-sized particle growth. Nanoparticles were grown in porous polysilsesquioxanes.	[43–52]
Sol-gel technology	Borosilicate glass matrix, silica and silicate glass with and without inorganic salt additives	Oxidation of cadmium chalcogenide microcrystals, doped in silica glasses, prepared by the sol-gel process was investigated. Quantum confinement effects of CdS nanocrystals in a sodium borosilicate glass, prepared by the sol-gel process, were studied. Effect of heat and gases on the photoluminescence of CdS quantum dots confined in silicate glasses prepared by the sol-gel method was inspected. Optical properties of CdS nanocrystals dispersed in a sol-gel silica glass were studied.	[53–61]
Protein monolayer	CdS/channel protein	Nanosized CdS particles were formed within a channel protein monolayer on water.	[62]

Table 18.2. Preparations of CdSe nanoparticles.

Method	Media/Stabilizer	Comments	References
Electrodeposition	Aqueous electrolytes/ Au single crystal film coated on electrode	Epitaxial size control was achieved by mismatch tuning in electrodeposited Cd(Se, Te) quantum dots on Au (111). Electrodeposited quantum dots were found to be epitaxial. High-resolution TEM and electronic absorption spectra were employed to characterize the epitaxial nanocrystals.	[36, 39, 63]
Colloid synthesis, arrest precipitation, and size-selective precipitations	Aqueous or nonaqueous/ small molecule stabilizers: aminocalixarene, TOPO, and water-soluble thiols	Organically capped, highly monodisperse quantum-sized particles were made by colloidal route followed by size-selective precipitation. TEM, XRD, fluorescence emission, optical absorption characterization were carried out. Size-dependent oscillator strength, temperature shift of the excitonic transition energy, and reversible absorbency shift were observed.	[19, 20, 35]
Chemical solution deposition	Aqueous electrolytes	Quantum size effects were studied in chemical solution deposition mechanisms of semiconductor films.	[38]

Table 18.3. Preparations of ZnS and ZnSe nanoparticles.

Method	Media/Stabilizer	Comments	References
Colloid synthesis	Organic solvents/thiophenol, thioglycerol	Thiophenol-capped ZnS quantum dots 0.7–1.5 nm in diameter were prepared. XRD and optical absorption characterization were carried out. Thioglycerol-capped ZnS particles 1.2–1.5 nm in diameter were prepared. ESR spectra of the particles were compared with those of bulk ZnS.	[64]
Langmuir–Blodgett film	Organized surfactant LB films	Size-confined ZnS particles were synthesized in Langmuir–Blodgett films by the reaction of metal fatty acid with H_2S.	[65]
Organometallic block polymer	Lamella or spherical microphased polymers	ZnS clusters and zinc particles of about 3.0 nm in diameter were synthesized within microphase-separated domains of organometallic block copolymers. TEM and optical absorption characterization were carried out.	[66]
Polymer	Aqueous/polymer	ZnS and CdS particles were prepared by the reaction of a metal alkyl polymer adduct with H_2S in solution. TEM showed the particle sizes ranging from 2 to 5 nm in diameter. Absorption band edge changes were investigated by photoacoustic spectroscopy.	[45]
Polymer	Aqueous/polymer or chitosan	ZnS nanocrystals were prepared in a polymer matrix and in chitosan film. Electroluminescence was observed. ZnS thin films were grown by SILAR on poly(vinyl chloride) and polycarbonate substrates.	[67–69]
Clay platelet compartmentalization	Ethanol-cyclohaxane/ layered silicate or colloidal particles	Size-quantized CdS and ZnS particle dispersions were prepared in nanophase reactors provided by binary liquids adsorbed at layered silicates and at colloidal silica particles. XRD, rheological, and calorimetric measurements were performed to characterize the dispersions.	[13, 28]
MOCVD	Vacuum	Intrazeolite topotactic MOCVD was used for three-dimensional structure-controlled synthesis of II–VI nanoclusters.	[70]
Sol-gel processing	Aqueous	Electroluminescent ZnS devices were produced by sol-gel processing. Sol-gel-derived ZnSe crystallites were prepared in glass films and optical properties were studied.	[71, 72]

Table 18.4. Preparations of ZnO nanoparticles.

Method	Media/Stabilizer	Comments	References
Colloid synthesis	Aqueous, ethanol, and 2-propanol	Transparent sol was prepared in water and ethanol or 2-propanol. Fluorescence spectra show that adsorbed electron relays are necessary to transport electrons from conduction band to deep traps.	[73]
Sol-gel technology	Aqueous/conducting glass	Transparent nanocrystalline ZnO film was prepared by sol-gel techniques. Visible-light sensitization by cis-bis(thiocyanato) bis(2,2′-bipyridyl-4,4′-dicarboxylato) ruthenium(II) on the film was shown to have high light-to-current conversion efficiency.	[74]
Polymer matrix	Aqueous/polymer	ZnO nanoparticles were synthesized with capsulation of polymer. Steady and time-resolved fluorescence, absorption spectroscopy were used to characterize the particles.	[75]
Electrodeposition	Aqueous/electrode	Cathodic electrodeposition was accomplished from aqueous solution to form dense or open-structured ZnO films.	[76]
Sublimation	Vacuum	Quantum size effects in zinc oxide nanoclusters were synthesized by reactive sublimation. TEM, XRD, and photoacoustic FTIR data were obtained.	[77]
Laser vaporation	Vacuum	Metal oxide nanoparticles were synthesized by using laser vaporization/condensation in a diffusion cloud chamber.	[78]

Table 18.5. Preparations of PbS nanoparticles.

Method	Media/Stabilizer	Comments	References
Polymer	Aqueous or organic solvents/polymer	Metal sulfide nanoparticles were prepared by the reaction of soluble metal alkyls and polymers containing pyridyl groups with H_2S. The particles were observed by TEM to be evenly distributed in the polymer matrix.	[79]
Polymer	Benzene/Pb-containing polymer film	PbS nanoclusters were synthesized within microphase-separated diblock copolymer films and within block copolymer nanoreactors. TEM, XRD, and electronic absorption were employed to characterize the particles embedded in the films.	[70, 80, 81]
Langmuir–Blodgett film	Fatty acid or amphiphilic polymers in ordered films	Size-confined metal sulfides were synthesized in Langmuir–Blodgett films. Control of distance and size of the nanoparticles was achieved by organic matrices in the LB films.	[82, 83]
Monolayer	Surfactant monolayer on water surfaces	Morphology control was accomplished for epitaxially grown PbS nanocrystallites under mixed surfactant monolayers. Morphology-dependent spectroelectrochemical behavior of the nanoparticulate films was studied on conducting glass electrode in electrochemical cell.	[84, 85]
Surfactant mesophase	Aqueous/AOT	Quantum-sized PbS particles were prepared in surfactant-based complex fluid media and in bicontinuous cubic phase of AOT. Structural nd morphological characterization by TEM, small-angle X-ray scattering were carried out.	[86, 87]
Sol-gel processing	Aqueous/silicon methoxide, boron ethoxide	PbS quantum dot materials were prepared by the sol-gel process using Pb-doped gel of silicon methoxide or boron ethoxide and lead acetate followed by sintering.	[88]

Table 18.6. Preparations of GaAs and other III–V nanoparticles.

Method	Media/Stabilizer	Comments	References
Colloid synthesis	Decane/methylsilane	3 nm GaAs crystallites were prepared by refluxing GaCl$_3$ with As(SiMe$_3$)$_3$ at 180°C. The particle growth was controlled by autoclaving. Electronic absorption spectra were studied.	[89]
Colloid synthesis	Organic solvents/ monoglyme or diglyme	Nanocrystalline GaAs and GaP were synthesized in organic solvents by a route of *in situ* formation of particles followed by metathetical reaction with coordinate solvents. UV, TEM, XRD, and elemental analysis were used to characterize the nanoparticles.	[90]
Colloid synthesis	Aromatic solvents/ chelating agents	5 nm-diameter GaAs nanoparticles were prepared via metathetical reaction of GaCl$_3$ with chelating agents bearing As(III). XRD, XPS, NMR, FTIR, high-resolution TEM characterization were carried out.	[91]
Colloid synthesis	Triethylene glycol ether/ methylsilyl	GaAs quantum dots were synthesized by the reaction of gallium acetylacetonate with trimethylsilylarsine at 216 K. The particles were characterized by XRD, XPS, UV, and elemental analysis. Transient pump-probe spectroscopy was used to examine charge carrier transport.	[92]
Colloid synthesis	Organic solvent/TOPO	2.6 nm InP nanocrystals were synthesized by reacting chlorindium oxalate complex with P(SiMe$_3$)$_3$ at 270°C. The nanoparticles were characterized by TEM and optical absorption.	[93]
Colloid synthesis	Triglyme	Size-quantized GaAs nanocrystals 1.5–9.0 nm in diameter were prepared by wet process. Photoinduced reduction of methyl viologen on the particles was investigated.	[94]
			[95–97]

persions of colloidal CdS resulted in the preferential precipitation of the larger nanoparticles, while the smaller ones remained in the supernatant. Centrifugation permitted the separation of the precipitate and the supernatant and thus it provided the means for isolating CdS nanoparticles in narrow size distributions.

- Structurally defined clusters (containing 15–45 molecules) of semiconductors [23–27] can now be chemically synthesized and crystallized in a superlattice structure.

The recently reported preparations of $Cd_{32}S_{14}(SC_6H_5)_{36}DMF_4$ [26] and $Cd_{17}S_4(SCH_2CH_2OH)_{36}$ [27] clusters illustrate the current activities in this area. The structure of the yellow cubic crystalline $Cd_{32}S_{14}(SC_6H_5)_{36}DMF_4$ was determined to consist of a 12 Å-diameter spherical 82-atom CdS core with dangling surface bonds terminating in hexagonal (wurtzitelike) CdS units at the four tetrahedral corners [26]. Two clusters of $Cd_{17}S_4(SCH_2CH_2OH)_{36}$ were observed to assemble to an interlaced diamondlike superlattice [27]. Importantly, both of these clusters remained intact upon dissolving them in dipolar aprotic solvents, and their absorption and emission spectra, both in the solid state and in colloidal dispersions, indicated size quantization.

- Nanoparticles can be generated *in situ* in (or at) such membrane mimetic systems [5, 31] as aqueous and reversed micelles, surfactant vesicles, proteins, monolayers, Langmuir–Blodgett (LB) films, bilayer lipid membranes (BLMs), polymers, polyelectrolytes, synthetic membranes, silicates (organoclay complexes or pillared silicates, for examples), and porous glass.

Aqueous micellar cobalt and iron(II) dodecyl sulfate have been employed for the preparation of nanosized cobalt and iron magnetic particles [98]. The size of the particles was controlled by the surfactant concentration. The average size of the particles, determined by transmission electron microscopy and by comparison with simulated Langevin curves, varied from 2 to 5 nm, with 30–35% polydispersity in the size distribution and thus they were superparamagnetic. The saturation magnetization was found to decrease and the surface anisotropy to increase with an increase in size of the nanoparticles.

Reversed micelles are surfactant-entrapped water pools in an organic solvent, and the water-to-surfactant molar ratio determines the size of the water pool. Aerosol–OT (sodium dialkylsulfosuccinate) has been the favored surfactant for reversed micelle formation since it can solubilize up to 50 moles of water per mole of surfactant and since its properties in organic solvents are well understood [99]. Sizes of nanoparticles, generated *in situ* in reversed micelles, have been controlled by the judicious selection of water-to-surfactant ratios, by using functionalized surfactant (replacing the sodium ion by the copper ion in Aerosol–OT for copper particle formation, for example [100, 101]) and by arresting the growth of nanoparticles by capping them by a nucleophilic reagent (thiophenol, for example) [20, 102].

Nanoparticle precursor ions can be selectively attracted to, and hence nanoparticles can be grown at, the inner or the outer surfaces of surfactant vesicles [103]. This arrangement has permitted the demonstration of vectorial photoelectron transfer in surfactant vesicles [103]. Similarly, nanoparticles have been prepared at one or both sides of bilayer lipid membranes [102].

Hydrophilic cavities in channel proteins provide a suitably confined space for nanoparticle generation. Cadmium and zinc sulfides have been formed, for example, within the cavity of a channel protein by a simple process in which a closely packed monolayer of the channel protein is formed on a neutral subphase, transported to a cadmium-chloride-containing subphase, transferred to a slide by Langmuir–Blodgett transfer, and exposed to hydrogen sulfide [62]. The size of the cluster formed is limited by the small number of ions capable of assimilation into the channel and, because each group of ions is compartmentalized, there is no possibility of spontaneous aggregation to produce macroscopic particulates inside of the protein channel.

Semiconducting, metallic, and magnetic nanoparticulate films have been grown under monolayers whose aqueous subphase contained the appropriate precursor ions (Cd^{2+}, Ag^+, Fe^{2+}/Fe^{3+}, for example) upon exposure to gaseous precursors (H_2S, formaldehyde, for example) [5, 105]. Silver ions, attached to negatively charged monolayers have also been reduced by *in situ* electrolysis [6, 105].

Advantage has also been taken of the hydrophilic interlayers of Langmuir–Blodgett films to incorporate or *in situ* generate nanoparticles [106–109]. Formation of CdS nanoparticles in cadmium arachidate Langmuir–Blodgett films has been monitored by an electrochemical quartz crystal microbalance (EQCM) and absorption spectrophotometry [106]. The formation of CdS nanoparticles was consistent with that expected for the quantitative conversion of Cd^{2+} ions in the films to CdS and the corresponding conversion of CdAr to arachidic acid. Subsequent reexposure of the film to H_2S increased the mole fraction of CdS in the film. Indeed, exposing a cadmium stearate Langmuir–Blodgett (LB) film to H_2S, then immersion in aqueous $CdCl_2$ was found to regenerate the cadmium stearate multilayer without the escape of CdS and the repetition of sulfidation–intercalation cycles allowed the size-quantized CdS to grow in a stepwise fashion within the hydrophilic interlayers [106]. This approach opens the door to the formation of nanoparticles in controllable thickness.

Polar liquids selectively adsorbed at solid interfaces from binary apolar–polar liquids have been shown to provide a suitable nanoreactor for the generation of nanoparticles [28]. Precise information on the volume of the nanoreactor, in a given system, can be obtained by the determination of excess adsorption isotherms. Thus, for example, in silica particle dispersions in ethanol–cyclohexane and in methanol–cyclohexane binary mixtures 0.5–5.0 nm-thick alcohol-rich adsorption layers have been shown to form at the silica–particle interface. This alcohol nanolayer has been used for the generation of controlled-sized CdS and ZnS particles [28].

A large variety of different nanoparticles have been prepared in the cavities provided by porous glasses, membranes, and zeolites [30, 74]. The current trend is to alter the pore sizes and chemical composition by derivatization or chemical synthesis.

- Using suitable templates it is possible to control the shapes of incipient nanoparticles and to grow them epitaxially [101].

The shape of the template has been shown to profoundly influence the shape of the
nanoparticles grown thereon (or therein). Evidence has been reported, for example,
for the formation of spherical and cylindrical copper nanoparticles in spherical and
cylindrical reversed micelles [101].

Monolayers spread on aqueous solution surfaces have been shown to provide
suitable templates for oriented crystallization of nanoparticles [85, 109–111]. Lead
sulfide (PbS) particulate films composed of highly oriented, equilateral triangular
crystals have been generated in situ by the exposure of arachidic acid (AA) mono-
layer-coated aqueous lead nitrate solutions to hydrogen sulfide. The AA-coated
PbS particulate films, at different stages in their growth, were transferred to solid
substrates and characterized by transmission electron microscopy (TEM), atomic
force microscopy (AFM), and electron diffraction measurements. Each individual
crystal had its (111) plane parallel and its (112), (121), and (211) plane perpendicular (arranged in threefold symmetry at 120° angles) to the AA monolayer surface.
This epitaxial growth has been rationalized in terms of an almost perfect fit between
the (111) plane of the cubic crystalline PbS and the (100) plane of the hexagonally
close-packed AA monolayer [110]. Interestingly electrical and spectroelectrical
properties of epitaxially grown PbS nanocrystals were found to be morphology dependent [89]. CdS [111, 112] and CdSe [113] have also been grown epitaxially under
Langmuir monolayers [111].

- Chemical and colloid chemical approaches can be extended to many other
  nanoparticle preparations.

It is not unfair to say that only our lack of imagination and ingenuity limits us
in preparing nanoparticles of any composition, size, and shape by suitable chemical
means. Indeed, new and innovative preparations of metallic, semiconducting,
magnetic, and ferroelectric nanoparticles appear with ever increasing frequency.

The desirable properties and potential applications (components of diode lasers
and nonlinear electro-optical devices, for example) of gallium arsenide, indium
arsenide, gallium phospide, and indium phosphide quantum wells and superlattices
prompted the ever increasing effort to design novel preparations of these nanoparticles. Indeed, several viable gallium arsenide [89, 94–96], gallium phosphide [90, 91],
indium arsenide [97], and indium phosphide [93] preparations have been reported.
Equally important is the ongoing intensive effort to prepare surface-oxidized silicon
[114–116] nanoparticles that have unique photoluminescence properties.

18.2.3 Preparation of Composite Nanoparticles, Nanoparticle Arrays, and Nanostructured Films

Recent activities are summarized in Table 18.7, some of which are highlighted
below.

Table 18.7. Preparations of composite nanoparticles.

Method	Particles Components	Comments	References
Heterosupramolecular synthesis	CdSe, diamine	Homodimers of CdSe nanocrystals were synthesized and characterized by TEM and electrophoresis.	[119]
Heterosupramolecular synthesis	TiO$_2$, cyanopyridino organics	Polyviologen-modified TiO$_2$ prepared by photocatalytic polymerization of bis(4-cyano-1-pyridinio)-p-xylene dibromide.	[122]
Heterosupramolecular synthesis	TiO$_2$, pyridnium organics	Photoinduced electron transfer processes were observed in organized redox-functionalized bipyridinium-polyethylenimine-TiO$_2$ colloids and particulate assemblies.	[123]
Solid solution	CdSe, molecular bridge	CdSe nanocrystallite networks were built with molecular connectors.	[124]
Electrospray MOCVD	CdSe/ZnSe	Luminescent thin-film CdSe/ZnSe quantum dot composites were synthesized by electrospray MOCVD using CdSe quantum dots passivated with an overlayer of ZnSe. The films were characterized by XRD and luminescence spectroscopy.	[125]
Polymer	Mn/ZnS	Mn-doped ZnS nanocrystals were precipitated with a poly(ethylene oxide) matrix. Photoluminescence excitation was employed to study the dopant effect on bandgap and the size quantization.	[117]
Colloid synthesis	CdS, HgS	A quantum dot quantum well CdS/HgS/CdS was combined in aqueous sol synthesis. TEM and fluorescence decay data were obtained. Chemistry and photophysics of mixed layered CdS/HgS colloids were investigated.	[126, 127]
Colloid synthesis	CdS, CdSe	Coupled composite CdS–CdSe and core-shell types of (CdS)CdSe and (CdSe)CdS nanoparticles were synthesized in aqueous sol form. TEM and optical absorption were used to characterize the composites. Luminescence spectral showed charge carrier transfer between CdS and CdSe moieties.	[118]
Colloid synthesis	CdS, SiO$_2$	Monodisperse colloidal silica CdS nanocomposites were prepared by controlled hydrolysis of tetraethyl orthosilicate. Tailored morphology was observed by TEM and SEM. XRD and light-scattering characterization were carried out to show the size of CdS to be 2.5 nm in diameter.	[33]
Colloid synthesis	TiO$_2$, SnO$_2$	Capped TiO$_2$–capped SnO$_2$ nanocrystallites were prepared in colloid dispersion. Photoelectrochemical behavior of the composite particles was studied.	[128]

Method	Materials	Description	Ref.
Nanoporous matrix	CdS, ZnS	Nanocomposite systems were grown *in situ* in porous silica-pillared layered phosphates.	[129]
Solid solution	CdS, CdSe	CdS_xSe_{1-x} nanocrystals were grown in silicate glass. Their structural and interfacial properties were studied by high-resolution TEM and optical absorption.	[130]
Solid solution	CdS, metals	Growth kinetics and quantum size effects of CdS nanocrystallites were examined in glasses. Metals were coated on semiconductor particles embedded in glass.	[131, 132]
Xerogel	Cr, CdS	Nanosized Cr clusters and intimate mixtures of Cr/CdS phases were synthesized in a porous hybrid xerogel by an internal doping method. New procedures were analyzed for the preparation of CdS and heterogeneous Cr/CdS phases in hybrid xerogel matrices' pore structure.	[133]
Photoreduction	Rh, TiO_2, SiO_2	Homogeneous Rh particles were generated by photoreduction of Rh(III) on TiO_2 colloids grafted on SiO_2.	[134]
Evaporation of metal in reaction gas	TiO_2, SiO_2	Quantum-sized TiO_2 supported on silica were prepared. UV, LEES, XPS, and REELS were used to characterize the composites. Spectroscopic features were used to examine the influence of size and support on photoemission of the composites.	[135]
Porous matrix	TiO_2, PbS, Ag_2S	Porous TiO_2 was sensitized by PbS quantum dots. Photoconduction properties were studied. Particle size and pH effects on the sensitization of nanoporous TiO_2 electrodes by Q-sized Ag_2S were investigated.	[136, 137]
Colloid synthesis	TiO_2, Fe, Mo, Ru, Os, Re, V	Role of metal ion dopants in quantum-sized TiO_2 was studied and correlation between photoreactivity and charge carrier recombination dynamics was established.	[138, 139]
Blending	CdSe, polymer	Electroluminescence was observed for CdSe quantum dot polymer composites.	[140]
Sol-gel processing	TiO_2, 4-(dimethylamino)-4'-nitrostilbene, 4-(2-(4-hydroxylphenyl) ethyl)-N-methylpyridium	TiO_2 film doped with organic compounds was made by two-dimensionally poled sol-gel processing for nonlinear optical activity.	[141]
Sol-gel processing	TiO_2, SiO_2	Selective epoxidation of α-isophorone was achieved with mesoporous titania–silica aerogels and tert-butyl hydroperoxide.	[142]
Sol-gel processing	RuO_2, TiO_2	Ultrafine RuO_2–TiO_2 binary oxide particles were prepared by a sol-gel process.	[143]

- Composite and core-shell types nanoparticles can now be routinely prepared.

Depending on the method of preparation, it is quite possible to vary the composition of composite nanoparticles. H_2S exposure of a monolayer floating on a mixture of aqueous cadmium nitrate and zinc nitrate solutions, for example, leads to the formation of composite CdS_xZnS_{1-x} nanoparticulate films [105]. The ratio of CdS to ZnS in the mixed nanoparticulate films could be varied by changing the ratio of the precursor ions in the subphase. A similar approach was used in forming mixed MnS_xZnS_{1-x} nanoparticulates in reversed micelles [117].

Sequential generation of two (or more) different nanoparticles results in the formation of core-shell-type structures [105, 118]. By paying careful attention to the different parameters (the order, the amount, and the rate of precursor introduction) it is quite feasible to achieve a fine level of thickness control of the different layers (core and shell, for example). Due attention has to be paid, however, to the solubility products of the nanoparticles. For example, the larger solubility product of ZnS than PbS permits the replacement of lead ions by zinc ions upon the immersion of PbS nanoparticles into an aqueous zinc ion solution. Conversely, because of their solubility products, lead ions cannot replace zinc ions in ZnS. Introduction of zinc ions into PbS is tantamount, of course, to doping. The extent of doping can be controlled, at least to some extent, by varying the time of exposure of the semiconductor nanoparticles to the dopant ions.

Sandwich layers of PbS and ZnS nanoparticulate films have been prepared under monolayers [105]. The method involved the following steps: (i) formation of the PbS nanoparticulate film by exposing a monolayer (floating on aqueous $Pb(NO_3)_2$ subphase) to H_2S, (ii) exchanging the aqueous $Pb(NO_3)_2$ subphase to aqueous $Zn(NO_3)_2$ subphase without perturbing the monolayer-supported PbS nanoparticulate films, and (iii) formation of the ZnS nanoparticulate film by exposing a monolayer-supported PbS (floating on aqueous $Zn(NO_3)_2$ subphase) to H_2S.

Core-shell-type (CdS)CdSe (or(CdSe)CdS) nanoparticles have been prepared by the sequential introduction of different amounts of H_2S and H_2Se (or H_2Se and H_2S) into aqueous $Cd(ClO_4)_2$ solutions containing $(NaPO_3)_6$ [118]. Conditions in these experiments were adjusted (by adding excess cadmium and hydroxide ions) for the maximization of excitonic fluorescence. Two emission bands were observed in the coupled and in the core-shell-type mixed semiconductor nanoparticles. The first one, centered around 470 nm, was attributed to the 1s(e)–1s(h) excitonic emission of CdS. The second, centered around 560 nm, was proposed to arise from charge transfer of CdS to core-shell-type (CdS)CdSe (or (CdSe)CdS) nanoparticles [118].

- Chemists have learned to treat nanoparticles as large macromolecules and to link them, by electrostatic interactions or covalent bonds, into heterostructured supramolecules, two-dimensional arrays, or three-dimensional networks.

The recognition that nanoparticles can be derivatized and treated like any other molecule is a major development in our quest for the full exploitation of chemistry for advanced materials synthesis. The following examples all illustrate our accomplishments in reacting and assembling nanoparticles.

Two thiol-capped cadmium selenide nanocrystals have been covalently linked by the addition of a bis(acyl hydrazine) derivative to give a stable homodimeric CdSe product [119]. CdS and TiO_2 nanoparticles have been bridged [120]. Phosphonated polypyridyl ligands have been used to anchor transition metal complexes to titanium dioxide nanoparticulate films [121]. Gold nanocrystals, each encapsulated by a monolayer of alkyl thiol molecules, have been cast from a colloidal solution onto a flat substrate to form a close-packed monoparticulate layer [144]. Adjacent clusters have then been covalently linked by aryl dithiols or aryl di-isonitriles, which displaced the alkyl thiol molecules in the monolayer and formed a two-dimensional superlattice of gold nanoparticles that exhibited nonlinear Coulomb-charging behavior when placed in the gap between two gold contacts [144]. An even simpler method of assembly of gold nanoparticle superlattices has been accomplished by an acid-facilitated phase transfer (from an aqueous dispersion to toluene) of thiol-derivatized Au, Pt, and Ag nanoparticles [145].

Of particular significance is the employment of DNA for the reversible assembly of nanoparticles into desired three-dimensional networks [146, 147]. The method involves the attachment of single-stranded DNA oligonucleotides of defined length and sequence to nanoparticles via thiol linkages and the assembling of the desired structures by Watson–Crick-type complementary base pairing. Thermal denaturation results in the disassembly of the supramolecular structure formed. The advantage of this approach is its versatility and diversity, which, at least in principle, permit the manufacturing of any tailor-made advanced nanoparticle-based materials by well-established biochemical protocols.

Self-assembly of alternative nanolayers of oppositely charged polyelectrolytes and nanoparticles provides a convenient means for the preparation of two-dimensional arrays and three-dimensional networks of simple and composite nanoparticles. Self-assembly of macromolecular species into larger units is well established in nature, of course. It is only recently that chemists have been able to mimic this process. They have layer-by-layer self-assembled oppositely charged polyelectrolytes and polyelectrolyte-nanoparticle films [148]. The layer-by-layer self-assembly of polyelectrolytes and nanoparticles onto substrates is deceptively simple. A well-cleaned substrate is primed by adsorbing a layer of surfactant or polyelectrolyte onto its surface. The primed substrate is then immersed into a dilute aqueous solution of a cationic polyelectrolyte, for a time optimized for adsorption of a monolayer, rinsed, and dried. The next step is the immersion of the polyelectrolyte monolayer-covered substrate into a dilute dispersion of surfactant-coated negatively charged nanoparticles, also for a time optimized for adsorption of a monoparticulate layer, rinsing, and drying. These operations complete the self-assembly of a polyelectrolyte monolayer–monoparticulate layer of semiconductor nanoparticle sandwich unit onto the primed substrate. Subsequent sandwich units are deposited analogously. Alternating polyelectrolye monolayers and CdS [148], PbS [149], TiO2 [149], BaTiO3 [149], lead–zirnonate–titanate, PZT [150], and Au [151] monoparticulate layer sandwich units have been prepared this way.

The method of self-assembly is extremely versatile. Sandwich units can be layered in any order for a variety of different nanoparticles. Furthermore, elimination of the polyelectrolyte layers by burning leads to ultrathin films in which the nanoparticles are arranged in a three-dimensional network with controllable interparticle distances. Such films are, of course, extremely useful for many practical applications. For example, the self-assembly of monodisperse gold and silver colloid particles into monolayers on polymer-coated substrates yields macroscopic surfaces that are highly active for surface-enhanced Raman scattering (SERS) [151] and optically nonlinear materials [152].

Spreading surfactant-stabilized nanoparticles on aqueous solutions in a Langmuir trough and transferring them to solid substrates by the Langmuir–Blodgett technique provide an alternative approach to superlattice construction [153, 154]. The technique can be regarded as analogous to monolayer formation from simple surfactants. There are many intrinsic benefits to this method. That the particles are prepared prior to their incorporation into the films enables their dimensions and physical properties and the particle size distribution to be precisely controlled. Spreading the particles in a Langmuir trough provides a means for defining the interparticle distances and facilitates subsequent transfer of the particulate films to a wide range of solid substrates by using standard Langmuir–Blodgett (LB) techniques. This may be contrasted with deposition techniques, where the quality of the film is highly dependent on the solid substrate itself. Thus, in essence, this method is highly versatile, facilitating film construction from a diverse range of materials and substrates; it is also extremely simple experimentally. Additionally, it is quite possible to affect the layer-by-layer transfer of different nanoparticles and thus form a composite three-dimensional structure.

Surfactant-coated cadmium sulfide [153, 154], TiO_2 [155], the magnetic iron oxide magnetite Fe_3O_4 [156], ferroelectric barium titanate [150], lead zirconium titanate [157], Pt [158], Pd [158], and Ag [159, 160] nanoparticles have been prepared to date in our laboratories by this Langmuir monolayer LB-film technique. In all cases, particulate films were spread on water in a Langmuir trough by dispersing solution aliquots from a Hamilton syringe. Physical properties of the monoparticulate films were characterized, *in situ*, on the water surface and subsequently transferred to solid substrates, by using a range of physical techniques. The structures of the films on the water subphase were examined on the micrometer scale by Brewster-angle microscopy (BAM) and at higher magnifications by transmission electron microscopy (TEM). Absorption spectroscopy and steady state fluorescence spectroscopy were applied where appropriate, and reflectivity measurements permitted the estimation of film thicknesses on water surfaces. Standard techniques were applied for the LB transfer. The surfactant coating of the nanoparticles can, of course, be burned off to produce the desired ultrathin nanoparticulate film.

The construction of nanoparticle arrays and nanostructured films is summarized in Table 18.8.

Table 18.8. Fabrications of nanoparticle arrays and nanostructured films.

Method	Particles/Substrates	Comments	References
Electrodeposition	TiO_2/electrode	Nanocrystalline TiO_2 thin films were prepared by cathodic electrodeposition.	[161]
Monolayer	TiO_2/PbS	Monolayers of TiO_2/PbS coupled nanoparticles were formed by electrostatic attraction. TEM, XPS, and absorption spectroscopy were used to characterize the films.	[162]
Langmuir–Blodgett deposition	CdSe, CdS, TiO_2/glass, TEM grids, metal electrode	Langmuir–Blodgett films of size-selected CdSe nanocrystallites were prepared. TEM, absorption and emission spectroscopy were used to characterize the film. LB films were made from fluorescence-activated, surfactant-capped, size-selected CdS nanoparticles. AFM, STM were employed to image the surfaces. Monoparticulate layer and layers of size-quantized CdS clusters were fabricated by LB techniques. Monoparticulate layers of TiO_2 nanocrystallites were deposited with controllable interparticle distances.	[21, 153, 154, 158–160]
Organized bilayer	CdS/bilayer cast film	CdS clusters were generated in an organic bilayer template to form an organic-inorganic superlattice structure.	[104]
Deposition	CdSe/solid supports	Thin and superthin photoconductive CdSe films were deposited on substrate at room temperature.	[166]
Self-assembly	Au, CdS	Self-assemblies of semiconductor nanocrystals were surveyed on thiol-functionalized Au surfaces. Self-assembling multilayer structures were studied by quartz crystal microgravimetry.	[167, 168]
Self-assembly	CdSe/thiol organics on ITO electrode	Light-emitting diodes were made from CdSe nanocrystals and a semiconducting polymer.	[170]
	CdS/porous film	Quantum-sized CdS particles were assembled to form a regularly nanostructured porous film.	[171]
Surfactant liquid crystal	WO_2, Sb_2O_5/surfactant organized layers	Organic molecules with inorganic molecular species were organized into nanocomposite biphase arrays.	[172]
Solution decomposition epitaxy	CdS/silicon wafer	CdS nanostructured films were deposited by decomposition of thiourea in basic solution.	[167]
Atomic layer epitaxy	TiO_2	Titanium isopropoxide was used as a precursor in atomic layer epitaxy of TiO_2 thin films. Atomic layer epitaxy growth of TiO_2 thin films from titanium ethoxide.	[168, 169]

Table 18.8. (cont.)

Method	Particles/Substrates	Comments	References
Aggregation	CdS	Ultrathin semiconductor films were formed by CdS nanostructure aggregation.	[170]
	CdO, CdS, ZnO, ZnS	Nanoclusters were embedded in siliceous faujasite.	[171]
Chemical solution deposition	CdS, TiO$_2$	CdS films were deposited by decomposition of thiourea in basic solutions. TiO$_2$ thin films were prepared from aqueous solution.	[172, 173]
Chemical solution deposition	ZnO	Transparent and conductive ZnO thin films were prepared by applying a solution of zinc alkoxid. Spectroelectrochemical investigations of nanocrystalline ZnO films.	[174, 175]
Chemical solution deposition	CdS	CdS thin films comprising nanoparticles were prepared by a solution growth technique.	[176]
Electrochemistry	TiO$_2$/polymer mold	TiO$_2$ nanotube array was formed by electrochemical deposition in a polymer mold.	[177]
Electrochemistry	TiO$_2$	Photoelectrochemical doping of TiO$_2$ particles and the effect of charge carrier density on the photocatalytic activity of microporous semiconductor electrode films were studied.	[178]
Adsorption	TiO$_2$	Photochemical quartz crystal microbalance was used to study the nanocrystalline TiO$_2$ semiconductor–electrode–water interfaces. Simultaneous photoaccumulations of electrons and protons were investigated.	[179]
Colloid synthesis	SnO$_2$/TiO$_2$	Photoelectrochemical behavior of SnO$_2$/TiO$_2$ composite systems and its role in photocatalytic degradation of a textile azo dye were investigated.	[128]
Sol-gel technology	TiO$_2$	TiO$_2$ ultrathin films were prepared by two-dimensional sol-gel process.	[180]
Nanochannel glass	GaAs, InAs/ nanochannel glass	GaAs and InAs nanowires, in the form of 3–15 nm crystallites, were prepared by reaction of organogallium or organoindium with arsine in porous nanochannel glass.	[181]
Chemical beam epitaxy	GaAs film	Nanostructured GaAs films were grown using tris(di-ter-butylarsino)gallane as single source precursor in chemical beam epitaxy.	[182]

18.3 Properties of Bulk Semiconductors and Semiconductor Nanoparticles Compared and Contrasted

The structural, optical, and electronic properties of intrinsic bulk CdS, CdSe, ZnS, ZnSe, PbS, TiO_2, GaAs, Ge, and Si semiconductors are summarized in Table 18.9. These properties are not expected to change for a given semiconductor with changes in size or shape. In contrast, both the optical and the electronic behaviors of semiconductor nanoparticles are distinctly size dependent. In fact, the major advantage of using semiconductor nanoparticles is that the optoelectronic properties can be tailored quite simply by size alteration without changing the chemical composition of the material. This behavior spectacularly manifests itself in size-dependent color changes. The black bulk PbS can be made, for example, red, yellow, and even colorless if prepared as progressively smaller- and smaller-diameter nanoparticles. The practical relevance of size-quantized semiconductor particles is that upon photoexcitation the undesirable electron–hole recombination diminishes since the smaller the particle the larger the surface-area-to-volume ratio, and hence the ease of electron transfer to surface-bound acceptors or donors. That the size quantization decreases the absorption edge and hence increases the required excitation energy is another matter.

A related, and often forgotten, issue is that for size-quantized particles morphological changes often influence electrical and optoelectrical behavior. For example, the differences in morphology between equilateral-triangular PbS (PbS-I), right-angle-triangular PbS (PbS-II), both epitaxially grown under monolayers (prepared from AA:ODA = 1:0 and AA:ODA = 1:1), and disk-shaped PbS (PbS-III, nonepitaxially grown under monolayers, prepared from hexadecylphosphonic acid), manifested themselves in different spectroelectrochemical behavior [85]. Specifically, marked differences were observed in the potential-dependent absorption spectra of PbS-I, PbS-II, and PbS-III. Biasing the epitaxially grown PbS-I and PbS-II nanoparticulate films to negative potentials (from 0.5 V to −1.1 V) increased the intensity of absorption in the ultraviolet region. In contrast, no change in the absorption at wavelengths longer than 700 nm was observed in the nonepitaxially grown PbS nanoparticulate film on changing the potential from 0 to −1.5 V. Absorption spectra of the optically transparent conductive glass (i.e., the control) remained unaltered upon biasing the potential between +0.5 and −1.5 V. The near-infrared absorption is likely to correspond to the spectrum of trapped charge carriers. Increase of this absorption resulted from the accumulation of trapped conduction band electrons at negative bias potentials in PbS-I and PbS-II. Indeed, absorbances for PbS-II at 750 nm were found to decrease with increasing applied positive potential linearly to −0.6 V, after which they remained unaltered. The point of inflection, $−0.50 \pm 0.05$ V, may be taken to correspond to the flat-band potential, V_{fb}, of the PbS-II nanoparticulate film [85]. Similarly, marked differences in capacitance vs. potential and photocurrent curves were observed between PS-I, PS-II, and PS-III [85]. Dependence of the absorbance on the applied potential, as well as the observed photocurrent and voltage-dependent capacitances, reflected a

Table 18.9. Solid state physical properties of selected semiconductors.[1]

Property	CdS	CdSe	ZnS	ZnSe	PbS
Lattice parameters (Å)	Hexagonal: $a = 4.136, c = 6.714$ Zincblende: $a = 5.818$ Rocksalt: $a = 5.42$	Hexagonal: $a = 4.299, c = 7.011$ Zincblende: $a = 6.052$ Rocksalt: $a = 5.49$	Hexagonal: $a = 3.8226, c = 6.2605$ Zincblende: $a = 5.4102$	Hexagonal: $a = 3.996, c = 6.540$ Zincblende: $a = 5.6676$	Orthorhombic: $a = 11.28, b = 3.98, c = 4.21$ Rocksalt: $a = 5.936$
Density (g cm^{-3})	4.82	5.81	4.075	5.266	7.597
Melting point (K)	1750	1514	Sublime before melting 2103	1793	1383
Bandgap (eV)	$\sim$2.554–2.599	$\sim$1.751–1.771	$\sim$3.9107–3.9407 (hex.) $\sim$3.68–3.74 (zinc.)	$\sim$2.70–2.82 (295 K)	0.286 (4.2 K) 0.42 (300 K)
Optical absorption edge (nm)	481–489	697–704	314–317 (hex.) 331–336 (zinc.)	$\sim$439–459	2951 (300 K)
Exciton energies (eV)	$\sim$2.5517–2.5528	$\sim$1.8249–1.8266	$\sim$3.698–3.801 (zinc., 273 K) $\sim$3.8715–3.8932 (hex, 77 K)	$\sim$2.793–2.803	
g factor	1.789, 1.774, 1.75, 1.23, 1.8, 0.7	$g_{c\parallel} = 0.6, g_{c\perp} = 0.51,$ $g_{v\parallel} = 1.41$	$g_c = 1.8846,$ $g_v = 0.93,$	$g_c = \sim1.18\text{–}1.37$ $g_v = \sim-0.12\text{--}-0.28$	$g_{c\parallel} = 12$ $g_{v\parallel} = 13$
Effective mass	$M_A = m_p + m_n$ $= \sim0.78\text{–}0.9,$ $M_A = m_p + m_n = 3.0$ $= 0.158$ $m_n = 0.210$ $m_p = 0.64$	$m_n = \sim0.112\text{–}0.13$ $m_{p\perp A} = 0.45,$ $m_{p\perp B} = 0.9$	$m_n = 0.34$	$m_n = 0.160$ $m_p = 0.6$	$m_{\perp,n} = 0.080$ $m_{\parallel,n} = 0.105$
Piezoelectric strain coefficient	$d_{15} = -11.91$ $d_{31} = -5.09$ $d_{33} = 9.70$	$d_{15} = -10.51$ $d_{31} = -3.93$ $d_{33} = 7.84$	(Hex.) $d_{15} = -4.37$ $d_{31} = -2.14$ $d_{33} = 3.66$ (Zinc) $d_{14} = 3.117$	(Zinc.) $d_{14} = 1.10$	

Piezoelectric stress coeff. coefficient		$\varepsilon_{15} = -0.138$ $\varepsilon_{31} = -0.160$ $\varepsilon_{33} = 0.347$	$\varepsilon_{15} = -0.118$ $\varepsilon_{31} = -0.238$ $\varepsilon_{33} = 0.265$ $\varepsilon_{14} = 0.140$	$\varepsilon_{14} = 0.049$	
Dielectric constants	$\varepsilon(0) = 9.38$ $\varepsilon_{11} = 8.7$ $\varepsilon_{33} = 9.25$ $\varepsilon_{11}(\infty) = 5.53$ $\varepsilon_{33}(\infty) = 5.5$	$\varepsilon(0) = 9.38$ $\varepsilon_{11} = 10.16$ $\varepsilon_{\perp}(0) = 9.29$ $\varepsilon_{11}(\infty) = 6.20$ $\varepsilon_{\perp}(\infty) = 6.30$	$\varepsilon_{\parallel}(0) = 8.31$ $\varepsilon_{33}(0) = 8.76$ $\varepsilon_{\parallel}(\infty) = 8.25$ $\varepsilon_{33}(\infty) = 8.59$	$\varepsilon(0) = 9.14$ $\varepsilon(\infty) = 6.3$	$\varepsilon(0) = \sim 160\text{–}181$ $\varepsilon(0) = \sim 16.2\text{–}18.2$
Nonlinear optical parameters	Two-photon absorption coefficient: $K_2 = \sim 0.56\text{–}0.7$	Two-photon absorption coefficient: $K_2 = 0.050$ at 1.06 μm	Two-photon absorption coefficient: $K_2 = 0.2 \times 10^{-4}$ at 1.06 μm	Two-photon absorption coefficient: $K_2 = 0.17 \times 10^{-4}$ at 1.06 μm $K_2 = 0.08 \times 10^{-6}$ at 3.56 eV	

1. Effective masses are given in the unit of electron mass, m_0. The subscripts of the g factor are: v = valence electron, c = conduction electron, ∥ = parallel component, ⊥ = perpendicular component, n = electron, p = hole. $\varepsilon(0)$ = low-frequency dielectric constant, $\varepsilon(\infty)$ = high-frequency dielectric constant.

Table 18.9. (cont.).[1]

Property	TiO$_2$	GaAs	Ge	Si
Lattice parameters (Å)	Tetragonal (Anatase) $a = 3.7845$, $c = 9.5143$ tetragonal (Rutile) $a = 4.5941$, $c = 2.9589$	Zincblende: stable at normal pressure $a = 5.653$ Orthorhombic distortion, high-pressure phase $a = 4.946$, $b = 4.638$, $c = 5.493$	I: diamond, $a = 5.6579$ II: hex. $a = 4.88$, $c = 2.692$	I: $a = 5.431$ (diamond) II: $a = 4.686$, $a = 2.258$ III: $a = 6.636$ IV: $a = 3.8$, $c = 6.28$
Density (g cm^{-3})	3.894 (Anatase) 4.259 (Rutile)	5.3176	5.3234	I: 2.329
Melting point (K)	Sublime before melting $T_m = 1513$	Sublime before melting $T_m = 1513$	1210	1685
Band gap (eV)	Direct: $\sim$3.033–3.062 Indirect: $\sim$3.049–3.101	1.424 (300K)	Direct: 0.805 (293K) Indirect: 0.664 (293 K)	Indirect: 1.110 (300 K) 1.17 (0 K)
Optical absorption edge (nm)	404–408 (direct)	870	1540	1117
Exciton energies (eV)	1.515 (1.8 K)	1.515 (1.8 K)	0.739 (1.7 K)	1.155 (1.8 k)
g factor		$g_c = -0.44$ (4 K)	$g_v = -1.8$ (1.7 K) -3.0 (30 K) $g_n = 1.5$, $g_p = 3.8$	$g_c = -1.999$
Effective mass		$m_{p,h} = 0.51$ $m_{p,l} = 0.082$ $m_{s,o} = 0.154$ $m_{p,d} = 0.53$	$m_{n,perp} = 0.0807$ $m_{n,para} = 1.57$	$m_{n,perp} = 0.1905$ $m_{n,para} = 0.9163$ $m_{n,ds} = 1.062$ $m_{n,opt} = 0.43$
Piezoelectric strain coefficient	$d_{11} - d_{12} = 111 \times 10^{-8}$ 3.5 μm $d_{44} = 140 \times 10^{-8}$ 10.39 μm			

Dielectric constants	$\varepsilon_\parallel(0) = 8.31$	$\varepsilon = 12.1$ (4.2 K)
	$\varepsilon_{33}(0) = 8.76$	$\varepsilon = 11.9$ (300)
	$\varepsilon_\parallel(\infty) = 8.25$	
	$\varepsilon_{33}(\infty) = 8.59$	
Nonlinear optical parameters	Two-photon absorption coefficient: $K_2 = 0.2 \times 10^{-4}$ at 1.06 μm	

1. Effective masses are given in the unit of electron mass, m_0. The subscripts of the g factor are: v = valence electron, c = conduction electron, $\parallel$ = parallel component, $\perp$ = perpendicular component, n = electron, p = hole. $\varepsilon(0)$ = low-frequency dielectric constant, $\varepsilon(\infty)$ = high-frequency dielectric constant.

complex interplay between the electron population in the electronic bands, in the traps (whose levels correspond to bulk imperfections), and in the available surface states, in addition to the ongoing interfacial electrochemical and photoelectro-chemical processes. The significance of this work is that it has unambiguously demonstrated morphology-dependent spectroelectric, electric, and electrochemical properties of PbS nanocrystallites grown under monolayers [85].

Metals behave like to semiconductors; in the macroscopic regime their properties are size independent, whereas in the nanodomains there is a recognized transition from nonmetallic to metallic behavior [183]. The situation for magnetic, let alone ferroelectric and piezoelectric, materials is more complex and as yet incompletely understood. Such questions as (i) what is the size of the smallest single-domain magnetic (or ferroelectric) particle?, and (ii) what are the properties of size-quantized and dimensionally reduced magnetic (or ferroelectric) particles? have not been adequately answered either theoretically or experimentally.

18.4 Current Trends and Future Directions

An attempt has been made to select contributions in the present book that represent the current trends and future directions in the preparation, characterization, and utilization of nanoparticles and nanostructured films.

Chemists have risen rather successfully to the challenge of making nanoparticles, as evidenced in the present chapter. It is well within their means to prepare dispersions of any desired simple or complex nanoparticles in a high degree of mono-dispersity. Furthermore, it is often possible to isolate the nanoparticles as solid powders and redisperse them without affecting their sizes and size distributions. Most of these preparations are based on well-established colloid chemical processes. The use of such relatively simple and flexible templates as monolayers (Chapter 2), reversed micelles (Chapter 4) and polymers (Chapter 7) are also based on colloid and surface chemical principles. Similarly, determinations of adsorption isotherms provided [13, 28, 184–189] the information necessary for the characterization of nanoreactors that spontaneously form upon the selective adsorption of a polar liquid onto a solid surface from binary polar–apolar mixtures. Generation of size-quantized semiconductor nanoparticles at dispersed organoclay complexes [13] and layered silicates [28] illustrates the use of the nanophase reactors provided by adsorbed binary liquids. The binary liquid pairs ethanol(1)–cyclohexane(2) and methanol(1)–cyclohexane(2) are highly suitable since the polar component of the liquid mixture (1) is preferentially adsorbed at the solid interface, and hence its mole fraction in the bulk (x_1) is negligible (i.e. $x_1^s \gg x_1$ and $x_2^s \ll x_2$), and since the semi-conductor precursors (Cd^{2+} and Zn^{2+}) are highly soluble in the liquid which pref-erentially adsorbed at the interface (methanol and ethanol), although they are in-soluble in the bulk phase (predominantly cyclohexane). These conditions effectively limited the nucleation and growth of the semiconductors to the nanophase reactor provided by the adsorption layer at the solid interface. By varying the mole fraction of the polar liquid (1) it is possible to control the volume of the nanophase reactor and hence the size of the semiconductor particles grown therein.

Much progress has been made in exploiting such rigid templates as zeolites (Chapter 17), opals (Chapter 13), and porous membranes (Chapter 10) for the preparation of nanoparticles, nanofibrils, and nanotubes. Versatility and reproducibility are distinct advantages of this approach. Electrochemistry is increasingly being employed for the generation of nanoparticles, nanostructured films, and superlattices (Chapters 1 and 3). Indeed, electrodeposition in the cavities of porous membranes can be considered to be a kind of templating in a nanobeaker (Chapter 10).

Nanoparticle research not only benefits from colloid and surface science but also contributes to it. Our need to construct ever more sophisticated supramolecular and heterosupramolecular structures (Chapter 16) demands a better understanding of intricate surface and interparticle colloid chemical interactions and reactions. Thus, it is not surprising that physicists, chemical physicists, and materials scientists are devoting increasingly more attention to the surface and colloid phenomena (Chapters 11 and 12, for example). Their efforts are made easier, of course, by the availability of a large variety of different techniques. Of these, special mention should be made of the surface force apparatus and the scanning force microscope [190–192], which permit both the imaging and the manipulation of nanoparticles and atomic clusters [193, 194] as well as the determination of interparticle interaction forces [195].

Considerable work is being directed toward examining the mechanism of nanoparticle-mediated electron and photoelectron transfers (Chapters 9 and 14). The importance of electrical double layers to electron transfer cannot be overemphasized (Chapter 12). It is quite remarkable that determination of nanoparticle-mediated single-electron transfer events are in our grasp (Chapter 15). This should lead to the fabrication of novel nanodevices.

Many new nanoparticles have been prepared by enterprising chemists. Preparation of porous silicon nanoparticles by a variety of different methods (see Chapters 5 and 8) is particularly significant since they have been shown to undergo photoluminescence and they can be readily integrated into silicon-based integrated circuitry.

Reference should be made to polymeric nanoparticles and dendrimers even though they have not been explicitly discussed in the present volume. Polymeric nanoparticles in the 1–100 nm-size regime, just like their metallic and semiconducting counterparts, exhibit optical and magnetic properties that are different than those observed in the bulk for the same polymers [196, 197]. Dendrimers are characterized by highly branched structures in which all bonds converge to a central core [198]. They are uniform in shape, size, and structure and can be spread on water and aqueous solution surfaces. Dendrimers also display properties that are different than those associated with polymers. These differences manifest themselves in altered chemical reactivities and in distinct nanoenvironments [198]. Many optical and electro-optical [197, 199], as well as biological [200], applications have been found for polymeric nanoparticles. Furthermore, polymeric nanoparticles [201], just like polyelectrolytes [202, 203], have been self-assembled in two-dimensional arrays and three-dimensional superlattices.

In spite of the extremely rapid progress in nanoparticles research, exploitation of the results to practical functioning economic devices has been disappointingly slow. We are confident that we shall witness a rapid progress in this direction in the very near future. The key lies, we believe, in using nanoparticles as molecules and constructing, or preferentially self-assembling, functional three-dimensional networks with a desired composition and topology.

Acknowledgments

We are grateful to the National Science Foundation, which supported most of the experimental work of our group. Janos H. Fendler thanks the donors of the Meyerhof Visiting Professorship for making his stay at the Weizmann Institute of Science, Rehovot, Israel possible and for allowing him to complete the writing of this chapter and the editing of this book. The true credit is due, of course, to the numerous scientists whose creative work has defined and stimulated our progress in nanoparticle research.

References

[1] D. F. Evans, H. Wennerstrom, *The Colloid Domain. Where Physics, Chemistry and Technology Meet*, VCH, New York **1994**.
[2] G. A. Ozin, *Acc. Chem. Res.* **1997**, *30*, 17–27.
[3] (a) S. Mann, *Biomimetic Materials Chemistry*, VCH, New York **1996**. (b) P. Calvert, P. Rieke, *Chem. Mater.* **1996**, *8*, 1715–1727.
[4] L. Addadi, S. Weiner, *Angew. Chem. Int. Ed. Eng.* **1992**, *31*, 153–169.
[5] J. H. Fendler, *Membrane Mimetic Approach to Advanced Materials*, Springer-Verlag, Berlin, **1992**.
[6] J. H. Fendler, F. C. Meldrum, *Adv. Mater.* **1995**, *7*, 607–632.
[7] A. P. Alivisatos, *Materials Research Society Bulletin* **1995**, *XX*, 23–31.
[8] A. Henglein, *Ber. Bunseng. Phys. Chem.* **1995**, *99*, 903–913.
[9] G. Hodes, *Solar Energy Materials and Solar Cells* **1994**, *32*, 323.
[10] H. Weller, *Angew. Chem. Int. Ed. Eng.* **1993**, *32*, 41–53.
[11] P. V. Kamat, *Progr. Reaction Kinetics* **1994**, *19*, 277–316.
[12] M. A. Marcus, W. Flood, M. Steigerwald, L. Brus, M. Bawendi, *J. Phys. Chem.* **1991**, *95*, 1572–1576.
[13] I. Dekany, L. Turi, E. Tombacz, J. H. Fendler, *Langmuir* **1995**, *11*, 2285–2292.
[14] A. P. Alivisato, *J. Phys. Chem.* **1996**, *100*, 13266–13239.
[15] R. W. Collins, P. M. Fauchet, I. Shimizu, J. C. Vial, T. Shimada, A. P. Alivisatos, *Advances in Microcrystalline and Nanocrystalline Semiconductors*, Materials Research Society, Pittsburgh **1997**.
[16] J. H. Fendler, I. Dekany, *Nanoparticles in Solids and Solutions*, NATO ASI Series, Kluwer, Dordrecht **1996**.
[17] (a) A. Ulman, *Adv. Mater.* **1993**, *5*, 55–57. (b) J. Liu, A. Kim, L. Q. Wang, B. J. Palmer, Y. L. Chen, P. Bruinsma, B. C. Bunker, G. J. Exarhos, G. L. Graff, P. C. Rieke, G. E. Fryxell, J. W. Virden, B. Tarasevich, L. A. Chick, *Adv. Colloid Interface Sci.* **1996**, *69*, 131–180.

[18] A. Chemseddine, H. Weller, *Ber. Bunseng. Phys. Chem.* **1993**, *97*, 636–637.

[19] T. Vossmeyer, L. Katsikas, M. Giersig, I. G. Popovic, K. Diesner, A. Chemseddine, A. Eychmuller, H. Weller, *J. Phys. Chem.* **1994**, *98*, 7665–7673.

[20] C. B. Murray, D. J. Norris, M. G. Bawendi, *J. Am Chem. Soc.* **1993**, *115*, 8706–8715.

[21] B. O. Dabbousi, C. B. Murray, M. F. Rubner, M. G. Bawendi, *Chem. Mater.* **1994**, *6*, 216–219.

[22] M. Nirmal, C. B. Murray, M. G. Bawendi, *Phys. Rev.* B, **1994**, *50*, 2293–2300.

[23] Y. Nosaka, H. Shigeno, T. Ikeuchi, *J. Phys. Chem.* **1995**, *99*, 8317–8322.

[24] Y. Wang, M. Harmer, N. Herron, *Israel J. Chem.* **1993**, *33*, 31–39.

[25] M. E. Brenchley, M. T. Weller, *Angew. Chem. Int. Ed. Engl.* **1993**, *32*, 1663–1665.

[26] N. Herron, J. C. Calabrese, W. E. Farneth, Y. Wang, *Science* **1993**, *259*, 1426–1428.

[27] T. Vossmeyer, C. Reck, L. Katsikas, E. T. K. Haupt, B. Schulz, H. Weller, *Science* **1995**, *267*, 1476–1479.

[28] I. Dekany, L. Nagy, L. Turi, Z. Kiraly, N. A. Kotov, J. H. Fendler, *Langmuir* **1996**, *12*, 3709–3715.

[29] Y. A. Barnakov, M. S. Ivanova, R. A. Zvinchuk, A. A. Obryadina, V. P. Petranovskii, V. V. Poborchii, Y. E. Smirnov, A. Shchukarev, Y. V. Ulashkevich, *Inorg. Mater.* **1995**, *31*, 752–755.

[30] G. A. Ozin, *Adv. Mater.* **1994**, *6*, 71–76.

[31] T. Sugimoto, G. E. Dirige, A. Muramatsu, *J. Colloid & Interface Sci.* **1996**, *180*, 305–308.

[32] S. A. Gurevich, A. I. Ekimov, I. A. Kudryavtsev, O. G. Lyublinskaya, A. V. Osinskii, A. S. Usikov, N. N. Faleev, *Semiconductors* **1994**, *28*, 486–493.

[33] S. Y. Chang, L. Liu, S. A. Asher, *J. Am Chem. Soc.* **1994**, *116*, 6739–6744.

[34] R. A. Hobson, P. Mulvaney, F. Grieser, *J. Chem. Soc., Chem. Commun.* **1994**, 823–824.

[35] (a) M. Kundu, A. A. Khosravi, S. K. Kulkarni, P. Singh, *J. Mater. Sci.* **1997**, *32*, 245–258. (b) H. Yoneyama, T. Torimoto, *Adv. Mater.* **1995**, *7*, 492–494.

[36] J. L. Coffer, R. R. Chandler, C. D. Gutsche, I. Alam, R. F. Pinizzotto, H. Yang, *J. Phys. Chem.* **1993**, *97*, 696–702.

[37] (a) T. Trindade, P. O'Brien, *J. Mater. Chem.* **1996**, *6*, 343–347. (b) T. Trindade, P. O'Brien, *Adv. Mater.* **1996**, *8*, 161–163.

[38] Y. Golan, L. Margulis, G. Hodes, I. Rubinstein, J. L. Hutchison, *Surface Sci* . **1994**, *311*, 633–640.

[39] U. Demir, C. Shannon, *Langmuir* **1994**, *10*, 2794–2799.

[40] S. Gorer, G. Hodes, *J. Phys. Chem.* **1994**, *98*, 5338–5346.

[41] G. Hodes, *Israel J. Chem.* **1993**, *33*, 95–106.

[42] T. Edamura, J. Muto, *Thin Solid Films* **1993**, *226*, 135–139.

[43] S. W. Haggata, X. C. Li, D. J. Cole-Hamilton, J. R. Fryer, *J. Mater. Chem.* **1996**, *6*, 1771–1780.

[44] S. R. Bigham, J. L. Coffer, *Colloids and Surfaces A – Physicochemical and Engineering Aspects* **1995**, *95*, 211–219.

[45] J. Lin, E. Cates, P. A. Bianconi, *J. Am. Chem. Soc.* **1994**, *116*, 4738–4745.

[46] M. Moffitt, A. Eisenberg, *Chem. Mater.* **1995**, *7*, 1178–1184.

[47] M. Moffitt, L. McMahon, V. Pessel, A. Eisenberg, *Chem. Mater.* **1995**, *7*, 1185–1192.

[48] H. Noglik, W. J. Pietro, *Chem. Mater.* **1995**, *7*, 1333–1336.

[49] V. S. Gurin, M. V. Artemyev, *J. Cryst. Growth* **1994**, Sep 13–17, 993–997.

[50] H. Yao, N. Kitamura, *Bull. Chem. Soc. Jpn.* **1996**, *69*, 1227–1232.

[51] K. M. Choi, K. J. Shea, *J. Phys. Chem.* **1994**, *98*, 3207–3214.

[52] K. M. Choi, K. J. Shea, *Chem. Mater.* **1993**, *5*, 1067–1069.

[53] M. Nogami, A. Kato, *J. Non-Cryst. Sol.* **1993**, *163*, 242–248.

[54] H. Mathieu, T. Richard, J. Allegre, P. Lefebvre, G. Arnaud, *J. Appl. Phys.* **1995**, *77*, 287–293.

[55] T. Fujii, Y. Hisakawa, E. J. Winder, A. B. Ellis, *Bull. Chem. Soc. Jpn.* **1995**, *68*, 1559–1564.

[56] A. Othmani, J. C. Plenet, E. Berstein, C. Bovier, J. Dumas, P. Riblet, P. Gilliot, R. Levy, J. B. Grun, *J. Cryst. Growth* **1994**, *144*, 141–149.

[57] K. M. Choi, J. C. Hemminger, K. J. Shea, *J. Phys. Chem.* **1995**, *99*, 4720–4732.

[58] M. Nogami, A. Kato, Y. Tanaka, *J. Mater. Sci.* **1993**, *28*, 4129–4133.

[59] D. Lincot, R. Ortegaborges, M. Froment, *Philos. Mag. B – Phys. Condens. Matter Struct. Electr. Opt. Magn. Prop.* **1993**, *68*, 185–194

[60] M. Nogami, A. Nakamura, *Physics and Chemistry of Glasses* **1993**, *34*, 109–113

[61] J. Butty, N. Peyghambarian, Y. H. Kao, J. D. Mackenzie, *Appl. Phys. Lett.* **1996**, *69*, 3224–3226.

[62] J. Y. Wang, R. A. Uphaus, S. Ameenuddin, D. A. Rintoul, *Thin Solid Films* **1994**, *242*, 127–131.

[63] Y. Golan, J. L. Hutchison, I. Rubinstein, G. Hodes, *Adv. Mater.* **1996**, 8, 631–633.

[64] (a) S. Mahamuni, A. A. Khosravi, M. Kundu, A. Kshirsagar, A. Bedekar, D. B. Avasare, P. Singh, S. K. Kulkarni, *J. Appl. Phys.* **1993**, *73*, 5237–5240. (b) Y. Nosaka, Y. Nosaka, *Langmuir* **1997**, *13*, 708–713.

[65] V. Sankaran, J. Yue, R. E. Cohen, R. R. Schrock, R. Silbey, *J. Chem. Mater.* **1993**, *5*, 1133–1142.

[66] I. Moriguchi, H. Nii, K. Hanai, H. Nagaoka, Y. Teraoka, S. Kagawa, *Colloids and Surfaces* **1995**, *103*, 173–181.

[67] Y. Yang, J. M. Huang, S. Liu, J. Shen, *J. Mater. Chem.* **1997**, *7*, 131–133.

[68] S. Lindroos, T. Kanniainen, M. J. Leskela, *Mater. Chem.* **1996**, *6*, 1497–1500.

[69] W. B. Sang, Y. B. Qian, W. M. Shi, D. M. Wang, J. H. Min, W. H. Wu, Y. F. Liu, J. D. Hua, J. Fang, Y. F. Yue, *J. Phys.-Condens. Matter.* **1996**, *8*, 499–504.

[70] G. A. Ozin, M. R. Steele, A. Holmes, *J. Chem. Mater.* **1994**, *6*, 999–1010.

[71] W. Tang, D. C. Cameron, *Thin Solid Films* **1996**, *280*, 221–226.

[72] G. M. Li, M. Nogami, *J. Appl. Phys.* **1994**, *75*, 4276–4278.

[73] D. W. Bahnemann, *Israel J. Chem.* **1993**, *33*, 115–136.

[74] G. Redmond, D. Fitzmaurice, M. Graetzel, *Chem. Mater.* **1994**, *6*, 686–691.

[75] S. Mahamuni, B. S. Bendre, V. J. Leppert, C. A. Smith, D. Cooke, S. H. Risbud, H. W. H. Lee, *Nanostructured Materials* **1996**, *7*, 659–666.

[76] S. Peulon, D. Lincot, *Adv. Mater.* **1996**, *8*, 166–170.

[77] J. Y. Ying, G. McMahon, *Mater. Res. Soc. Symp. Proc.* **1993**, *286*, 73–79.

[78] M. S. El–Shall, W. Slack, W. Vann, D. Kane, D. Hanley, *J. Phys. Chem.* **1994**, *98*, 3067–3070.

[79] X. C. Li, J. R. Fryer, D. J. Colehamilton, *J. Chem. Soc., Chem. Commun.* **1994**, 1715–1716

[80] R. Tassoni, R. R. Schrock, *Chem. Mater.* **1994**, *6*, 744–749.

[81] R. S. Kane, R. E. Cohen, R. Silbey, *Chem. Mater.* **1996**, *8*, 1919–1924.

[82] X. Peng, R. Lu, Y. Zhao, L. Qu, H. Chen, T. Li, *J. Phys. Chem.* **1994**, *98*, 7052–7055.

[83] M. Gao, Y. Yang, B. Yang, F. Bian, J. Shen, *J. Chem. Soc., Chem. Commun.* **1994**, 2779–2780.

[84] J. Yang, J. H. Fendler, H. Janos, *J. Phys. Chem.* **1995**, *99*, 5505–5511.

[85] Y. C. Tian, C. J. Wu, N. Kotov, J. H. Fendler, *Adv. Mater.* **1994**, *6*, 959–962.

[86] J. P. Yang, S. B. Qadri, B. R. Ratna, *J. Phys. Chem.* **1996**, *100*, 17255–17259.

[87] A. J. Ward, E. C. Sullivan, J.-C. Rang, J. Nedeljkovic, R. C. Patel, *J. Colloid & Interface Sci.* **1993**, *161*, 316–320.

[88] T. Takada, T. Yano, A. Yasumori, M. Yamane, *J. Ceramic Soc. Japan* **1993**, *101*, 73–75.

[89] L. Butler, G. Redmond, D. Fitzmaurice, *J. Phys. Chem.* **1993**, *97*, 10750.

[90] S. S. Kher, R. L. Wells, *Chem. Mater.* **1994**, *6*, 2056–2062.

[91] S. S. Kher, R. L. Wells, *Nanostructured Materials* **1996**, *7*, 591–603.

[92] H. Matsumoto, H. Uchida, H. Yoneyama, T. Sakata, H. Mori, *Res. Chem. Intermed.* **1994**, *20*, 723–733.

[93] O. I. Micic, C. J. Curtis, K. M. Jones, J. R. Sprangue, A. J. Nozik, *J. Phys. Chem.* **1994**, *98*, 4966–4969.

[94] (a) H. Uchida, C. J. Curtis, A. J. Nozik, *J. Phys. Chem.* **1991**, *95*, 5382–5384. (b) H. Uchida, C. J. Curtis, P. V. Kamat, K. M. Jones, A. J. Nozik, *J. Phys. Chem.* **1992**, *96*, 1156–1160.

[95] O. V. Salata, P. J. Dobson, P. J. Hull, J. L. Hutchison, *Appl. Phys. Lett.* **1994**, *65*, 189–191.

[96] A. J. Nozik, H. Uchida, P. V. Kamat, C. Curtis, *Israel J. Chem.* **1993**, *33*, 15–20.

[97] R. L. Wells, S. R. Aubuchon, S. S. Kher, M. S. Lube, P. S. White, *Chem. Mater.* **1995**, *7*, 793–800.

[98] N. Moumen, M. P. Pileni, *J. Phys. Chem.* **1996**, *100*, 1867–1873.

[99] H. Sato, T. Hirai, I. Komasawa, *Industrial & Engineering Chem. Res.* **1995**, *34*, 2493–2498.

[100] J. Tanori, T. Gulik-Krzywicki, M. P. Pileni, *Langmuir* **1997**, *13*, 632–638.

[101] J. Tanori, M. P. Pileni, *Langmuir* **1997**, *13*, 639–646.

[102] L. Motte, F. Billodet, M. P. Pileni, *J. Mater. Sci.* **1996**, *31*, 38–42.

[103] O. Horvath, J. H. Fendler, *J. Phys. Chem.* **1992**, *96*, 9591–9594.

[104] I. Ichinose, N. Kimizuka, T. Kunitake, *J. Phys. Chem.* **1995**, *99*, 3736–3742.

[105] K. C. Yi, Ph.D. Thesis, Syracuse University **1994**.

[106] R. S. Urquhart, D. N. Furlong, H. Mansur, F. Grieser, K. Tanaka, Y. Okahata, *Langmuir* **1994**, *10*, 899–904.

[107] I. Moriguchi, K. Hosoi, H. Nagaoka, I. Tanaka, Y. Teraoka, S. Kagawa, *J. Chem. Soc., Faraday Trans.* **1994**, *90*, 349–354.

[108] Y. C. Tian, C. J. Wu, J. H. Fendler, *J. Phys. Chem.* **1994**, *98*, 4913–4918.

[109] Z. Pan, G. Shen, L. Zhang, J. Liu, *J. Mater. Chem.* **1997**, *7*, 531–535.

[110] J. P. Yang, F. C. Meldrum, J. H. Fendler, *J. Phys. Chem.* **1995**, *99*, 5500–5504.

[111] J. P. Yang, J. H. Fendler, *J. Phys. Chem.* **1995**, *99*, 5505–5511.

[112] Z. Y. Pan, G. J. Shen, L. G. Zhang, Z. H. Lu, J. Z. Liu, *J. Chem. Mater.* **1997**, *7*, 531–535.

[113] F. Grieser, D. N. Furlong, D. Scoberg, I. Ichinose, N. Kimizuka, T. Kunitake, *J. Chem. Soc., Faraday Trans.* **1992**, *88*, 2207–2214.

[114] A. Fojtik, A. Henglein, *Chem. Phys. Lett.* **1994**, *221*, 363–367.

[115] K. A. Littau, P. J. Szajowski, A. J. Muller, A. R. Kortan, L. E. Brus, *J. Phys. Chem.* **1993**, *97*, 1224–1230.

[116] L. E. Brus, P. F. Szajowski, W. L. Wilson, T. D. Harris, S. Schuppler, P. H. Citrin, *J. Am. Chem. Soc.* **1996**, *117*, 2915–2922.

[117] D. Gallagher, W. E. Heady, J. M. Racz, R. N. Bhargava, *J. Mater. Res.*, **1995**, *10*, 870–876.

[118] Y. C. Tian, T. Newton, N. A. Kotov, D. M. Guldi, J. H. Fendler, *J. Phys. Chem.* **1996**, *100*, 8927–8939.

[119] X. G. Peng, T. E. Wilson, A. P. Alivisatos, P. G. Schultz, *Angew. Chem., Internat. Ed. Engl.* **1997**, *36*, 145–147.

[120] D. Lawless, S. Kapoor, D. Meisel, *J. Phys. Chem.* **1995**, *99*, 10329–10335.

[121] P. Pechy, F. P. Rotzinger, M. K. Nazeeruddin, O. K. S. Zakeeruddin, R. Humphry-Baker, M. Gretzel, *J. Chem. Soc., Chem. Commun.* **1995**, 65–66.

[122] T. Saika, T. Iyoda, T. Shimidzu, *Chem. Lett.* **1993**, 2025–2028.

[123] I. Willner, Y. Eichen, A. J. Frank, M. A. Fox, *J. Phys. Chem.* **1993**, *97*, 7264–7271.

[124] S. A. Majetich, A. D. Carter, R. D. McCullough, *Mat. Res. Soc. Symp. Proc.* **1993**, *286*, 87–92.

[125] M. Danek, K. F. Jensen, C. B. Murray, M. G. Bawendi, *Chem. Mater.* **1996**, *8*, 173–180.

[126] A. Eychmüller, A. Mews, H. Weller, *Chem. Phys. Lett.* **1993**, *208*, 59–62.

[127] A. Hässelbarth, A. Eychmüller, R. Eichberger, M. Giersig, A. Mews, H. Weller, *J. Phys. Chem.* **1993**, 97, 5333–5340.

[128] K. Vinodgopal, I. Bedja, P. V. Kamat, *Chem. Mater.* **1996**, *8*, 2180–2187.

[129] T. Cassagneau, G. B. Hix, D. J. Jones, P. Mairelestorres, M. Rhomari, J. Roziere, *J. Mater. Chem.* **1994**, *4*, 189–195.

[130] M. Gandais, M. Allais, Y. Zheng, M. Chamarro, *J. Physique IV Ed. Phys.* **1994**, 47–56.

[131] V. Sukumar, R. H. Doremus, *Physica Status Solidi B – Basic Research* **1993**, *179*, 307–314.

[132] V. Sukumar, S. C. Kao, P. G. N. Rao, R. H. Doremus, *J. Mater. Res.* **1993**, *8*, 2686–2693.

[133] K. M. Choi, K. J. Shea, *J. Am Chem. Soc.* **1994**, *116*, 9052–9060.

[134] A. Fernández, A. R. González-Elipe, C. Real, A. Caballero, G. Munuera, *Langmuir* **1993**, *9*, 121–125.

[135] G. Lassaletta, A. Fernandez, J. P. Espinos, A. Gonzalez-Elipe, *J. Phys. Chem.* **1995**, *99*, 1484–1490.

[136] P. Hoyer, R. Konenkamp, *Appl. Phys. Lett.* **1995**, *66*, 349–351.

[137] P. Hoyer, H. Weller, *Chem. Phys. Lett.* **1994**, *224*, 75–80.

[138] W. Choi, A. Termin, M. R. Hoffmann, *J. Phys. Chem.* **1994**, *98*, 13669–13679.

[139] S. T. Martin, C. L. Morrison, M. R. Hoffmann, *J. Phys. Chem.* **1994**, *98*, 13695–13704.

[140] B. O. Dabbousi, M. G. Bawendi, O. Onitsuka, M. F. Rubner, *Appl. Phys. Lett.* **1995**, *66*, 1316–1318.

[141] Y. Nosaka, N. Tohriiwa, T. Kobayashi, N. Fujii, *Chem. Mater.* **1993**, *5*, 930–932.
[142] R. Hutter, T. Mallat, A. Baiker, *J. Chem. Soc., Chem. Commun.* **1995**, 2487–2488.
[143] K. Kameyama, S. Shohji, S. Onoue, K. Nishimura, K. Yahikozawa, Y. Takasu, *J. Electrochem. Soc.* **1993**, *140*, 1034–1037.
[144] R. P. Andres, J. D. Bielefeld, J. I. Henderson, D. B. Janes, V. R. Kolagunta, C. P. Kubiak, W. J. Mahoney, R. G. Osifchin, *Science* **1996**, *273*, 1690–1693.
[145] K. V. Sarathy, G. U. Kulkarni, C. N. R. Rao, *Chemical Communications* **1997**, 537–538.
[146] A. P. Alivisatos, P. K. Johnsson, X. Peng, E. T. Wilson, J. C. Loweth, P. M. Bruchez, C. P. Schultz, *Nature* **1996**, *382*, 609–611.
[147] C. A. Mirkin, R. L. Letsinger, R. C. Mucic, J. J. Storhoff, *Nature*, **1996**, *382*, 607–609.
[148] N. A. Kotov, I. Dekany, J. H. Fendler, *J. Phys. Chem.* **1995**, *99*, 13065–13069.
[149] J. H. Fendler, N. A. Kotov, I. Dékány, in: *Fine Particles Science and Technology From Micro to Nanoparticles* (Ed.: E. Pelizzetti), Kluwer, Netherlands **1996**, pp. 557–577.
[150] N. A. Kotov, G. Zavala, J. H. Fendler, *J. Phys. Chem.* **1995**, *99*, 12375–12378.
[151] R. C. Freeman, K. C. Grabar, K. J. Allison, R. M. Bright, J. A. Davis, A. Gruthrie, M. B. Hommer, M. A. Jackson, P. C. Smith, D. G. Walter M. J. Natan, *Science*, **1995**, *267*, 1629–1632.
[152] N. Satoh, H. Hasegawa, K. Tsujii, K. Kimura, *J. Phys. Chem.*, **1994**, *98*, 2143–2147.
[153] (a) N. A. Kotov, F. C. Meldrum, C. Wu, J. Fendler, *J. Phys. Chem.* **1994**, *98*, 2735–2738. (b) M. Brust, R. Etchenique, E. J. Calvo, G. J. Gordillo, *J. Chem. Soc., Chem. Commun.* **1996**, 1949–1950.
[154] Y. C. Tian, J. H. Fendler, *Chem. Mater.* **1996**, *8*, 969–974.
[155] N. A. Kotov, F. C. Meldrum, J. H. Fendler, *J. Phys. Chem.* **1994**, *98*, 8827–8830.
[156] (a) F. C. Meldrum, N. A. Kotov, J. H. Fendler, *J. Phys. Chem.* **1994**, *98*, 4506–4510. (b) T. Nakaya, Y. J. Li, K. Shibata, *J. Mater, Chem.* **1996**, *6*, 691–697.
[157] N. A. Kotov, G. Zavala, J. H. Fendler, *J. Phys. Chem.* **1995**, *99*, 12375–12378.
[158] F. C. Meldrum, N. A. Kotov, J. H. Fendler, *Langmuir* **1994**, *10*, 2035–2040.
[159] F. C. Meldrum, N. A. Kotov, J. H. Fendler, *J. Chem. Soc. Faraday Trans.* **1995**, *91*, 673–680.
[160] F. C. Meldrum, N. A. Kotov, J. H. Fendler, *Chem. Mater.* **1994**, *7*, 1112–1116.
[161] C. Natarajan, G. Nogami, *J. Electrochem. Soc.* **1996**. *143*, 1547–1550.
[162] Y. P. Sun, E. C. Hao, X. Zhang, B. Yang, M. Y. Gao, J. Shen, *J. Chem. Soc., Chem. Commun.* **1996**, 2381–2382.
[163] D. Nesheva, D. Arsova, R. J. Ionov, *Mater. Sci.* **1993**, *28*, 2183–2186.
[164] V. L. Colvin, M. C. Schlamp, A. P. Alivisatos, *Nature* **1994**, *370*, 354–357.
[165] P. Hoyer, N. Baba, H. Masuda, *Appl. Phys. Lett.* **1995**, *66*, 2700–2702.
[166] Q. S Huo, D. I. Margolese, U. Ciesla, D. G. Demuth, P. Y. Feng, T. E. Gier, P. Sieger, A. Firouzi, B. F. Chmelka, F. Schuth, G. D. Stucky, *Chem. Mater.* **1994**, *6*, 1176–1191.
[167] P. C. Rieke, S. B. Bentjen, *Chem. Mater.* **1993**, *5*, 43–53.
[168] M. Ritala, M. Leskelä, L. Niinistö, P. Haussalo, *Chem. Mater.* **1993**, *5*, 1174–1181.
[169] M. Ritala, M. Leskela, E. Rauhala, *Chem. Mater.* **1994**, *6*, 556–561.
[170] P. Facci, V. Erokhin, A. Tronin, C. Nicolini, *J. Phys. Chem.* **1994**, *98*, 13323–13327.
[171] A. Jentys, R. W. Grimes, *J. Chem. Soc., Faraday Trans.* **1996**, *92*, 2093–2097.
[172] (a) S. Kohtani, A. Kudo, T. Sakata, *Chem. Phys. Lett.* **1993**, *206*, 166–170. (b) N. Serpone, P. Maruthamuthu, P. Pichat, E. Pelizzetti, H. Hidaka, *J. Photochem. Photophys. A*, **1995**, *85*, 247–255.
[173] S. Deki, Y. Aoi, O. Hiroi, A. Kajinami, *Chem. Lett.* **1996**, 433–434.
[174] T. Isago, S. Sonobe, T. Ohkawa, H. Sunayama, *J. Ceram. Soc. Japan* **1996**, *104*, 1052–1055.
[175] P. Hoyer, R. Eichberger, H. Weller, *Ber. Bunsenges. Phys. Chem.* **1993**, *97*, 630–635.
[176] I. Yu, T. Isobe, M. Senna, *Mater. Res. Bull.* **1995**, *30*, 975–980.
[177] P. Hoyer, *Langmuir* **1996**, *12*, 1411–1413.
[178] T. Torimoto, R. J. Fox, M. A. Fox, *J. Electrochem. Soc.* **1996**, *143*, 3712–3717.
[179] B. I. Lemon, J. T. Hupp, *J. Phys. Chem.* **1996**, *100*, 14578–14580.
[180] I. Moriguchi, H. Maeda, Y. Teraoka, S. S. Kagawa, *J. Am. Chem. Soc.* **1996**, *117*, 1139–1140.
[181] R. J. Tonucci, M. Fatemi, *Appl. Phys. Lett.* **1996**, *69*, 2846–2848.

[182] V. Lakhotia, J. M. Heitzinger, A. H. Cowley, R. A. Jones, J. G. Ekerdt, *Chem. Mater.* **1994**, *6*, 871–874.

[183] A. Henglein, *Berichte Der Bunsen-Gesellschaft – Physical Chemistry Chemical Physics* **1995**, *99*, 903–913.

[184] I. Dékány, in: *Fine Particle Science and Technology from Micro to Nanoparticles* (Ed.: E. Pelizzetti), Kluwer, Netherlands **1996**, pp. 293–322.

[185] I. Dékány, L. G. Nagy, G. Schay, *J. Colloid and Interface Sci.* **1978**, *66*, 197–199.

[186] I. Dékány, F. Szántó, A. Weiss, G. Lagaly, *Berichte der Bunsen-Gesellschaft für Physikalische Chemie* **1985**, *89*, 62–67.

[187] I. Dékány, F. Szántó, A. Weiss, G. Lagaly, *Berichte der Bunsen-Gesellschaft für Physikalische Chemie* **1986**, *90*, 427–431.

[188] I. Dékány, F. Szántó, A. Weiss, G. Lagaly, *Berichte der Bunsen-Gesellschaft für Physikalische Chemie* **1986**, *90*, 422–427.

[189] I. Dékány, F. Szántó, A. Weiss, *Colloids and Surfaces* **1989**, *41*, 107–121.

[190] J. N. Israelachvili, *Chemtracts – Analytical and Physical Chemistry* **1989**, *1*, 1–12.

[191] P. M. Claesson, T. Ederth, V. Bergeron, M. W. Rutland, *Advances in Colloid and Interface Science* **1996**, 119–183.

[192] M. C. Lieber, J. Liu, P. E. Sheehan, *Angew. Chemie, Int. Edn. English* **1996**, *35*, 687–704.

[193] W. A. Hayes, H. Kim, X. H. Yue, S. S. Perry, C. Shannon, *Langmuir* **1997**, *13*, 2511–2518.

[194] D. M. R. Kolb, T. Will, *Science* **1997**, *275*, 1097–1099.

[195] Y. K. Leong, D. V. Boger, P. J. Scales, T. W. J. Healy, *Colloid and Interface Sci.* **1996**, *181*, 605–612.

[196] D. Y. Godovski, Y. K. P. V. P. E. Godovsky, *Advances in Polymer Science,* **1995**, *119*, 79–122.

[197] C. Barbero, R. Kötz, *J. Electrochem. Soc.* **1994**, *141*, 859–865.

[198] J. M. J. Frechet, C. J. Hawker, I. Gitsov, J. W. J. Leon, *Macromolecular Sci. – Pure and Applied Chemistry* **1996**, *a33*, 1399–1425.

[199] M. Granstrom, M. Berggren, O. Inganas, *Science* **1995**, *267*, 1479–1481.

[200] S. Margel, I. Burdygin, V. Reznikov, B. Nitzan, O. Melamed, M. Kedem, S. Gura, G. Mandel, M. Zuberi, L. Boguslavsky, in *Recent Research Developments in Polymer Science* (Ed.: S. G. Patidali), Transworld Research Network **1997**.

[201] O. Karthaus, K. Ijiro, M. Shimomura, *Chemistry Letters* **1996**, 821–822.

[202] L. H. Radzilowski, B. O. Carragher, S. I. Stupp, *Macromolecules* **1997**, *30*, 2110–2119.

[203] G. Decher, in: *Comprehensive Supramolecular Chemistry*, Vol. 9 (Ed.: J.-P. Sauvage), Pergamon, Oxford **1996**, pp. 507–528.